Public Health Microbiology

METHODS IN MOLECULAR BIOLOGY™

John M. Walker, Series Editor

293. **Laser Capture Microdissection:** *Methods and Protocols,* edited by *Graeme I. Murray and Stephanie Curran, 2005*
292. **DNA Viruses:** *Methods and Protocols,* edited by *Paul M. Lieberman, 2005*
291. **Molecular Toxicology Protocols**, edited by *Phouthone Keohavong and Stephen G. Grant, 2005*
290. **Basic Cell Culture**, *Third Edition,* edited by *Cheryl D. Helgason and Cindy Miller, 2005*
289. **Epidermal Cells**, *Methods and Applications,* edited by *Kursad Turksen, 2004*
288. **Oligonucleotide Synthesis**, *Methods and Applications,* edited by *Piet Herdewijn, 2004*
287. **Epigenetics Protocols**, edited by *Trygve O. Tollefsbol, 2004*
286. **Transgenic Plants:** *Methods and Protocols,* edited by *Leandro Peña, 2004*
285. **Cell Cycle Control and Dysregulation Protocols:** *Cyclins, Cyclin-Dependent Kinases, and Other Factors,* edited by *Antonio Giordano and Gaetano Romano, 2004*
284. **Signal Transduction Protocols,** *Second Edition,* edited by *Robert C. Dickson and Michael D. Mendenhall, 2004*
283. **Bioconjugation Protocols,** edited by *Christof M. Niemeyer, 2004*
282. **Apoptosis Methods and Protocols,** edited by *Hugh J. M. Brady, 2004*
281. **Checkpoint Controls and Cancer, Volume 2:** *Activation and Regulation Protocols,* edited by *Axel H. Schönthal, 2004*
280. **Checkpoint Controls and Cancer, Volume 1:** *Reviews and Model Systems,* edited by *Axel H. Schönthal, 2004*
279. **Nitric Oxide Protocols,** *Second Edition,* edited by *Aviv Hassid, 2004*
278. **Protein NMR Techniques,** *Second Edition,* edited by *A. Kristina Downing, 2004*
277. **Trinucleotide Repeat Protocols,** edited by *Yoshinori Kohwi, 2004*
276. **Capillary Electrophoresis of Proteins and Peptides,** edited by *Mark A. Strege and Avinash L. Lagu, 2004*
275. **Chemoinformatics,** edited by *Jürgen Bajorath, 2004*
274. **Photosynthesis Research Protocols,** edited by *Robert Carpentier, 2004*
273. **Platelets and Megakaryocytes, Volume 2:** *Perspectives and Techniques,* edited by *Jonathan M. Gibbins and Martyn P. Mahaut-Smith, 2004*
272. **Platelets and Megakaryocytes, Volume 1:** *Functional Assays,* edited by *Jonathan M. Gibbins and Martyn P. Mahaut-Smith, 2004*
271. **B Cell Protocols,** edited by *Hua Gu and Klaus Rajewsky, 2004*
270. **Parasite Genomics Protocols,** edited by *Sara E. Melville, 2004*
269. **Vaccina Virus and Poxvirology:** *Methods and Protocols,*edited by *Stuart N. Isaacs, 2004*
268. **Public Health Microbiology:** *Methods and Protocols,* edited by *John F. T. Spencer and Alicia L. Ragout de Spencer, 2004*
267. **Recombinant Gene Expression:** *Reviews and Protocols, Second Edition,* edited by *Paulina Balbas and Argelia Johnson, 2004*
266. **Genomics, Proteomics, and Clinical Bacteriology:** *Methods and Reviews,* edited by *Neil Woodford and Alan Johnson, 2004*
265. **RNA Interference, Editing, and Modification:** *Methods and Protocols,* edited by *Jonatha M. Gott, 2004*
264. **Protein Arrays:** *Methods and Protocols,* edited by *Eric Fung, 2004*
263. **Flow Cytometry,** *Second Edition,* edited by *Teresa S. Hawley and Robert G. Hawley, 2004*
262. **Genetic Recombination Protocols,** edited by *Alan S. Waldman, 2004*
261. **Protein–Protein Interactions:** *Methods and Applications,* edited by *Haian Fu, 2004*
260. **Mobile Genetic Elements:** *Protocols and Genomic Applications,* edited by *Wolfgang J. Miller and Pierre Capy, 2004*
259. **Receptor Signal Transduction Protocols,** *Second Edition,* edited by *Gary B. Willars and R. A. John Challiss, 2004*
258. **Gene Expression Profiling:** *Methods and Protocols,* edited by *Richard A. Shimkets, 2004*
257. **mRNA Processing and Metabolism:** *Methods and Protocols,* edited by *Daniel R. Schoenberg, 2004*
256. **Bacterial Artifical Chromosomes, Volume 2:** *Functional Studies,* edited by *Shaying Zhao and Marvin Stodolsky, 2004*
255. **Bacterial Artifical Chromosomes, Volume 1:** *Library Construction, Physical Mapping, and Sequencing,* edited by *Shaying Zhao and Marvin Stodolsky, 2004*
254. **Germ Cell Protocols, Volume 2:** *Molecular Embryo Analysis, Live Imaging, Transgenesis, and Cloning,* edited by *Heide Schatten, 2004*
253. **Germ Cell Protocols, Volume 1:** *Sperm and Oocyte Analysis,* edited by *Heide Schatten, 2004*
252. **Ribozymes and siRNA Protocols,** *Second Edition,* edited by *Mouldy Sioud, 2004*
251. **HPLC of Peptides and Proteins:** *Methods and Protocols,* edited by *Marie-Isabel Aguilar, 2004*
250. **MAP Kinase Signaling Protocols,** edited by *Rony Seger, 2004*

METHODS IN MOLECULAR BIOLOGY™

Public Health Microbiology

Methods and Protocols

Edited by

John F. T. Spencer

Alicia L. Ragout de Spencer

Planta Piloto de Procesos Industriales Microbiológicos
(PROIMI)–CONICET
San Miguel de Tucumán, Argentina

[redacted]

HUMANA PRESS TOTOWA, NEW JERSEY

999 Riverview Drive, Suite 208
Totowa, New Jersey 07512

www.humanapress.com

This publication is printed on acid-free paper. ∞
ANSI Z39.48-1984 (American Standards Institute)

Permanence of Paper for Printed Library Materials.

Cover design by Patricia F. Cleary.

For additional copies, pricing for bulk purchases, and/or information about other Humana titles, contact Humana at the above address or at any of the following numbers: Tel.: 973-256-1699; Fax: 973-256-8341; E-mail: humana@humanapr.com; or visit our Website: www.humanapress.com

E-ISBN 1-59259-766-1

Library of Congress Cataloging-in-Publication Data

Public health microbiology : methods and protocols / edited by John F.T. Spencer, Alicia L. Ragout de Spencer.

p. ; cm. -- (Methods in molecular biology, ISSN 1064-3745 ; v. 268)

Includes bibliographical references and index.

ISBN 1-58829-117-0 (alk. paper)

1. Diagnostic microbiology. [DNLM: 1. Environmental Microbiology. 2. Microbiological Techniques. QW 55 P976 2004] I. Spencer, J. F. T. II. Ragout de Spencer, Alicia L. III. Methods in molecular biology (Clifton, N.J.) ;v. 268.

QR67.P83 2004
616.9'041--dc22 [redacted]

2003025016

Preface

Public Health Microbiology: Methods and Protocols is focused on microorganisms that can present a hazard to human health in the course of everyday life. There are chapters dealing with organisms that are directly pathogenic to humans, including bacteria, viruses, and fungi; on organisms that produce toxins during growth in their natural habitats; on the use of bacteriocins produced by such organisms as lactobacilli and bifidobacteria; as well as several chapters on hazard analysis, the use of disinfectants, microbiological analysis of cosmetics, and microbiological tests for sanitation equipment in food factories. Additional chapters look at the use of animals (mice) in the study of the various characteristics of milk and their relationships with lactic acid bacteria in particular. Other chapters focus on special methods for determining particular components of milk.

In particular, in Parts I and II, on bacterial and viral pathogens, special attention is given to methods for PCR detection of genes with resistance to tetracycline, as well as to *Salmonella enterica*; for identification and typing of *Campylobacter coli;* for detection of the abundance of enteric viruses, hepatitis A virus, and rotaviruses in sewage, and of bacteriophages infecting the O157:H7 strain of *Escherichia coli.*

Part III offers methods for computerized analysis and typing of fungal isolates, for isolation and enumeration of fungi in foods, and for the determination of aflatoxin and zearalenone.

Important pathogens discussed in Part IV include *Legionella,* amoebae, fungal conidia, and yeasts (by flow cytometry), *Cryptosporidium parvum,* and *Erysipelothrix rhusopathiae.* Protocols for the separation of pathogenic organisms from environmental matrices by immunomagnetic methods are also included here.

Part V describes methods for spectrophotometric determination of histamine in fish processing and the antimicrobial effects of flavonoids from Tagetes species in Argentina.

The uses of animals (mice) in studies of various characteristics of lactobacilli are covered in the six chapters of Part VI, and Part VII includes special methods of analysis.

The final part of *Public Health Microbiology: Methods and Protocols* includes three very important reviews, one on the microflora of the intestinal tract, another on the relation of the spread of pathogens from livestock and poultry production by disposal of manure and similar wastes on agricultural land, and finally, a review of the threat to public health by the prion-based diseases, the several forms of transmissible spongiform encephalopathies, that is, mad cow disease, scrapie in sheep, wasting disease in elk and mule deer, and Creutzfeld-Jacob disease in humans. In actuality, all three animal conditions may be transmissible to humans. Chapters 51 and 52, in particular, should be required reading for all public officials, at all levels of government, who have a responsibility for public health.

The editors wish to thank Dr. Faustino Siñeriz, Director of PROIMI and Head of Secretaría de Ciencia y Técnica (CIUNT) in Tucumán for his consideration in providing facilities and supplies for this work; Pharm. Laura Tereschuk, and Dra. Alejandra Martinez, for valuable assistance in the compilation of this work; Dr. Javier Ochoa, for much useful technical assistance; and finally all the investigator/authors who contributed chapters to our book.

John F. T. Spencer
Alicia L. Ragout de Spencer

Contents

Contributors

FRANCESC-XAVIER ABAD • *Enteric Virus Laboratory, Department of Microbiology, School of Biology, University of Barcelona, Barcelona, Spain*

LIDIA R. ABDALA • *Facultad de Ciencias Naturales e Instistuto Miguel Lillo, Universidad Nacional de Tucumán, Tucumán, Argentina*

DONALD G. AHEARN • *Environmental Research Center, Department of Biology, Georgia State University, Atlanta, GA*

MARÍA R. ALBERTO • *Centro de Referencia para Lactobacilos (CERELA)–CONICET, Tucumán, Argentina*

JUAN ALVAREZ • *Departamento de Inmunología, Microbiología y Parasitología, Facultad de Farmacia, Universidad del País Vasco, Vitoria-Gasteiz, España*

RUSTAM I. AMINOV • *Rowett Research Institute, Aberdeen, UK*

MARIO E. ARENA • *Facultad de Bioquímica, Química y Farmacia, Universidad Nacional de Tucumán; CONICET, Tucumán, Argentina*

JEAN-MARC BERJEAUD • *Laboratoire de Microbiologie Fondamentale et Appliquée, Université de Poitiers, Poitiers, France*

MARIO D. BAIGORÍ • *Laboratorio de Microbiología, Planta Piloto de Procesos Industriales Microbiológicos (PROIMI)–CONICET, Tucumán, Argentina*

ALBERT BOSCH • *Enteric Virus Laboratory, Department of Microbiology, School of Biology, University of Barcelona, Barcelona, Spain*

SONIA BRENA • *Departamento de Inmunología, Microbiología y Parasitología, Facultad de Farmacia, Universidad del País Vasco, Lejona, España*

ELENA BRU • *Centro de Referencia para Lactobacilos (CERELA)–CONICET, Tucumán, Argentina*

DANTE JAVIER BUENO • *Centro de Referencia para Lactobacilos (CERELA)–CONICET, Tucumán; Estacíon Experimental Concepción del Uruguay, Instituto Nacional de Tecnología Agropecuaria (INTA), Entre Riós, Argentina*

BRENDAN P. BURNS • *School of Biotechnology and Biomolecular Sciences, Australian Center for Astrobiology, University of New South Wales, Sydney, Australia*

AINARA CADAVAL • *Departamento de Inmunología, Microbiología y Parasitología, Facultad de Farmacia, Universidad del País Vasco, Vitoria-Gasteiz, España*

ROSA CANGEMI DE GUTIERREZ • *Facultad de Bioquímica, Química y Farmacia, Universidad Nacional de Tucumán, Tucumán, Argentina*

PATRICIA CASTELLANO • *Centro de Referencia para Lactobacilos (CERELA)–CONICET, Tucumán, Argentina*

YVES CENATIEMPO • *Laboratoire de Microbiologie Fondamentale et Appliquée, Université de Poitiers, Poitiers, France*

ADRIANA PÉREZ CHAIA • *Centro de Referencia para Lactobacilos (CERELA)–CONICET; Universidad Nacional de Tucumán, Tucumán, Argentina*

JOANNE C. CHEE-SANFORD • *Invasive Weed Managment Unit, Agricultural Research Service, United States Department of Agriculture (USDA–ARS), Urbana, IL*
JAUME COMAS • *Flow Cytometry Unit, Serveis Cientificotecnics, University of Barcelona, Spain*
MAIRÉAD DALY • *Molecular Diagnostics Unit, Cork Institute of Technology, Bishoptown, Cork, Ireland*
LUCIA I. C. DE FIGUEROA • *Facultad de Bioquímica, Química y Farmacia, Universidad Nacional de Tucumán; Laboratorio de Levaduras y Hongos Filamentosos, Planta Piloto de Procesos Industriales Microbiológicos (PROIMI)–CONICET, Tucumán, Argentina*
MARÍA DEL CARMEN AHUMADA OSTENGO • *Centro de Referencia para Lactobacilos (CERELA)–CONICET, Tucumán, Argentina*
DIANA DRAKSLER • *Centro de Referencia para Lactobacilos (CERELA)–CONICET, Tucumán, Argentina*
SILVINA FADDA • *Centro de Referencia para Lactobacilos (CERELA)–CONICET, Tucumán, Argentina*
SÉAMUS FANNING • *Centre for Food Safety, Faculties of Agriculture, Medicine, and Veterinary Medicine, University College, Belfiold, Dublin, Ireland*
XIAOCHUAN FENG • *Division of Infectious Diseases, Tufts University School of Veterinary Medicine, North Grafton, MA*
SILVANA G. FIDALGO • *Department of Microbiology, University of Western Australia, Queen Elizabeth II Medical Centre, Western Australia*
GRACIELA FONT DE VALDEZ • *Centro de Referencia para Lactobacilos (CERELA)–CONICET, Tucumán, Argentina*
CECILIA FONTANA • *Centro de Referencia para Lactobacilos (CERELA)–CONICET, Tucumán, Argentina*
JOSÉ D. FONTANA • *Director Técnico, Instituto Técnico do Paraná (TECPAR), Universidad Federal do Paraná, Paraná, Brasil*
JAVIER GARAIZAR • *Departamento de Inmunología, Microbiología y Parasitología, Facultad de Farmacia, Universidad del País Vasco, Vitoria-Gasteiz, España*
NATALIE GARRIGUES • *Department of Veterinary Pathobiology, University of Illinois at Urbana–Champaign, Urbana, IL*
SILVIA GONZÁLEZ • *Centro de Referencia para Lactobacilos (CERELA)–CONICET; Universidad Nacional de Tucumán, Tucumán, Argentina*
JIEWEN GUAN • *Animal Diseases Research Institute/Centre for Plant Quarantine Pests, Canadian Food Inspection Agency, Ottawa, Ontario, Canada*
CARLOS GUSILS • *Centro de Referencia para Lactobacilos (CERELA)–CONICET; Universidad Nacional de Tucumán, Tucumán, Argentina*
WEI-LONG HAO • *Department of Microbiology, National University of Singapore, Singapore*
ANAVELLA GAITAN HERRERA • *Planta Piloto de Procesos Industriales Microbiológicos (PROIMI)–CONICET, Tucumán, Argentina*

JIANLIN JIANG • *Division of Parasitic Diseases, Centers for Disease Control and Prevention, Public Health Service, United States Department of Health and Human Services, Atlanta, GA*

JUAN JOFRE • *Department of Microbiology, Faculty of Biology, University of Barcelona, Barcelona, Spain*

MARTA BEATRIZ KATZ • *Centro de Referencia para Lactobacilos (CERELA)–CONICET, Tucumán, Argentina*

IDOIA LACONCHA • *Departamento de Inmunología, Microbiología y Parasitología, Facultad de Farmacia, Universidad del País Vasco, Vitoria-Gasteiz, España*

ALTAF A. LAL • *Division of Parasitic Diseases, Centers for Disease Control and Prevention, Public Health Service, United States Deparment of Health and Human Services, Atlanta, GA*

YUAN-KUN LEE • *Department of Microbiology, National University of Singapore, Singapore*

NURIA LÓPEZ-MOLINA • *Departamento de Inmunología, Microbiología y Parasitología, Facultad de Farmacia, Universidad del País Vasco, Vitoria-Gasteiz, España*

BRIGID LUCEY • *Molecular Diagnostics Unit, Cork Institute of Technology, Bishoptown, Cork, Ireland*

RODERICK I. MACKIE • *Department of Animal Sciences, University of Illinois at Urbana-Champaign, Urbana, IL*

MARÍA C. MANCA DE NADRA • *Facultad de Bioquímica, Química y Farmacia, Universidad Nacional de Tucumán, Tucumán, Argentina*

MARTA MÉDICI • *Centro de Referencia para Lactobacilos (CERELA)–CONICET, Tucumán, Argentina*

ROXANA BEATRIZ MEDINA • *Centro de Referencia para Lactobacilos (CERELA)–CONICET; Universidad Nacional de Tucumán, Tucumán, Argentina*

ASMA MEHBOOB • *Department of Animal Sciences, University of Illinois at Urbana-Champaign, Urbana, IL*

MICHELLE C. MOFFITT • *School of Biotechnology and Biomolecular Sciences, University of New South Wales, Sydney, Australia*

MARÍA CECILIA MONFERRAN • *Universidad Nacional de Catamarca, Catamarca, Argentina*

VILMA MORATA • *Universidad Nacional de Cuyo, Mendoza, Argentina*

MAITE MUNIESA • *Department of Microbiology, Faculty of Biology, University of Barcelona, Barcelona, Spain*

MARÍA ELENA NADER-MACÍAS • *Centro de Referencia para Lactobacilos (CERELA)–CONICET, Tucumán, Argentina*

BRETT A. NEILAN • *School of Biotechnology and Biomolecular Sciences, Australian Center for Astrobiology, University of New South Wales, Sydney, Australia*

BIRGID NEUMEISTER • *Department Transfusion Medicine, University Hospital of Tuebingen, Tuebingen, Germany*

Judith A. Noble-Wang • *Environmental Research Center, Department of Biology, Georgia State University, Atlanta, GA*
Martha Núñez de Kairúz • *Centro de Referencia para Lactobacilos (CERELA)–CONICET, Tucumán, Argentina*
Virginia Ocaña • *Centro de Referencia para Lactobacilos (CERELA)–CONICET, Tucumán, Argentina*
Fiona O'Halloran • *Molecular Diagnostics Unit, Cork Institute of Technology, Bishoptown, Cork, Ireland*
Rubén Oliszewski • *Centro de Referencia para Lactobacilos (CERELA)–CONICET, Universidad Nacional de Tucumán, Tucumán, Argentina*
Guillermo Oliver • *Centro de Referencia para Lactobacilos (CERELA)–CONICET, Tucumán, Argentina*
María Claudia Otero • *Centro de Referencia para Lactobacilos (CERELA)–CONICET, Tucumán, Argentina*
Gabriela Perdigón • *Centro de Referencia para Lactobacilos (CERELA)–CONICET, Tucumán, Argentina*
Marcos L. Pessatti • *Universidade do Vale do Itajaí, Santa Catarina, Brasil*
Tânia L. P. Pessatti • *Universidade do Vale do Itajaí, Santa Catarina, Brasil*
Rosa M. Pintó • *Enteric Virus Laboratory, Department of Microbiology, School of Biology, University of Barcelona, Barcelona, Spain*
José Pontón • *Departamento de Inmunología, Microbiología y Parasitología, Facultad de Farmacia, Universidad del País Vasco, Lejona, España*
Daniel Price • *Interface Research Center, Kennesaw, GA*
Suzette A. Priola • *Laboratory of Persistent Viral Diseases, National Institute of Allergy and Infectious Diseases, Rocky Mountain Laboratories, Hamilton, MT*
Guillermo Quindós • *Departamento de Inmunología, Microbiología y Parasitología, Facultad de Farmacia, Universidad del País Vasco, Lejona, España*
Aitor Rementeria • *Departamento de Inmunología, Microbiología y Parasitología, Facultad de Farmacia, Universidad del País Vasco, Lejona, España*
Rosario Rey • *Facultad de Bioquímica, Química y Farmacia, Universidad Nacional de Tucumán, Tucumán, Argentina*
Kelly A. Reynolds • *University of Arizona, Environmental Research Laboratory, Tuscon, AZ*
Thomas V. Riley • *Department of Microbiology, University of Western Australia, Queen Elizabeth II Medical Centre, Western Australia*
María Teresa Ruesga • *Departamento de Inmunología, Microbiología y Parasitología, Facultad de Farmacia, Universidad del País Vasco, Lejona, España*
Lucila Saavedra • *Centro de Referencia para Lactobacilos (CERELA)–CONICET, Tucumán, Argentina*
Martin L. Saker • *Departamento de Zoologia e Antropologia, Faculdade de Ciencias, Universidade do Porto, Porto, Portugal*

GLÒRIA SÁNCHEZ • *Enteric Virus Laboratory, Department of Microbiology, School of Biology, University of Barcelona, Barcelona, Spain*
VIVIANA M. SANTOS • *Facultad de Bioquímica, Química y Farmacia, Universidad Nacional de Tucumán, Tucumán, Argentina*
FERNANDO SESMA • *Centro de Referencia para Lactobacilos (CERELA)–CONICET, Tucumán, Argentina*
CLARA SILVA • *Facultad de Bioquímica, Química y Farmacia, Universidad Nacional de Tucumán, Tucumán, Argentina*
JULIO OSCAR SILVA • *Facultad de Bioquímica, Química y Farmacia, Universidad Nacional de Tucumán, Tucumán, Argentina*
J. LLOYD SPENCER • *Animal Diseases Research Institute/Centre for Plant Quarantine Pests, Canadian Food Inspection Agency, Ottawa, Ontario, Canada*
KATHLEEN L. STADTERMAN-KNAUER • *The Knauer Group, Pittsburgh, PA*
SULTAN TANRIVERDI • *Division of Infectious Diseases, Tufts University School of Veterinary Medicine, North Grafton, MA; Department of Pediatric Hematology-Oncology, Study of Medicine, Cukurova University, Adana, Turkey*
MARÍA PÍA TARANTO • *Centro de Referencia para Lactobacilos (CERELA)–CONICET, Tucumán, Argentina*
MARÍA L. TERESCHUK • *Facultad de Ciencias Naturales e Instituto Miguel Lillo, Universidad Nacional de Tucumán, Tucumán, Argentina*
MARÍA SILVINA JUÁREZ TOMÁS • *Centro de Referencia para Lactobacilos (CERELA)–CONICET, Tucumán, Argentina*
GRACIELA VIGNOLO • *Centro de Referencia para Lactobacilos (CERELA)–CONICET, Tucumán, Argentina*
CRISTINA VILLENA • *Enteric Virus Laboratory, Department of Microbiology, School of Biology, University of Barcelona, Barcelona, Spain*
ELISA VINTIÑI • *Centro de Referencia para Lactobacilos (CERELA)–CONICET, Tucumán, Argentina*
ANA BELÉN VIVANCO • *Departamento de Inmunología, Microbiología y Parasitología, Facultad de Farmacia, Universidad del País Vasco, Vitoria-Gasteiz, España*
INA VORBERG • *Institute for Virology, Technical University of Munich, Munich, Germany*
GIOVANNI WIDMER • *Division of Infectious Diseases, Tufts University School of Veterinary Medicine, North Grafton, MA*
LIHUA XIAO • *Division of Parasitic Diseases, Centers for Disease Control and Prevention, Public Health Service, United States Deparment of Health and Human Services, Atlanta, GA*
GARY P. YAKUB • *North Huntington, PA*
GABRIELA ZÁRATE • *Centro de Referencia para Lactobacilos (CERELA)–CONICET, Tucumán, Argentina*
SHANGTONG ZHANG • *Environmental Research Center, Department of Biology, Georgia State University, Atlanta, GA*

I

BACTERIA

1

Detection of Tetracycline Resistance Genes by PCR Methods

Rustam I. Aminov, Joanne C. Chee-Sanford, Natalie Garrigues, Asma Mehboob, and Roderick I. Mackie

1. Introduction

Rapid, accurate, and sensitive determination of antibiotic resistance profiles of various human and animal pathogens becomes a vital prerequisite for successful therapeutic intervention in the face of the increased occurrences of drug-resistant bacterial infections. The current methods, which are dependent on cultivation of pathogens and phenotypic expression of antibiotic resistance, usually require excessive time, special microbiological equipment, and qualified personnel. However, even with all these requisites, for example, no bacteria can be grown from more than 80% of all clinical samples sent to clinical microbiology laboratories *(1)*. Besides the cultivation limitations, the cultivation-based determination of an antibiotic resistance profile lacks the genotypic information, which is essential for understanding the epidemiology and routes of transmission of antibiotic resistance genes. These genes often reside on mobile genetic elements and can move freely between commensal and pathogenic microbiota, occurring even between taxonomically distant clinical and environmental microbiota. Therefore, development of genotyping methods for detection of antibiotic resistance genes is highly desirable for fast, accurate, and sensitive detection of antibiotic resistance genes in a broad range of pathogenic and commensal bacteria in both clinical and environmental samples.

As a model for our studies we have chosen the genes conferring resistance to tetracyclines. Tetracyclines belong to a family of broad-spectrum antibiotics that include tetracycline, chlortetracycline, oxytetracycline, demeclocycline, methacycline, doxycycline, minocycline, and a number of other semisynthetic derivatives *(2)*. These antibiotics inhibit protein synthesis in Gram-positive and Gram-negative bacteria by preventing the binding of aminoacyl-tRNA molecules to the 30S ribosomal subunit *(3)*. The antibiotics of this group were introduced in the late 1950s and since then have been widely used in clinical and veterinary medicine, as well as for prophylaxis and

From: *Methods in Molecular Biology, vol. 268: Public Health Microbiology: Methods and Protocols*
Edited by: J. F. T. Spencer and A. L. Ragout de Spencer © Humana Press Inc., Totowa, NJ

growth promotion in food animals. Because of the possible misuse and overuse of these drugs, resistance to this class of antibiotics is widespread among many clinical isolates, thus limiting the utility of tetracyclines in treating infections. Despite this shortcoming, antibiotics of this class still remain in the active arsenal for dermatologists to treat skin infections such as acne ***(4)*** and rosacea ***(5)***. The local application of high concentrations of doxycycline has been proposed for clinical management of patients with periodontitis ***(6)***. Currently, the third generation of tetracyclines, the so-called glycylcyclines, are in phase II clinical trials ***(2)***, but microbial resistance to these novel antibiotics can already be seen among *Salmonella* isolates ***(7)***. Interestingly, this resistance to glycylcyclines is mediated by the well-known tetracycline resistance gene, *tet*(A), which has acquired an additional mutation conferring resistance to novel antibiotics that have not yet been introduced into clinical practice. Thus, routine monitoring for existing and novel tetracycline resistance genes may aid in evaluating possible decreases in efficacy of third-generation tetracyclines caused by spreading resistances.

Bacterial resistance to tetracycline is mediated mainly by two mechanisms including protection of ribosomes by the synthesis of ribosomal protection proteins (RPPs), which share homology with the GTPases involved in protein synthesis, namely, EF-Tu and EF-G ***(8–14)***, and by the energy-dependent efflux of tetracycline from the cell ***(3,12,15)***. A third mechanism, enzymatic inactivation of tetracycline, is relatively uncommon and functions only during heterologous expression in *E. coli* ***(16)***. This reaction requires the presence of oxygen and NADPH and is not functional in the natural anaerobic host (*Bacteroides*). The first nomenclature for tetracycline resistance determinants was proposed in 1989 ***(17)***, and a recent update appeared in 1999 ***(18)***. Phylogeny-based classifications of the RPP genes allowed the identification of nine classes: TetB P, TetM, TetO, TetQ, TetS, TetT, TetW, Tet 32, and *otrA* ***(19,20)***. The second mechanism of tetracycline resistance, efflux of tetracycline from the cell, is mediated by transporters, which share a common structure with the 12 or 14 *trans*-membrane segments (12-TMS and 14-TMS) and belong to the major facilitator superfamily (MFS) ***(21,22)***. The 12-TMS permeases are found almost exclusively in Gram-negative bacteria and they uniformly catalyze drug/H^+ antiport ***(21)***. Recent phylogenetic analysis has suggested that the *tet* subcluster includes 11 genes catalyzing the efflux of tetracycline from the cell: *tet*(A), *tet*(B), *tet*(C), *tet*(D), *tet*(E), *tet*(G), *tet*(H), *tet*(J), *tet*(Y), *tet*(Z), and *tet*(30) ***(23)***.

Genotyping of antibiotic resistance genes can be done by hybridization with any of the known *tet* genes or by polymerase chain reaction (PCR). A hybridization approach is time-consuming and is less sensitive when one is working with environmental or clinical samples. Based on the phylogenetic analysis of tetracycline resistance genes, we have developed PCR primer sets suitable for detection and tracking of these genes in various bacteria and environmental samples ***(19,23,24)*** as well as for quantification of these genes in a variety of bacterial isolates and samples (unpublished data). This PCR-based approach can be easily automated and scaled up, and a similar detection approach can be adapted for detection of other, more clinically relevant, antibiotic resistance genes.

2. Materials

1. Depending on the type of samples used for PCR analysis, genomic DNA can be isolated using the kits from Mo Bio Laboratories (Solana Beach, CA, www.mobio.com):
   a. From tissues, bones, and cell cultures: UltraClean™ Tissue DNA Kit.
   b. From whole blood: UltraClean BloodSpin Kit or UltraClean™ Blood DNA Kit.
   c. From plant tissue: UltraClean Plant DNA Kit.
   d. From soil samples: UltraClean Soil DNA Kit.
   e. From water and urine samples: UltraClean Water DNA Kit.
   f. From fecal samples: UltraClean DRY Soil DNA Kit for fecal samples.
   g. From microorganisms: UltraClean Microbial DNA Kit.
   h. From forensic samples: UltraClean Forensic DNA Kit.

   Additional equipment for genomic DNA isolation using these kits includes a vortex with a flat-bed pan for processing of multiple samples and a microcentrifuge.
2. PCR primers for detection of the RPP genes (**Table 1**) or tetracycline efflux pump genes of Gram-negative bacteria (TEPGNB) (**Table 2**) are synthesized commercially on a 50-nmol scale (high performance liquid chromatography [HPLC] purified). For PCR-denaturing gradient gel electrophoresis (DGGE) analysis, the forward or reverse primer is synthesized with a GC clamp (CGCCCGGGGCGCGCCCCGGGCGGGGCGGGGGC ACGGGGGG) attached to the 5'-end.
3. Takara Ex Taq™ DNA polymerase kit (Takara Shuzo, Japan).
4. Agarose gel electrophoresis buffer: 45 m*M* tris-borate, 1 m*M* EDTA, pH 8.0. This buffer should not require pH adjustment. The 10X stock is stored at room temperature and is used at a working strength of 0.5X.
5. Agarose gel sample loading buffer (6X): 40% (w/v) sucrose in water and 0.5% (w/v) Orange G (Sigma). This dye migrates as a DNA fragment of approx 50 bp.
6. Agarose gel stain solution: 0.5 mg/L (w/v) ethidium bromide in water. Alternatively, for faster results, the fluorescent dye GelStar® (FMC Bioproducts, Rockland, ME) can be incorporated into the agarose gels at a 1:10,000 dilution (v/v).
7. 100-bp DNA Ladder (Promega).
8. DGGE buffer: 40 m*M* Tris-acetate, 1 m*M* EDTA, pH 8.0. The 50X stock solution is stored at room temperature and is used at a working strength of 1X.
9. DGGE sample loading buffer: 0.05% bromophenol blue, 0.05% xylene cyanol, and 70% glycerol in sterile nanopure water.
10. GelBond PAG gel support films (FMC Bioproducts).
11. Freshly prepared silver stain solution: 0.1% $AgNO_3$ in ddH_2O.
12. Developing solution: 0.01% $NaBH_4$ and 0.4% v/v formaldehyde in 1.5% w/v NaOH.
13. QuantiTect SYBR Green PCR kit (Qiagen, Germany) or other high-sensitivity real-time PCR kits based on quantification of amplified DNA by binding the SYBR Green fluorescent dye.
14. Control strains to generate the standard curves for quantification of tetracycline resistance genes in the environmental samples or for determining the copy number of tetracycline resistance genes in microbial cells (**Table 3**). These strains are also used as controls in PCR and PCR-DGGE.
15 QIAquick Gel Extraction Kit (Qiagen).

Table 1
PCR Primers Targeting the Ribosomal Protection Protein Genes

Primer pair	Class targeted	Sequence 5'→3'	PCR annealing temperature (°C)	Amplicon size (bp)
TetB/P-FW	TetB P	AAAACTTATTATATTATAGTG	46	169
TetB/P-RV		TGGAGTATCAATAATATTCAC		
TetM-FW	TetM	ACAGAAAGCTTATTATATAAC	55	171
TetM-RV		TGGCGTGTCTATGATGTTCAC		
TetO-FW	TetO	ACGGARAGTTTATTGTATACC	60	171
TetO-RV		TGGCGTATCTATAATGTTGAC		
TetQ-FW	TetQ	AGAATCTGCTGTTTGCCAGTG	63	169
TetQ-RV		CGGAGTGTCAATGATATTGCA		
TetS-FW	TetS	GAAAGCTTACTATACAGTAGC	50	169
TetS-RV		AGGAGTATCTACAATATTTAC		
TetT-FW	TetT	AAGGTTTATTATATAAAAGTG	46	169
TetT-RV		AGGTGTATCTATGATATTTAC		
TetW-FW	TetW	GAGAGCCTGCTATATGCCAGC	64	168
TetW-RV		GGGCGTATCCACAATGTTAAC		
Tet32-FW	Tet32	TCGACCTACAGCGTGTTTACC	62	277
Tet32-RV		CTAATAGTTCATCGCTTCCGG		

3. Methods

3.1 Genomic DNA Isolation

Genomic DNA of sufficient purity for PCR amplification can be isolated using the corresponding kits from Mo Bio (*see* Materials and **Notes 1** and **2**). Alternatively, if there are difficulties in obtaining these kits, total genomic DNA can be isolated using the following protocol:

1. To 200 mg of material (microbial or cell culture pellet, soil, plant tissue, or other sample), in a 2-mL screw-top cryogen storage propylene plastic tube, add 0.1 g zirconium beads (0.1-mm diameter), 700 μL lysis solution (50 m*M* Tris-HCl, 10 m*M* EDTA, 2% sodium dodecyl sulfate [SDS], pH 8.0), and 700 μL Tris-buffered phenol, pH 8.0.
2. Shake the tubes ("homogenize" setting on a Mini Bead-Beater-8™ [Biospec Products, Bartlesville, OK]) for 1 min and chill on ice for 2 min.
3. Repeat the previous step twice.
4. Centrifuge the tubes in a tabletop centrifuge at 10,000*g* for 5 min.
5. Transfer the upper water phase to a 2-mL microcentrifuge tube and extract with phenol-chloroform-isoamyl alcohol (25:24:1) twice or until the interphase is clear.
6. Precipitate with ethanol, wash with 70% ethanol, dry, and dissolve in 50 μL of TE buffer (10 m*M* Tris-HCl, 1 m*M* EDTA, pH 8.0) with 10 μg/mL RNaseA.

3.2. Conventional PCR Detection of RPP Genes

1. On ice, combine the following components of the PCR mixture in a 200-μL thin-walled PCR tube: 25 pmol of each primer, 1X ExTaq reaction buffer, 100 μM of each

Table 2
PCR Primers Targeting tet Efflux Pumps of Gram-Negative Bacteria

Primer pair	Class targeted	Sequence 5'→ 3'	PCR annealing/ extension temperature (°C)	Amplicon size (bp)
TetA-FW	TetA	GCG CGA TCT GGT TCA CTC G	61	164
TetA-RV		AGT CGA CAG YRG CGC CGG C		
TetB-FW	TetB	TAC GTG AAT TTA TTG CTT CGG	61	206
TetB-RV		ATA CAG CAT CCA AAG CGC AC		
TetC-FW	TetC	GCGGGATATCGTCCATTCCG	68	207
TetC-RV		GCGTAGAGGATCCACAGGACG		
TetD-FW	TetD	GGA ATA TCT CCC GGA AGC GG	68	187
TetD-RV		CAC ATT GGA CAG TGC CAG CAG		
TetE-FW	TetE	GTT ATT ACG GGA GTT TGT TGG	61	199
TetE-RV		AAT ACA ACA CCC ACA CTA CGC		
TetG-FW	TetG	GCA GAG CAG GTC GCT GG	68	134
TetG-RV		CCY GCA AGA GAA GCC AGA AG		
TetH-FW	TetH	CAG TGA AAA TTC ACT GGC AAC	61	185
TetH-RV		ATC CAA AGT GTG GTT GAG AAT		
TeJ-FW	TetJ	CGA AAA CAG ACT CGC CAA TC	61	184
TetJ-RV		TCC ATA ATG AGG TGG GGC		
TetY-FW	TetY	ATT TGT ACC GGC AGA GCA AAC	68	181
TetY-RV		GGC GCT GCC GCC ATT ATG C		
TetZ-FW	TetZ	CCT TCT CGA CCA GGT CGG	61	204
TetZ-RV		ACC CAC AGC GTG TCC GTC		
Tet30-FW	Tet30	CAT CTT GGT CGA GGT GAC TGG	68	210
Tet30-RV		ACG AGC ACC CAG CCG AGC		

deoxynucleoside triphosphate, 1.0 U of ExTaq DNA polymerase, and 200 ng of the purified DNA template, adjusted to a total volume of 20 μL.

2. Perform PCR amplification (25 cycles) with a GeneAmp PCR system 2400 (Perkin-Elmer, Norwalk, CT) or a DNA Engine Thermocycler (MJ Research, Waltham, MA) using initial denaturation at 94°C for 5 min, followed by cycling at 94° for 30 s, 30 s of annealing (annealing temperatures are shown in **Table 2**), 30 s of extension at 72°C, with final extension at 72°C for 7 min. Include positive (**Table 1**) and negative (e.g., a noncomplementary template) controls for each pair of primers used.
3. A second, nested PCR can be performed using 1 μL of product from the first PCR as a template and amplifying for 25 cycles as described above if PCR amplification fails owing to the presence of unidentified PCR-inhibiting substances.

3.3. Conventional PCR Detection of TEPGNB Genes

1. On ice, combine the components of the following PCR mixture in a 200-μL PCR tube: 25 pmol of each primer, 1X ExTaq reaction buffer, 100 μ*M* of each deoxynucleoside triphosphate, 1.0 U of ExTaq DNA polymerase, and 125 ng of the purified DNA template, adjusted to a total volume of 25 μL.

Table 3
Control Strains and Plasmids

Strains and plasmids	*tet* gene	Reference/source
RPP genes		
Clostridium perfringens JIR4202	*tet*B(P)	***25***
pFD310	*tet*(M)	***26***
pGEM-tetO	*tet*(O)	***27***
pBT-1	*tet*(Q)	***28***
pVP2	*tet*(S)	***29***
pAT451	*tet*(S)	***30***
Streptococcus pyogenes A498	tet(T)	***31***
pGEM-tetW	*tet*(W)	***27***
pGEM-tet32	*tet*(32)	***20***
Efflux genes		
RP1	*tet*(A)	***32***
pRT11	*tet*(B)	***33,34***
pBR322	*tet*(C)	***35***
pUC119D	*tet*(D)	***36***
pSL1504	*tet*(E)	***37***
pUC119G	*tet*(G)	***38,39***
pVM111	*tet*(H)	***40***
PVM6	*tet*(J)	***41***
pIE1122	*tet*(Y)	Dr. E.
Tietze		
pAGHD1	*tet*(Z)	***42***
pZLE4.5	*tet*(30)	***43***

2. For detection of *tet*(C), *tet*(D), *tet*(G), *tet*(Y), and *tet*(30), perform a two-step PCR amplification consisting of initial denaturation at 94°C for 5 min followed by 25 cycles at 94° for 5 s and 10 s of annealing/extension at 68°C, with a final extension at 68°C for 7 min. Include positive (**Table 2**) and negative (e.g., a noncomplementary template) controls for each pair of primers used.
3. For detection of *tet*(A), *tet*(B), *tet*(E), *tet*(H), *tet*(J), and *tet*(Z), perform a two-step PCR amplification consisting of initial denaturation at 94°C for 5 min followed by 25 cycles at 94° for 5 s and 30 s of annealing/extension at 61°C, with a final extension at 61°C for 7 min. Include positive (**Table 2**) and negative (e.g., a noncomplementary template) controls for each pair of primers used.
4. A second, nested PCR can be performed using 1 μL of product from the first PCR as a template and amplifying for 25 cycles as just described if amplification fails owing to the presence of unidentified PCR-inhibiting substances.
5. Analyze the PCR products by electrophoresis of 5-μL aliquots on a 2.5% (w/v) agarose gel (NuSieve®, FMC Bioproducts) containing the fluorescent dye GelStar® (FMC Bioproducts). If running the gel without the dye, stain it in the ethidium bromide or GelStar solution after the run. The expected amplicon sizes are shown in **Tables 1** and **2**.

3.4. PCR-DGGE

1. Perform PCR reaction as described earlier for the RPP and TEPGNB genes, but using a primer pair with a GC-clamp on one primer (*see* **Note 3**).
2. Analyze a 5-μL aliquot by agarose gel electrophoresis to confirm that sufficient quantities of the desired amplicon are synthesized.
3. Assemble the vertical gel casting module of the Bio-Rad D-Code System (Hercules, CA) including a GelBond PAG gel supporting film.
4. Form a 15–60% urea/formamide gradient (100% denaturant is equivalent to 7 *M* urea and 40% deionized formamide) in a 8% polyacrylamide gel (0.5X TAE buffer) using a Bio-Rad Gradient Former.
5. Insert the comb and allow the gel to polymerize for 2 h.
6. Remove the comb and wash the pockets with the electrophoresis buffer.
7. Apply samples (1–10 μL), mount the gel module in the preheated electrophoresis tank of the Bio-Rad D-Code System, and perform electophoresis at 60°C and 150 V for 2 h followed by 200 V for 1 h.
8. After the run, remove the supporting film with the gel on it, rinse in ddH_2O, and fix in a solution of 10% ethanol and 0.5% acetic acid for 4 h to overnight with gentle shaking.
9. Rinse gel briefly in ddH_2O.
10. Shake gently in a solution of freshly prepared silver stain solution for 20 min.
11. Rinse gel briefly in ddH_2O.
12. Incubate the gel in developing solution with gentle shaking until the desired intensity of bands is obtained.
13. Rinse gel in ddH_2O.
14. The gel can now be photographed or scanned. We usually capture and digitize the gel images using the Bio-Rad system, which includes a GS-710 Calibrated Imaging Densitometer connected to a G3 Macintosh computer with the Diversity Database™ software. Images can be saved in formats compatible with a number of other image software such as Adobe® Photoshop®.

3.5. Quantitative Real-Time PCR of TEPGNB Genes

1. Amplify the control templates using the corresponding primer pairs (*see* **Subheading 3.3.**, **Tables 2** and **3**, and also **Notes 4** and **5**).
2. Run a preparative agarose gel (TAE buffer, 0.8% agarose) electrophoresis with the whole PCR mix applied.
3. Stain the gel with ethidium bromide.
4. Excise the band from the gel corresponding to the expected size amplicon while viewing under long-wave UV light.
5. Extract DNA with the QIAquick Gel Extraction Kit.
6. Determine the molar concentration of the purified DNA spectrophotometrically. Serial dilutions of this DNA will be amplified in parallel with the experimental samples using a real-time PCR thermocycler and will be used for generating standard curves to quantify the corresponding target genes in experimental samples.
7. For each sample: to QuantiTect SYBR Green master mix, add primers (at final concentration of 5 m*M*) and template, and then add RNase-free water from the kit to a final volume of 100 μL. Distribute 25-μL aliquots to four tubes in a 96-well PCR plate.
8. Prepare other samples and control mixes as described in **step 7**. Seal the plate and perform PCR with initial denaturation/enzyme activation at 94°C for 10 min, followed by 45 cycles of two-step PCR consisting of 15-s denaturation at 94°C and 60-s annealing/extension at 61°C.

9. After the run, perform control analyses such as melting curve analysis and, if necessary, analyze the samples by agarose gel electrophoresis to verify the amplicon sizes.
10. Quantify the target gene concentration in experimental samples based on standard curves generated from the control *tet* templates.

4. Notes

1. Some DNA preparations obtained with the Mo Bio kits may still contain the substances inhibitory to PCR. If the concentration of your genomic DNA preparation is sufficiently high, then the sample may be diluted to decrease the concentration of these substances. Additional DNA purification may include phenol-chloroform-isoamyl alcohol (25:24:1) extraction and gel filtration. As a control for the suitability of DNA for PCR amplification, a PCR reaction with the universal bacterial primers 27f (5'AGAGTTTGATCM TGGCTCAG) and 1525r (AAGGAGGTGWTCCARCC) ***(44)*** can be performed.
2. In some cases, PCR amplification directly from samples (e.g., microbial isolates, tissue, and other biomasses) may be used, omitting the DNA extraction step. In our hands, successful amplification was done with the fresh colony biomass of laboratory *Escherichia coli* strains, *Streptococcus* spp., *Enterococcus* spp., *Salmonella* spp., and lactic acid bacteria. For this, resuspend one-half of a 1–2-mm individual colony in 20 μL of sterile water and use 1–2 μL of this suspension as a template for PCR amplification. The initial denaturation time at 94°C should to be extended to 10 min to allow better denaturation of cellular proteins and more complete release of genomic DNA into a PCR mix. This labor- and time-saving shortcut may be especially important when a large number of samples need to be examined.
3. The DNA markers for PCR-DGGE of tetracycline resistance genes can be generated from control strains (**Table 3**).
4. Real-time PCR is extremely sensitive, allowing detection of several target molecules in a reaction, and care should be taken to reduce the possibility of contamination during all steps, beginning with purity of reagents, preparation of genomic DNA, and mixing of the PCR components. First, all solutions, water, and equipment for genomic DNA isolation must be free of tetracycline resistance gene contamination. Many molecular biology labs perform routine cloning with plasmids and strains that possess tetracycline resistance markers, the most widely used being *tet*(C), *tet*(B), and *tet*(A). These genes may circulate in the lab environment and air and may contribute to false-positive signals in real-time PCR. All surfaces, pipets, tube stands, and so on have to be cleaned with 3% H_2O_2 and ethanol and then UV-irradiated. The use of aerosol-protected filter tips is highly recommended. Second, genomic DNA isolation from samples and control strains should preferably be carried out separately to avoid crosscontamination. Third, mixing the components for real-time PCR should, if possible, be performed in a designated UV-treated box.
5. Albeit suitable for conventional PCR and PCR-DGGE, the TetC primer set (**Table 2**) produced an additional low-molecular-weight amplicon in real-time PCR. Therefore, this pair cannot be recommended for quantitative real-time PCR.

References

1. Bergeron, M. G. and Quellette, M. (1995) Diagnosing bacterial infectious diseases in one hour: an essential upcoming revolution. *Infection* **23,** 69–72.
2. Chopra, I. and Roberts, M. (2001) Tetracycline antibiotics: mode of action, applications, molecular biology, and epidemiology of bacterial resistance. *Microbiol. Mol. Biol. Rev.* **65,** 232–260.

3. Schnappinger D. and Hillen, W. (1996) Tetracyclines: antibiotic action, uptake, and resistance mechanisms. *Arch. Microbiol.* **165,** 359–369.
4. Leyden, J. J. (2001) Current issues in antimicrobial therapy for the treatment of acne. *J. Eur. Acad. Dermatol. Venereol.* **15(suppl 3),** 51–55.
5. Del Rosso, J. Q. (2000) Systemic therapy for rosacea: focus on oral antibiotic therapy and safety. *Cutis* **66(4 suppl),** 7–13.
6. Johnson, L. R. and N. H. Stoller (1999) Rationale for the use of Atridox therapy for managing periodontal patients. *Compend. Contin. Educ. Dent.* **20,** 19–25.
7. Tuckman, M., Petersen, P. J., and Projan, S. J. (2000) Mutations in the interdomain loop region of the *tet*A(A) tetracycline resistance gene increase efflux of minocycline and glycylcyclines. *Microb Drug Resist.* **6,** 277–282.
8. Burdett, V. (1986) Streptococcal tetracycline resistance mediated at the level of protein synthesis. *J. Bacteriol.* **165,** 564–569.
9. Burdett, V. (1991) Purification and characterization of Tet(M), a protein that renders ribosomes resistant to tetracycline. *J. Biol. Chem.* **266,** 2872–2877.
10. Manavathu E. K., Hiratsuka, K., and Taylor, D. E. (1988) Nucleotide sequence analysis and expression of a tetracycline resistance gene from *Campylobacter jejuni. Gene* **62,** 17–26.
11. Manavathu E. K., Fernandez, C. L., Cooperman, B. S., and Taylor, D. E. (1990) Molecular studies on the mechanism of tetracycline resistance mediated by Tet(O). *Antimicrob. Agents Chemother.* **34,** 71–77.
12. Roberts, M.C. (1996) Tetracycline resistance determinants: mechanisms of action, regulation of expression, genetic mobility, and distribution. *FEMS Microbiol. Rev.* **19,** 1–24.
13. Sanchez-Pescador, R., Brown, J. T., Roberts, M. C., and Urdea, M. S. (1988) Homology of the Tet(M) with translational elongation factors: implication for potential modes of tet(M)-conferred tetracycline resistance. *Nucleic Acids Res.* **16,** 1218.
14. Taylor D. E. and Chau, A. (1996) Tetracycline resistance mediated by ribosomal protection. *Antimicrob. Agents Chemother.* **40,** 1–5.
15. Levy, S. B. (1992) Active efflux mechanisms for antimicrobial resistance. *Antimicrob. Agents Chemother.* **36,** 695–703.
16. Speer, B. S. and Salyers, A. A. (1989) Novel aerobic tetracycline resistance gene that chemically modifies tetracycline. *J. Bacteriol.* **171,** 148–153.
17. Levy, S. B., McMurry, L. M., Burdett, V., et al. (1989) Nomenclature for tetracycline resistance determinants. *Antimicrob. Agents Chemother.* **33,** 1373–1374.
18. Levy, S. B., McMurry, L. M., Barbosa, T. M., et al. (1999) Nomenclature for new tetracycline resistance determinants. *Antimicrob. Agents Chemother.* **43,** 1523–1524.
19. Aminov, R. I., Garrigues-Jeanjean, N., and Mackie. R. I. (2001) Molecular ecology of tetracycline resistance: development and validation of primers for detection of tetracycline resistance genes encoding ribosomal protection proteins. *Appl. Environ. Microbiol.* **67,** 22–32.
20. Melville, C. M., Scott, K. P., Mercer, D. K., and Flint, H. J. (2001) Novel tetracycline resistance gene, *tet*(32), in the *Clostridium*-related human colonic anaerobe K10 and its transmission *in vitro* to the rumen anaerobe *Butyrivibrio fibrisolvens. Antimicrob Agents Chemother.* **45,** 3246–3249.
21. Pao, S. S., Paulsen, I. T., and Saier, M. H. (1998) Major facilitator superfamily. *Microbiol. Mol. Biol. Rev.* **62,** 1–34.
22. Paulsen, I. T., Brown, M. H., and Skurray, R. A. (1996) Proton-dependent multidrug efflux systems. *Microbiol. Rev.* **60,** 575–608.

23. Aminov R. I., Chee-Sanford, J. C., Garrigues, N., et al. (2002) Development, validation, and application of primers for detection of tetracycline resistance genes encoding tetracycline efflux pumps in gram-negative bacteria. *Appl. Environ. Microbiol.* **68,** 1786–1793.
24. Chee-Sanford, J., Aminov, R. I., Garrigues, N., Krapac, I., and Mackie, R. (2001) Occurrence and diversity of tetracycline resistance genes in lagoons and groundwater underlying two swine production facilities. *Appl. Environ. Microbiol.* **67,** 1494–1502
25. Sloan, J., McMurray, L. M., Lyras, D., Levy, S. B., and Rood, J. I. (1994) The *Clostridium perfringens* Tet P determinant comprises two overlapping genes: *tetA*(P), which mediates active tetracycline efflux, and *tetB*(P), which is related to the ribosomal protection family of tetracycline-resistance determinants. *Mol. Microbiol.* **11,** 403–415.
26. Smith C. J., Rogers, M. B., and McKee, M. L. (1992) Heterologous gene expression in *Bacteroides fragilis*. *Plasmid* **27,** 141–154.
27. Barbosa, T. M., Scott, K. P., and Flint, H. J. (1999) Evidence for recent intergenic transfer of a new tetracycline resistance gene, *tet*(W), isolated from *Butyrivibrio fibrisolvens*, and the occurrence of *tet*(O), in ruminal bacteria. *Environ. Microbiol.* **1,** 53–64.
28. Nikolich, M. P., Shoemaker, N. B., and Salyers, A. A. (1992) A *Bacteroides* tetracycline resistance gene represents a new class of ribosome protection tetracycline resistance. *Antimicrob. Agents Chemother.* **36,** 1005–1012.
29. Perreten, V., Schwarz, F., Cresta, L., Boeglin, M., Dasen, G., and Teuber, M. (1997) Antibiotic resistance spread in food. *Nature* **389,** 801–802.
30. Charpentier, E., Gerbaud, G., and Courvalin, P. (1993) Characterization of a new class of tetracycline-resistance gene tet(S) in *Listeria monocytogenes* BM4210. *Gene* **131,** 27–34.
31. Clermont, D., Chesneau, O., De Cespedes, G., and Horaud, T. (1997) New tetracycline resistance determinants coding for ribosomal protection in streptococci and nucleotide sequence of *tet*(T) isolated from *Streptococcus pyogenes* A498. *Antimicrob. Agents Chemother.* **41,** 112–116.
32. Waters, S.H., Grinsted, J., Rogowsky, P., Altenbuchner, J., and Schmitt, R. (1983) The tetracycline resistance determinants of RP1 and Tn1721: nucleotide sequence analysis. *Nucleic Acids Res.* **11,** 6089–6105.
33. Jorgensen, R. A. and Reznikoff, W. S. (1979) Organisation of structural and regulatory genes that mediate tetracycline resistance in transposon Tn*10*. *J. Bacteriol.* **138,** 705–714.
34. Hillen, W. and Schollmeier, K. 1983. Nucleotide sequence of the Tn*10* encoded tetracycline resistance gene. *ucleic Acids Res.* **11,** 525–539.
35. Bolivar, F., Rodriguez, R. L., Greene, P. J., et al. (1977) Construction and characterization of new cloning vehicles, II. A multipurpose cloning system. *Gene* **2,** 95–113.
36. Kim, E. H. and Aoki, T. (1994) The transposon-like structure of IS26-tetracycline, and kanamycin resistance determinant derived from transferable R plasmid of fish pathogen, *Pasteurella piscicida*. *Microbiol. Immunol.* **38,** 31–38.
37. Marshall, B., Morrissey, S., Flynn, P., and Levy, S. B. (1987) A new tetracycline-resistance determinant, class E, isolated from Enterobacteriaceae. *Gene* **50,** 111–117.
38. Aoki, T., Satoh, T., and Kitao, T. (1987) New tetracycline resistance determinant on R plasmid from *Vibrio anguillarum*. *Antimicrob. Agents Chemother.* **31,** 1446–1449.
39. Zhao, J. and Aoki, T. (1992) Nucleotide sequence analysis of the class G tetracycline resistance determinant from *Vibrio anguillarum*. *Microbiol. Immunol.* **36,** 1051–1060.
40. Hansen, L.M., McMurry, L. M., Levy, S. B., and Hirsh, D. C. (1993) A new tetracycline resistance determinant, Tet H, from *Pasteurella multocida* specifying active effux of tetracycline. *Antimicrob. Agents Chemother.* **37,** 2699–2705.

41. Magalhaes, V. D., Schuman, W., and Castilho, B. A. (1998) A new tetracycline resistance determinant cloned from *Proteus mirabilis. Biochim. Biophys. Acta* **1443,** 262–266.
42. Tauch, A., Puhler, A., Kalinowski, J., and Thierbach, G. (2000) TetZ, a new tetracycline resistance determinant discovered in gram-positive bacteria, shows high homology to gram-negative regulated efflux systems. *Plasmid* **44,** 285–291.
43. Luo, Z.-Q. and Farrand, S. K. (1999) Cloning and characterization of a tetracycline resistance determinant present in *Agrobacterium tumefaciens* C58. *J. Bacteriol.* **181,** 618–626.
44. Lane, D. J. (1991) 16S/23S rRNA sequencing, In: *Nucleic Acid Techniques in Bacterial Systematics* (Stackebrandt, E. and Goodfellow, M., eds.). John Wiley & Sons, New York, pp. 115–175.

2

Integron Analysis and Genetic Mapping of Antimicrobial Resistance Genes in *Salmonella enterica* serotype Typhimurium

Mairéad Daly and Séamus Fanning

1. Introduction

Antimicrobial resistance determinants may be transferred among bacteria via mobile genetic elements including plasmids, transposons, and the more recently explored integrons *(1)*. Integrons are naturally occurring genetic elements found as part of the Tn*21* transposon family or located on various broad host-range plasmids *(2)*. The fundamental integron structure consists of a 5'-conserved segment (5'-CS) of 1.4-kbp and a 2-kbp 3'-CS *(3)*. Between these conserved regions are DNA sequences of variable length and molecular complexity. These intervening sequences are known as *gene cassettes*, and several have now been characterized (**Fig. 1**). Acquisition and dissemination of these genes located within the integron structure, results in an increase in antimicrobial resistance *(4)*.

Three classes of integron structure have been described. Class 1 integrons are of principal importance in clinical isolates. The 5'-CS of class 1 integrons includes an *intI* 1 gene of 1358 bp, which encodes a specific recombinase, a member of the DNA integrase family *(5)*. This gene contains the *attI* recombination site, required for specifically integrating gene cassettes (**Fig. 1**). Classes 2 and 3 also contain integrase genes (*intI* 2 and *intI* 3), with the former showing 40% sequence identity to those of class 1, and the latter showing 61% *(1)*. All three classes of integrons contain similar gene cassettes from the same families, which suggests the existence of a common pool of gene cassette with cross-specificity between the classes *(1)*.

When the 3'-CS region is examined in detail, it contains several open reading frames (ORFs). These include *qac*EΔ1, which confers resistance to quaternary ammonium compounds, often associated with antiseptics, along with a *sulI* gene expressing resistance to sulphonamide antimicrobial agents. The *sulI* gene encodes the enzyme hydopteroate synthase. Transcription of the *sulI* gene begins at a promoter located in

From: *Methods in Molecular Biology, vol. 268: Public Health Microbiology: Methods and Protocols*
Edited by: J. F. T. Spencer and A. L. Ragout de Spencer © Humana Press Inc., Totowa, NJ

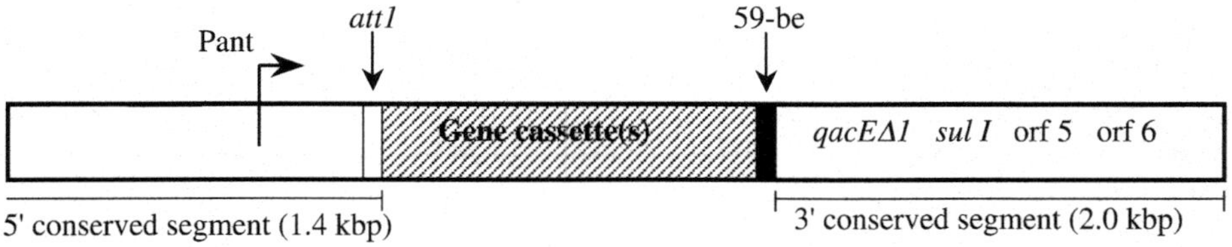

Fig.1. Schematic representation of a class 1 integron structure. Several gene cassettes (hatched box) have been found located between the 5'- and 3'-CS. Often, more than one open reading frame (ORF) gene cassette can be found within the same integron, in the same orientation, and transcribed from a common promoter located proximal to the 5'-site of insertion.

the 5'-CS. The latter is also responsible for the transcription of the inserted gene cassette(s) ***(6)***. Two additional ORFs, ORF-5 and -6, are located toward the distal end of the 3'-CS. The gene product of ORF 5 appears to share some sequence similarity with puromycin acetyltransferase, and this feature suggests a possible role in antimicrobial resistance ***(7)***. A biological function has yet to be ascribed to ORF-6.

Gene cassettes are discrete mobile DNA elements, which may exist as free, circular, nonreplicating DNA molecules (possibly) in transit from one genetic locus to another ***(8)***. They usually contain one gene along with a recognition sequence known as the 59-base element (59-be). The latter is a necessary component required for their integration into the larger mobile elements (i.e., integrons). Examples of gene cassettes commonly found in integrons include the aminoglycoside-modifying enzyme-encoding genes *ant (3")-1a*, *ant (2")-1a*, *aac (3')-1*, and *aac (6')-1b*, the β-lactamase-encoding genes, $bla_{\text{PSE-1}}$, $bla_{\text{OXA-1}}$, $bla_{\text{OXA-2}}$, $bla_{\text{OXA-5}}$, and $bla_{\text{CARB-3}}$, genes conferring resistance to chloramphenicol (*cat*, *cml*), and *dfr* genes resulting in resistance to trimethoprim. Gene cassettes lack a functional promoter ***(9)*** and therefore cannot be transcribed until they are correctly positioned within an integron ***(10)***. Pant, the integron promoter responsible for controlling gene expression is located toward the 5'-end of the *attI* site, 214 bases from the inner boundary of the 5'-CS (**Fig. 1**). Thus these unusual genetic elements act as natural expression vectors ***(1,11)***.

Salmonalla enterica serotype Typhimurium is recognized as a significant human pathogen. It is currently estimated that, of the 40,000 *Salmonella* isolates reported to the Centers for Disease Control and Prevention (CDC), 8.5% are identified as serotype Typhimurium ***(12)***. These organisms are often simultaneously resistant to five or more common antimicrobial agents including ampicillin (A), chloramphenicol (C), streptomycin (S), sulphomamides (Su), and tetracycline (T). The increasing spectrum of resistance among *Salmonella* Typhimurium, definitive type (DT)104, has led several investigators to investigate the mechanism(s) of resistance acquisition. Antimicrobial resistance in *S.* Typhimurium DT204c and DT193 has been entirely attributed to the presence of plasmids; however, DT104 represents an unusual case in that some of the resistance determinants have become chromosomally integrated ***(13)***.

Resistance to sulfonamides when encountered is suggestive of the involvement of a class 1 integron. This is a useful phenotypic indicator, as it is a property attributed to class 1 integrons (as outlined above). Multiple drug-resistant (MDR)-DT104 of resistance (R)-type ACSSuT are often associated with the presence of two class 1 integrons, containing gene cassettes of 1.0 and 1.2 kbp. The former gene cassette contains *ant(3')-1a*, conferring resistance to aminoglycoside antibiotics, streptomycin, and spectinomycin; the 1.2-kbp gene cassette contains a bla_{PSE-1} gene, encoding a β-lactamase enzyme conferring resistance to β-lactam antibiotics including ampicillin. Fine genetic mapping studies have located these structures to the chromosome of *S.* Typhimurium DT104 within a highly conserved 10-kbp *XbaI* DNA fragment ***(13-15)***. This chromosomal region has been described as a multiresistant gene cluster consisting of two integrons flanking an R-plasmid ***(16)***. Molecular analysis of antimicrobial resistance indicated that the same gene cassettes accounted for MDR-DT104 isolates from diverse geographical regions ***(14,17-20)***.

The inserted gene cassettes described above may be recovered by DNA amplification using the previously characterized Int1 F and Int1 R primer set ***(2)*** (**Table 1**). In a standard polymerase chain reaction (PCR), the complete amplification of any inserted gene cassette(s) within a class 1 integron structure can be successfully achieved. A typical result, for a DT104 isolate of R-type ACSSuT, is shown in **Fig. 2** (lane 1). Two intensely ethidium bromide-stained DNA fragments corresponding to the 1.0- and 1.2-kbp amplicons outlined previously are detected after conventional agarose gel electrophoresis. These structures accounted for three (ASSu) of the five resistance traits normally associated with MDR-DT104 isolates (see below). Larger amplicons (approx 1.6 kbp [**Fig. 2**, lane 2]) have also been identified in *S.* Typhimurium isolates including phage types DT170a and -193. The amplicon shown in **Fig. 2** (lane 2) contained two resistance genes, *dfrI* and *aadA*, fused in a classical head-to-tail fashion ***(15)***.

Chloramphenicol acetyl transferases (CATs) are the enzymes most frequently reaponsible for resistance to chloramphenicol. However nonenzymatic mechanisms have also been described including efflux pumps, membrane transporter impermeability, and ribosomal modifications. A *cmlA* gene found in MDR-DT104 and recently described by Briggs and Fratamico ***(14)*** encodes an exporter enzyme and not the more common acetyltransferase. The *floR* gene, which belongs to the MDR efflux pump family of proteins, was described ***(16)*** and is responsible for cross-resistance to chloramphenicol and florfenicol. The *floR* gene is located downstream of the integron containing the *ant (3")-1a* gene cassette. Distal to the *floR* gene are the *tetRA* genes (class G tetracyclines). Tetracycline resistance arises as a result of the production of an efflux pump. These resistance determinants are located on the integrated R-plasmid, outlined above.

The R-type pattern ACSSuT can therefore be accounted for by the presence of this conserved MDR locus on a 10-kpb *Xba1* digested DNA fragment. However, additional resistances to trimethoprim and fluoroquinolones are now being encountered in MDR-DT104 isolates ***(15)***. Trimethoprim resistance can arise owing to the presence of a nonconjugative but mobilizable 4.6-MDa plasmid, which can also encode resistance to sulphonamides (SuTp). Genes associated with trimethoprim resistance (including those of the *dfr* gene family), expressing a dihydrofolate reductase enzyme, have also been found in integrons ***(15)***. Considering the nature of integrons, this presents an opportunity for trimethoprim resistance determinants to become chromosomally inte-

Table 1
Sequence and Characteristics of Oligonucleotide Primers Used for PCR

Primer	Sequence 5' →3'	Conc. per reaction (pmol)	% GC	PCR T-annealing (°C)
IntI1 F	GGC ATC CAA GCA GCA AGC	25	61	55
IntI1 R	AAG CAG ACT TGA CCT GAT	25	44	55
sul1 F	CTT CGA TGA GAG CCG GCG GC	25	70	65
sul1 R	GCA AGG CGG AAA CCC GCG CC	25	75	65
qac EΔ1 F	ATC G2CA ATA GTT GGC GAA GT	25	45	55
qac EΔ1 R	CAA GCT TTT GCC CAT GAA GC	25	50	55
ant(3")-1a F	GTG GAT GGC GGC CTG AAG CC	25	70	65
ant(3")-1a R	ATT GCC CAG TCG GCA GCG	25	67	65
pse-1 F	CGC TTC CCG TTA ACA ACT AC	25	55	55
pse-1 R	CTG GTT CAT TTC AGA TAG CG	25	45	55
dfrI F	GTG AAA CTA TCA CTA ATG GTA GCT	25	37	65
dfrI R	ACC CTT TTG CCA GAT TTG GTA ACT	25	42	65
flo F	ACC CGC CCT CTG GAT CAA GTC AAG	25	58	70
flo R	CAA ATC ACG GGC CAC GCT GTA TC	25	54	70

grated. A larger second plasmid of 60-MDa is also found in DT104s of R-type ACSSuT; however, this plasmid has not been linked with antimicrobial resistance but rather contains virulence determinants ***(13,21)***. *Salmonella* plasmid virulence (*spv*) genes are carried on this plasmid.

The presence of putative resistance islands can be investigated using a combination of molecular approaches including pulsed-field gel electrophoresis (PFGE), Southern blotting and probe hybridization experiments. PFGE is a useful molecular typing protocol, wherein the complete bacterial chromosome is investigated with a rare cutting restriction endonuclease ***(22)***. The macrodigested DNA fragments produced must be resolved by PFGE. These can then be denatured to yield single-stranded DNA fragments and transferred by Southern blotting to nylon membranes. Selected molecular probes for hybridization can be conveniently prepared by PCR, with the specific primer sets outlined in **Tables 2** and **3**.

For example, when making the $bla_{PSE\text{-}1}$ probe, the primer set given was used to amplify a segment of the *pse* gene from the gene cassette. In this case the *int1* primer set is not used for probe generation purposes, as the flanking sequences are common among many gene cassettes and may lead to false-positive hybridization signals. Furthermore, the PFGE profiles may also provide epidemiological data and can be analyzed to indicate the genomic relationships of isolates (**Fig. 3A**) prior to Southern blotting. Suitably labeled molecular weight markers must be included for size determination, and these must be chosen depending on the type of hybridization and detection system used. When a digoxigenin (DIG)-labeling and detection system is used, prelabeled molecular weight markers can be included in two or three gel lanes. Once

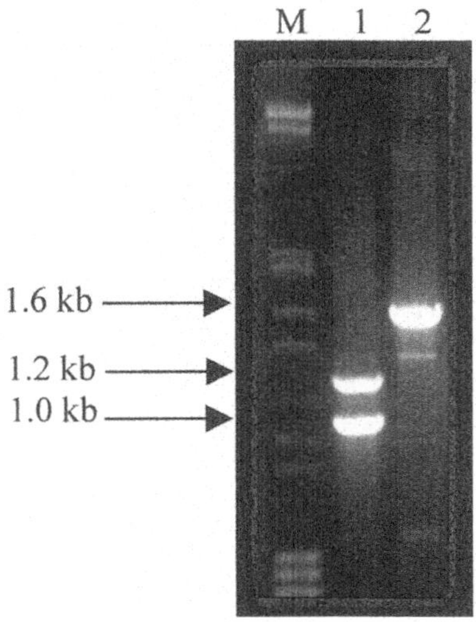

Fig.2. Amplified gene cassettes of two representative strains of *Salmonella enterica* serotype Typhimurium. After PCR, 10 μL of the amplified reaction mixture was loaded into a 1% (w/v) agarose gel in 1X Tris-EDTA-acetate (TAE) buffer containing 0.1 μg of ethidium bromide per mL. Samples were horizontally electrophoresed at 80 V for 60 min. The lanes marked M contain an equal mixture of molecular weight markers grade III (ranging in size from 0.56 to 21.2 kb) and molecular weight markers grade V (ranging in size from 8 to 597 bp; Roche Diagnostics). Lane 1, CIT-F45, DT 104 isolate, containing two integrons with gene cassettes of 1.0 and 1.2 kbp; lane 2, CIT-F41, DT 193 isolate, containing one integron carrying a gene cassette with two ORFs, producing a PCR amplicon of 1.6 kbp.

Table 2
Thermocycler Programs, Step Design, Cycle Number, and Corresponding Citation Used With Oligonucleotide Primers[a]

PCR program	Step 1 Predenaturation	Step 2 Denaturation	Step 3 Annealing	Step 4 Extension	No. of cycles	Ref.
IntI1	95°C, 5 min	95°C, 1 min	55°C, 1 min	72°C, 2 min	30	***2***
sulI	95°C, 5 min	95°C, 1 min	65°C, 1 min	72°C, 2 min	30	***24***
qac EΔ1	95°C, 5 min	95°C, 1 min	55°C, 1 min	72°C, 2 min	30	***6***
ant(3")-Ia	95°C, 5 min	95°C, 1 min	65°C, 1 min	72°C, 2 min	30	***25***
pse-1	95°C, 5 min	95°C, 1 min	55°C, 1 min	72°C, 2 min	30	***26***
dfrI	95°C, 5 min	95°C, 1 min	68°C, 1 min	72°C, 1 min	30	***15***
flo	95°C, 5 min	95°C, 1 min	70°C, 1 min	72°C, 1 min	30	***16***

[a]A final extension step of 72°C for 5 min was incorporated into each program; the samples are then maintained at 4°C until analysis.

Table 3
Southern Blotting of *Salmonella* Typhimurium Isolates and Probe Hybridization Using Selected Gene Cassettes

Phage type	R-type	10-kb *Xba*I fragment	Probe hybridization *ant (3")-Ia*	*pseI*	*dfrI*
DT104					
CIT-V38	ACSSuT	+	+	+	–
CIT-H164	ACSSuT	+	+	+	–
CIT-H176	ACSSuTTp	+	+	+	–
CIT-H183	SSu	–	–	–	–
CIT-V115	ACSSuT	+	+	+	–
CIT-F44	ASu	–	–	–	–
DT104 related					
CIT-F107 (104b)	ACSSuTK	+	+	+	–
CIT-H144 (104b)	ACSSuTTp	+	+	+	+
CIT-V37 (U302)	ACSSuTTp	+	+	+	–
DT193					
CIT-F34	ASSuTN	–	+	+	+
CIT-F41	ASSuTTp	–	+	+	+
DT195					
CIT-V60	SuTTp	–	–	–	–
DT208					
CIT-V75	T	–	–	–	–
DT170a					
CIT-V127	ASSuTTp	–	+	+	+
CIT-V129	SuTTp	–	–	–	–
Nontypable					
CIT-F40	Sensitive	–	–	–	–
CIT-F105	SuTp	–	–	–	–
CIT-H195	SuTp	–	+	+	–
Controls					
E. coli R100.1		–	+	+	+
E. coli R751		–	+	+	–

ant (3")-1a, aminoglycoside modifying enzyme coding gene; *pse*I, β-lactamase gene; *dfrI*, dihydrofolate reductase gene; R-type, antimicrobial resistance pattern; A, ampicillin; C, chloroamphenicol; S, streptomycin; Su, sulphonamides; T, tetracycline; N, nalidixic acid; K, kanamycin; V, veterinary; H, human; F, food.

Fig. 3. (*opposite page*) The lanes marked M contain DIG-labeled DNA molecular weight marker grade II (Roche Diagnostics). Unlabeled mid-range PFG markers (New England BioLabs) in lane M* were included for fragment sizing before Southern transfer. Lane 1, CIT-V38; lane 2:CIT-F45; lane 3, CIT-H164; lane 4, CIT-H176; lane 5, CIT-H183; lane 6, CIT-V115; lane 7, CIT-F107; lane 8, CIT-H144; lane 9, CIT-V37; lane 10, CIT-F34; lane 11, CIT-F41; lane 12, CIT-F44; lane 13, CIT-V60; lane 14, CIT-V75; lane 15, CIT-V127; lane 16, CIT-V129; lane 17, CIT-F40; lane 18, CIT-F105; lane 19, CIT-H195; lane 20, *E. coli* R100.1 (control isolate); and lane 21, *E. coli* R751 (control isolate). (**A**) Pulsed-field gel electrophoresis

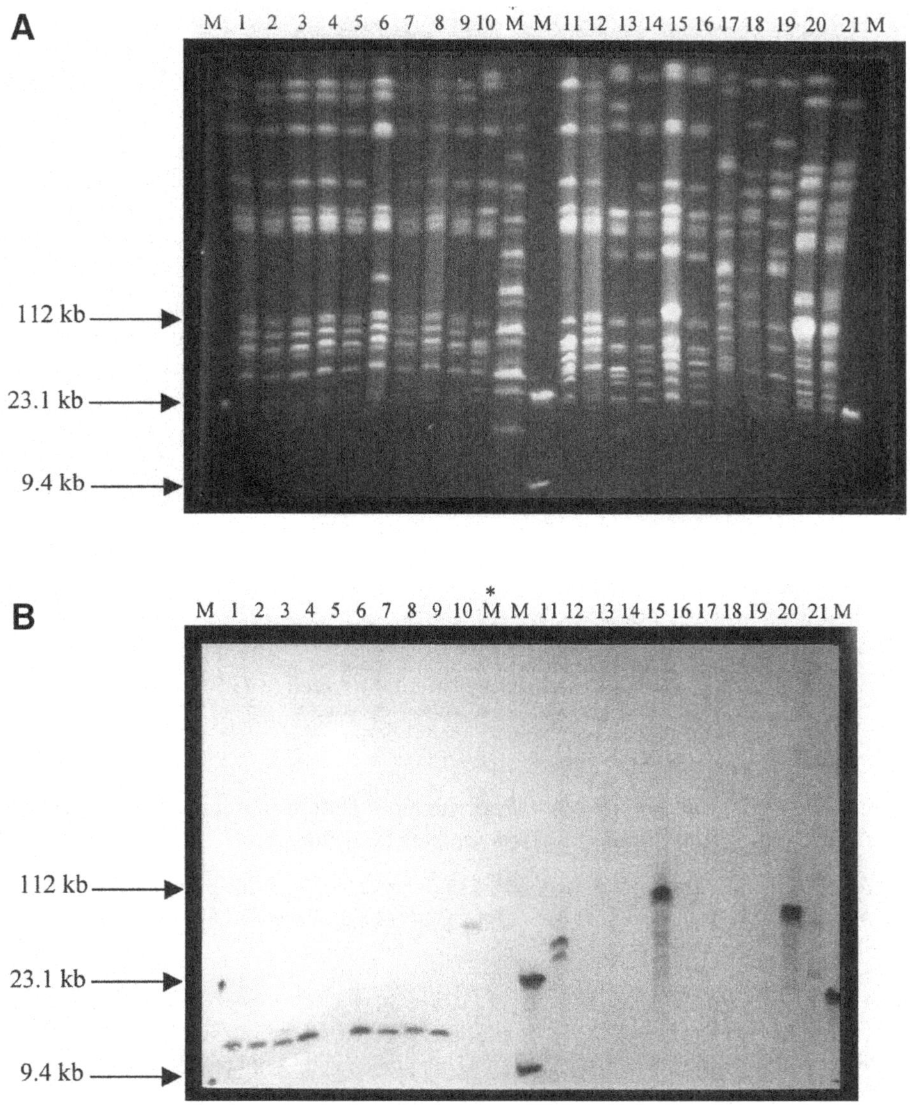

(*Fig. 3. continued*) profiles of *Salmonella* Typhimurium isolates digested with *Xba*I following electrophoresis through 1% (w/v) SeaKem Gold agarose in 0.5X TBE for 18 h at 10.5°C. Electrophoresis was performed at 200 V using a Gene Navigator system (Pharmacia) in the interpolation mode pulsing from 1 through 40 s. (**B**) Following pulsed-field gel electrophoresis the *Xba*I macrorestricted DNA fragments were transferred to nylon membranes. Individual gene cassettes were then used to probe the membranes. Southern blot using the $bla_{PSE\text{-}1}$ gene as a probe.

the PFGE patterns are analyzed, DNA is transferred as an exact pattern replica onto a nylon membrane or other suitable solid phase. The membrane can then be hybridized with a specific antimicrobial resistance gene probe and the location of any hybridizing signal(s) generated detected visually (**Fig. 3B**). This entire procedure must be repeated for each new probe and the position of the hybridizing signal recorded each time. One possible limitation to this approach is the time required, as several replica sets of PFGE gels must be produced, transferred to nylon membranes, and subsequently probed. However, it is an advantage, if several agarose slices are digested simultaneously for PFGE (*see* **Note 1**).

Figure 3B shows the results of a hybridization experiment using the $bla_{PSE\text{-}1}$ gene probe with the isolates listed in **Table 3**. The corresponding PFGE patterns are shown in **Fig. 3A**. A signal was detected for the DT104 isolates of R-type ACSSuT at approx 10 kbp. A similar result was obtained when using the *ant (3")-1a* and the *flo* gene probes sequentially. This indicated that these genes are located within the same region on the *S.* Typhimurium genome. Larger fragments in non-DT104s can also be seen, the largest of which is 112 kbp (**Fig. 3B**, lane 15). This isolate (CIT-V127, **Table 3**) also had a hybridization signal at this position when probed with *ant (3")-1a* and *dfrI* (**Table 1**). Analysis of hybridization signals from the various gene probes can therefore identify multiresistant gene clusters in bacterial isolates.

2. Materials

All reagents marked with an asterisk (*) are autoclaved at 121°C for 15 min prior to use.

2.1. DNA Extraction

This following protocol is a general purification method for Gram-negative organisms *(23)*, yielding good-quality DNA templates (*see* **Note 2**).

1. All *S.* Typhimurium isolates used in this study are listed in **Table 3**.
2. Tryptone soya broth (TSB*; Oxoid, Hampshire, UK): dispensed in 5-mL volumes into sterile universal containers.
3. 1 *M* NaCl*.
4. TE solution*: 50 m*M* Tris-HCl, pH 8.0, 50 m*M* EDTA.
5. Lysozyme: obtained lyophilized from Sigma (Poole, UK), dissolved in sterile distilled water to a final concentration of 2 mg/mL, aliquoted, and stored at –20°C.
6. 20% (w/v) Sodium dodecyl sulfate (SDS).
7. Proteinase K: obtained lyophilized from Sigma, reconstituted in sterile distilled water to a final concentration of 10 mg/mL, aliquoted, and stored at –20°C.
8. Phenol/chloroform: isoamylalcohol (25:24:1; Sigma), saturated with 10 m*M* Tris-HCl, pH 8.0, 1 m*M* EDTA.
9. 3 *M* Ammonium acetate, filter-sterilized, and stored at room temperature.
10. 70 and 100% (v/v) ethanol stored at –20°C.
11. 1X TE solution*: 10 m*M* Tris-HCl, pH 8.0, 1 m*M* EDTA.

2.2. Polymerase Chain Reaction

1. Deoxyribonucleoside triphosphates (dNTPs; Promega, Madison, WI), stock concentration of 100 m*M*. Prepare a working concentration of 1.25 m*M* dNTPs by mixing 2.5 μL of

each (dATP, dCTP, dGTP, and dTTP) with 190 µL of sterile distilled water (final volume 200 µL). Store at –20°C until required.

2. 2.5 U*Taq* DNA polymerase (Promega). The enzyme is supplied with 25 m*M* $MgCl_2$ and 10X reaction buffer (100 m*M* Tris-HCl, pH 9.0, 500 m*M* KCl, 1% Triton X-100).
3. Sterile distilled water.
4. Oligonucleotide primers: all primers used for integron and gene cassette analysis were synthesized by Oswel (Southampton, UK) and purified by high-performance liquid chromatography (HPLC). These primer sequences together with their relevant characteristics are listed in **Table 1**.

2.3. Conventional Agarose Gel Electrophoresis

1. *ultra* PURE agarose, electrophoresis grade from GibcoBRL Life Technologies (Paisley, Scotland). Gelling temperature 36-42°C.
2. 10 mg/mL Ethidium bromide in distilled water (stored in a dark bottle).
3. 10X TAE buffer: 400 m*M* Tris-HCl, pH 7.8, 400 m*M* glacial acetic acid, and 2 m*M* EDTA; working concentration 1X.
4. Loading dye: prepared by mixing 100 µL of 10% (w/v) bromophenol blue with 6.6 mL glycerol and 3.3 mL 10X TAE.
5. Molecular weight markers: several preparations of molecular weight markers are available, and the choice of a suitable marker is made based on the expected size of the amplified DNA fragment(s). Usually grades III and V molecular weight markers from Roche Diagnostics (East Sussex, UK) are used, and these have the following known DNA fragments:
   a. DNA molecular weight marker III: consists of 13 DNA fragments ranging in size from 0.12 to 21.2 kbp; prepared by cleaving λ DNA with *Eco*RI and *Hin*dIII (250 µg/mL).
   b. DNA molecular weight marker V: consists of 22 DNA fragments ranging in size from 8 to 587 bp; prepared by cleaving pBR322 with *Hae*III (250 µg/mL).

2.4. Pulsed-Field Gel Electrophoresis

2.4.1. Preparation of Agarose Plugs

1. Nutrient agar plates (Oxoid).
2. 1-mL Sterile plastic syringes with the nozzle removed.
3. Saline 0.85% (w/v) NaCl.
4. TEN buffer: 100 m*M* Tris-HCl, pH 7.5 100 m*M* EDTA.
5. Incert® agarose (FMC BioProducts, Vallesbaek Strand, Denmark).
6. 20 mg/mL Proteinase K in distilled water. Store at –20°C.
7. Lysis buffer: 6 m*M* Tris-HCl, pH 7.6, 1 *M* NaCl, 100 m*M* EDTA, pH 7.6, 0.5% (w/v) Brij-58 (polyoxyethylene-20-cetyl-ether), 0.2% (w/v) sodium deoxycholate, 0.5% (w/v) sarkosyl. Filter-sterilize and store at 4°C.
8. 400 m*M* EDTA, pH 9.3, 1% (w/v) sarkosyl.
9. 1X TE*: 100 m*M* Tris-HCl, pH 7.6, 100 m*M* EDTA, pH 7.6, 150 m*M* NaCl.

2.4.2. Restriction Endonuclease Digestion of Agarose Slices

1. DNS buffer: 100 m*M* Tris-HCl, pH 8.0, 5 m*M* $MgCl_2$.
2. RNAse super stock: RNAse obtained from Sigma, diluted to 10 mg/mL with sterile distilled water, and frozen at –20°C in 1-mL aliquots.

3. RNase A 1:100 dilution: prepared by mixing; 200 μL RNAse A super stock, 200 μL Tris-HCl, pH 7.5, 75 μL 4 *M* NaCl, and 19.5 mL sterile distilled water in a sterile glass universal. Boil this mixture for 15 min, allow to cool gradually to room temperature in a water bath, and freeze 1-mL aliquots at –20°C.
4. 1 *M* $MgCl_2$ (Sigma).
5. BNE (buffer with no restriction enzyme): mix 5 μL 10X restriction enzyme buffer with 5 μL 1:100 RNAse A dilution and 40 μL sterile distilled water, per sample.
6. BWE (buffer with restriction enzyme): same as for BNE (in **step 5** above) but with 2 μL of the appropriate restriction enzyme and only 38 μL of sterile distilled water.
7. 10 U/μL *Xba*I restriction endonuclease.

2.4.3. Electrophoresis of Pulsed-Field Gels

1. SeaKem® Gold agarose (FMC BioProducts).
2. 5X Tris-borate-EDTA (TBE): 450 m*M* Tris-HCl, pH 8.3; 450 m*M* borate, pH 8.3; 10 m*M* EDTA (working concentration of 0.5X TBE).
3. 10 mg/mL Ethidium bromide.
4. Molecular weight markers: low-range and mid-range PFG markers embedded in 1% (w/v) low melting point (LMP) agarose supplied by New England BioLabs (Hertfordshire, UK). In addition, a DIG-labeled marker is also used to facilitate size determination from the colored developed blots: DNA molecular weight marker grade II, DIG-labeled, consisting of eight fragments ranging in size from 125 to 23,130 bp (Roche Diagnostics).

2.4.4. Visualization of DNA Bands

Ethidium bromide solution at a final concentration of 0.5 μg/mL in distilled water.

2.5. Southern Blotting of Pulsed-Field Gels

2.5.1. DNA Transfer From Pulsed-Field Gels to Nylon Membranes

1. Depurination buffer*: 0.5 *M* HCl.
2. Denaturing buffer*: 0.5 *M* NaOH, 1.5 *M* NaCl.
3. 20X SSC*: 3 *M* NaCl, 0.3 *M* sodium citrate, pH 7.0.
4. Neutralizing buffer*: 0.5 *M* Tris-HCl, pH 7.0, 1.5 *M* NaCl.

DNA transfer of pulsed-field gels was achieved using a Vacu-Aid Blot processing Pump (Hybaid, Middlesex, UK) onto positively charged nylon membranes (Roche Diagnostics) (*see* **Note 3**).

2.6. Preparation and Detection of Digoxigenin (DIG)-Labeled Probes

2.6.1. Preparation of DIG-Labeled Probes

1. DIG-labeled dUTP, at a concentration of 1 nmol/μL (DIG-11-dUTP; Roche Diagnostics).
2. 4 *M* Lithium chloride.
3. Cold 100% ethanol.
4. Cold 70% (v/v) ethanol.

2.6.2. Hybridization and Detection

A DIG-DNA labeling and detection kit obtained from Roche Diagnostics was used for probe hybridization and for color detection of hybridization events.

The solutions required are as follows:

1. Hybridization buffer: 5X SSC (1:4 dilution of 20X SSC), 1% blocking solution (from kit), 0.1% (w/v) *N*-lauroyl-sarcosine, 0.02% (w/v) SDS.
2. Wash solution A: 2X SSC, 0.1% (w/v) SDS (for 500 mL combine 50 mL 20X SSC, 450 mL sterile distilled water, and 0.5 g SDS).
3. Wash solution B: 0.1% SSC, 0.1% (w/v) SDS (for 500 mL combine 2.5 mL 20X SSC, 497.5 mL sterile distilled water, and 0.5 g SDS).
4. Buffer 1: maleic acid buffer; 0.1 *M* maleic acid, 0.15 *M* NaCl; adjust pH with NaOH to pH 7.5 (*see* **Note 4**).
5. Buffer 2: blocking buffer; dilute 10X blocking solution in buffer 1 (maleic acid buffer) to 1X final concentration.
6. Buffer 3: detection buffer; 0.1 *M* Tris-HCl, pH 9.5, 0.1 *M* NaCl, 50 m*M* $MgCl_2$.

3. Methods

3.1. DNA Extraction

This extraction method is a modification of the procedure described by Versalovic et al. *(23)*.

1. Grow bacterial cells overnight in 5 mL TSB at 37°C and then pellet by centrifugation at 15,000*g* for 2 min.
2. Wash the cell pellet with 1 mL of 1 *M* NaCl followed by 1 mL TE solution (50 m*M* Tris-HCl [pH 8.0], 50 mM EDTA) and then resuspend in 0.7 mL TE solution (*see* **Note 5**). Transfer the cells to 1.5-mL Eppendorf tubes.
3. Add 100 μL lysozyme (2 mg/mL) to the cells and incubate at 37°C for 30 min. To complete lysis, add 30 μL of 20% (w/v) SDS and incubate at 65°C for 10 min. This is followed by the addition of 60 μL proteinase K (10 mg/mL) and returning the cells to 37°C for 1 h (*see* **Note 6**)
4. The protein must then be extracted from this crude nucleic acid preparation by performing two phenol/chloroform extractions: Add 800 μL phenol/chloroform/isoamylalcohol (25:24:1) to the lysate mixture and invert the tubes vigorously for 5 min. Centrifuge the samples for 5 min at 20,000*g* and then transfer the *top* aqueous phase (containing the DNA) to a new 1.5-mL tube. Repeat the phenol/chloroform extraction on the aqueous phase.
5. After the second extraction, precipitate the DNA by adding a 1:10 volume of 3 *M* ammonium acetate and cold 100% ethanol to fill the 1.5-mL Eppendorf tube and place at –20°C to precipitate overnight (*see* **Note 7**).
6. After precipitation, centrifuge the DNA samples at 18,000*g* for 30 min at 4°C. Wash the DNA pellet with 70% ethanol, remove all ethanol with a pipet tip, and dry the DNA by leaving the tubes open on the bench for 5 min. Finally, resuspend in 200 μL 1X TE solution.

3.2. Polymerase Chain Reaction

3.2.1. Integron and Gene Cassette Amplification

1. Amplification of gene cassettes located between the 5' conserved segment and the 3' conserved segment of class I integrons is performed in a final volume of 50 μL.

2. For each reaction combine 5 µL 25 m*M* $MgCl_2$, 5 µL 10X reaction buffer, 8 µL working stock dNTPs to give a final concentration of 0.2 m*M* each (dATP, dCTP, dGTP, and dTTP), 100 ng of template DNA, and 2.5 U *Taq* DNA polymerase (5 U/µL).
3. All primers are used at a concentration of 25 pmol per reaction (**Table 1**), and reaction conditions are given in **Table 2**.

A wide range of thermocyclers are available for PCR; the MiniCycler™ (MJ Research, Watertown, MA) is compact (if laboratory space is limited) and has proved very reliable.

3.3. Agarose Gel Electrophoresis

1. Prepare a 1.5% (w/v) agarose gel by mixing 1.5 g agarose and 100 mL 1X TAE (*see* **Note 8**). Heat gently in a microwave oven until the agarose bubbles, mix briefly, reduce the heat, and simmer for 30 s. Allow the gel to cool to 60°C and then add 1 µL of ethidium bromide (10 mg/mL). Mix in carefully and pour into the gel mold with the well comb in place.
2. Allow the gel to solidify for 20–30 min and then pour 1X TAE buffer containing 0.5 µg/mL Ethidium Bromide into the electrophoresis tank until it just covers the surface of the gel. Remove the comb and mold barriers (if present).
3. Prepare the samples by combining 10 µL each PCR with 3 µL loading dye and load into the preformed wells.
4. Electrophoresis times vary from approx 1 h at 80 V for smaller gels (6.5 × 8.0 cm) to 1.5–2 h at 100 V for larger gels (11.5 × 14 cm).
5. The ethidium bromide-stained fragments can be viewed and photographed with a UV transilluminator, e.g., from UVP (Ultra-Violet-Products, Cambridge, UK), using Phoretix software (Newcastle-upon-Tyne, UK).

3.4. Pulsed-Field Gel Electrophoresis

3.4.1. Preparation of Samples in Agarose Plugs for Pulsed-Field Gel Electrophoresis

Day 1:

1. For each strain to be analyzed plate out a single colony onto a nutrient agar plate and incubate overnight at 37°C.
2. Take one 1.5-mL Eppendorf tube per isolate and weigh the tube to 2 decimal places. Record the weights and add 1 mL of saline to each tube. Leave on the bench overnight.

Day 2:

1. Using a disposable inoculating loop, remove a large inoculum of each isolate from the agar plates and place in the saline. Mix well and centrifuge at 8500*g* for 2 min.
2. Remove the saline by pipeting and repeat this step twice. After the last wash fully aspirate any remaining saline and reweigh each Eppendorf tube to get the precise weight of the pellet. Resuspend the pellet in an equal volume of saline, to the nearest 10 (i.e., if the pellet weighs 0.038 g, add 40 µL saline). These tubes now represent stock samples.
3. To prepare a working suspension, mix 5 µL of each stock sample with 10 µL of saline and then to each of these working suspensions add 225 µL of TEN and mix.
4. Prepare 250 µL of 2% (w/v) Incert agarose for each sample (plus 250 µL extra), by combining the required amount of agarose with sterile distilled water in a heat-proof capped tube. Boil the agarose for 15 min, mixing occasionally, and then allow to cool to approx 50°C.

5. Working quickly, add 230 μL of agarose to each sample, mix well using a 1-mL pipete tip (*see* **Note 9**) and aspirate into a syringe barrel. Seal the opening of the syringe with Parafilm and place on ice for 15–20 min.
6. When plugs have solidified, push them out into 10-mL tubes, add 2 mL of EC buffer to each tube, and put gently shaking at 37°C for 5 h.
7. After 5 h aspirate the EC buffer and add 3 mL TE buffer. Aspirate the TE buffer and add another 3 mL of TE buffer and then place the tubes on a gently shaking platform for 10 min. Aspirate the TE buffer, and then to each sample add 2 mL freshly prepared ES buffer and 100 μL proteinase K (20 mg/mL). Place in a shaking waterbath at 50°C overnight.

Day 3:

1. Remove all samples from the waterbath and place them on ice for 10 min. Aspirate the ES buffer from each plug and wash the plugs in 3 mL TE buffer. Immediately replace the TE buffer with fresh TE and incubate with gentle shaking for 1 h.
2. Repeat this procedure for a total of four washes. The plugs can now be stored in fresh TE buffer at 4°C until ready for restriction digestion.

3.4.2. Restriction Endonuclease Digestion of Genomic DNA Embedded in Agarose Plugs

1. Aliquot 300 μL of DNS buffer into the required number of Eppendorf tubes.
2. Chill the samples to be digested on ice for 10 min. Remove the surrounding TE buffer and aspirate the solid plug back into the syringe barrel originally used for preparing the plug.
3. Using a sterile scalpel, previously flamed with alcohol, slice 1-mm-thick pieces from the plug into the tubes containing the DNS buffer (*see* **Note 1**). Immediately replace the buffer with fresh DNS. Incubate the samples at room temperature for 1 h. Repeat this step for a total of four washes.
4. After aspirating of final wash, add 50 μL BNE to all tubes and incubate for 1 h at 4°C. Replace the BNE with 50 μL BWE, refrigerate for 2 h (at 4°C), and then place at 37°C for 12–20 h.
5. After incubation, add 250 μL 0.5X TBE to all tubes to stop the enzyme digestion. Samples can now be stored at 4°C or run immediately.

3.4.3. Preparation and Loading of Pulsed-Field Gels

1. Prepare a 1% (w/v) agarose gel by mixing 1.1 g of SeaKem Gold agarose with 110 mL 0.5X TBE buffer in a conical flask. Record the weight of the flask along with its contents.
2. Heat the agarose on full power in a microwave oven until boiling. Reduce the heat until the agarose bubbles very gently for 3 min. Mix the agarose by gently swirling the flask (*see* **Note 10**) and reheat for a further 3 min. Reweigh the flask and replace the moisture loss with hot distilled water.
3. Allow the agarose to cool to 60°C before casting the gel mold. Leave to solidify for 30 min at room temperature.
4. Finally, pre-electrophorese the gel (prior to loading the samples) for 1.5 h under sample run conditions to improve resolution of DNA bands. Alternatively, place the gel in the electrophoresis tank and allow the 0.5X TBE buffer to wash the gel overnight.
5. Load the prechilled, digested agarose slices into the corresponding wells in the agarose gel with the aid of a glass coverslip and an inoculating loop.
6. Seal the wells with molten 1% (w/v) agarose before applying electric current. Samples are electrophoresed through the homogenous field using a Gene Navigator® System (Pharmacia Biotech, Uppsala, Sweden) or other similar PFGE system, with a hexagonal

(HEX) electrode array. The running buffer consists of 2.5 L of 0.5X TBE, and the following run conditions are applied: 200 V at 10.5°C for 18 h with pulse times ramped from 1 to 40 s.

3.4.4. Visualization of DNA Bands

1. Following electrophoresis, stain the gel for 30 min with an ethidium bromide solution (0.5 μg/mL). This may be followed by destaining in distilled water for 1–2 h.
2. All ethidium bromide-stained DNA fragments can then be directly viewed and photographed over a UV transilluminator (Ultra-Violet-Products). A permanent record is maintained either electronically by creating a TIFF file using suitable capture software such as Phoretix software, or a standard photograph can be taken.

3.5. Southern Blotting

3.5.1. Transfer of DNA From Agarose Gels to Nylon Membranes

DNA transfer from the PFGE gels to nylon membranes can be achieved efficiently using vacuum blotting with the Hybaid Vacu-aid apparatus (*see* **Note 3**).

1. Cut a piece of 3MM filter paper such that it is approx 2–4 cm larger than the gel to be blotted, and prewet the filter paper in 2X SSC (prepared by diluting 20X SSC 1:10 with sterile distilled water).
2. Being careful not to touch the membrane, cut it so that it is slightly (0.5 cm) larger than the gel. Again, prewet it in transfer buffer for 3–5 min.
3. Assemble the Vacu-aid (or similar device) as outlined in the manufacturer's instructions, and overlay the membrane with the gel.
4. The blotting process is greatly accelerated with the *in situ* pretreatment steps. First, apply the vacuum and add *depurination solution*. Allow this solution to penetrate the gel for 10 min and then replace it with *denaturation solution* for 10 min followed by the *neutralization solution*, also for 10 min. Start the transfer by overlaying the gel with 20X SSC and allow transfer to proceed for at least 1 h (*see* **Note 11**).
5. Once the transfer is complete, rinse the membrane in 2X SSC and place it on fresh filter paper to air-dry completely.
6. Fix the DNA to the membrane either by crosslinking in a UV crosslinker or by wrapping the membrane in aluminium foil and baking at 80°C for 1 h. The membrane is now ready for DNA probe hybridization.

3.6. Preparation and Detection of DIG-Labeled Probes

3.6.1. Preparation of DIG-DNA Gene Probes

1. Incorporation of a DIG label can be achieved using 1 μL of DIG-labeled dUTP (DIG-11-dUTP) in the integron PCR reaction mixture as previously outlined in **Subheading 3.2.**
2. After the labeling reaction, recover the probe by precipitating in a mixture containing 2 μL 4 *M* LiCl and 50 μL cold ethanol (100%).
3. Allow the probe to stand for 2 h at –20°C, followed by a centrifugation step at 18,000*g* for 15 min.
4. Decant the surrounding ethanol and wash the pellet (*see* **Note 12**) with 70% (v/v) cold ethanol. Centrifuge for 5 min at 18,000*g*, fully aspirate the ethanol, and allow the pellet to air dry for 5 min.
5. Resuspend the pellet in 50 μL 1X TE and solubilize the probe at 37°C for 30 min. Place the probe in 10 mL hybridization solution and store at –20°C until required.

3.6.2. Hybridization With DIG-DNA Probes

1. Prewarm the hybridization buffer to 58°C for Southern blots of PFGE gels. Place the membrane in a hybridization bottle with approx 50 mL prewarmed hybridization solution and incubate in a hybridization oven at 58°C for 1 h (*see* **Note 13**).
2. Boil the probe for 10 min to ensure that the DNA is completely denatured and cool rapidly in ice water for an additional 10 min. Remove the hybridization solution from the membrane, add the denatured probe, and allow to hybridize overnight at 58°C with constant agitation of the hybridization-probe solution. This is achieved in the hybridization bottle simply by rotating it. This step should continue for a period of 18 h.
3. Following probe-hybridization, remove the probe and freeze it at –20°C for later use (*see* **Note 14**).
4. Wash the membrane twice with *wash solution A* at room temperature for 5 min per wash, and wash twice with *wash solution B* at 68°C for 15 min per wash, with constant shaking.

3.6.3. Color Detection of the Hybridized DIG Label

1. Rinse the membrane briefly in *buffer 1* (maleic acid buffer), and then incubate the membrane for 30 min with 300 mL *buffer 2* (blocking buffer) at room temperature.
2. Dilute anti-DIG-AP conjugate to 150 mU/mL (1:5000 dilution) in *buffer 2* per the manufacturers instructions (8-µL conjugate in 40 mL *buffer 2*) and incubate the membrane for a further 30 min with the antibody conjugate.
3. Remove unbound antibody conjugate by washing twice with 300 mL *buffer 1* for 15 min for each wash at room temperature. Then equilibrate the membrane in approx 60 mL detection buffer for 2–5 min followed by incubation with 30 mL freshly prepared color solution (600 µL NBT/BCIP in 30 mL *buffer 3*), in a sealed plastic bag (*see* **Note 15**).
4. Once the color solution is added, place the membrane in the dark and do not shake. Allow the color to develop, which could take from several minutes to several hours.
5. After color development, wash the membrane in 1X TE or in sterile distilled water for a few minutes and then allow the membrane to dry before photographing and recording results.

4. Notes

1. When cutting the gel slices for restriction digests, it is useful to cut several slices together and digest them all at once in the same tube. This is particularly advantageous if the samples are to be electrophoresed more than once for hybridization with several probes.
2. The DNA extraction protocol outlined here can also be used for Gram-positive bacteria. However, note that lysozyme is not effective against Gram-positive bacterial cells and therefore must be substituted for by using mutanolysin (50–100 U per sample) or lysostaphin (50–100 U per sample) for staphylococcal isolates.
3. The transfer of immobilized DNA fragments by Southern blotting can be greatly accelerated when using a vacuum-based blotting device. Denaturation, depurination, neutralization, and transfer of DNA to the solid-phase membrane can be achieved in under 2 h. Transfer can also be achieved using more traditional and slower methods including capillary action through multiple layers of filter paper.
4. When preparing the maleic acid buffer, to achieve the required pH 7.5, approx 7 g of NaOH per liter are required. Add the sodium hydroxide pellets carefully and allow them to dissolve, check the pH, and then add less concentrated NaOH until the desired pH is reached. **Caution:** *Be aware that granular NaOH can cause skin burns and must be handled carefully.*

5. It is very important to resuspend the bacterial pellet completely before adding the lysozyme. Repeated pipeting proves more efficient than vortexing in this respect.
6. After incubation with proteinase K for 1 h at 37°C, it is often difficult to manipulate the DNA in solution, owing to its "stickiness," typically occurring following the release of chromosomal DNA. Incubating the samples for an additional hour at 65°C often improves this situation by reducing the "stickiness."
7. As soon as the cold ethanol is added, strands of DNA should become visible as they precipitate. Allow the DNA to precipitate for at least 2 h or preferably overnight at –20°C, to increase the yield further.
8. Agarose gels can be made up in several different sizes; a gel of dimensions 11.5 × 14 cm requires 100 mL of agarose TAE buffer mix, whereas smaller gels approx 6.5 × 8.0 cm require only 30 mL.
9. When preparing plugs for PFGE, it is very important to prepare a homogenous mixture of bacterial cell culture and warm agarose. Maintain the solubilized agarose at a constant temperature of 50°C and dispense the 230-μL aliquot directly into each of the bacterial cultures. Mix by gently pipeting up and down, but avoid introducing air bubbles into the samples. When the samples are thoroughly mixed, immediately aspirate into the barrel of the syringe, again avoiding air bubble formation. If air bubbles form, gently flick the syringe barrel until the air bubbles escape by floating to the surface of the gel/culture mixture.
10. **Caution:** *a fully heated flask of molten agarose is very likely to boil over when disturbed, resulting in a severe skin burn. Care should be taken when gently swirling molten agarose to avoid injury. Make sure that gloves are worn.*
11. When performing the Southern blot with a vacuum blotter, allow the transfer to proceed for a minimum of 1 h, but as the macrorestricted DNA fragments are quite large, an additional 30 min often ensures complete transfer of the DNA. As the transfer proceeds, ensure that there is adequate 20X SSC buffer for the transfer. Generally the 20X SSC must be topped up every 20–30 min.
12. A pellet may not always be visible when the DIG-labeled DNA probe is precipitated; therefore it is reassuring if 5 μL of the PCR product for the probe is analyzed by agarose gel electrophoresis prior to the precipitation step.
13. If several Southern blotting and hybridization experiments are to be performed in the laboratory, it is wise to invest in a hybridization oven. The oven will maintain a constant temperature for hybridization, while continually rotating the membrane in a hybridization bottle to ensure even distribution of probe.
14. After hybridization, the probe can be frozen at –20°C and used for up to five hybridization experiments, depending on the strength of the original DIG-labeled PCR product. However, it is important to remember that probes must be denatured each time before use.
15. Using a sealed plastic bag has the advantage of increasing the membrane-solution contact by reducing the volume of the container.

References

1. Recchia, G. D. and Hall, R. M. (1995) Gene cassettes; a new class of mobile element. *Microbiology* **141,** 3015–3027.
2. Lévesque, C., Pyché, L., Larose, C., and Roy, P.H. (1995) PCR mapping of integrons reveals several novel combinations of resistance genes. *Antimicrob. Agents Chemother.* **39,** 185–191.

3. Hall, R. M., Brookes, D. E., and Stokes, H. W. (1991) Site-specific insertion of genes into integrons: role of the 59-base element and determination of the recombination cross-over point. *Mol. Microbiol.* **5,** 1941–1959.
4. Recchia, G. D. and Hall, R. M. (1997) Origins of the mobile gene cassettes found in integrons. *Trends Microbiol.* **5,** 389–394.
5. Sallen, B., Rajoharison, A., Desvarenne, S., and Malibat, C. (1995) Molecular epidemiology of integron associated antibiotic resistance genes in clinical isolates of Enterobactericeae. *Microb. Drug Resist.* **1,** 195–202.
6. Stokes, H. W. and Hall, R. M. (1989) A novel family of potentially mobile DNA elements encoding site-specific gene-integration functions: integrons. *Mol. Microbiol.* **3,** 1669–1683.
7. Paulsen, I.T., Littlejohn, T.G., Rådström, P., et al. (1993) The 3' conserved segment of integrons contains a gene associated with multidrug resistance to antisepics and disinfectants. *Antimicrob. Agents Chemother.* **37,** 761–768.
8. Bennett, P. M. (1999) Integrons and gene cassettes: a genetic construction kit for bacteria. *J. Antimicrob. Chemother.* **43,** 1–4.
9. Collis, C. M. and Hall, R. M. (1995) Expression of the antibiotic resistance genes in the integrated cassettes of integrons. *Antimicrob. Agents Chemother.* **39,** 155–162.
10. Stokes, H. W., Ogorman, D. B., Recchia, G. D., Parsekhian, M., and Hall, R. M. (1997) Structure and function of 59-base element recombination sites associated with mobile gene cassettes. *Mol. Microbiol.* **26,** 731–745.
11. Collis, C. M., Grammaticopoulous, G., Briton, J., Stokes, H. W., and Hall, R. M. (1993) Site specific insertion of gene cassettes into integrons. *Mol. Microbiol.* **9,** 41–52.
12. Glynn, M. K., Bopp, W., Dewitt, P., Mokhtar, M., and Angulo, F. J. (1998) Emergence of multidrug resistant *Salmonella enterica* serotype Typhimurium DT104 infections in the United States. *N. Engl. J. Med.* **338,** 1333–1338.
13. Threlfall, E. J., Frost, J. A., Ward, L. R., and Rowe, B. (1994) Epidemic in cattle and humans of *Salmonella typhimurium* DT104 with chromosomally integrated multiple drug resistance. *Vet. Rec.* **134,** 577.
14. Briggs, C. E. and Fratamico, P. M. (1999) Molecular characterization of an antibiotic resistance gene cluster of *Salmonella typhimurium* DT104. *Antimicrob. Agents Chemother.* **43,** 846–849.
15. Daly, M. and Fanning, S. (2000) Characterization and chromosomal mapping of antimicrobial resistance genes in *Salmonella enterica* serotype Typhimurium. *Appl. Environ. Microboiol.* **66,** 4842–4848.
16. Arcangioli, M. A., Leroy-Setrin, S., Martel, J. L., and Chaslus-Dancla, E. (2000) Evolution of chloramphenicol resistance, with emergence of cross resistance to florfenicol, in bovine *Salmonella* Typhimurium strains implicates definitive phage type (DT) 104. *J. Med. Microbiol.* **49,** 103–110.
17. Daly, M., Buckley, J., Power, E., et al. (2000) Molecular characterisation of Irish *Salmonella enterica* serotype Typhimurium: detection of class 1 integrons and assessment of genetic relationships by DNA amplification fingerprinting. *Appl. Environ. Microbiol.* **66,** 614–619.
18. Ng, L. K., Mulvey, M. R., Martin, I., Peters, G. A., and Johnson, W. (1999) Genetic characterization of antimicrobial resistance in Canadian isolates of *Salmonella* serovar Typhimurium DT104. *Antimicrob. Agents Chemother.* **43,** 3018–3021.
19. Ridley, A. and Threlfall, E. J. (1998) Molecular epidemiology of antibiotic resistance genes in multiresistant epidemic *Salmonella typhimurium* DT104. *Microb. Drug Res.* **4,** 113–118.

20. Sandvang, D., Aarestrup, F. M., and Jensen, L. B. (1998) Characterisation of integrons and antibiotic resistance genes in Danish multiresistant *Salmonella enterica* Typhimurium DT104. *FEMS Microbiol. Lett.* **160,** 37–41.
21. Threlfall, E. J., Frost, J. A., Ward, L. R., and Rowe, B. (1996) Increasing spectrum of resistance in multiresistant *Salmonella typhimurium. Lancet* **347,** 1053–1054.
22. Maslow, J. N., Slutsky, A. M., and Arbeit, R. D. (1993) Application of pulsed field gel electrophoresis to molecular epidemiology. In: *Diagnostic Molecular Microbiology, Principles and Application*, (Persing, D. H., Smith, T. F., Tenover, F. C., and White, T. J., eds.) American Society for Microbiology, Washington, DC, pp. 563–572.
23. Versalovic, J. T., Koeuth, T., and Lupski, J. R. (1991) Distribution of repetitive DNA sequences in eubacteria and application to fingerprinting of bacterial genomes. *Nucleic Acids Res.* **24,** 6823–6831.
24. Sundström, L., Rådström, P., Swedberg, G., and Sköld, O. (1988) Site specific recombination promotes linkage between trimethoprim and sulfonamide resistance genes. Sequence characterisation of *dhfrV* and *sulI* and a recombination active locus of Tn21. *Mol. Gen. Genet.* **213,** 191–201.
25. Hollingshead, S. K. and Vapnek, D. (1985) Nucleotide sequence analysis of a gene encoding a streptomycin/spectinomycin adenyltransferase. *Plasmid* **13,** 17–30.
26. Huovinen, P. and Jacoby, G. A. (1991) Sequence of PSE-1 beta-lactamase gene. *Antimicrob. Agents Chemother.* **35,** 2428–2430.

3

Molecular-Based Identification and Typing of *Campylobacter jejuni* and *C. coli*

Brigid Lucey, Fiona O'Halloran, and Séamus Fanning

1. Introduction

Thermophilic *Campylobacter* spp., mainly *Campylobacter jejuni* and to a lesser extent *C. coli* are recognized as the most common bacteriological causes of gastroenteritis in humans *(1)*. As enteric infection with *Campylobacter* organisms cannot be distinguished from that caused by other enteric pathogens, a definitive diagnosis can only be made by isolating or detecting the organism from the feces. The epidemiology of *Campylobacter* enteritis has been complicated by the ubiquitous nature of the organism (commonly found as a commensal in the intestines of domestic animals, in milk, and in water). Furthermore, identification is carried out only to genus level by most clinical laboratories.

Because of the biochemical similarity known to exist between *C. jejuni* and *C. coli*, the hippurate hydrolysis test is often used as the only phenotypic test capable of differentiating the two species. This test, however, has some acknowledged technical limitations and is dependent on inoculum size; results can be difficult to interpret accurately *(2)*. Furthermore, almost all *C. jejuni* isolates possess the hippuricase gene *(3)*, fewer *C. jejuni* isolates express the hippuricase gene. For this reason, certain polymerase chain reaction (PCR)-based species identification methods, for both *C. jejuni* and *C. coli*, and for the other thermophilic species, provide more reliable identification; they also help to highlight mixed species cultures, should they occur. However, even with these methods, false negatives or nonspecifically amplified product(s) can occur in a minority of isolates tested owing to genomic anomalies. Thus a second molecular identification method may be required in these circumstances.

Gonzalez et al. *(4)* developed a species-specific PCR assay for the identification of *C. jejuni* and *C. coli* based on the *ceuE* gene, which is involved in siderophore transport. Using this method two primer sets are employed in separate PCR amplification reactions. Another method, developed by Eyers et al. *(5)*, performs PCR amplification of 23S rRNA gene fragments, based on regions specific for *C. jejuni*, *C. coli*, *C. lari*,

From: *Methods in Molecular Biology, vol. 268: Public Health Microbiology: Methods and Protocols*
Edited by: J. F. T. Spencer and A. L. Ragout de Spencer © Humana Press Inc., Totowa, NJ

and *C. upsaliensis*. In addition, Hani and Chan *(6)* developed a PCR assay that detected and amplified the hippuricase gene. This molecular approach may offer a more reliable means of identifying *C. jejuni* strains compared with the phenotypic hippurate hydrolysis test alone.

Epidemiological typing methods currently in use for *Campylobacter* spp. include phenotypic methods such as serotyping and genotypic methods including DNA amplification fingerprinting (DAF), pulsed-field gel electrophoresis (PFGE), and ribotyping and restriction fragment length polymorphism (RFLP) analysis of the flagellin A gene (*flaA*). Serotyping remains primarily a reference laboratory protocol because of the time and expense needed to maintain high-quality antisera. Furthermore, there are a high percentage of untypable strains, and the standard serotyping systems such as the Lior and Penner schemes are time-consuming and technically demanding *(7)*.

DAF is a PCR-based typing method that typically uses randomly designed 10-mer primers, under conditions that allow some base-pair mismatches, to increase the number of primer binding sites. Primer binding throughout the genome generates an array of DNA amplicons with varying lengths and intensities, which typify a genomic fingerprint. Interstrain variation is dependent on the number, location, and degree of mismatches tolerated when the primer binds to the genome. Advantages of this technique include a high degree of typability, ease of performance, cost effectiveness, and ready availability of reagents and equipment. In a recent study of 378 *Campylobacter* isolates, from human and poultry sources, only one isolate proved untypable with DAF analysis (Lucey et al., unpublished observations), using the primer HLWL85 *(8)*. DAF analysis has proved to be highly discriminatory when primers are carefully selected (*see* **Note 1**).

PFGE is a molecular typing technique that uses rare cutting restriction endonucleases to cleave bacterial genomes producing a small number of very large DNA fragments. These DNA fragments can only be resolved and visualized by PFGE methods. Despite some obvious technical considerations such as the labor-intensive process involved, PFGE is one method that offers a high degree of reproducibility and typability, with the production of highly discriminatory profiles and the possibility of interlaboratory comparisons if methods are standardized. Compared with DAF, PFGE requires specialized equipment and can be more expensive to run on a routine basis *(7)*.

A study by Møller Nielsen et al. *(9)* compared the effectiveness of six typing methods for use on a population of 90 *Campylobacter* isolates of human, cattle, and poultry origins (comprising outbreak and non-outbreak-related strains). The methods used were Penner heat-stable serotyping, automated ribotyping, DAF analysis, PFGE, RFLP of *flaA*, and denaturing gradient gel electrophoresis of *flaA* (*flaA*-DGGE). DAF and PFGE were shown to be the most discriminating methods, which the authors ascribed to their ability to determine polymorphisms across the entire bacterial genome. Serotyping was found to be the least discriminatory.

Bacterial resistance to antimicrobial agents often results from the acquisition of new genes, as well as from mutations. These new genes can be acquired by several mechanisms including genetic elements, such as resistance (R-)plasmids, transposons, and integrons. The latter group are a class of genetic elements involved in the dissemination of antimicrobial resistance-encoding genes. This is mediated by the integration of a gene cassette containing the resistance genes via a site-specific recombinational event ***(10)***. Integrons have been found to be associated with antimicrobial resistance in many Gram-negative species, including *Salmonella*, *Pseudomonas*, and *Klebsiella*. A recent report suggests that *Campylobacter* spp. also possess integron-like structures ***(11)***. When the plasticity of the *Campylobacter* genome is considered, this has interesting implications for the continued evolution of this species. Therefore, it is reasonable to suggest that complete characterization of these species should include investigations for the presence of unique integron structures.

Most cases of clinical *Campylobacter* enteritis are sufficiently mild or self-limiting not to require antimicrobial chemotherapy ***(12)***. Nevertheless, in severe or recurrent cases for which antibiotics are required, susceptibility testing is important to ensure appropriate and timely treatment. Macrolides remain the agents of choice for such cases, and resistance rates remain comparatively low ***(13)***. Since the 1980s, however, the development of the fluoroquinolones, which are effective against most enteric pathogens, offered an effective therapy to treat acute bacterial diarrhea; ciprofloxacin becoming used extensively as prophylaxis for travellers ***(14)***. The emergence of resistance to these agents, however, has since then made their efficacy less certain. Resistance was reported to develop among patients after treatment with fluoroquinolones ***(15)*** and was also found to coincide with the introduction of these agents in veterinary medicine ***(16)***. In *Campylobacter* and other Gram-negative bacteria, fluoroquinolones operate by interfering with the type II topoisomerase (DNA gyrase) and topoisomerase IV ***(17)***. The predominant mechanism of resistance to ciprofloxacin in *C. jejuni* and *C. coli* has been shown to result from a mutation in the *gyrA* gene, whereby all isolates tested demonstrated a Thr-86-Ile substitution in the A-subunit of DNA gyrase ***(18,19)***. A specific PCR assay, the mismatch amplification mutation assay (MAMA-PCR), has been demonstrated to be a useful screening method for ciprofloxacin resistance among each of these isolates ***(18,19)***. This method uses a conserved, forward primer and a reverse, diagnostic primer, which together generate a 265-bp product that is a positive indication of the presence of the Thr-86-Ile amino acid substituion consistent with resistance to ciprofloxacin.

2. Materials

All materials marked with an asterisk (*) were autoclaved at 121°C for 15 min prior to use.

2.1. DNA Extraction

This protocol is a modified version of a purification method for *Campylobacter* spp. ***(20)*** to yield purified DNA. Cultures were treated prior to DNA extraction with formaldehyde to inhibit any DNAse activity (*see* **Note 2**), according to the method of Gibson et al. ***(21)***.

1. Preston agar Campylobacter Agar Base, Oxoid CM689, containing Modified Campylobacter Selective Supplement Sr204E and 5% (v/v) lysed horse blood. Prepare according to the manufacturer's instructions.
2. Distilled water*.

3. 0.85% (w/v) NaCl*.
4. Formaldehyde solution (37–40% v/v).
5. 5X TE solution*: 50 m*M* Tris-HCl, pH 8.0, 50 m*M* EDTA.
6. Lysozyme: obtained lyophilized from Sigma (Poole, UK), reconstituted in 5X TE solution to a final concentration of 5 mg/mL, aliquot, and store at –20°C.
7. 20% (w/v) Sodium dodecyl sulfate (SDS), heat to 35–45°C to dissolve.
8. Proteinase K: obtained lyophilized from Sigma, reconstituted in 5X TE solution to a final concentration of 10 mg/mL, aliquot, and store at –20°C.
9. Phenol:chloroform:isoamylalcohol (25:24:1) solution (Sigma), saturated with 10 m*M* Tris-HCl, pH 8.0, 1 m*M* EDTA.
10. 3 *M* Ammonium acetate. Filter sterilize and store at room temperature.
11. 70 and 100% (v/v) ethanol. Store at –20°C.
12. 1X TE solution*: 10 m*M* Tris-HCl, pH 8.0, 1 m*M* EDTA.

2.2. PCR Amplification Methods

1. Sterile water.
2. Deoxyribonucleoside triphosphates (dNTPs; Promega, Madison WI). Provided at a concentration of 100 m*M* each dNTP (dATP, dCTP, dGTP, dTTP). Prepare a working solution of 1.25 m*M* dNTP by adding 2.5 µL of each dNTP to 190 µL of sterile H_2O. Store at –20°C until required.
3. *Taq* DNA polymerase (Promega; 5 U/µL). Supplied with 25 m*M* $MgCl_2$ and 10X reaction buffer (100 m*M* Tris-HCl, pH 9.0, 500 m*M* KCl, 1% [w/v] Triton X-100).
4. Oligonucleotide primers diluted to optimal working concentration in sterile H_2O. All primers used for speciation, DAF analysis, integron analysis, and ciprofloxacin resistance-determining MAMA-PCR were purchased from Oswel DNA Services (Southampton, UK) and were purified by high-performance liquid chromatography (HPLC) prior to use. Primer sequences along with their relevant characteristics are listed in **Table 1**.

2.3. Agarose Gel Electrophoresis

1. Conventional agarose gel electrophoresis was used to characterize PCR amplicons, and a list of the reagents used is given in Chapter 2 of this volume.
2. Molecular weight markers: Three DNA molecular weight marker mixtures were employed to analyze PCR products. These included *DNA marker III* (generates fragments from 0.12 to 21.2 kbp, *DNA marker V* (generates fragments from 8 to 587 bp), and *DNA marker XIV* (generates fragments from 100 to 1500 bp in length). All markers were supplied by Roche Diagnostics (East Sussex, UK).

2.4. Pulsed-Field Gel Electrophoresis

2.4.1. Preparation of Agarose Plugs and Restriction Endonuclease Digestion

1. Preston agar (Campylobacter Agar Base, Oxoid CM689, containing Modified Campylobacter Selective Supplement Sr204E and 5% [v/v] Lysed Horse Blood), prepared according to the manufacturer's instructions.
2. 0.85% (w/v) NaCl*.
3. Formaldehyde solution (37–40% v/v).
4. 1-mL Sterile plastic syringes with the nozzle removed.
5. Sterile microcentrifuge tubes (1.5 and 2.0 mL).
6. Sterile clear polystyrene capped 17 × 120-mm (Sarstedt) tubes (or equivalent).

Table 1
Sequence and Characteristics of Oligonucleotide Primers Used for PCR

Primer	Sequence 5'→3'	Conc. per reaction (pmol)	%GC	PCR T-annealing (°C)
*ceuE*jej F	CCT GCT ACG GTG AAA GTT TTG C	25	50	57
*ceuE*jej R	GAT CTT TTT GTT TTG TGC TGC	25	38.1	57
*ceuE*coli F	ATG AAA AAA TAT TTA GTT TTT GCA	25	16.7	57
*ceuE*coli R	ATT TTA TTA TTT GTA GCA GCG	25	28.6	57
Hip F	GAA GAG GGT TTG GGT GGT G	25	57.9	66
Hip R	AGC TAG CTT CGC ATA ATA ACT TG	25	39.1	66
Int1F	GGC ATC CAA GCA GCA AGC	25	61	55
Int1R	AAG CAG ACT TGA CCT GAT	25	44	55
HLWL85	ACA ACT GCT C	50	50	40
MAMA*gyrA*1	TTT TTA GCA AAG ATT CTG AT	20	25	50
MAMA*gyrA*5	CAA AGC ATC ATA AAC TGC À	20	36.8	50
GZ*gyrA*coli3F	TAT GAG CGT TAT TAT CGG TC	10	40	50
MAMA*gyr*8	TAA GGC ATC GTA AAC AGC CA	10	45	50

7. SeaKem® Gold (SKG) agarose (FMC BioProducts, Vallesbaek Strand, Denmark).
8. 10 mg/mL Ethidium bromide.
9. Molecular weight markers: Low-range and mid-range PFG markers embedded in 1% (w/v) low melting point (LMP) agarose (New England BioLabs, Hertfordshire, UK).
10. 1 *M* Tris-HCl, pH 8.0*.
11. 10 *M* NaOH*.
12. 0.5 *M* EDTA adjusted to pH 8.0*.
13. 10 mg/mL Lysozyme stock solution.
14. 20% (w/v) SDS.
15. 20 mg/mL Proteinase K stock solution.
16. 10% (w/v) sarcosyl (*N*-lauroylsarcosine, sodium salt).
17. 10X TBE buffer, pH 8.0*.
18. Tris-EDTA buffer (TE), pH 8.0*.
19. Cell suspension buffer: 100 m*M* Tris-HCl, pH 8.0, 100 m*M* EDTA.
20. Cell lysis buffer: 50 m*M* Tris-HCl, pH 8.0, 50 m*M* EDTA, 1% (w/v) sarcosyl, 20 mg/mL Proteinase K (add just before use).
21. Restriction endonucleases: 12 U/μL *Sma*I or 10 U/μL *Sac*II (Promega).

3. Methods

3.1. DNA Extraction With Formaldehyde Treatment (see *Note 3*)

Perform the complete extraction at 4°C except for those steps for which an alternative temperature is indicated.

1. To establish purity of culture, examine a fixed smear with 10% carbol fuchsin (*see* **Note 4**). Allow to stain for 2 min, wash with tap water, and dry. Examine by oil immersion microscopy and note the presence of typical "seagull-shaped" bacilli.

2. Emulsify pure plate culture of *Campylobacter* to a turbidity of ≥ McFarland 6 in a sterile 2-mL microcentrifuge tube containing 1.0 mL of 0.85% (w/v) NaCl (*see* **Note 5**).
3. Wash twice in 1 mL 0.85% (w/v) sterile NaCl.
4. After the second wash, decant the NaCl solution and resuspend the cell pellet in 0.9 mL fresh 0.85% (w/v) NaCl.
5. Add 100 μL of formaldehyde solution to cell suspension. Mix thoroughly and incubate for 1 h at room temperature.
6. Following this incubation wash cells *three* times in 1 mL of 0.85% (w/v) NaCl and then once in 1 mL of 5X TE. Centrifuge at 11,000*g* for 5 min and decant supernatant.
7. Resuspend cell pellet in 540 μL of 5X TE containing lysozyme (5 mg/mL).
8. Invert tube several times to mix the contents thoroughly and incubate at 37°C for 30 min.
9. Add 60 μL 20% (w/v) SDS, inverting tube several times, and incubate at 37°C for 15 min.
10. Add 150 μL of 5X TE containing Proteinase K (10 mg/mL), mix, and incubate at 37°C overnight.
11. Reincubate tubes at 65°C for 1 h.
12. Centrifuge at 15,000*g* for 15 min and transfer the supernatant to a sterile 2-mL Eppendorf tube.
13. Add an approx equal volume of phenol chloroform (*see* **Note 6**), shake vigorously for 5 min, and centrifuge for 5 min at 15,000*g*. Transfer the top aqueous phase (containing DNA) to a new tube (*see* **Note 7**).
14. Repeat **steps 11** and **12** for the phenol chloroform.
15. Add 1:10 volume of 3 *M* ammonium acetate and mix by inversion. To this mixture add cold ethanol to the top of each tube (approx 1 mL). To precipitate the DNA, incubate the mixture at –20°C overnight.
16. Centrifuge at 15,000*g* for 30 min and decant the supernatant.
17. Wash the DNA pellet twice with 70% (v/v) ethanol, fully aspirate the ethanol, and allow to air dry until no moisture remains (5–10 min). Add 300 μL of 1X TE and resuspend the DNA pellet.
18. Examine preparations of chromosomal DNA by gel electrophoresis (to ensure that the DNA is intact). Determine the DNA concentration in each sample using absorbance spectrophotometry at 260 nm: 1 absorbance unit = 50 ng/μL DNA.
19. Dilute DNA suspensions to give a stock template DNA concentration of 100 ng/μL. DNA suspensions may be stored at 4°C (*see* **Note 8**).

3.2. PCR Amplification Methods

3.2.1. Campylobacter Species Identification

1. Success of this species-specific PCR is greatly facilitated by the sequence divergence (of approx 13%) that exists between the *ceuE* genes of *C. jejuni* and *C. coli*. All PCR reactions were performed in a final volume of 50 μL (*see* **Fig. 1**).
2. For each reaction combine 7 μL 25 m*M* $MgCl_2$, 5 μL 10X reaction buffer, 8 μL working stock dNTPs to give a final concentration of 0.2 m*M* for each dNTP, 200 ng purified template DNA, and 1 U *Taq* DNA polymerase (5 U/μL).
3. Each isolate was analyzed by two primer sets, *CeuE*jej forward and reverse and *CeuE*coli forward and reverse (**Table 1**), in two separate PCR reactions. Each primer was used at a concentration of 25 pmol in a final reaction volume of 50 μL.
4. The reaction conditions required for amplification are listed in **Table 2**. Expected product sizes are as follows: *CeuE*jej: 793 bp; *CeuE*coli: 894 bp.

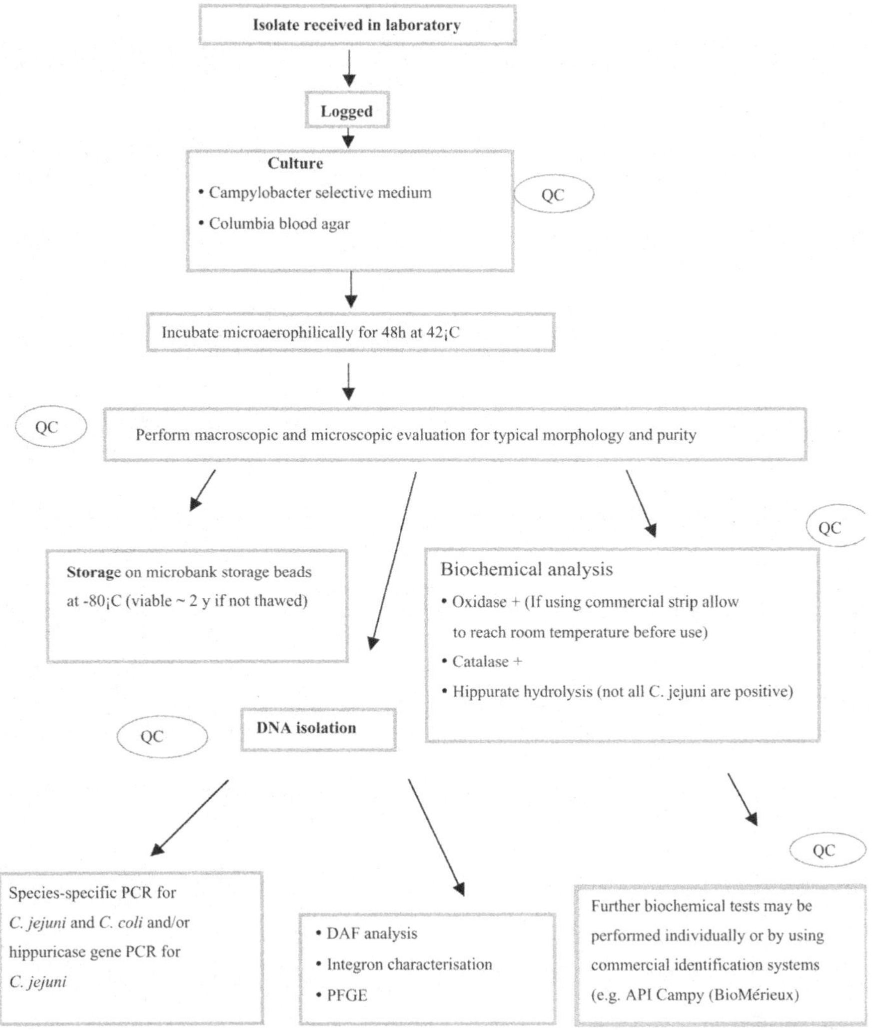

Fig. 1. Procedure for the isolation and identification of *Campylobacter* spp. DAF, DNA amplification finger printing; PFGE, pulsed-field gel electrophoresis; QC, quality control step.

3.2.2. Amplification of the Hippuricase Gene

1. For each reaction combine 2.5 µL of 25 m*M* $MgCl_2$, 2.5 µL of 10X reaction buffer, 4 µL working stock dNTPs to give a final concentration of 0.2 m*M* for each dNTP, 25 ng template DNA and 2.5 U *Taq* DNA polymerase (5 U/µL).
2. Primer pairs *Hip* F and *Hip* R were used at a concentration of 25 pmol per reaction (**Table 1**) in a final reaction volume of 25 µL.
3. The reaction conditions for successful amplification are given in **Table 2**, and expected product size is 735 bp.

Table 2
Thermocycling Reaction Conditions for PCR Programs to Analyze *Campylobacter* Isolates

PCR program	Step 1 Initial denaturation	Step 2 Denaturation	Step 3 Annealing	Step 4 Extension	Step 5 Final extension	No. of cycles	Ref.
*Ceu*Ejej	94°C, 3 min	94°C, 30 s	57°C, 30 s	72°C, 1 min	72°C, 5 min	30	***4***
*Ceu*Ecoli	94°C, 3 min	94°C, 30 s	57°C, 30 s	72°C, 1 min	72°C, 5 min	30	***4***
Hippuricase gene	—	94°C, 1 min	66°C, 1 min	72°C, 1 min	—	25	***24***
int	95°C, 5 min	95°C, 1 min	55°C, 1 min	72°C, 2 min	22°C, 5 min	30	***10***
HLWL85	94°C, 5 min	94°C, 30 s	40°C, 1 min	72°C, 1 min	72°C, 5 min	40	***8***
MAMA PCR	94°C, 3 min	94°C, 30 s	50°C, 30 s	72°C, 20 s	—	30	***18,19***

3.2.3. DNA Amplification Fingerprinting (DAF) Analysis

Each DNA template was analyzed in duplicate by DAF to ensure the reproducibility of the DNA profile and to provide reliable data quality.

1. Each PCR assay incorporated the 10-mer oligonucleotide primer HLWL85 for DAF analysis.
2. For each reaction combine 5.0 µL 25 m*M* $MgCl_2$, 5.0 µL 10X PCR buffer, 8 µL working stock dNTPs to give a final concentration of 0.2 m*M* dNTP, 200 ng genomic DNA, and 2.5 U *Taq* DNA polymerase.
3. The primer HLWL85 (**Table 1**) was used at a concentration of 50 pmol per reaction in a final reaction volume of 50 µL.
4. The reaction conditions used for amplification are listed in **Table 2**. Expected product sizes are variable but may range from 180 to 6000 bp (*see* **Fig. 2** and ref. ***22***).

3.2.4. Molecular Detection of Ciprorofloxacin Resistance in Campylobacter *Isolates*

1. Amplifications were performed in 50-µL reaction volumes containing 20 pmol of *C. jejuni* and 10 pmol of *C. coli* the correct primer pair (**Table 1**), 200 µ*M* each dNTP, and 1.5 m*M* $MgCl_2$.
2. A 2.5-µL volume of 10X *Taq* DNA polymerase buffer and 2.5 U *Taq* DNA polymerase were added.
3. Approximately 33 ng of DNA template, corresponding to 1 µL were added to the reaction mixture.
4. The reaction conditions used for amplification are listed in **Table 2**.
5. Expected product sizes: *C. coli*: 192-bp product, (A*C*T→A*T*T) mutation (**Fig. 3A**); *C. jejuni*: 265-bp product, (A*C*A→A*T*A) mutation (**Fig. 3B**).

3.2.5. Integron PCR Analysis

1. The integron primer sequences ***(10)*** used to identify the presence of integron structures are listed in **Table 1**.
2. In a final reaction volume of 50 µL, the following was added; 5 µL 10X PCR buffer, 5 µL 25 m*M* $MgCl_2$, 8 µL dNTP working stock solution, 25 pmol of each IntF and IntR primer (**Table 1**), 200 ng template DNA, and 2.5 U *Taq* DNA polymerase.

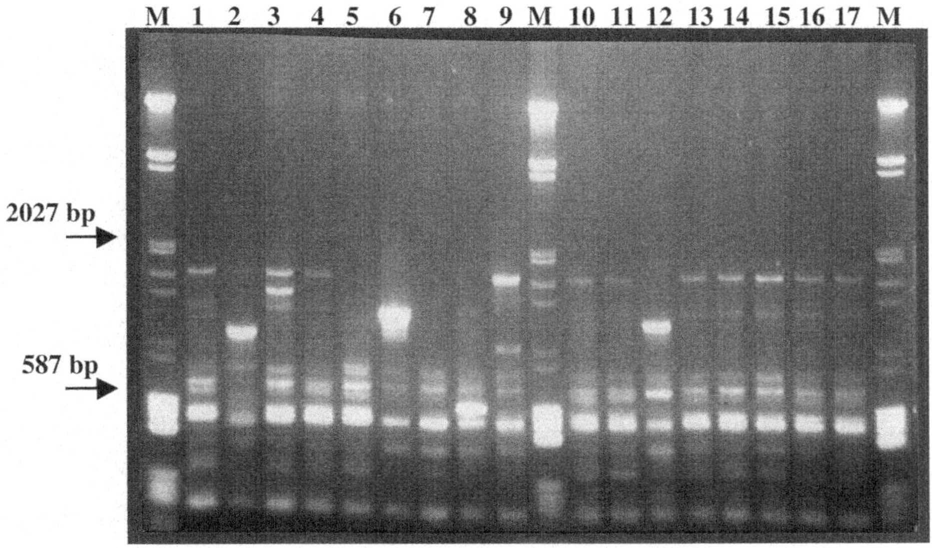

Fig. 2. An agarose gel (1.5% w/v) showing the DNA profiles of a collection of unrelated *Campylobacter* spp. isolates.

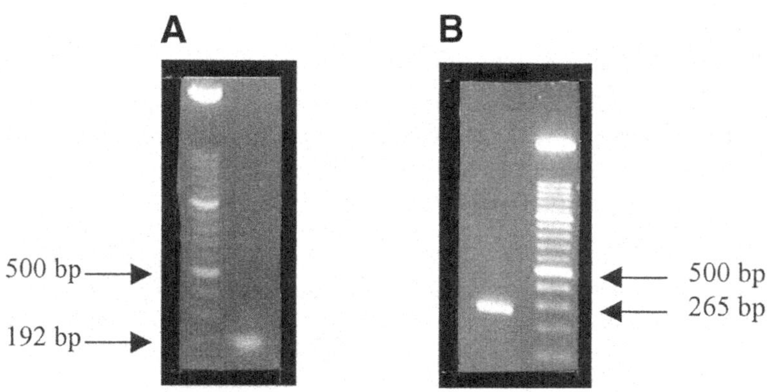

Fig. 3. MAMA-PCR showing a 192-bp DNA fragment corresponding to the wild-type DNA gyrase in *C. coli* (**A**) and a 265-bp fragment corresponding to the mutated gene in *C. jejuni* (**B**).

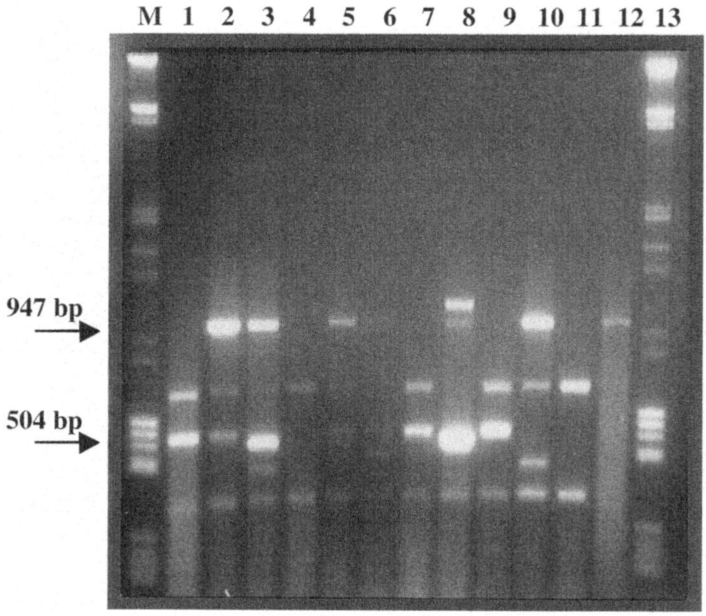

Fig. 4. Amplified gene cassettes from a collection of *Campylobacter* spp. isolates showing a range of potential antimicrobial-resistance encoding genes.

3. The reaction conditions used for amplification are listed in **Table 2**.
4. Integron PCR products were analyzed by conventional agarose gel (1.5% w/v) electrophoresis (**Fig. 4**).

3.3. Pulsed-Field Gel Electrophoresis

A one-day standardized laboratory protocol for the molecular subtyping of *Campylobacter* spp. by PFGE (adapted for *Campylobacter* spp. from the PulseNet Method The National Molecular Subtyping Network for Foodborne Disease Surveillance for nontyphoidal *Salmonella*) was applied in this study.

3.3.1 Preparation of Agarose Plugs for Pulsed-Field Gel Electrophoresis

1. Cultures of *Campylobacter* spp. to be investigated should be well grown (48–72 h) on Preston medium and checked microscopically for purity (*see* **Note 4** and **Fig. 1**).
2. Emulsify the plate contents of each isolate in a 1-mL volume of 0.85% (w/v) NaCl in sterile 2-mL microcentrifuge tubes. Centrifuge at 11,000*g* for 5 min.
3. Decant the supernatant and resuspend isolates in 0.9 mL 0.85% (w/v) NaCl.
4. Add 100 µL of formaldehyde (37–40% v/v), mix, and leave at room temperature for 1 h.
5. Centrifuge for 5 min at 11,000*g* and then decant the supernatant. Wash the pellet three times in 1 mL 0.85% (w/v) NaCl.
6. Resuspend the pellet in 2 mL of cell suspension buffer.

7. Adjust the volume of the cell suspension buffer to give an optical density of 1.8–2.0 at 610 nm.
8. Prepare agarose plugs as follows: weigh 0.25 g SKG into a 50-mL screw-capped tube (Sarstedt). Add 23.5 mL TE buffer and mix gently. Dissolve the agarose completely by microwaving.
9. Place tubes in a 50°C water bath for 5 min and then add 1.25 mL of 20% (w/v) SDS (preheated to 50°C). Mix well by gentle inversion of the capped tube.
10. For each isolate, transfer 0.4 mL of the pure bacterial culture suspension (at room temperature) to a 1.5-mL Eppendorf tube, add 20 μL of Proteinase K (20 mg/mL) to each tube, and mix gently by pipeting.
11. Add 0.4 mL of the SKG/SDS mixture to each bacterial culture and again mix by pipeting gently up and down several times. Place in a 50°C waterbath.
12. Aspirate or pipet each sample into the barrel of a prelabeled syringe. Cover the nozzle of each syringe with parafilm and place on ice for 15 min to allow the plugs to solidify.

3.3.2. Cell Lysis

1. Label 17 × 120-mm sterile tubes with appropriate isolate numbers.
2. Add 5 mL of the cell lysis buffer to each tube, containing 25 μL of the stock Proteinase K. Calculate the total volume of the cell lysis/Proteinase K buffer required and aliquot 5 mL of this mixture into the prelabeled tubes.
3. Trim the agarose plugs with a sterile scalpel and add the whole plug/or plug cut in two segments to the lysis buffer.
4. Incubate in a 54°C waterbath for 1.5–2 h, with constant shaking at 175–200 rpm. Ensure that the water level in the bath is above the level of the lysis buffer in the tubes.
5. Heat sufficient sterile distilled H_2O to 50°C to allow plugs to be washed twice with 10–15 mL of water.
6. Remove tubes from waterbath and carefully pour off lysis buffer into an appropriate discard container.
7. Add 10–15 mL of preheated sterile distilled H_2O to each tube and shake the tubes vigorously in a 50°C waterbath for 10–15 min.
8. Again pour off the water from the plugs and repeat this wash step with preheated water, ensuring that no lysis buffer remains on the rims or caps of tubes.
9. Pour off water, add 10–15 mL preheated sterile TE buffer, and shake the tubes vigorously in a 50°C waterbath for 10–15 min.
10. Pour off TE and repeat wash step with preheated TE three more times.
11. Decant last wash and add 5 mL sterile TE. At this stage it is possible to continue with the restriction digest, or plugs can be stored at 4°C until ready to proceed.

3.3.3. Restriction Digest of DNA in Agarose Plugs With Sma*I*

1. Label the appropriate number of 1.5-mL microcentrifuge tubes.
2. Prepare the *Sma*I digest reactions (*see* **Note 9**) according to the table below and add the mixtures into the microcentrifuge tubes:

Reagent	mL/sample
Sterile distilled H_2O	87.5
10X digest buffer J	10
Enzyme (10 U/μL)	2.5
Total volume	100

3. Cut 1.5–2-mm wide slices from each plug with a sterile scalpel blade and transfer to the tubes containing the digest mixture (*see* **Note 10**).
4. Mix by tapping gently and ensure that plug slices are fully submerged in the mixture.
5. Incubate samples in 25–30°C waterbath for a minimum of 2 h (*see* **Note 11**).

3.3.4. Preparation of Agarose Gels and Sample Loading

1. Heat a waterbath to 55–60°C.
2. Prepare 3 L of 0.5X TBE buffer by mixing 150 mL of the 10X TBE stock buffer with 2850 mL distilled water.
3. Add 1.1 g SKG agarose to 110 mL of the 0.5X TBE buffer in a 250-mL flask and mix gently (*see* **Note 12**).
4. Cover loosely with parafilm and microwave for 60 s; mix gently and repeat for 15-s intervals until the agarose is completely dissolved. Cool to approx 60°C and pour into a precasting gel apparatus. Allow to solidify for approx 30 min and then gently remove the comb.
5. Place the gel into the electrophoresis tank and overlay completely with 0.5X TBE buffer. Pre-electrophorese the gel for 1 h.
6. After the pre-electrophoresis step, remove the gel from the electrophoresis tank.
7. Remove digested plug slices from 37°C water bath. Gently decant the enzyme/buffer solution, taking care not to score the plug slices with the pipet tip, and add 200 μL 0.5X TBE to stop the digestion reaction. Incubate at room temperature for 5 min.
8. Remove one plug slice from each tube at a time and load into the preformed wells in the agarose gel (*see* **Note 13**).
9. Prepare 20 mL of a mixture containing 1% (w/v) SKG agarose in 0.5X TBE as follows: Add 0.2 g agarose to 20 mL 0.5X TBE and microwave until dissolved. Seal the gel wells by overlaying with this molten agarose solution (*see* **Note 14**).
10. Replace the loaded gel back into the gel tank, submerging it in 0.5X TBE electrophoresis buffer, and apply an electric current of 200 V to perform the electrophoresis. DNA fragments are separated using a ramped pulse switching every 0.5–25 s for 20 h *(23)*, at 10.5°C (**Fig. 5**).

3.3.5. Staining and Documentation of PFGE Agarose Gel

Ethidium bromide solution is at a final concentration of 0.5 μg/mL in distilled water. Stain the gel for 20–30 min in a covered container. Destaining may be performed if required in distilled water.

4. Notes

1. To ensure reproducibility of DAF patterns, the same thermocycler and reagents should be used. The interpretation of weak bands can be problematic, and typing should be duplicated in separate runs to help overcome this problem (*see* **Subheading 1.**).
2. A minority of *Campylobacter* isolates may be nontypable by DAF or PFGE owing to the production of DNAses. This can be overcome by routinely treating bacterial suspensions with formaldehyde before proceeding with the typing method (*see* **Subheading 2.1.**).
3. **Caution:** Formaldehyde is toxic and harmful if inhaled or absorbed through skin or mucosa. This reagent should only be handled in a safety cabinet with extraction ventilation (*see* **Subheading 3.1.**).

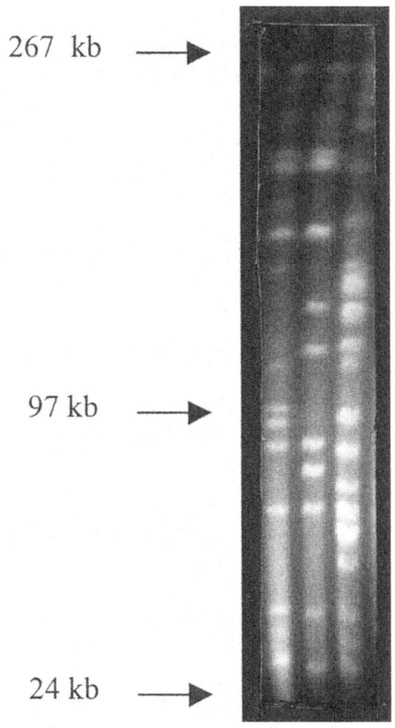

Fig. 5. PFGE profiles from three *Campylobacter* spp. *Sma*I macrorestricted fragments ranging from 24 to 267 kbp are resolved by this method.

4. Microscopic examination by dilute carbol fuchsin staining. Prepared by dissolving 20 g of carbol fuchsin (BDH Chemicals, Poole, England) and liquefying 100 g of phenol (BDH Chemicals) together in a 3-L conical flask. Add 200 mL of absolute ethanol and 1715 mL of distilled water. Filter the resultant solution and dilute in sterile distilled water to give a 10% working carbol fuchsin solution (*see* **Subheadings 3.1.** and **3.3.1.**).
5. Round-bottomed microcentrifuge tubes facilitate emulsification of organisms (*see* **Subheading 3.1.**)
6. **Caution:** care should be taken when handling phenol-chloroform. As phenol is an acid, it can cause skin burns. This reagent should only be handled in a safety cabinet with extraction ventilation (*see* **Subheading 3.1.**).
7. Ensure that none of the white layer at the interface of the phenol chloroform and DNA solution is disturbed (*see* **Subheading 3.1.**).
8. DNA suspensions may be stored at 4°C for up to 3 mo, or, alternatively, may be aliquoted and frozen at –20°C and thawed once prior to use (*see* **Subheading 3.1.**).
9. Optimal temperature for *Sma*I is 25–30°C (*see* **Subheading 3.3.3.**).
10. Up to four gel slices may be cut for digestion in 100 μL total volume of restriction solution (*see* **Subheading 3.3.3.**).

11. Alternatively, agarose slices may be allowed to digest overnight (*see* **Subheading 3.3.3.**).
12. Prepare sufficient gel volumes according to the PFGE apparatus used. Quoted volume applies to the Gene Navigator System (Pharmacia Biotech, Uppsala, Sweden), having an actual gel size of 152 mm^2 (*see* **Subheading 3.3.4.**).
13. Use blotting paper to blot dry the surface of wells, facilitating the loading of plug slices into the gel (*see* **Subheading 3.3.4.**).
14. Unused agarose can be allowed to solidify and can be reheated several times as needed (*see* **Subheading 3.3.4.**).

References

1. Jackson, C. J., Fox, A. J., and Jones, D. M. (1996) A novel polymerase chain reaction assay for the detection and speciation of thermophilic *Campylobacter* spp. *J. Appl. Bacteriol.* **81,** 467–473.
2. On, S. L. W. and Holmes, B. (1991) Effect of inoculum size on the phenotypic characterisation of *Campylobacter* species. *J. Clin. Microbiol.* **33,** 923–926.
3. Chan, V. L., Hani, E. K., Joe, A., Lynett, J., Ng, D., and Steele, M. (2000) The hippurate hydrolase gene and other unique genes of *Campylobacter jejuni.* In: *Campylobacter* (Nachamkin, I. and Blaser, M., eds.). American Society for Microbiology, Washington DC, pp. 455-463.
4. Gonzalez, I., Grant, K. A., Rirchardson, P. T., Park, S. F., and Collins M. D. (1997) Specific identification of the enteropathogens *Campylobacter jejuni* and *Campylobacter coli* by using a PCR test based on the ceuE gene encoding a putative virulence determinant. *J. Clin. Microbiol.* **35,** 759–763.
5. Eyers, M., Chapelle, S., Van Camp, G., Goossens, H., and De Wachter, R. (1993) Discrimination among thermophilic *Campylobacter* species by polymerase chain reaction amplification of 23s rRNA gene fragments. *J. Clin. Microbiol.* **31,** 3340–3343.
6. Hani, E. K. and Chan, V. L. (1995) Expression and characterisation of *Campylobacter jejuni* benzoylglycine amidohydrolase (hippuricase) gene in *Escherichia coli. J. Bacteriol.* **177,** 2396–2402.
7. Waassenaar, T. M. and Newell, D. (2000) Genotyping of *Campylobacter* spp. *Appl. Environ. Microbiol.* **66,** 1–9.
8. Mazurier, S., Van de Giessen, A., Heuvelman, K., and Wernars, K. (1992) RAPD analysis of *Campylobacter* isolates: DNA fingerprinting without the need to purify DNA. *Lett. Appl. Microbiol.* **14,** 260–262.
9. Møller Nielsen, E., Engberg, J., Fussing, V., Petersen, L., Brogren, C.-H., and On, S. L. W. (2000) Evaluation of phenotypic and genotypic methods for subtyping *Campylobacter jejuni* isolates from humans, poultry and cattle. *J. Clin. Microbiol.* **38,** 3800–3810.
10. Lévesque, C., Pyché, L., Larose, C., and Roy, P. H. (1995) PCR mapping of integrons reveals several novel combinations of resistance genes. *Antimicrob. Agents Chemother.* **39,** 185–191.
11. Lucey, B., Crowley, D., Moloney, P., et al. (2000) Integronlike structures in *Campylobacter* spp. of human and animal origin. *Emerg. Infect. Dis.* **6,** 50–55.
12. Allos, B. M. and Blaser, M. J. (1995) *Campylobacter jejuni* and the expanding spectrum of related infections. *Clin. Infect. Dis.* **20,** 1092–1099.
13. Nachamkin, I., Engberg, J., and Møller Aarestrup, F. (2000) Diagnosis and antimicrobial susceptibility of *Campylobacter* species. In: *Campylobacter* (Nachamkin, I. and Blaser, M., eds). American Society for Microbiology, Washington DC, pp. 45–66.

14. Gibreel, A., Sjögren, E., Kaijser, B., Wretland, B., and Sköld, O. (1998) Rapid emergence of high-level resistance to quinolones in *Campylobacter jejuni* associated with mutational changes in *gyrA* and *parC*. *Antimicrob. Agents Chemother.* **42,** 3276–3278.
15. Adler-Mosca, H., Lüthy-Hottenstein, J., Martinetti-Lucchini, G., Burnens, A., and Altwegg, H. (1991) Development of resistance to quinolones in five patients with campylobacteriosis treated with norfloxacin or ciprofloxacin. *Eur. J. Clin. Microbiol. Infect. Dis.* **10,** 953–957.
16. Endtz, H. P., Ruijs, G. J., van Klingeren, B., Jansen, W. H., van der Reyde ,T., and Mouton, R. P. (1991) Quinolone resistance in *Campylobacter* isolated from man and poultry following the introduction of fluoroquinolones in veterinary medicine. *J. Antimicrob. Chemother.* **27,** 199–208.
17. Drlica, K. and Zhao, X. (1997) DNA gyrase, topoisomerase and the 4-quinolones. *Microbiol. Mol. Biol. Rev.* **61,** 377–392.
18. Zirnstein, G., Swaminathan, B., and Angulo, F. (1999) Ciprofloxacin resistance in *Campylobacter jejuni* isolates: detection of *gyrA* resistance mutations by mismatch amplification mutation assay PCR and DNA sequence analysis. *J. Clin. Microbiol.* **37,** 3276–3280.
19. Zirnstein, G., Helsel, L., Li, Y., Swaminathan, B., and Besser, J. (2000) Characterisation of *gyrA* mutations associated with fluoroquinolone resistance in *Campylobacter coli* by DNA sequence analysis and MAMA PCR. *FEMS Microbiol. Lett.* **190,** 1–7.
20. Lind, L., Sjögren, E., Melby, K., and Kaijser, B. (1996) DNA fingerprinting and serotyping of *Campylobacter jejuni* isolates from epidemic outbreaks. *J. Clin. Microbiol.* **34,** 892–896.
21. Gibson, J. R., Sutherland, K., and Owen, R. J. (1994) Inhibition of DNAse activity in PFGE analysis of DNA from *Campylobacter jejuni*. *Lett. Appl. Microbiol.* **19,** 357–358.
22. Lucey, B., Feurer, C., Greer, P., Moloney, P., Cryan, B., and Fanning, S. (2000) Antimicrobial resistance profiling and DNA amplification fingerprinting (DAF) of thermophilic *Campylobacter* spp. in human, poultry and porcine samples from the Cork region of Ireland. *J. Appl. Microbiol.* **89,** 727–734.
23. Hänninen, M-L., Pajarre, S., Klossner, M-L., and Rautelin, H. (1998) Typing of human *Campylobacter jejuni* isolates in Finland by Pulsed-Field Gel Electrophoresis. *J. Clin. Microbiol.* **36,** 1787–1789.
24. Linton, D., Lawson, A. J., Owen, R. J., and Stanley, J. (1997) PCR detection, identification to species level and fingerprinting of *Campylobacter jejuni* and *Campylobacter coli* direct from diarrheic samples. *J. Clin. Microbiol.* **35,** 2568–2572.

4

Molecular Genotyping Methods and Computerized Analysis for the Study of *Salmonella enterica*

Ana Belén Vivanco, Juan Alvarez, Idoia Laconcha, Nuria López-Molina, Aitor Rementeria, and Javier Garaizar

1. Introduction

Salmonella enterica is widely recognized as a major cause of foodborne diseases in humans and animals and has been isolated from environmental sources in increasing numbers worldwide *(1,2)*. Conventional typing methods such as serotyping and phage typing have been and still are the mainstay in descriptive epidemiology of this microorganism. Nevertheless, limitations on the availability of phage reagents circumscribes the performance of such technique in reference laboratories.

The resolving power of epidemiological typing has been expanded during recent years through the molecular analysis of microbial DNA. The broader availability of the reagents and equipment is accelerating their generalized use in clinical and public health laboratories. Important differences in the performance criteria of the genotyping techniques (typability, reproducibility, stability, and discriminatory power) and the convenience criteria (flexibility, accessibility, and ease of use) exist between them, and there is no ideal typing system for universal use *(3,4)*.

Most of these powerful strain-discriminative techniques are based on comparison of electrophoretic patterns or fingerprints, for which computer-assisted strategies and software packages have been developed to help in construction and analysis of microbial databases. Several initiatives, such as PulseNet (http://www.cdc.gov/pulsenet) or Harmony (http://www.phls.org.uk/inter/harmony), have arisen during recent years for international construction of such fingerprinting databases, which will allow the rapid detection of new strains and the spread of pathogenic clones of bacteria through different regions or countries. Nevertheless, complete consensus has not yet been achieved on the techniques to use or the criteria for interpretation of the results, but these goals may be reached soon.

From: *Methods in Molecular Biology, vol. 268: Public Health Microbiology: Methods and Protocols*
Edited by: J. F. T. Spencer and A. L. Ragout de Spencer © Humana Press Inc., Totowa, NJ

The purpose of this chapter is to describe some of the fingerprinting methods that are currently available to differentiate strains of *Salmonella enterica* and the computer-assisted strategies that can be used for their analysis. We describe here methods based on DNA restriction such as pulsed-field gel electrophoresis (PFGE) ***(2,5)***, methods based on DNA amplification such as polymerase chain reaction (PCR) fingerprinting ***(6)*** methods, based on DNA restriction and amplification such as infrequent restriction site PCR (IRS-PCR) ***(7)***, and an approach to their computerized analysis and generation of fingerprinting databases. When the purpose of the study is the analysis of a limited number of epidemiologically related isolates, a visual analysis of the gel is faster and more efficient. However, if the objective of the analysis is surveillance of a pathogen over time or requires the analysis of higher numbers of isolates, then computerized analysis is more convenient. We use PCR fingerprinting and IRS-PCR methods when a quick analysis of a limited number of isolates in a gel is required. We use the PFGE method for more permanent epidemiological studies, including those that require interlaboratory comparisons, and we suggest their use for construction of computerized libraries of fingerprints with software packages.

2. Materials

2.1. Pulsed-Field Gel Electrophoresis

1. TES buffer: 10 m*M* Tris-HCl, pH 8.0, 10 m*M* Na_2EDTA, 1 *M* NaCl. Sterilize by autoclaving and store at 4°C.
2. Certified Low-Melt agarose (Bio-Rad).
3. Lysis buffer: 1 *M* NaCl, 10 m*M* Tris-HCl, pH 8.0, 200 m*M* Na_2EDTA, 0.5% w/v *N*-laurylsarcosine, and 0.2% w/v sodium deoxycholate. Make up 100 mL of solution. Filter sterilize and store at room temperature. Add 1 mg/mL lysozyme, and 2 µg/mL RNAse prior to use (*see* **Note 1**).
4. ESP buffer: 0.5 *M* Na_2EDTA, pH 8.0, 1% w/v *N*-laurylsarcosine. Filter-sterilize and store at room temperature. Add 1 mg/mL Proteinase K prior to use.
5. TE buffer: 10 m*M* Tris-HCl, pH 8.0, 10 m*M* Na_2EDTA. Sterilize by autoclaving and store at 4°C.
6. PMSF solution: dissolve 17 mg phenylmethylsulfonyl fluoride (PMSF; Sigma) in 1 mL of isopropanol (*see* **Note 2**). Store at –20°C. If necessary, redissolve before using warming up to 56°C.
7. 10X TBE buffer: 890 m*M* Tris-HCl, 890 m*M* boric acid, 20 m*M* Na_2EDTA. Sterilize by autoclaving and store at room temperature.
8. Electrophoresis buffer: 75 µ*M* thiourea (Sigma) in 0.5X TBE buffer (*see* **Note 3**). Prepare fresh and discard after use.
9. Pulsed-field Certified Agarose (Bio-Rad).
10. Rare-cutting restriction endonucleases *Xba*I, *Spe*I, and *Bln*I and appropriate buffers (Boehringer Mannheim).
11. Ethidium bromide solution: prepare a concentrate stock solution of ethidium bromide (200 µg/mL solution) using ultrapure water. Ethidium bromide is sensitive to light: use an opaque bottled to store the stock solution. **Caution:** Ethidium bromide is also carcinogenic. *Always* wear gloves when handling materials containing ethidium bromide. Follow decontamination instructions using activated charcoal or incineration.
12. Lambda ladder concatamers (Bio-Rad).

13. CHEF-DR II Pulsed-field Electrophoresis System (Bio-Rad).
14. Long-wave UV transilluminator for visualization of agarose gels (UltraLum).
15. MP4 Land camera (Polaroid).

2.2. PCR Fingerprinting

1. Tissue culture grade sterile water (Sigma).
2. 10X PCR buffer: 500 m*M* KCl, 100 m*M* Tris-HCl, pH 8.3 (Applied Biosystems).
3. 25 m*M* $MgCl_2$ (Applied Biosystems).
4. Ampli*Taq* Polymerase 5 U/μL (Applied Biosystems).
5. PCR dNTPs solution: 10 m*M* each of dATP, dCTP, dGTP, and dTTP (Applied Biosystems).
6. PCR primers: ERIC2 (5'-AAG TAA GTG ACT GGG GTG AGC G-3'), M13 (5'-GTA AAA CGA CGG CCA GT-3'), and OPS-19 (5'-GAG TCA GCA G-3') (Applied Biosystems; *see* **Note 4**).
7. Thermal cycler for PCR: Robocycler gradient 96 (Stratagene).
8. 1X TBE buffer.
9. Molecular grade agarose (Bio-Rad).
10. Gel loading solution (Sigma).
11. pGEM DNA marker (Promega).
12. 200 μg/mL Ethidium bromide stock solution.
13. Standard apparatus for the electrophoresis of agarose gels and power supply (Bio-Rad).
14. Long-wave UV transilluminator for visualization of agarose gels (UltraLum).
15. MP4 Land camera (Polaroid).

2.3. Infrequent Restriction Site PCR

1. Lysis solution: mix 5 *M* guanidine thiocyanate with 0.1 *M* Na_2EDTA in 20 mL of distilled water. Heat to 65°C to dissolve and add 1.7 mL of 30% (v/v) Sarkosyl. When cold, make up to 100 mL with distilled water. Keep at room temperature and light protected.
2. 7.5 *M* Ammoniun acetate. Keep at 4°C.
3. Chloroform/2-pentanol (24:1 v/v). Keep at 4°C.
4. Isopropanol.
5. 70% v/v Ethanol. Keep at 4°C.
6. TE buffer: 10 m*M* Tris-HCl, pH 8.0, 10 m*M* Na_2EDTA. Sterilize by autoclaving and store at 4°C.
7. Restriction endonucleases *Xba*I, *Hha*I, and *Taq*I and appropriate buffers (Boehringer Mannheim).
8. T4 DNA ligase 10 U/μL and appropriate buffer (Bioline).
9. ATP solution: 10 m*M* ATP, 50 m*M* Tris-HCl, pH 7.5 (Bioline).
10. DNA adapters and primers (Applied Biosystems) (**Tables 1** and **2**).
11. Tissue culture grade sterile water (Sigma).
12. 10X PCR buffer: 500 m*M* KCl, 100 m*M* Tris-HCl, pH 8.3 (Applied Biosystems).
13. 25 m*M* $MgCl_2$ (Applied Biosystems).
14. Ampli*Taq* Polymerase 5 U/μL (Applied Biosystems).
15. PCR dNTPs solution: 10 m*M* of either dATP, dCTP, dGTP, or dTTP (Applied Biosystems).
16. 30% (w/v) Acrylamide/bis solution, 37.5:1 (Bio-Rad).
17. 10% (w/v) Ammonium persulfate solution (Bio-Rad).
18. TEMED (Bio-Rad).
19. Gel loading solution (Sigma).
20. 1X TBE buffer.

Table 1
Adapters for Ligation in IRS-PCR

Restriction enzyme	Adapters	Sequence
*Hha*I	AH1	5'-AGA ACT GAC CTC GAC TCG CAC G-3'
	AH2	5'-TGC GAG T-3'
*Xba*I	AX1	5'-PO_4CTA GTA CTG GCA GAC TCT-3'
	AX2	5'-GCC AGT A-3'
*Taq*I	AT1	5'-CCT GAT GAG TCC TGA C-3'
	AT2	5'-CGG TCA G-3'

IRS-PCR, infrequent restriction site polymerase chain reaction.

Table 2
IRS-PCR Primers

Primers	Sequence
PX (*Xba*I)	5'-AGA GTC TGC CAG TAC TAG A-3'
PAH1 (*Hha*I)	5'-AGA ACT GAC CTC GAC TCG CAC G-3'
PAT1 (*Taq*I)	5'-CCT GAT GAG TCC TGA C-3'

IRS-PCR, infrequent restriction site polymerase chain raction.

21. pGEM marker.
22. Mini Protean electrophoresis cell and power supply (Bio-Rad).
23. Long-wave UV transilluminator for visualization of polyacrylamide gels (UltraLum).
24. MP4 Land camera (Polaroid).

2.4. Computerized Analysis of Electrophoretical Patterns

1. Scanner ScanJet IICX (Hewlett Packard).
2. Software GelCompar 4.0 or superior (Applied Maths).
3. PC personal computer.
4. Laser printer.

3. Methods

3.1. Pulsed-Field Gel Electrophoresis

1. Grow the isolates on trypticasein soy agar plates overnight at 37°C. Inoculate cultures into 6 mL of nutrient broth and incubate at 37°C with shaking until growth reaches an absorbance of 0.12 at 650 nm (1 × 10^9 cells/mL).
2. Spin 5 mL of cells in a refrigerated centrifuge (Beckman GS-6R or similar) at 3000*g* for 10 min and decant supernatant. Wash the cells with 5 mL ice-cold TES buffer and spin once again. Discard supernatant and resuspend pellet in 0.3 mL TES buffer.
3. Prepare by boiling a solution of 1% (w/v) low-melt agarose in distilled water and maintain at 56°C in a water bath. Mix 700 µL of warm agarose with the 300 µL of bacterial suspension and dispense into a block mold, allowing the agarose blocks to solidify by placing them at 4°C for 15–30 min.

4. Add lysozyme and RNAse to the lysis buffer and incubate the blocks for 2 h at 37° C in 3 mL of this lysis buffer.
5. Discard the lysis buffer and add 3 mL ESP buffer. Add the Proteinase K to ESP buffer just before use in a final concentration of 1 mg/mL. Incubate overnight at 56°C in a waterbath.
6. Discard the ESP buffer carefully, washing the blocks with 3 mL distilled water for 15 min with gentle agitation at room temperature. Discard the water and add 2 mL TE buffer and 30 μL of 100 m*M* PMSF solution for each sample (*see* **Note 2**). **Caution:** This step has to be performed in a fume hood and wearing gloves. Agitate gently at room temperature for 30 min. Repeat this step with fresh TE buffer and PMSF solution.
7. Remove TE buffer and PMSF solution, add 3 mL of distilled water, and incubate samples at room temperature for 15 min with gentle agitation.
8. Remove water, add 3 mL of TE buffer, and incubate at room temperature for 30 min with gentle agitation. Repeat this step twice.
9. Remove the buffer and add 3 mL of new TE buffer. Keep the samples refrigerated.
10. Cut a small piece of the block, place it in an Eppendorf tube with 100 μL of the restriction enzyme buffer, and incubate at room temperature for 1 h. Remove buffer carefully with a pipet and replace with 20 U of the enzyme, buffer, and distilled water in a final volume of 50 μL. Incubate at 37°C for 4 h.
11. Prepare 110 mL of 1.2% (w/v) Pulsed-Field Certified Agarose in 0.5X TBE buffer and cast the gel. When gel hardens, cast the gel wells with the agarose blocks, making sure they are placed flat against the leading edge of the well. Insert polymerized phage lambda DNA agarose block as a molecular size marker. Seal the wells with warm agarose.
12. Transfer the gel together with 2 L of cold 0.5X TBE buffer with thiourea to the machine (CHEF-DRII) and leave for 15 min for equilibrium.
13. Program the pulse machine and run the electrophoresis for the established voltages, times, and pulses, with a constant electrophoresis buffer temperature of 12°C (**Table 3**).
14. Dilute the stock solution of ethidium bromide to reach a concentration of 2 μg/mL. Incubate the gel in obscurity for 30 min. Wash the gel with distilled water and photograph the gel under 254-nm UV transillumination.
15. Interpret the gel visually or use the computerized analyzer (*see* **Note 5**). Tenover et al. *(8)* describe several criteria for strain typing through the use of PFGE DNA restriction patterns.

3.2. PCR Fingerprinting

1. Grow the isolates overnight at 37°C in trypticasein soy agar plates.
2. Resuspend in an Eppendorf tube a small loopful of bacteria in 500 μL of sterile water. Boil for 10 min and spin at 11,600*g* for 10 min. Measure the DNA concentration of the supernatant by spectrophotometry and prepare DNA stock solutions of 50 ng/μL. Keep at –20°C.
3. Carry out a PCR reaction with 10 m*M* Tris-HCl, pH 8.3, 50 m*M* KCl, 3 m*M* $MgCl_2$, 200 μ*M* of each dNTPs, 1 U Ampli*Taq* polymerase, 250 ng template DNA, and primer (50 p*M* for ERIC2, 50 p*M* for M13, or 24 p*M* for OPS-19) in a 50-μL final volume.
4. Program the thermal cycler and run the PCR as described in **Table 4**.
5. Prepare 2% (w/v) molecular grade agarose gels in 1X TBE buffer, load the samples and DNA marker, and run the electrophoresis for 1 h, 30 min at 150 V.
6. Dilute the stock solution of ethidium bromide to reach a concentration of 0.5 μg/mL. Incubate the gel in the dark for 40 min. Wash the gel with distilled water and photograph the gel under 254-nm UV transillumination.
7. Interpret the banding patterns visually or with the help of a computerized analyzer (*see* **Note 6**).

Table 3
PFGE Electrophoretic Conditions for Some *Salmonella* Serotypes

		Restriction endonuclease			
		*Xba*I			
Serotype	Conditions	(1)	(2)	*Bln*I	*Spe*I
Enteritidis	Voltage	170 V	230 V	230 V	230 V
	Time	23 h	22,5 h	22,5 h	22 h
	Pulses	10 –35 s	(a) (17–26 s)–17.5 h (b) (50–60 s)–5 h	(a) (17–26 s)–17.5 h (b) (50–60 s)–5 h	(a) (8–20 s)–18 h (b) (30–40 s)–4 h
Typhimurium	Voltage	170 V	230 V	230 V	230 V
	Time	23 h	24 h	24 h	24 h
	Pulses	10–35 s	(a) (10–26 s)–18 h (b) (50–60 s)–6 h	(a) (10–26 s)–18 h (b) (50–60 s)–6 h	(a) (8–20 s)–18.5 h (b) (30–40 s)–5.5 h
California	Voltage	230 V		230 V	
	Time	22.5 h		22.5 h	
	Pulses	(a) (17–26 s)–17.5 h (b) (50–60 s)–5 h		(a) (17–26 s)–17.5 h (b) (50–60 s)–5 h	
[4,5,12:i:-]	Voltage	200V			
	Time	26 h			
	Pulses	(a) (5–15 s)–7 h (b) (15–60 s)–19 h			

PFGE, pulsed-field gel electrophoresis.

(a) and (b), different electrophoretical conditions evaluated for *Salmonella* serotypes Enteritidis and Typhimurium.

3.3 Infrequent-Restriction-Site PCR

1. Grow the isolates in 20 mL of trypticasein soy broth at 37°C overnight with shaking.
2. Spin at 3000*g* for 20 min. Resuspend the pellet in 10 mL TE buffer to wash the cells and spin again.
3. Resuspend the pellet with 500 μL of lysis solution and transfer to an Eppendorf tube.
4. Add 250 μL ammonium acetate, mix, and leave on ice for 10 min.
5. Add 500 μL of chloroform/2-pentanol and shake by hand vigorously for 10 min.
6. Spin at 11,600*g* for 10 min and transfer 740 μL of supernatant to a fresh Eppendorf tube. Add 400 μL of cold isopropanol, mix by inverting the tube for 1 min, and spin at 11,600*g* for 5 min.
7. Wash the DNA three times with 1 mL of ice-cold 70% (v/v) ethanol and vacuum dry.
8. Rehydrate the DNA in 100 μL of TE buffer and keep at 4°C overnight.
9. Double-restrict 1 μg of bacterial DNA with 10 U of *Xba*I and 10 U of *Hha*I endonucleases (or the same amount of *Taq*I), with the appropriate buffer in a 12.5-μL final volume for 1 h at 37°C. If *Taq*I restriction endonuclease is used, carry out a second restriction step for 1 h at 65°C.
10. Add the following reagents to the reaction: 2 U T4 DNA ligase and corresponding buffer, 1 m*M* ATP, 20 p*M* of each adapter (*see* **Note 7**) in a final reaction volume of 20 μL. Carry out the DNA ligation at 16°C for 1 h.

Table 4
PCR Fingerprinting Amplification Temperatures

Temperature (°C)	Time (min)	No. of cycles
ERIC2 primer		
95	7	1
94	1	
52	1	30
72	3	
72	3	1
M13 primer		
94	5	1
93	1	
50	1	30
72	2	
72	6	1
OPS19 primer		
94	10	1
94	1	
34	1	45
72	2	
72	2	1

PCR, polymerase chain reaction.

11. To inactivate the T4 DNA ligase, heat the samples at 65°C for 20 min.
12. Carry out a second round of DNA double restriction to cleave any restriction site reformed by ligation with 5 U of each restriction enzyme and the corresponding buffer for 15–30 min.
13. Carry out a PCR reaction with the following reagents: 10 m*M* Tris-HCl, pH 8.3, 50 m*M* KCl, 2.5 m*M* $MgCl_2$, 200 μ*M* each dNTPs, 1 μ*M* of each of the corresponding primers (**Table 2**), 1 U of Ampli*Taq* polymerase, and 1 μL of the restricted-ligated DNA. The final reaction volume is 50 μL.
14. Perform the PCR reaction in the thermal cycler at the following temperatures: 94°C for 6 min and one cycle and 94°C for 1 min, 60°C for 1 min, 72°C for 2 min, for 30 cycles.
15. Prepare an 8% (w/v) polyacrylamide gel adding 2.66 mL acrylamide/bis solution to 7.34 mL 1X TBE buffer. Add 75 μL of 10% (w/v) ammonium persulfate and 10 μL TEMED and let it polymerize for 1 h in a Mini Protean electrophoresis cell.
16. Mix samples and the molecular weight standard with loading buffer and run an electrophoresis with 1X TBE buffer at a constant 150 V for 45 min.
17. Stain the polyacrylamide gel with ethidium bromide (3 μg/mL) for 15 min, wash the gel with distilled water for 30 min, and photograph under UV transillumination.
18. Interpret the banding patterns visually or with help of a computerized analyzer.

3.4. Computerized Analysis of Fingerprinting Patterns

1. Scan and save images of gels in a TIFF file and then analyze them by GelCompar version 4.0 or superior software (Applied Maths; *see* **Note 5**).
2. After conversion and normalization of gels, determine the degrees of similarity of DNA fingerprints by Dice or related coefficients (*see* **Note 8**).
3. Generate dendrograms using the UPGMA (unweighted pair group method using arithmetic averages) algorithm.
4. To construct the libraries with the PFGE fingerprints, follow the instructions of the software manufacturer. Such libraries allow the user to identify new eletrophoretic patterns and to incorporate them in the database.

4. Notes

1. *N*-laurylsarcosine may precipitate when lysozyme and RNAse are added. If this happens, place the reaction product at 50°C in a water bath to dissolve and then warm it at room temperature.
2. PMSF is highly toxic by inhalation. Its use for Proteinase K inactivation could be avoided, but more washing with distilled water is then required.
3. In some *Salmonella* serotypes, extensive DNA degradation is observed during electrophoresis. Addition of thiourea to the electrophoresis buffer prevents such events.
4. Several primers are available in the literature for such PCR fingerprinting procedures, and different levels of strain discrimination could be achieved depending on the serotype studied.
5. GelCompar is licenced in the United States and several countries to Bio-Rad laboratories. Other software such as Molecular Analyst run on an Apple platform. The Bio Image Whole Band Analyzer runs on a Unix platform. We have compared and evaluated such software to determine whether results generated by these programs correlated adequately with visual interpretation of DNA patterns *(9)*.
6. The interassay reproducibility of PCR fingerprinting is inherently low owing to the low annealing temperatures and influences of minor changes in the amounts of reagent in the PCR reaction *(10)*. Intra-assay reproducibility is higher, allowing for analysis of a group of isolates in a single gel looking for band differences.
7. Choose the adapters in relation to the restriction enzyme used as shown in **Table 1**.
8. The software packages try to imitate the automatic band alignments that our eyes perform and usually accomplish it successfully when gels are of good quality. Computerized alignments of the internal control bands in heavily distorted gels are frequently performed inaccurately; therefore such gels should be eliminated from the comparison or repeated.

Acknowledgments

This work was supported by The University of Basque Country and the Basque Government. A. B. Vivanco is supported by a grant from The University of Basque Country and J. Alvarez by a grant from the Basque Government, Spain.

References

1. Rodrigue, D. C., Tauxe, R. V., and Rowe, B. (1990) International increase in *Salmonella enteritidis*: a new pandemic? *Epidemiol. Infect.* **105,** 21–27.
2. Laconcha, I., Baggesen, D. L., Rementeria, A., and Garaizar, J. (2000) Genotypic characterisation by PFGE of *Salmonella enterica* serotype Enteritidis phage types 1, 4, 6,

and 8 isolated from animal and human sources in three European countries. *Vet. Microbiol.* **75,** 155–165.

3. Struelens, M. J., Bauernfeind, A., van Belkum, A., et al. and members of ESGEM (ESCMID) (1996) Guidelines for appropriate use and evaluation of microbial epidemiological typing systems. *Clin. Microbiol. Infect.* **2,** 2–11.
4. Van Belkum, A., Struelens, M., De Visser, A., Verbrugh, H., and Tybayrenc, M. (2001) Role of genomic typing in taxonomy, evolutionary genetics, and microbial epidemiology. *Clin. Microbiol. Rev.* **14,** 547–560.
5. Laconcha, I., López-Molina, N., Rementeria, A., Audicana, A., Perales, I., and Garaizar, J. (1998) Phage typing combined with pulsed-field gel electrophoresis and random amplified polymorphic DNA increases discrimination in the epidemiological analysis of *Salmonella enteritidis* strains. *Int. J. Food Microbiol.* **40,** 27–34.
6. López-Molina, N., Laconcha, I., Rementeria, A., Audicana, A., Perales, I., and Garaizar, J. (1998) Typing of *Salmonella* serotypes including *Salmonella enteritidis* of different phage types by PCR fingerprinting. *J. Appl. Microbiol.* **84,** 877–882.
7. Garaizar, J., López-Molina, N., Laconcha, I., et al. (2000) Suitability of PCR fingerprinting, infrequent-restriction-site PCR, and pulsed-field gel electrophoresis, combined with computerised gel analysis, in library typing of *Salmonella enteritidis*. *Appl. Environ. Microbiol.* **66,** 5273–5281.
8. Tenover, F. C., Arbeit, R. D., Goering, R. V., et al. (1995) Interpreting chromosomal DNA restriction patterns produced by pulsed-field gel electrophoresis: criteria for bacterial strain typing. *J. Clin. Microbiol.* **33,** 2233–2239.
9. Rementeria, A., Gallego, L., Quindós, G., and Garaizar, J. (2001) Comparative evaluation of three commercial software packages for analysis of DNA polymorphism patterns. *Clin. Microbiol. Infect.* **7,** 331–336.
10. Schweder, M. E., Shatters, Jr, R. G., West, S. H., and Smith, R. L. (1995) Effect of transition interval between melting and annealing temperatures on RAPD analyses. *Biotechniques* **19,** 38–42.

II

VIRUSES

5

Detection of Infectious Rotaviruses by Flow Cytometry

Albert Bosch, Rosa M. Pintó, Jaume Comas, and Francesc-Xavier Abad

1. Introduction

Human rotaviruses are considered the main cause of viral gastroenteritis in infants and young children throughout the world *(1)*. Their transmission is through the fecal-oral route, mostly after ingestion of contaminated water and food *(2)*. Since an extremely high number of virus particles are present in the feces during the acute gastroenteritis, methods based on electron microscopy, passive particle agglutination tests, or enzyme-linked immunosorbent assays are readily employed for clinical diagnosis. However, the sensitivity of these procedures is not high enough to detect the low number of viral particles sometimes present in the environment *(3)*. In the case of environmental samples, amplification of viral nucleic acids by polymerase chain reaction assays coupled to reverse transcription (RT-PCR) has been increasingly applied to detect rotaviruses in water *(4)* and shellfish samples *(5)*. However, procedures based on molecular approaches have to face the drawback that they do not differentiate between infectious and noninfectious particles, which is of major relevance from the public health point of view.

Virus propagation in cell culture prior to detection by immunological or molecular procedures accomplishes the dual purpose of increasing the amount of target material and incorporating an infectivity assay as well.

Wild-type rotaviruses present difficulties in their in vitro replication, although some of them may be adapted to grow in several cell lines such as the monkey kidney cell line MA104 or the human intestinal cell line CaCo-2 *(6,7)*. More than a decade ago, an assay for the specific detection of infectious rotaviruses in environmental samples, involving an indirect immunofluorescence test (IIF) and optical microscopy (OM) counting of infected foci in infected MA-104 cell monolayers, was described *(4)*. On the other hand, CaCo-2 cells have been successfully employed in our laboratory for infectivity assays of several fastidious enteric virus strains present in water samples *(8)*.

From: *Methods in Molecular Biology, vol. 268: Public Health Microbiology: Methods and Protocols*
Edited by: J. F. T. Spencer and A. L. Ragout de Spencer © Humana Press Inc., Totowa, NJ

Flow cytometry (FCM) is a method to quantify components or study the structural characteristics of cells, mainly by optical means. The term *flow cytometry* derives from the measurement of single cells as they flow in a fluid stream through a measuring point surrounded by several detectors. FCM involves the use of a beam of light projected through a liquid stream that contains cells. As the cells cross the focused light, they emit signals that are captured by detectors. These signals are then converted for computer storage and data analysis and can provide information about cellular properties. These biophysical properties are then correlated with biological and biochemical properties of interest. To make the measurement of biological properties of concern, the cells are usually stained with fluorescent dyes that specifically bind to cellular constituents. Sorting instruments may also isolate specific cells according to their cytometric profile. FCM can be thus employed for counting and specific cell sorting with several cell types.

The use of FCM for the detection of virally infected cells presents several advantages with respect to OM detection and has been used for different viruses ***(9–11)***. The IIF-OM detection method, although efficient, is cumbersome and requires well-trained personnel, whereas FCM is an automatable procedure that allows the processing of a large number of samples. Additionally, most major hospitals and public health institutions in developed countries possess a flow cytometer. However, standardization of an FCM method requires the study of several critical points, such as fixation (type of fixative, contact time, and so on) or minimization of background noise (lowering the antibody working solution concentration, assaying different blocking solutions, and so on).

2. Materials

2.1. Infectivity Assay

1. Horizontal laminar flow hood.
2. Vertical laminar flow hood.
3. CO_2 incubator.
4. CaCo-2 cell line, a human colon adenocarcinoma cell line, used at passage level 80–100.
5. A cytopathogenic strain of human rotavirus, i.e., Wa or Ito^r P13, used as positive control.
6. Eagle's minimum essential medium (MEM) with Earle's salts; Auto-Pow® (ICN, Costa Mesa, CA, cat. no. 1110024) as the cell culture medium. Dissolve and sterilize according to the manufacturer's instructions, and supplement with the following reagents at final concentrations: 0.15% $NaHCO_3$, 15 m*M* HEPES, 2 m*M* L-glutamine, 100 U/mL penicillin, 100 μg/mL streptomycin.
7. Fetal calf serum (FCS; BioWhittaker, Walkerville, MD).
8. Phosphate buffer solution (PBS): dissolve 8 g NaCl, 0.2 g KCl, 1.15 g Na_2HPO_4, and 0.2 g KH_2PO_4 in 1000 mL deionized water. After complete dissolution, adjust the pH to 7.1–7.2 by 1 *M* HCl addition. Sterilize by autoclaving and store at 4 ± 1°C.
9. Trypsin for cell passage: dissolve 2.5 g of trypsin (tissue culture grade 1:250; DIFCO, Sparks, MD, cat. no. 0152-13) and 0.2 g of EDTA in 1000 mL PBS. After complete dissolution sterilize by filtration through 0.22-μm GS-type filters (Millipore, Bedford, MA; cat. no. GSWP047S0) and store at –20°C until use.

10. Trypsin for rotavirus activation: dissolve 0.1 g of trypsin grade IX (Sigma, St. Louis, MO; cat. no. T-0134) in 10 mL PBS. Sterilize by filtration. Prepare a stock solution of 1 mg/mL, by subsequent 1:10 dilution with PBS. Store in 1-mL aliquots at –20°C.
11. Sterile plastic ware: 5- and 10-mL plastic pipets, 58 × 17-mm Petri dishes (Nunc, Roskilde, Denmark; cat. no. 150288), sterile 0.2- and 1.0-mL tips, and 10-mL plastic tubes.

2.2. Indirect Immunofluorescence

1. Orbital vertical mixer (Selecta, Barcelona, Spain) or similar.
2. 1 *M* Sucrose solution in deionized water. Sterilize by filtration through 0.22-μm filters and store at room temperature.
3. *p*-Formaldehyde solution (20%): dissolve 20 g *p*-formaldehyde (Merck, cat. no. 1.04005.) in 70 mL deionized water. Heat the mixture for 1 h at 80°C in a thermostatic bath inside an extraction cabinet. Add 2–3 drops of 1 *M* NaOH. After 1 h more of heating, filter the mixture through Whatman filter paper. Bring to a final volume of 100 mL with deionized water and store at –20°C.
4. 1 *M* Dipotassium phosphate solution in deionized water: filter through a 0.45-μm filter and autoclave. Store at room temperature.
5. 1 *M* Sodium phosphate solution in deionized water: filter through a 0.45-μm filter and sterilize by autoclaving. Store at room temperature.
6. 1 *M* Phosphate buffer: adjust the pH of 80 mL 1 *M* dipotassium phosphate solution to pH 7.4 with 1 *M* sodium phosphate solution.
7. 0.2 *M* Phosphate buffer: add 20 mL of 1 *M* phosphate buffer to 80 mL of sterile deionized water.
8. Saline solution: add 0.8 g of NaCl in 100 mL deionized water. Sterilize by autoclaving and store at 4 ± 1°C.
9. Fixative: mix, at room temperature, 3 mL of 20% *p*-formaldehyde solution, 1.2 mL of 1 *M* sucrose, 10 mL of 0.2 *M* phosphate buffer, and 5.8 mL of deionized water.
10. Blocking solution: the day of IIF assay, dissolve 5 g powdered skim milk and 0.2 mL of a 10% Triton X-100 solution in 100 mL of saline solution.
11. Rotavirus positive control serum (Institute Virion, Rüschlikon, Switzerland; cat. no. 3193) as primary antibody: Reconstitute the lyophilized powder with 0.4 mL of deionized water. Transfer the volume to a plastic tube and add blocking solution to reach the working dilution recommended by the manufacturer (between 1:32 and 1:110). Prepare the same day of experiment.
12. Fluorescein isothiocyanate (FITC)-labeled rabbit anti-human IgG (Sigma; cat. no. F-4512) as secondary antibody: Prepare a 1:400 working dilution with blocking solution. Prepare the same day of experiment.

2.3. FCM Detection

1. Coulter Epics XL flow cytometer (Beckman-Coulter, Miami, FL) or similar.

3. Methods

3.1. Infectivity Assay

1. Prepare CaCo-2 cell monolayers in a laminar flow hood by trypsinization at a split ratio of 1:3. Cells are used when they are confluent or near confluence. In normal growing conditions, a maximum of 20 58 × 17-mm Petri dishes may be produced from a 175-cm^2

culture flask after trypsinization (*see* **Subheading 2.1., item 9**). The cells are grown in cell culture medium with 10% (v/v) FCS.

2. Samples (*see* **Note 1**) are pretreated with 10 µg/mL trypsin (*see* **Note 2**) for 30 min at 37°C .
3. Wash the cells monolayers twice, using 5 mL of cell culture medium per Petri dish to remove all traces of FCS (*see* **Note 3**).
4. Remove the washing medium and add 200 µL of each trypsin-treated sample per Petri dish. As positive controls inoculate 200 µL of a viral suspension of a cytopathogenic rotavirus strain with an original titer of 10^3–10^4 infectious units per mL in each of two Petri dishes. As negative controls inoculate 200 µL serum-free medium over each of four Petri dishes.
5. Incubate for 60 min at 37°C, with two or three gentle swirlings of the inoculum over the monolayer.
6. Add 5 mL per plate of postinfection overlay medium consisting of serum-free cell culture medium supplemented with 5 µg/mL trypsin (*see* **Note 4**).
7. Incubate Petri dishes for 4 d at 37°C in an atmosphere of 5% CO_2-air.

3.2. Indirect Immunofluorescence

1. Recover the cells by vigorously pipeting up and down the postinfection medium, and transfer the whole volume to a 10-mL plastic tub.
2. Centrifuge at low speed, around 900*g* for 10 min.
3. Carefully discard the supernatant.
4. Resuspend cell pellet in 1 mL of saline.
5. Transfer the cell suspension to a 1.5-mL microtube.
6. Centrifuge microtubes in an Eppendorf 5415C microcentrifuge (or similar) at 10,000*g* for 90 s, at room temperature (*see* **Note 5**).
7. Discard the supernatant by aspiration using a micropipet and resuspend again the cell pellet with another 1 mL of saline.
8. Spin the microtube again at the same conditions described above.
9. Discard the supernatant and resuspend cell pellet with 1 mL of freshly prepared (the same day) fixative solution.
10. Incubate in gentle agitation for 30 min at room temperature.
11. Centrifuge at 10,000*g* for 90 s, at room temperature. Carefully discard the fixative and add 1 mL of saline to each microtube (*see* **Note 6**).
12. Perform two more washes with saline as described above (1 mL of saline solution per microtube, centrifuge at previous described conditions, discard the supernatant, add 1 mL more of saline solution, centrifuge another time, and pipet out the supernatant).
13. Permeabilize the fixed cells by a 15-min treatment with 0.1% Triton X-100 at room temperature, with constant gentle agitation.
14. Add 300 µL of primary antibody solution to the cell pellet of each microtube, with gentle mixing (*see* **Note 7**). Place the microtubes in an orbital vertical mixer.
15. Keep in agitation for 45 min at room temperature (*see* **Note 8**).
16. Centrifuge the microtubes at 10,000*g* for 90 s, and discard the primary antibody solution by aspiration with a pipet.
17. Wash the cellular pellet five times with the blocking solution, using for each wash 1 mL of blocking solution per microtube (*see* **Note 7**), centrifuge after each wash at 10,000*g* for 90 s, and gently discard the supernatant by careful aspiration with a pipet.

18. Add 300 μL of the FITC-labeled secondary antibody solution. Place the microtubes in an orbital vertical mixer.
19. Keep in agitation for 45 min in the dark at room temperature (*see* **Note 8**).
20. Centrifuge at 10,000*g* for 90 s, and discard the secondary antibody solution by careful aspiration with a pipette.
21. Wash the cell pellet at least four times with saline solution, using for each wash 1 mL per microtube (*see* **Note 7**), centrifuge after each wash at 10,000*g* for 90 s, and gently discard the supernatant by careful aspiration with a pipet.
22. Resuspend the cell pellet in 1 mL of saline.
23. Store at 4 ± 1°C in dark conditions to await the FCM assay (*see* **Note 9**).

3.2. FCM Detection

The cellular suspensions are analyzed with a Coulter Epics XL flow cytometer equipped with the standard 488-nm argon-ion laser at 15 mW power.

1. Select logarithmic forward angle light scatter (FSC), logarithmic side-angle light scatter (SS), logarithmic green fluorescence (FL1), and time as the acquisition parameters.
2. Adjust cytometer settings according to the values given in **Table 1**.
3. Define an FS log vs SS log dotplot (resolution: 128 × 128 channels) (**Fig. 1A**) and a 1024-channel histogram using the FL1 parameter (**Fig. 1B**).
4. Check instrument stability, analyzing 10-μm calibration beads (Flowcheck, Coulter). Record the channel position of these beads for FSC, SSC, and FL1 as a daily quality control of the instrument.
5. Acquire a minimum of 100,000 cells. To avoid coincidences, injection flow rate must be adjusted to keep total events/s below 800.
6. Draw a polygonal region on the cellular population of the FSC vs SS dotplot. FSC is used to select cell size, and SS is used to select shape and structure in order to restrain the readings to the population of intact eukaryotic cells and not the cell debris.
7. Create a gate using the polygonal region of the FSC/SS dotplot. Apply the gate on the fluorescence histogram (**Fig. 1 B**) in order to represent fluorescence of the cells inside this region. Fluorescence intensity (*x*-axis) is expressed in log scale, which means that in the higher channels small differences in channel number represent large differences in the amount of dye per cell.
8. Define a cursor on the FL1 histogram from 59.7 to 1024. This cursor is used to quantify positive events. The XL analysis software give us the percentage of cells under the cursor. The position of this cursor is established after analysis of a pool of negative control samples *(11)*. For quality assurance, a large number (30–50 *n*) of mock-infected samples should be processed before attempting the detection of wild-type rotaviruses in natural environmental samples. An arbitrary cursor (A) is drawn at the right end of their fluorescence curves (channels 10–1024). This cursor included some of the negative cell population counts. The mean fluorescence of each of the A cursors from the negative controls is calculated. These mean values follow a normal distribution. The mean and standard deviation of this latter curve are figured, and a second cursor (B) is then defined starting at the point obtained by adding 2 standard deviations to the mean fluorescence (channel 60), and ending at channel 1024. The ratio of cells present in cursor B in respect to the total counted cells is calculated for each negative sample, and the mean plus 2 standard deviations of these ratios in the negative samples is established as the threshold of positivity .

Table 1
XL Cytometer Settings

Signal	Volts	Gain	Total gain
FS	0	1.0	1.00
SS	0	1.0	1.00
FL1	500	1.0	
Discriminator: Sensor SS, value 7			

FS, forward scatter; SS, side scatter; FL1, logarithmic green fluorescence.

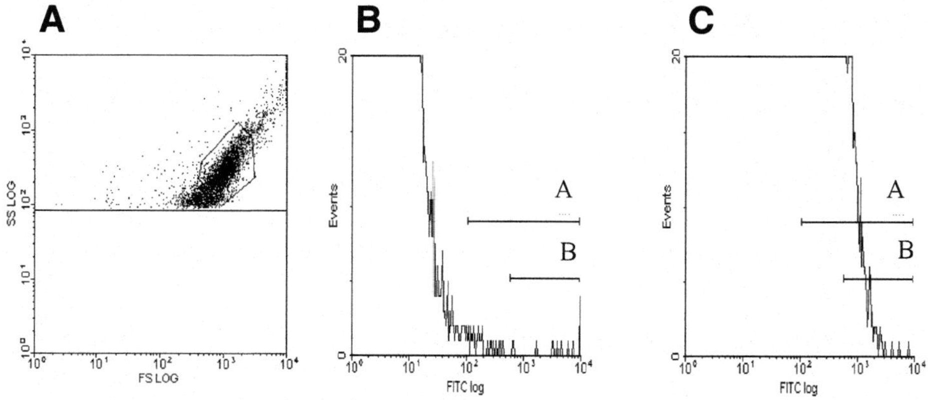

Fig. 1. (**A**) Dot plot of FSC vs SS, used to identify intact cells (inside the polygonal region) and to exclude debris and aggregates. (**B**) Green fluorescence histogram of a negative control. (**C**) Green fluorescence histogram of a positive sample. Events included inside the polygonal region are gated and represented in these two histograms according to its fluorescence.

The procedure must be ascertained using 10-fold dilutions of a cell-adapted rotavirus strain to infect CaCo-2 cell monolayers. Different infectious doses are assayed and each of them determined at least in triplicate. When wild-type rotaviruses are assayed, negative and positive control samples should be included in each assay.

4. Notes

1. In environmental studies, a water concentration step is frequently required to reduce the sampled volume to an amount able to be analyzed in the laboratory. Additionally, we regularly lyophilyze (freeze-dry) 2 mL of the water concentrate and resuspend the lyophilyzate in 200 μL of saline solution, which adds a further 10X concentration factor. This procedure contributes to detoxification of the sample prior to its inoculation onto the cell monolayers.
2. Trypsin treatment enhances rotavirus infectivity by cleaving capsid protein VP4 *(12)*.

3. Alternatively, cell monolayers may be kept overnight with the serum-free cell culture medium at 37°C to remove all traces of fetal bovine serum, which is inhibitory to rotavirus infectivity.
4. Addition of different concentrations of trypsin (1–20 µg/mL) to media overlays is generally employed for the propagation of rotavirus in cell cultures. However, the CaCo-2 cell line is somewhat sensible to this enzyme, and in our studies the maximum permissive concentration was 5 µg/mL. Moreover, although the cells may detach from the plates, they keep alive and show no deleterious effect on virus infectivity.
5. This centrifugation speed usually allows us to recover all the cells from the suspension. Sometimes, however, the cell pellet is somewhat fluffy. In this case, it is recommended to repeat this centrifugation step before discarding the saline.
6. Do not exceed the fixation time. A contact time between 25 and 30 min does not affect the final outcome of fixation. The fixation works best on monodispersed cells.
7. When any solution is added, cells have to be dispersed by gentle agitation prior to centrifugation. This is specially crucial in the final washing step, to reduce the presence of the blocking agent (skim milk) and FITC-labeled secondary antibody. The visual aspect of the final cell suspensions should be near transparence. If not, perform additional washes. Too much blocking agent can affect proper sample processing by the flow cytometer.
8. One of the most critical points in IIF is the contact time of antibodies and cell suspension. From our experience, it is not recommended to exceed 45 min of contact time, including centrifugation time. Longer contact time, particularly in secondary antibody incubation, result in high backgrounds, which causes problems in discriminating between negative and positive samples.
9. Although it is better to perform the flow cytometry detection the same day as the immunofluorescence staining, no significant loss in fluorescence signal has been observed 2 wk after immunofluorescent staining.

References

1. Bern, C., Martines, J., de Zoysa, I., and Glass, R. I. (1992) The magnitude of the global problem of diarrhoeal disease: A ten-year update. *Bull. WHO* **70,** 705–714.
2. Craun, G .F. (1991) Causes of waterborne outbreaks in the United States. *Wat. Sci. Tech.* **24,** 17–20.
3. Bosch, A., Pinto, R. M., Blanch, A. R., and Jofré, J. T. (1988) Detection of human rotavirus in sewage through two concentration procedures. *Wat. Res.* **22,** 343–348.
4. Gajardo, R., Bouchriti, N., Pintó, R. M., and Bosch, A. (1995) Genotyping of rotaviruses isolated from sewage. *Appl. Environ. Microbiol.* **61,** 3460–3462.
5. Bosch, A., Sánchez, G., Le Guyader, F., Vanaclocha, H., Haugarreau, L., and Pintó, R. M. (2001) Human enteric viruses in coquina clams associated with a large hepatitis A outbreak. *Wat. Sci. Tech.* **43,** 61–65.
6. Kitamoto, N. K., Ramig, R. F., Matson, D. O., and Estes, M. K. (1991) Comparative growth of different rotavirus strains in differentiated cells (MA104, HepG2 and CaCo-2). *Virology* **184,** 729–737.
7. Pintó, R. M., Diez, J. M., and Bosch, A. (1994) Use of the colonic carcinoma cell line CaCo-2 for in vivo amplification and detection of enteric viruses. *J. Med. Virol.* **44,** 310–315.
8. Pintó, R. M., Gajardo, R., Abad, F. X., and Bosch, A. (1995) Detection of fastidious infectious enteric viruses in water. *Environ. Sci. Tech.* **29,** 2636–2638.

9. Qvist, P., Aasted, B., Bloch, B., Mayling, A., Rosholt, L., and Houe, H. (1991) Comparison of flow cytometry and virus isolation in cell culture for identification of cattle persistently infected with bovine viral diarrhea virus. *J. Clin. Microbiol.* **29,** 660–666.
10. Snoeck, R., Schols, D., Sadzot-Delvaux, C., et al. (1992) Flow cytometric method for the detection of gpI antigens of varicella zoster virus and evaluation of anti-VZV agents. *J. Virol. Methods* **38,** 243–254.
11. Abad, F. X., Pintó, R. M., and Bosch, A. (1998) Flow cytometry detection of infectious rotaviruses in environmental and clinical samples. *Appl. Environ. Microbiol.* **64,** 2392–2396.
12. Estes, M. K. (2001) Rotaviruses and their replication. In: *Field's Virology*, 4th ed., vol 1 (Fields, B. N., Knipe. D. M., and Howley, P. M., eds.). Lippincott-Williams & Wilkins, New York, pp. 1747–1785.

6

Integrated Cell Culture/PCR for Detection of Enteric Viruses in Environmental Samples

Kelly A. Reynolds

1. Introduction

Recently, an integrated cell culture/polymerase chain reaction (ICC/PCR) technique has been developed for the detection of viruses in environmental samples providing a reliable method for practical analysis and direct monitoring of environmental samples for viral pathogens ***(1,2)***. CC/PCR allows for detection of infectious viruses in hours to days compared with the days or weeks necessary with cell culture alone. Bacterial indicator organisms are commonly used to evaluate environmental samples with respect to fecal contamination and potential public health impacts. These organisms do not correlate well with the presence of viruses, but a rapid, reliable method was not previously available for direct virus testing. Using ICC/PCR, environmental samples may be directly surveyed for pathogenic viruses, in a timely manner. Direct virus analysis will lead to better assessment of the presence and risk of human enteric viruses in the environment, so that control measures may be developed with true virus occurrence data. The ICC/PCR approach combines two previously applied virus detection methods, conventional cell culture and PCR amplification, utilizing the major advantages and overcoming the major limitations of each methodology when used alone.

Cell culture assay is the standard method for the detection of viable human viruses (i.e., poliovirus, coxsackievirus, echovirus, adenovirus, hepatitis A virus, reovirus, and rotavirus) in environmental samples, serving as the method against which all newer technologies are evaluated. Although cell culture is theoretically capable of detecting a single viable virus in relatively large volumes of sample, the time required for confirmed results with conventional cell culture makes it an impractical method for routine monitoring of environmental samples. Furthermore, cell culture does not detect noncytopathogenic viruses (viruses that are viable, infecting cells, and continually spreading to neighboring cells but that do not cause a visible cytopathogenic effect [CPE] on the cell monolayer). Rotavirus and most wild-type hepatitis A viruses (HAV) are infectious to cell cultures but do not produce a clear CPE.

From: *Methods in Molecular Biology, vol. 268: Public Health Microbiology: Methods and Protocols*
Edited by: J. F. T. Spencer and A. L. Ragout de Spencer © Humana Press Inc., Totowa, NJ

Another disadvantage to conventional cell culture is that a single cell line may support the growth of a variety of different viruses, preventing specific identification of the viruses present and allowing less fastidious viruses to proliferate, masking the presence of other, potentially pathogenic, viruses. Finally, conventional cell culture is subject to toxicity, meaning that substances such as toxic chemicals, waste products, suspended particles, and humic and fulvic acids can inactivate the cell monolayer. The presence of bacteria and fungi may also inactivate the cell monolayer, prior to virus infection. Cell culture toxins are common to environmental samples and visually mimic viral CPE, leading to false-positive results. Toxic samples must be purified, risking loss of original virus populations, and reassayed, at an increased cost of analysis.

Molecular methods such as PCR were developed to detect a wide range of viruses directly, including noncytopathogenic strains, and they have been used extensively for the sensitive detection and identification of viruses in clinical and environmental samples. The use of specifically designed nucleotide primers allows the sequences of target viruses to be amplified rapidly and specifically. Definitive results are available in less than 1–2 d.

There are some major disadvantages, however, to direct PCR amplification of viruses from environmental samples, including: (1) the presence of PCR-inhibitory compounds in sample concentrates; (2) the small reaction volumes; and (3) the detection of noninfectious virus particles. PCR-inhibitory compounds are naturally present in the environment (i.e., humic and fulvic acids, proteins, metals, and salts) and, when present, prevent reaction enzymes from amplifying a target sequence, producing a false-negative result. PCR inhibition can be very difficult, if not impossible, to overcome without loss of the target viral sequence. In addition, owing to the expense of the PCR reagents and the limitations of the thermal cycling equipment currently available, only small reaction volumes (approx 10 μL/reaction) can be examined by PCR.

Finally, PCR cannot distinguish between infectious and noninfectious viral targets. A commonly observed phenomenon with viruses is the presence of a particle-to-plaque-forming unit (PFU) ratio. A plaque-forming unit possesses the nucleic acid, capsid, and receptor components and is able to infect host cells; however, viral nucleic acid can be present in the absence of a complete, infectious virus. Depending on the environment, the particle/PFU ratio may be as high as 1000. Virus cultures grown under low-stress laboratory conditions typically have a particle/PFU ratio of 10. PCR does not differentiate between noninfectious and infectious viruses and thus expectedly overestimates the risk of virus infection in exposed populations. Since the actual ratio of particles to PFU is not known for individual environments, the level of overestimation of risk cannot be easily determined. Detection of noninfectious particles is particularly important in treated water supplies, in which viruses originally present are rendered noninfectious by the treatment process, such as chlorine disinfection, while their nucleic acid sequences persist. These particles are detectable by PCR but are not expected to be a public health risk.

The ICC/PCR method combines both cultural and molecular techniques for rapid detection of viable human viruses and thus is able to detect viruses in large equivalent volume concentrates, without the limitations of cell culture toxicity or PCR inhibition.

Inhibition is inherently overcome by the dilution of the sample concentrate with cell culture media. In addition, viruses are grown to much higher numbers in the cell culture flask, further ensuring the success of the PCR amplification. Therefore, using ICC/PCR, a large population of viruses is being detected with a low concentration of inhibitors, eliminating the uncertainty of true vs false negative results with direct PCR alone. Cell culture toxicity is minimized since the assays can be stopped and viruses detected prior to cell death owing to toxicity, eliminating the uncertainty of true vs false positive results with conventional cell culture alone.

Previous studies have shown the effective application of ICC/PCR methodology on environmental sample concentrates that were inhibitory to direct PCR ***(1,2)*** and at virus concentrations as low as 0.001 most probable number (MPN) per liter of original sample concentrate ***(3)***. The ICC/PCR method was able to detect enterovirus concentrations of ≥10 PFU per cell culture flask inoculum, after only 5 h of incubation. Detection of 1 PFU/flask occurred after only 20 h of incubation, compared with a minimum of 5–7 d using conventional cell culture alone ***(2)***. Finally, ICC/PCR has been used to evaluate the effectiveness of virus inactivation protocols (i.e., chlorine disinfection) ***(4)***. Compared with results from single-passage cell culture, ICC/PCR determined that the contact time for chlorine disinfection of poliovirus is up to five times greater than previously thought.

The primary limitation of the ICC/PCR procedure is that both cell culture and molecular laboratories must be maintained and viruses must be culturable to be detected. Some of the more significant public health viruses, i.e., Norwalk agent and enteric caliciviruses, cannot be grown in conventional cell culture, leaving direct molecular methods one of the only currently available options for their detection in environmental samples.

2. Materials

2.1. Cell Culture

Virus stocks, cell cultures, and environmental sample concentrates should be considered positive for human pathogenic viruses, handled with gloves, and contained in a biosafety level 2 (BL2) laboratory, or higher, equipped with appropriate BL2 safety equipment.

1. Virus control stocks (i.e., poliovirus strain LSc-2ab).
2. Cell culture monolayers (i.e., BGM).
3. Sample concentrate (stored at –70°C until use).
4. Cell culture maintenance medium with serum warmed to 37°C (i.e., minimum essential medium [MEM]/L-15 with 2% fetal bovine serum).
5. Inverted microscope (100X magnification).

2.2. RT-PCR Amplification

1. 10X PCR buffer: 500 m*M* potassium chloride, 100 m*M* Tris-HCl, pH 8.3. Store at –20°C, and thaw just prior to use.
2. 25 m*M* $MgCl_2$. Store at –20°C, and thaw just prior to use.
3. 10 m*M* Deoxynucleoside triphosphates (dNTPs). Combine all four dNTPs in one master mix, store at –20°C, and thaw just prior to use.

4. Mineral oil, PCR grade.
5. Thermal cycler (Perkin Elmer; Applied Biosystems, Forest City, CA).
6. Random hexamers. Store at –20°C, thaw on ice just prior to use, and keep on ice at all times.
7. Avian myeloblastosis virus reverse transcriptase (AMV-RT; Life Sciences, St. Petersburg, FL). Store at –20°C, and keep on ice at all times.
8. Ribonuclease inhibitor (RNasin; Promega, Madison, WI). Store at –20°C, and keep on ice at all times.
9. Upstream, downstream, and internal primer set. Store at –20°C, and thaw just prior to use. Specific primers, used to amplify a 149- and 192-bp target sequence on enteroviruses, include a downstream enterovirus primer, nucleotides 577–594: 5'-TGT CAC CAT AAG CAG CC-3'; with an upstream enterovirus primer, nucleotides 445–465: 5'-TCC GGC CCC TGA ATG CGG CT-3' *(5)*. The seminested primer used to yield a 105-bp product for enteroviruses is an internal downstream enterovirus primer from nucleotides 530–550: 5'-CCCAAAGTAGTCGGTTCCGC-3') *(2)*.
10. High-performance liquid chromatography (HPLC) grade sterile water. Store at –20°C, and thaw just prior to use.
11. AmpliTaq and Amplitaq gold enzymes (Applied Biosystems). Store at –20°C, and keep on ice at all times.
12. Metaphor agarose (FMC Bioproducts, Rockland, ME).
13. Loading buffer: 20% ficoll, 1% sodium dodecyl sulfate, 0.25% bromphenol blue, 0.1 *M* disodium EDTA, pH 8.0. Store at room temperature.
14. Gel electrophoresis power source, gel trays, and buffer trays, and recirculating pump with cooling ice bath.
15. TAE buffer (1% Tris-acetate/EDTA). Store at room temperature, or at 4°C.
16. Ethidium bromide solution (0.5 μg/mL). Store at room temperature, shielded from light. Prepare new solution every five staining cycles. Ethidium bromide is a known mutagen. Always wear gloves when handling.
17. 123-bp Ladder (Gibco BRL, Gaithersburg, MD).
18. 302-nm UV-transilluminator. Wear safety goggles and shield exposed skin when using UV light source.

3. Methods

Methods for the filtration, elution, and concentration of viruses from a variety of environmental samples including water, sewage, sludge, and soil have been reviewed elsewhere *(6,7)*. The following sections detail protocols specific to the ICC/PCR method of enteric virus detection.

3.1. Cell Culture

Various cell types may be used for the isolation of enteric viruses (*see* **Note 1**). A common continuous cell line, known as BGM, derived from African green monkey kidneys, is highly susceptible to enteroviruses (polioviruses, coxsackieviruses, and echoviruses) and is easily grown in culture flasks. A detailed protocol for the growth, maintenance and quality control requirements of BGM cells is supplied by the US Environmental Protection Agency *(8)*.

1. Briefly, cells are grown into confluent monolayers in tissue culture grade plastic, or glass vessels, bathed in warmed (37°C) MEM/L-15 (equal parts Eagle's MEM/

Leibovitz medium) supplemented with 10% (v/v) fetal bovine serum, 7.5% (w/v) stock sodium bicarbonate solution, and antibiotics (100 U/mL penicillin, 30 µg/mL amikacin sulfate, 100 µg/mL streptomycin sulfate).

2. Every 7 d, the cells are dislodged from the flask using minimal contact with EDTA-trypsin reagent.
3. The suspended cells are collected via centrifugation at 1000*g* for 10 min and resuspended in growth medium.
4. The cell suspension is counted using a hemocytometer and diluted to allow for an inoculum of approx 10^5 cells/25 cm^2 flask.
5. Eight milliliters of growth medium is then added to each 25-cm^2 flask.
6. Within several days, the BGM cells attach to the bottom surface of the flask and form a continuous monolayer of cells.
7. To assay an environmental sample for viruses, cells are used between d 3 and 6 of their growth stage (*see* **Note 2**). The growth media is poured off and the cell monolayer rinsed with serum-less media prior to being overlaid with sample concentrate.
8. The sample inoculum volume is calculated at <0.04 mL/cm^2 of monolayer surface area.
9. The test vessels are gently rocked to distribute the sample evenly and incubated at 37°C for 80–120 min.
10. The monolayer with the sample inoculum is then covered with warmed (37°C) maintenance medium (MEM/L-15 with 2% serum) and incubated at 37°C for the designated length of time (*see* **Note 3**). Positive (20 PFU/inoculum) and negative (rinse medium) controls should be assayed with environmental samples for quality control purposes.

Cytopathogenicity assays are based on the appearance of characteristic morphological changes in the cell monolayer. These visible changes are known as cytopathogenic effects (CPE). In conventional cell culture, the cells are monitored daily for CPE, noted microscopically over time by the rounding and detachment of the cells from the surface monolayer, initially as plaques, and often followed by total destruction. CPE may be visible in ≥2–14 d, depending on a variety of factors including the virus type and the initial virus concentration in the inoculum (*see* **Note 4**).

3.2. Integrated Cell Culture/PCR

1. Using ICC/PCR, environmental samples are incubated in cell culture for a minimum of 5 h (*see* **Note 3** for description of incubation conditions).
2. Freezing the cells at –70°C stops virus replication.
3. The cells are then thawed at room temperature, or 37°C, with gentle shaking to allow the ice crystals to scrape the cell monolayer from the culture flask.
4. The cell lysate, resulting from freezing and thawing, is centrifuged at 3000*g* to remove the cellular debris.
5. The supernatant is collected and a 10-µL aliquot is frozen for PCR amplification (*see* **Note 5**).

3.3. RT-PCR Amplification

PCR, used in combination with a culture assay, eliminates the problems of small reaction volumes, inhibitory compounds, and detection of noninfectious sequences, normally associated with molecular techniques. PCR provides a rapid, sensitive, and specific method for the detection of DNA (adenoviruses) or RNA (enteroviruses) viruses (*see* **Note 6** for DNA virus detection protocol and PCR kit information).

Direct PCR amplification of RNA viruses is known as reverse transcriptase-PCR (RT-PCR). RT-PCR involves: (1) heat extraction of the viral nucleic acid genome; (2) transcription of the RNA to cDNA; (3) amplification of the cDNA to yield multiple copies of the target genome using DNA polymerases; and (4) confirmational analysis of the cDNA using additional amplification cycles and primer sequences internal to the original target genome (seminested PCR). A confirmational step is necessary since PCR primers may nonspecifically amplify a nontarget organism, producing a false-positive result (*see* **Note 7**).

For single PCR reactions, two primers are designed to amplify a specific region (*see* **Note 8**). One primer is *sense* (of the same sequence) and one *antisense* (having a complementary sequence) to the original target RNA genome. For seminested PCR, one of the original primers is used with a third primer, designed internal to, (nested within) the single PCR product genome. If the sense primer that is used in the original PCR is used again for seminested PCR, the internal primer must be designed in an antisense orientation or vice versa. The seminested product, therefore will be smaller than the single PCR product and will serve as a further confirmation that the correct target sequence was indeed amplified. Positive and negative controls should always be used to validate PCR primers and reaction conditions (*see* **Note 9**).

1. Heat extraction of RNA: 3.0 µL of 10X buffer II, 7.0 µL of 25 m*M* $MgCl_2$, and 8.0 µL of 10 m*M* dNTP to total 18.0 µL.
   a. If processing multiple reactions, make a master mix by combining all the above reagents.
   b. Dispense 18.0 µL into each microfuge tube.
   c. Overlay reaction volume with 3 drops of mineral oil.
   d. Add 10.0 µL of environmental sample concentrate under mineral oil.
   e. To extract viral RNA, heat mixture at 95°C for 5 min.
2. Reverse transcriptase reaction: 1.0 µL of random hexamers, 1.0 µL of AMV reverse transcriptase, and 1.0 µL of RNase inhibitor to total 3.0 µL.
   a. Make a master mix of the above reagents, and add 3.0 µL per reaction tube.
   b. Using an automated thermal cycler, complete one cycle at 24°C, 10 min; 44°C, 50 min; 99°C, 5 min; 5°C, 5 min.
3. cDNA amplification: 7.0 µL of 10X Buffer II, 3.0 µL of 25 m*M* $MgCl_2$ 0.5 µL of Upstream primer (50 pmol/reaction), 0.5 µL of Downstream primer (50 pmol/reaction), 57.5 µL of distilled water, and 0.5 µL of AmpliTaq enzyme (2.5 U/reaction) to total 69.0 µL.
   a. Make a master mix of the above reagents, and add 69.0 µL per reaction tube.
   b. Complete one cycle at 94°C, 1 min; 50°C, 45 s; 72°C, 1 min.
   c. Link to 30 cycles at 94°C, 1 min; 55°C, 45 s; 72°C, 1 min.
   d. Link to one cycle at 72°C, 7 min. Store at 4°C.

3.4. PCR Confirmation

Seminested PCR is performed with an internal primer to increase the specificity while simultaneously increasing the sensitivity of the reaction. PCR results are evaluated based on the visual presence or absence of amplified product bands using gel electrophoresis.

1. Seminested PCR amplification: 10.0 µL of 10X buffer II, 10.0 µL of 25 m*M* $MgCl_2$, 8.0 µL of 10 m*M* DNTP, 0.5 µL Upstream primer (50 pmol/reaction), 0.5 µL of downstream primer

(50 pmol/reaction), 60.5 μL of distilled water, and 0.5 μL AmpliTaq Gold enzyme (2.5 U/reaction) to total 90.0 μL.

a. Make a master mix of the above reagents. Dispense 90.0 μL into a new microfuge tube.
b. Overlay with 3 drops of mineral oil.
c. Add 10.0 μL of sample from the original RT-PCR under the mineral oil.
d. Complete 20–25 cycles at 94°C, 1 min; 55°C, 45 s; 72°C, 1 min.
e. Link to one cycle at 72°C, 7 min. Store at 4°C.

All the above protocols require about 7 h to complete using a Gene Amp DNA thermal cycler 480 or 4 h with a Gene Amp 9600 series thermal cycler (*see* **Note 10**). Reagents may be adjusted in equal proportions for smaller volume reactions.

2. Gel electrophoresis. Amplicons of PCR are visualized by horizontal agarose gel electrophoresis using a 3% intermediate melting temperature agarose gel (Metaphor agarose; FMC Bioproducts).

a. Combine 15.0 μL of PCR product with 3.0 μL Ficoll loading buffer (20% Ficoll, 1% sodium dodecyl sulfate, 0.25% bromphenol blue, and 0.1 *M* disodium EDTA, pH 8.0).
b. Load samples into wells of premade gel slab submerged in 1% tris-acetate/EDTA (TAE) buffer.
c. Subject to electrophoresis at 100 V for 2–3 h with cooling recirculation.
d. Remove gel and submerge in an ethidium bromide DNA staining solution for 15 min.
e. Resubmerge gel in distilled water for 30 min to remove background stain.
f. Visualized stained amplicons by exposure to a 302-nm UV-transilluminator.

4. Notes

1. The ICC/PCR method can be easily adapted to other environmental viruses. Additional cell lines include: (1) Hep-2 from human epidermoid carcinoma cells for the growth of human adenoviruses; (2) PLC/PRF/5 cells from a human primary hepatocellular carcinoma for the isolation of human enteric adenovirus types 40 and 41; and (3) FRhK cells from fetal rhesus monkey kidneys for the growth of HAV, and (d) $CaCO_2$ cells from a human colorectal adenocarcinoma for the growth of multiple enteric viruses. Since the development of ICC/PCR ***(1)***, a variety of similar protocols have been developed for the detection of additional human viruses ***(2,6,9,10)***. A variety of cell lines can be purchased from the American Type Culture Collection (Manassas, VA).
2. To prevent potential contamination of cell lines, cell growth and sample inoculation should be performed in separate rooms with designated equipment. Some viruses, such as noncytopathogenic strains, may grow unnoticed with the cell line but may interfere with subsequent detection methods.
3. Positive ICC/PCR detection is a function of: (1) initial virus inoculum; (2) incubation time; and (3) the number of replicate flasks (equivalent volume) examined. Higher concentrations (≥10 PFU/flask) of poliovirus strain LSc-2ab, a cytopathogenic virus, may be detected in seven of eight replicate flasks after only 5 h of incubation. For 100% positive detection, at a poliovirus concentration of >10 PFU/flask, a minimum of 10 h is required. For the target concentration of 1 PFU/flask inoculum, >20 h is recommended. Similarly, higher levels (100 PFU/flask) of HAV strain HM175, a noncytopathogenic virus, may be detected as soon as 12 h post incubation, but this fastidious virus requires a minimum of two to three replicate flasks to evaluate its presence or absence effectively. Infectious HAV levels of 10 and 1 PFU/flask are detectable after 48 and 72 h and require two to three replicate flasks, respectively. Therefore, for a given virus and a desired

detection limit, one must evaluate the minimum incubation time and equivalent volume required to ensure reliable results using ICC/PCR.

4. The ICC/PCR method is most effective when cell cultures are stopped prior to complete lysis of the cell monolayer. Over time, PCR-inhibitory substances build up in the culture flasks, requiring sample purification. If inhibition does occur, reduce the culture assay time or use a resin column purification technique ***(11,12)*** as follows:
   a. Test samples for inhibition by seeding low levels of target virus into the PCR (1–10 PFU).
   b. If inhibition is evident, apply 50.0 μL of sample to a layered Sephadex G-25 coarse/ Chelex-100 resin column in a Silane-treated, glass wool plugged, 1-mL syringe (approx 0.5 mL vol of each resin are used with a wet particle size range of 87–510 and 150–300 μm, respectively).
   c. Allow sample to adsorb for 10 min.
   d. Centrifuge column at 3000*g*, and collect samples for RT-PCR analysis.
   e. Retest for inhibition as in **step a**.
   f. If inhibition remains, pass the sample through a second layered column or through one with greater size exclusion capacity, such as Sephadex G-200/Chelex-100 (G-200 wet particle size range of 30–380 μm).
5. A negative ICC/PCR result at incubation time zero is an assurance of viable virus detection. For the time zero replicates, samples are placed on the cell monolayer and incubated, as described in **Note 3**, to allow virus attachment to the cells. After the cells are covered with maintenance media, the assay is stopped, prior to production of progeny virus. Therefore, if all replicate flasks are negative at time zero and positive at time one, virus growth was necessary for positive integrated CC/PCR results, indicating the detection of viable viruses only.
6. Amplification protocols for a DNA virus are identical to the reaction conditions described above for cDNA targets. PCR kits offer time-saving advances, providing all reagents needed for complete reactions, ensuring high-quality control standards, and offering of single-tube RT-PCR amplification kits and premeasured PCR optimization kits (Promega; Roche Molecular Systems).
7. Cytotoxic effects, not caused by viruses but rather sample toxins, typically inactivate the cell monolayer within 1–3 d of the assay. If the target virus is fastidious, and growth does not occur within this time, cytotoxic effects may be reduced by: (1) changing the maintenance medium at the first sign of cytotoxicity; (2) diluting the sample inoculum with distilled water or cell culture media (1:2 or 1:4 dilution); (3) reducing the sample adsorption time to 15 or 30 min; (4) after the initial sample adsorption with the cells, thoroughly rinsing the sample inoculum from the cell monolayer with rinse media; or (5) pretreating the cells with a solution of 8.5 g NaCl in a final volume of 980 mL of reagent-grade water, autoclaved and cooled to room temperature. Add 20 mL of serum and mix thoroughly. Add 0.25 mL/cm^2 of cell surface area. This technique reduces the virus titer and is used only when necessary.
8. Primer design is also very important to ensure a reliable PCR result. In general, primers should have the following characteristics:
   a. Primers are typically between 17 and 30 bp long.
   b. Ideally, the amplification product should be between 100 and 500 bp (although possible, amplification of longer products is more difficult to optimize).
   c. Primers must not be complementary to one another and should uniquely amplify the intended target.

   d. The GC content should be >40% to enable higher annealing temperatures.
   e. Internal secondary structures within the primers should be avoided.

9. Laboratories using PCR on a routine basis must develop quality control procedures to prevent false positive results and maintain the integrity of the PCR ***(13)***. False-positive results owing to contamination can be reduced by the following precautions:
   a. Establish pre- and post-PCR work stations. These areas should be in separate rooms or as far away from one another as possible. By doing this, the PCR-amplified product is separated from prereaction preparations.
   b. If possible, dedicate sets of unique supplies and pipeting devices for each set of primers.
   c. Autoclave buffer solutions and use HPLC-grade water. (Primers, dNTPs, and enzymes cannot be autoclaved.)
   d. Aliquot reagents to minimize the number of repeated samplings from a single stock.
   e. Use disposable gloves and change frequently, taking care not to cross-contaminate pre- and post-PCR stations.
   f. Spin down tubes prior to opening to reduce aerosolization of the sample.
   g. Use positive displacement pipets or aerosol-resistant tips to avoid contamination via aerosols.
   h. Premix reagents before dividing into aliquots, i.e., use master mixes.
   i. Add all reagents before sample DNA, capping each tube before opening the next.
   j. Use hot start methodologies, or heat-activated enzymes (i.e., AmpliTaq Gold; Roche Molecular Systems) to reduce preamplification mispriming. (Hot start is a procedure whereby the PCR is first set up at 80°C without one critical component, the Taq polymerase, heated to 94°C to denature the template, and then cooled to around 80°C. The polymerase is then added after the chance for mispriming is reduced. Heat-activated enzymes cause a similar effect and do not start performing until the higher temperatures are reached for a certain period).
   k. Use the minimum number of PCR cycles possible for a given sample.
   l. Contamination can be checked by the use of negative controls and should be run at the beginning and end of the set of PCR tubes as a minimum requirement. Negative controls should be made from the same master mix as the other sample reactions but contain no added RNA or DNA template. Standard protocols may need to be optimized for various types of samples. The number of amplification cycles can range from 15 to 30; primer annealing temperatures typically range from 35°C to 60°C. Mg^{2+} is required in the reaction, but the optimal concentration range is from 1.5 to 4 m*M*.

10. Reaction times may be decreased by as much as one-half using the thin-walled PCR tubes and an accommodating thermal cycler (i.e., Gene Amp 9600 PCR thermal cycler, Perkin Elmer, Roche Molecular Systems). This equipment eliminates the need for mineral oil in the reaction and allows for more rapid heat transfer in the tubes and thus shorter reaction times. Furthermore, real-time PCR assays offer a more rapid completion of PCR amplification of samples, with an internal detection system using double-stranded DNA-specific dyes or internal probes. As the PCR product accumulates, the product is detected simultaneously, eliminating the need for gel electrophoresis. Although most efficient, real-time PCR equipment is very costly (upwards of $50,000).

References

1. Reynolds, K. A., Gerba, C. P., and Pepper, I. L., (1996) Detection of infectious enteroviruses by an integrated cell culture-PCR procedure. *Appl. Environ. Microbiol.* **62,** 1424–1427.

2. Reynolds, K. A., Gerba, C. P., Abbaszadegan, M., and Pepper, I. L., (2001). ICC/PCR detection of enteroviruses and hepatitis A virus in environmental samples. *Can. J. Microbiol.* **47,** 153–157.
3. Reynolds, K. A., Gerba, C. P., and Pepper, I. L., (1997) Rapid PCR-based monitoring of infectious enteroviruses in drinking water. *Wat. Sci. Tech.* **35,** 423–427.
4. Blackmer, F., Reynolds, K. A., Gerba, C. P., and Pepper, I. L., (2000) Use of integrated cell culture-PCR to evaluate the effectiveness of poliovirus inactivation by chlorine. *Appl. Environ. Microbiol.* **66,** 2267–2268.
5. Abbaszadegan, M., Huber, M. S., Gerba, C. P. and Pepper, I. L. (1993) Detection of enteroviruses in groundwater with the polymerase chain reaction. *Appl. Environ. Microbiol.* **59,** 1318–1324.
6. Hurst, C. J. and Reynolds, K. A., (2002) Detection of viruses in environmental waters, sewage and sewage sludges. In: *Manual of Environmental Microbiology* (Hurst, C. J., ed.). ASM Press, Washington, DC, pp. 244–253.
7. Hurst, C. J. and Reynolds, K. A., (2002) Sampling viruses from soil. In: *Manual of Environmental Microbiology* (Hurst, C. J., ed.). ASM Press, Washington, DC, pp. 527–534.
8. United States Environmental Protection Agency, (1995) *Virus Monitoring Protocol for the Information Collection Requirements Rule.* Document number: EPA/814-B-95-002. June 1995. USEPA, Cincinnati, OH, pp. 1–70.
9. Chapron, C. D., Ballester, N. A., Fontaine, J. H., Frades, C. N., and Margolin, A. B., (2000) Detection of astroviruses, enteroviruses, and adenovirus types 40 and 41 in surface waters collected and evaluated by the information collection rule and an integrated cell culture-nested PCR procedure. *Appl. Environ. Microbiol.* **66,** 2520–2525.
10. Greening, G. E., Hewitt, J., and Lewis, G. D. (2002) Evaluation of integrated cell culture-PCR (C-PCR) for virological analysis of environmental samples. *J. Appl. Microbiol.* **93,** 745–750.
11. Straub T. M., Pepper, I. L. and Gerba, C. P. (1994) Detection of naturally occurring enteroviruses and hepatitis A virus in undigested and anaerobically digested sludge using the polymerase chain reaction. *Can. J. Microbiol.* **40,** 884–888.
12. Reynolds, K. A., Roll, K., Fujioka, R. S., Gerba, C. P., and Pepper, I. L., (1998) Incidence of enteroviruses in Mamala Bay, Hawaii using cell culture and direct polymerase chain reaction methodologies. *Can. J. Microbiol.* **44,** 598–604.
13. Pepper, I. L. and Pillai, S. D., (1994) Detection of specific DNA sequences in environmental samples via polymerase chain reaction. In: *Methods of Soil Analysis, Part 2.* Microbiological and Biochemical Properties. ASA, CSSA, SSSA, Madison, WI, pp. 707–726.

7

Abundance in Sewage of Bacteriophages Infecting *Escherichia coli* O157:H7

Maite Muniesa and Juan Jofre

1. Introduction

Bacterial virulence factors such as toxins are often encoded by bacteriophages. Among other examples, factors encoded by phages have been described in some of the emerging or re-emerging pathogens, including the pyrogenic exotoxin A production in group A streptococci *(1)*, the cholera toxin in *Vibrio cholerae* *(2)*, or enterotoxin production in enterohemorrhagic (EHEC) strains of *E. coli (3)*.

Most described virulence factors in Shiga toxin (Stx)-producing *E. coli* strains are located in mobile genetic elements such as plasmids and bacteriophages. Stx, which are one of the most important virulence elements in Shiga toxin-producing *E. coli* (STEC), are encoded in the genome of temperate bacteriophages infecting *E. coli* and other Enterobacteriaceae *(3–5)*.

Studies on Stx phages indicate that they are transmitted between different bacteria in vivo *(6)* and in vitro *(7)*. Phages could also be transmitted extraintestinally, hence the observed presence of infectious Shiga toxin phages in sewage and in fecally contaminated rivers. Stx phages also show a higher persistence under natural inactivation and disinfectant treatments in aquatic environments *(8–10)*.

This background shows that phages or lysogenic strains carrying Stx2 phages might be the natural reservoir of *Stx2* genes and that lysogenization could be the main cause of the emergence of STEC strains, as suggested by several authors *(3,7,11)*. It has also been suggested that lysogenization/conversion processes could take place in food and water and probably inside the human and animal gut. Ingestion of Stx2 phages could produce conversion of non-Stx2-*E. coli* strains, present inside the gut and producing new pathogenic strains. To control these phenomena, it is first necessary to gain more information about the distribution of Stx phages in the environment.

From: *Methods in Molecular Biology, vol. 268: Public Health Microbiology: Methods and Protocols*
Edited by: J. F. T. Spencer and A. L. Ragout de Spencer © Humana Press Inc., Totowa, NJ

For this purpose, a method of detecting Stx2 phages present in environmental water samples has been developed. The particularity of this method is that it allows detection of all (infectious and noninfectious) Stx2 phages in a water sample; in a second stage, the method allows detection of those phages able to infect and replicate on *E. coli* O157:H7. Although this method has been applied to Stx2 phages able to infect *E. coli* O157:H7 *(8)*, it is also applicable to detection in the natural environment of other genes carried by other bacteriophages and other bacteria.

For direct isolation of phages carrying the gene from a water sample, phages are partially purified by ultracentrifugation of different volumes of the sample; the phage DNA is extracted and amplified by polymerase chain reaction (PCR)-nested PCR and then confirmed by hybridization and/or sequencing. For detection of infectious bacteriophages carrying the gene, a water sample (e.g. sewage or river water sample) is used to infect a culture of *E. coli* O157:H7 (ATCC 43888, which does not contain the *Stx2* gene). Phages present in the water sample infect the strain and multiply and can be recovered from the supernatant of the enrichment culture. After recovery of phages, phage DNA is extracted, and the *Stx2* gene is amplified by PCR and nested PCR.

This approach allows determination of the presence of converting phages in water as well as their quantification. To enumerate Stx2 phages from a water sample, several enrichment cultures are performed, each one infected with a different volume of treated water samples. Results of positive or negative nested PCR amplification for each volume of sample allow enumeration of Stx2 phages by applying the most probable number (MPN) technique. Although the number obtained is a statistic, this approach yields information about the amount of Stx2 phages in a given sample.

The method allows analysis of the supernatant of the enrichment culture to evaluate the process of transduction of the genetic character understudy (Stx or other proteins) carried by phages present in the sample. Detecting the presence of the protein codified by the gene being studied in the supernatant of a culture after infection with phages present in the environment is another extension of the protocol described here, which could help to evaluate the rate of phage conversion in natural environments.

2. Materials

2.1. Obtaining Phages Directly from Sewage

1. 0.25 *M* Glycine buffer, pH 9.5.
2. 2X Phosphate-buffered saline (PBS): 8 g NaCl, 0.2 g KCl, 0.2 g KH_2PO_4, 1.15 g Na_2HPO_4.
3. Ultracentrifuge.

2.2. Culture of the Host Strain and Phage Infection

1. A host strain negative for toxin production (e.g., *E. coli* O157:H7, ATCC 43888).
2. Water bath at 44°C.
3. Luria broth (LB) basal medium: 10 g peptone, 5 g yeast extract, 10 g NaCl in 1 L distilled water; or Luria agar (LA) made by addition of 7 or 15 g agar-agar. Adjust the pH so that after sterilization it will be 7.2 ± 0.2. Sterilize in the autoclave at 121 ± 1°C for 15 min.
4. 0.1 *M* Calcium chloride solution: dissolve $CaCl_2$ in the water while heating gently. Cool to room temperature (RT) and filter-sterilize through a 0.22-µm pore size membrane filter. Store in the dark at 5 ± 3°C for not longer than 6 mo.

5. Samples: although the protocol is developed for environmental water samples, other kinds of samples could be used.
6. Filters of 0.22-μm pore size of polyvinylidene difluoride (PVDF; a low-protein-binding membrane; Millex-GV, Millipore).
7. DNase I (final concentration 10 U/mL) and RNase (final concentration 30 U/mL).

2.3. DNA Extraction

1. 20 m*M* EDTA.
2. 50 μg/mL Proteinase K.
3. 0.5% Sodium dodecyl sulfate (SDS).
4. Phenol/chloroform/isoamyl alcohol (25:24:1, v/v).
5. 100% Ethanol.
6. 3 *M* Sodium acetate.

2.4. Polymerase Chain Reaction

1. PCR amplification kit.
2. Custom-made oligonucleotides:
   a. First-round PCR
      Upper primer, 5'-GCGTTTTGACCATCTTCGT-3'.
      Lower primer, 5'-ACAGGAGCAGTTTCAGACAG-3'.
   b. Nested PCR
      Inner upper primer, 5'-TAATACGGCAACAAATACT-3'.
      Inner lower primer, 5'-TGATGAAACCAGTGAGTGA-3'.
3. 2% Agarose.
4. 1X TBE buffer: 90 m*M* Tris-borate, 0.2 m*M* EDTA.
5. 2 mg/L Ethidium bromide.
6. Gel loading buffer: 40% sucrose, 0.25% bromophenol blue.
7. Standard molecular weight markers.

2.5. Labeling of the Probe With DIG

1. 1 m*M* Digoxigenin (DIG)-11-dUTP.
2. Nucleotide mix for labelling: 10 m*M* dATP, 10 m*M* dGTP, 10 m*M* m*M* dCTP, 7 m*M* dTTP.
3. PCR amplification kit and primers.
4. PCR Purification kit (e.g., QIAquick™, Qiagen, Hilden, Germany).

2.6. Hybridization

1. Nylon-N+ membranes.
2. Custom-made probe: 5'-ATGACAACGGACAGCAGTTATACCAC-3'. The PCR amplimer obtained either with first-step PCR or nested PCR could also be used as a probe.
3. Buffer 1: 100 m*M* maleic acid, 150 m*M* NaCl, pH 7.5, with solid NaOH. Sterilize by autoclave. Store at RT.
4. 20X SSC: 3 *M* NaCl, 0.3 *M* sodium citrate, pH 7.0. Sterilize by autoclave. Store at RT.
5. Hybridization buffer: 6X SSC, 50 m*M* $NaPO_4$, 0.5% SDS.
6. Washing buffer: buffer 1 + 0.3% Tween-20. Keep at RT.
7. Blocking reagent stock solution: 10% (w/v) of blocking reagent (Boehringer Mannheim) in buffer 1 with several 30-s heat pulses in the microwave (3 or 4 min total). Sterilize by autoclave. Store at RT or at 4C.
8. Buffer 2: blocking reagent stock solution diluted 1:10 in buffer 1. Store at –20C for no more than 2 mo.

9. AntiDIG antibody.
10. Buffer 3: 100 m*M* Tris-HCl, pH 9.5, 100 m*M* NaCl, 50 m*M* $MgCl_2$.
11. NBT/BCIP (4-nitroblue tetrazolium chloride/5-bromo-4-chloro-3-indolyl-phosphate; Boehringer Mannheim).
12. Plastic bags and hybridization tubes.
13. Hybridization oven or thermal water bath.
14. Thermal sealer.

2.7. Detection of the Protein Encoded by the Gene Carried by Bacteriophages in the Supernatant of the Enrichment Cultures

1. 10-K Cutoff filtration membrane microconcentrators (Microsep, Millipore).
2. Material for the preparation of the polyacrylamide gel (**Table 1**).
3. Electrophoresis supply and vertical electrophoresis apparatus.
4. SDS gel loading buffer: 50 m*M* Tris-HCl, pH 6.8, 2% SDS, 10% glycerol, 0.05% bromophenol blue, 0.1% 2-β-mercaptoethanol. Store at RT.
5. Running buffer: 25 m*M* Tris-HCl, 250 m*M* glycine, 0.1% SDS. Dissolve and store at 4°C.
6. Low-range stained SDS-PAGE standard, commercially available.
7. Semidry transfer cell (Trans-blot Semi Dry, Bio-Rad).
8. Nitrocellulose membranes (PROTRAN, Schleicher & Schuell).
9. Transfer buffer: 24 m*M* tris base, 192 m*M* glycine, and 20% methanol.
10. Blot absorbent filter paper (thick).
11. Blocking reagent: 2 m*M* Tris-HCl, 28 m*M* NaCl containing 0.02% Tween-20 and 3% serum albumin.
12. Antibody against Stx2-A subunit. Antibody is obtained from hybridoma cell line 11E10 (ATCC CRL1907).
13. Anti-mouse alkaline phosphatase-conjugated antibody.
14. Washing buffer: 2 m*M* Tris-HCl, 28 m*M* NaCl containing 0.02% Tween-20.
15. 5X Base buffer number 3: 1 *M* Tris-HCl, pH 9.6, 5 *M* NaCl.
16. Detection buffer for immunoblot: 20% base buffer #3, 2% 2 *M* $MgCl_2$. Dissolve and use immediately.
17. NBT/BCIP (Boehringer Mannheim).

3. Methods

The method described here can be applied to determine the presence of the gene either in DNA extracted from phages directly isolated from sewage (*see* **Subheading 3.1.**) or in phages that have infected a bacterial strain and are isolated from the supernatant of the enrichment cultures after infection (*see* **Subheading 3.2.**).

3.1. Obtaining Phages Directly From Sewage

1. Bacteriophages were recovered from sewage and partially purified as follows.
2. First, 100, 10, and 1 mL of sewage were centrifuged for 3 h at 48,000*g*.
3. The pellet was resuspended in 0.25 *N* glycine buffer, pH 9.5, shaken at 4°C for 30 min to disrupt the bacteriophage clumps, and then buffered to pH 7.4 with PBS 2X and again centrifuged at 12,000*g* for 20 min.
4. The supernatant was further centrifuged for 1 h at 171,000*g*.
5. The pellet was finally dissolved in 100 μL of PBS.

Table 1
Material for Preparing the Polyacrylamide Gel

Reactive	Separating gel (12% of polyacrylamide)	Compacting gel (4% of polyacrylamide)
Bidistilled water	1.658 mL	1.525 mL
Solution B: 1.5 *M* Tris-HCl, pH 8.8	1.25 mL	—
Solution C: 0.5 *M* Tris-HCl, pH 6.8	—	0.625 mL
Solution D: 10% SDS	50 µL	25 µL
Solution A: acrylamide/bis (30% T/27% C)	2 mL	0.325 mL
10% APS (0.05%)	37 µL	27.5 µL
TEMED (0.05%)	2.5 µL	2.5 µL

APS, ammonium persulfate; SDS, sodium dodecyl sulfate; TEMED, tetramethylethylenediamine.

6. To ensure that recovered DNA belonged exclusively to bacteriophages and was not free DNA, the phage suspension was treated with 10 U of DNase and RNase per mL for 30 min as previously described *(**12**)*.

3.2. Obtaining Phages After Infection of a Host Strain

3.2.1. Culture of the Host Strain and Phage Infection

1. Typical host strains recommended are *E. coli* DH5α, *E. coli* C600, or *E. coli* O157:H7, ATCC 43888 (*Stx2*⁻). However, other host strains could be used for a given experiment.
2. The host strain is inoculated in LB until an optical density of 0.1 is attained at 600 nm (containing approx 2×10^8 cells/mL).

3.2.2. Preparation of Enrichment Cultures

In order to determine the presence in sewage of bacteriophages carrying the *Stx2* gene, which is detectable by DNA amplification and able to infect the host strain, bacteriophage enrichment cultures are performed as follows.

1. Sewage samples are centrifuged at 12,000*g* for 30 min to remove particulate material, filtered through a 0.22-µm low-protein-binding membrane (Millex-GV, Millipore) to remove bacteria, and treated with DNase and RNase for 30 min at 37°C to eliminate free DNA and RNA.
2. Then volumes of 100, 10, 1, 0.1, 0.01, and 0.001 mL of the sample are added to 100 mL of liquid cultures in logarithmic growth phase of the host strain, which does not possess the gene for the Stx2 (*see* **Notes 1** and **2**).
3. LB culture medium is then added to a final volume of 250 mL.
4. After overnight incubation, aliquots of the enrichment cultures are centrifuged at 12,000*g* for 30 min.
5. The bacteriophages present in the supernatants are filtered through a 0.22-µm low-protein-binding membrane and then treated once more with 10 U of Dnase/mL for 30 min. All these steps are focused on eliminating bacteria, cell debris, and free DNA present in the supernatant of the enrichment cultures, which could interfere with the following steps. In this step, 10 ml of the supernatant may be stored at –20°C for posterior analysis of the proteins present in the supernatant (*see* **Note 3**).

3.3. DNA Extraction

DNA is extracted from bacteriophages according to Sambrook et al. *(12)*. The same protocol is used for phages directly isolated from sewage or for phages obtained from the supernatant of an enrichment culture.

1. Bacteriophages present in a 1-mL sample are digested by addition of 20 m*M* EDTA, 0.5% SDS, and 50 μg/mL of proteinase K and incubated at 45°C for 1 h.
2. Extraction of DNA is performed with phenol/chloroform/isoamyl alcohol (25:24:1) and centrifugation at 16,000*g* for 5 min (*see* **Note 4**).
3. After centrifugation, the upper part is recovered and the DNA present is precipitated with 10% 3 *M* sodium acetate and 2 vol of frozen absolute ethanol. Precipitation is allowed for at least 2 h at –70°C (*see* **Note 5**).
4. After precipitation, a pellet containing phage DNA is recovered by centrifugation at 16000*g* for 30 min.
5. DNA is washed with 200 μL of 70% ethanol and centrifuged at 16,000*g* for 15 min.
6. After centrifugation, the supernatant is discarded. The washing step is repeated twice (*see* **Note 6**).
7. The washed DNA is dried (air-dried or dried with a speed vacuum if available) and dissolved in 20 μL of double-distilled water. One microliter of the DNA solution is used for the PCR amplification.

3.4. Polymerase Chain Reaction

1. First-round PCR DNA amplification for specific detection of the *Stx2* gene is performed with the two primers described above at a concentration of 30 m*M* each, at an annealing temperature of 55°C for 1 min for 30 cycles. Primers were selected from the DNA sequence of the *Stx2* gene, aligned with previously published sequences, and evaluated against the sequences of the EMBL data bank by the FastA program of the Genetics Computer Group package. Other primers can be used for different purposes.
2. One microliter of the first-round PCR is used for the nested PCR at the same conditions as the first-round PCR (*see* **Note 7**).
3. Five microliters of the amplified DNA mixture is analyzed for amplification products by electrophoresis on a 2% agarose gel stained with ethidium bromide.
4. To estimate the number of phages carrying the *Stx* gene, the MPN method *(13)* could be used. For each given volume, all positive results of nested PCR are considered in relation to the total number of experiments assayed for the same volume. For example, from three different samples from site A, detection was made of three positives of 10 mL, two positives of 1 mL, one of 0.1 mL, and no positives for 0.01 and 0.001. The combination to be applied to the MPN is 3:2:1:0:0. These values will give an estimation of 147 phages containing Shiga toxin 2 in a 100-mL sample. Performance of more replicas of the same volume will give a more accurate final estimation.

3.5. Labeling of the Probe With DIG

Several methods could be used to label a probe. Short oligonucleotides could be purchased directly labeled. Another useful method is labeling with DIG-11-dUTP. For this purpose the amplimer of a PCR (or nested PCR) reaction is directly labeled with DIG.

1. From a positive control DNA, a standard PCR reaction is performed but adding 1 μL of the nucleotide mix for labeling, which has a lower concentration of dTTP (10 m*M* dATP, 10 m*M* dGTP, 10 m*M* dCTP, 7 m*M* dTTP).

2. To substitute for this lack of dTTP, 3 µL of DIG-11-dUTP (1 m*M*) are added.
3. The other PCR reagents are used in the standard concentrations, and the final volume is adjusted with double-distilled water.
4. The PCR reaction is carried on by using the corresponding PCR program according to the primers.
5. The DIG-labeled PCR product will be a bit longer than the nonlabeled product, thus confirming efficiency of labeling. The probe should be purified before use with a suitable PCR purification kit.

3.6. Hybridization

Hybridization to confirm PCR and/or nested PCR results could be performed by dot blot or Southern blot ***(14)*** of the amplification mixture. If a custom-made oligonucleotide is used, it should be selected from the DNA sequence of the *Stx2* gene and placed inside the sequence of the obtained amplimer.

1. For dot blot, 5 µL of each PCR product is blotted onto a nylon membrane and fixed by exposure of the membrane to UV light for 3 min at each side.
2. For Southern blot, DNA is transferred to the nylon membrane from the agarose gel by capillary blotting ***(12)***. Once transferred, DNA is fixed to the membrane by exposure of the membrane to UV light as described in **step 1** just above.

3.6.1. Hybridization and Probe Detection

1. Introduce the membrane in a plastic bag or hybridization tube with forceps, and add 1 vol of hybridization buffer (7 mL buffer/10 cm^2 of the nylon membrane).
2. Equilibrate the membrane with hybridization buffer at the hybridization temperature for 1 h. Close the bag with a thermal sealer.
3. Add 3.5 µL of DIG-labeled probe per 7 mL of hybridization buffer. Incubate at the hybridization temperature overnight.
4. Place the membrane in a Petri dish and wash the membrane twice by gently shaking in 2X SSC + 0.01% SDS for 5 min at room temperature.
5. Place the membrane into the hybridization tube or bag again. Wash the membrane twice in 0.04X SSC + 0.01% SDS for 15 min at the hybridization temperature.
6. Equilibrate the membrane for 5 min at RT in washing buffer by gently shaking.
7. Equilibrate the membrane for 30 min at RT with buffer 2.
8. Add 10 µL of the antibody anti-DIG 10 mL of buffer 2. Incubate the membrane for 30 min at RT with gentle shaking.
9. Wash the membrane twice by shaking with washing buffer for 15 min at RT.
10. Equilibrate the membrane for 5 min in buffer 3 (*see* **Note 8**).
11. Place the membrane in a plastic bag and add 10 mL of buffer 3 and 40 µL of NBT/BCIP. Seal the bag with a thermal sealer. Store the membrane in the dark at RT until the signal is visible. Time is dependent on the intensity of the reaction. The reaction is stopped by washing the membrane with distilled water.

3.7. Detection of the Protein Encoded by the Gene Carried by Bacteriophages in the Supernatant of the Enrichment Cultures

After infection of the host strain, the phages may have transferred the gene coding for the toxin to the chromosome of the bacterial cell. Thus, the bacteria were able to produce the protein after the infection. This phenomenon is known as *lysogenic con-*

version. The protein produced can be detected in the supernatant of the enrichment culture by using the specific antibody against the toxin.

The presence of the toxin in the supernatants of bacteriophage enrichment cultures could be analyzed by Immunoblot or Western blot ***(15)***. Immunoblot is easy and faster and is recommended for positive/negative results. If the objective is to identify the protein, then a Western blot is recommended. In both cases, protein should be recovered from the supernatant of the enrichment cultures.

3.7.1. Recovery of the Protein Fom the Supernatant of the Enrichment Cultures

1. Ten milliliters of each supernatant are concentrated 10-fold by passage through 10-K cutoff filtration membrane microconcentrators (Microsep, Millipore) by centrifugation at 16,000*g* at 4C.
2. Then 15 µL of each concentrated supernatant could either be directly blotted onto the nitrocellulose membrane for immunoblot or subjected to SDS-PAGE in slab gels for Western blot.

3.7.2. Protein Electrophoresis

1. A 12% polyacrylamide slab gel is prepared and samples are mixed 1:1 with SDS loading buffer.
2. The samples should be boiled for 5–7 min before loading in the gel. Keep one of the lanes for a prestained marker.
3. Meanwhile, prepare the vertical electrophoresis and fill it with running buffer. Run the electrophoresis at 100 V for around 1.5 h (*see* **Note 9**).

3.7.3. Protein Electrotransfer to the Membrane (Western Blot)

Transference is achieved by using a semidry transfer to the nitrocellulose membrane (0.45 µm) following the manufacturer's instructions.

1. Briefly, cut the nitrocellulose membrane and two extra pieces of thick blotting paper the same size as the gel.
2. Equilibrate the gel, the paper, and the membrane by soaking them for 15 min in the transfer buffer.
3. Place one piece of paper on the surface of the cell, then the membrane, then the gel on top of the transfer membrane, and finally another sheet of the presoaked filter paper.
4. Add 10 mL of transfer buffer between each element and close the cell.
5. Run the transference for 1 h at 100 V.

3.7.4. Immunodetection

1. After transference, the membrane is blocked at 4C overnight with blocking reagent.
2. Next, the nitrocellulose sheet is incubated by swirling in an orbital incubator for 1.5 h at RT with undiluted hybridoma culture supernatant (11E10, ATCC CRL 1907, which produces an IgG [Mab 11E10] toward the A subunit of *Stx2* [primary antibody]).
3. This should be followed by washing three times with washing buffer, rocking for 10 min at RT.
4. Bound MAb is detected after incubation for 1.5 h with a 1:1000 dilution of an anti-mouse alkaline phosphatase-conjugated antibody.

5. After three washings, secondary antibody is detected by adding NBT/BCIP at 40 μL for each 2 mL of the immunodetection buffer.
6. Store the membrane in the dark at RT until the signal is visible on the membrane. Time is dependent on the intensity of the reaction.
7. The reaction is stopped by washing the membrane with distilled water (*see* **Notes 10** and **11**).

4. Notes

1. Enrichment cultures performed with high volumes of sample may sometimes give negative results although smaller volumes give positive results. This is because other lytic bacteriophages present in the sample could cause total lysis of the host cell, interfering with the multiplication of the *Stx2* phages.
2. The density of the host strain is important. An overgrowth of the host strain will make multiplication of the phages difficult.
3. In case the proteins produced after phage conversion need to be analyzed, it is important to keep aliquots of the supernatant of the enrichment culture at –20°C until detection of the protein; otherwise proteases could produce degradation of the protein.
4. For DNA extraction, the extraction steps with phenol/chloroform/isoamyl alcohol should be repeated if a thick white interphase caused by protein debris is observed.
5. For DNA extraction: If recovery of DNA is not high enough, it could be increased by addition of 6 μL glycogen for each mL of sample together with sodium acetate and absolute ethanol.
6. For DNA extraction: Washing steps with 70% ethanol should be repeated if a large white precipitate is observed (caused by salt precipitation, which could interfere with the PCR).
7. Nested PCR should be handled with care because of contamination problems. False-positive results may be avoided by using several negative controls, such as negative controls from the DNA extraction process, from the first step of PCR, and from the nested PCR itself as well as from the enrichment cultures of the host strain without phage infection. Material used for nested PCR should be sterilized before use and the reaction mixture performed in a sterile bench.
8. For hybridization with DIG-labeled probes: If spotty background is observed, do not include $MgCl_2$ in buffer 3. Change the plates at each washing step.
9. Protein detection: **Caution:** Acrylamide used for the polyacrylamide gels is neurotoxic. **Handle with care**.
10. Immunodetection: To develop the signal with NBT/BCIP, a long incubation may be used if necessary. However, longer incubation could also increase the background, disturbing signal detection.
11. Immunodetection: signal obtained with alkaline phosphatase may disappear from the membrane after some weeks. Thus, the results should be photographed as soon as they are obtained.

References

1. Johnson, L. P., Tomai, M. A., and Schlievert, P. M. (1986) Bacteriophage involvement in group A streptococcal pyrogenic exotoxin A production. *J. Bacteriol.* **166,** 623–627.
2. Waldor, M. K. and Mekalanos, J. J. (1996) Lysogenic conversion by a filamentous phage encoding cholera toxin. *Science* **272,** 1910–1914.

3. O'Brien A. D., Newland, J. W., Miller, S. F., Holmes, R. K., Williams Smith, H., and Formal, S. B. (1984) Shiga like toxin-converting phages from *Escherichia coli* strains that cause hemorragic colitis or infantile diarrhea. *Science* **226,** 694–696.
4. Watarai, M., Sato, T., Kobayashi, M., et al. (1998) Identification and characterization of a newly isolated Shiga toxin-2-converting phage from Shiga toxin-producing *Escherichia coli. Infect. Immun.* **66,** 4100–4141.
5. Muniesa, M., Recktenwald, J., Bielaszewska, M., Karch, H., and Schmidt, H. (2000) Characterization of a Shiga toxin 2e-converting bacteriophage from an *Escherichia coli* strain of human origin. *Infect. Immun.* **68,** 4850–4855.
6. Acheson, D. W. K., Reidl, J., Zhang, X., Keusch, G. T., Mekalanos, J. J., and Waldor, M. K. (1998) In vivo transduction with Shiga toxin 1 encoding phage. *Infect. Immun.* **66,** 4496–4498.
7. Schmidt, H., Bielaszewska, M., and Karch, H. (1999) Transduction of enteric *Escherichia coli* isolates with a derivative of Shiga-toxin 2-encoding bacteriophage ϕ3538 isolated from *E. coli* O157:H7. *Appl. Environ. Microbiol.* **65,** 3855–3861.
8. Muniesa, M. and Jofre, J. (1998) Abundance in sewage of bacteriophages that infect *Escherichia coli* O157:H7 and that carry the Shiga toxin 2 gene. *Appl. Environ. Microbiol.* **64,** 2443–2448.
9. Muniesa, M. and Jofre, J. (2000) Occurrence of phages infecting *Escherichia coli* O157:H7 carrying the Stx2 gene in sewage from different countries. *FEMS Microbiol. Lett.* **183,** 197–200.
10. Muniesa, M., Lucena, F., and Jofre, J. (1999) Comparative survival of free Shiga toxin2 encoding phages and *Escherichia coli* strains outside the gut. *Appl. Environ. Microbiol.* **65,** 5615–5618.
11. Beutin, L., Geier, D., Zimmermann, S., Aleksic, S., Gillespi, H. A., and Whittam, T. S. (1997) Epidemiological relatedness and clonal types of natural populations of *Escherichia coli* strains producing Shiga toxins in separate populations of cattle and sheep. *Appl. Environ. Microbiol.* **63,** 2175–2180.
12. Sambrook, J., Fritsch, E., and Maniatis, F. (1989) *Molecular Cloning, A Laboratory Manual*, 2nd ed. (Nolan, C., ed.). Cold Spring Harbor Laboratory Press, Cold Spring Harbor, NY.
13. De Man, J. C. (1975) The probability of the most probable number. *Eur. J. Appl. Microbiol.* **1,** 72–77.
14. Southern, E. M. (1975) Detection of specific sequences among DNA fragments separated by gel electrophoresis. *J. Mol. Biol.* **98,** 503–517.
15. Harlow, E. and Lane, D. (1988) *Antibodies, a Laboratory Manual.* Cold Spring Harbor Laboratory Press, Cold Spring Harbor, NY.

8

Molecular Genotyping of Irish Rotavirus Strains

Fiona O'Halloran and Séamus Fanning

1. Introduction

Rotavirus is the primary etiological agent of gastroenteritis in infants and young children worldwide *(1)*. In developing countries, it is estimated that rotavirus is responsible for one-third of all diarrhea-associated hospitalizations and 873,000 deaths annually *(2,3)*. In industrialized countries, where mortality from rotavirus is low, infection is widespread, and nearly all children experience an episode of rotavirus diarrhea in the first 3–5 yr of life *(4,5)*.

Rotaviruses have important antigenic specificities including serogroup and serotype, and all viruses are classified accordingly. They are divided into seven morphologically indistinguishable but antigenically defined serogroups, delineated A through G *(6,7)*. The human infecting rotaviruses include groups A, B, and C, and it is well documented that group A rotaviruses are the major causative agents of diarrheal diseases in children *(8,9)*. They are responsible for 125 million cases of diarrhea annually *(10,11)*.

Within each serogroup, distinct serotypes exist. In group A rotavirus, serotype is specified by two viral proteins, VP4 and VP7. The neutralizing antibody response that is evoked by the antigenic determinants on VP4 and VP7 play an important role in protective immunity *(12)*. The rotavirus genome consists of 11 double-stranded (ds) RNA segments, and each genomic segment encodes a different protein. A dual system of reporting rotavirus serotype exists because the VP4 and VP7 proteins are encoded by different genes and thus can segregate independently *(13)*. The serotypes derived from VP7 are defined as G-serotypes. Currently 14 G-serotypes have been identified, and only 10 of these have been recovered from humans *(12)*. The predominating G-types worldwide are G1, G2, G3, and G4, with G1 being the most prevalent type *(14–16)*. Serotypes G5, G6, G8–G10, and G12 are rarely identified in humans and are usually recovered from animals. However, some of these unconventional types are now being frequently reported in humans, including G5, G8, and G9 *(17–20)*.

From: *Methods in Molecular Biology, vol. 268: Public Health Microbiology: Methods and Protocols*
Edited by: J. F. T. Spencer and A. L. Ragout de Spencer © Humana Press Inc., Totowa, NJ

The serotypes derived from VP4 are designated P-types (P from protease-sensitive protein) *(21)*. Currently 13 P-serotypes have been reported, and 9 of these have been identified in humans *(22)*. The two main P-types that predominate include P[8] and P[4] *(23)*, and an association between G- and P-type has been observed *(12)*. Extensive epidemiological studies in several health care systems characterizing rotavirus strains have identified the prevalent serotypes circulating in children within different populations. The predominating types recognized were G1P[8], G2P[4], G3P[8], and G4P[8] *(10,24,25)*.

Several laboratory strategies are available to identify and characterize rotavirus isolates. Owing to the segmented nature of the rotavirus genome, gel electrophoresis is often useful and can identify genetic heterogeneity within isolates as well as monitor virus transmission *(26)*. Migration of the dsRNA segments in polyacrylamide gels is the most frequently used gel electrophoresis method, producing distinct electropherotype patterns. A classification scheme was devised for group A rotavirus strains, based on their migration patterns *(27)*. All 11 gene segments are divided into four groups based on their migration to one of four characteristic regions in a nondenaturing polyacrylamide gel. Group I includes gene segments 1–4; group II contains segments 5 and 6; group III contains segments 7–9; and group IV contains the remaining segments, 10 and 11 (**Fig. 1**). Within each group distinct dsRNA banding patterns are observed resulting from differences in the rates of migration of cognate gene segments. These variations (or heterogeneity, as it is often referred to) are attributed to base sequence polymorphisms, together with, distinctive secondary and tertiary structure of each dsRNA fragment *(28,29)*. Sequence and corresponding structural differences can be detected based on the altered migration pattern(s) of the dsRNA segments under native conditions. The migration rate of segments 10 and 11 (group IV) broadly divides the migration patterns into two groups. Fast migration of these two gene segments generates a *long* electropherotype, and slow migration is referred to as a *short* electropherotype (**Fig. 2A**). Some authors have reported an association between specific G-types and a given electropherotype. For example, G1 and G4 strains predominantly demonstrate a *long* electropherotype, whereas G2 strains usually have *short* electropherotypes *(30)*.

This classification scheme is specific to the genome profiles characteristic of group A rotavirus, which are the most prevalent rotavirus strains found in both humans and animals. The migration patterns demonstrated by nongroup A rotavirus are markedly different, as the gene segments do not migrate to the same four distinct regions of the gel. Electropherotyping may therefore be used to differentiate between different rotavirus serogroups. The presence of more than 11 segments in the electropherotype indicates that more than one strain has infected the cells, and this is indicative of a mixed infection (**Fig. 2B**). Mixed infections can potentially lead to the formation of novel reassortant strains. These strains often demonstrate atypical migration patterns and thus can be identified by their particular array of genome segments in polyacrylamide gels *(31)*.

Rotavirus serotypes can be defined by antigen-based methods (e.g., enzyme-linked immunoassays) or molecular protocols, including reverse transcriptase (RT)-mediated

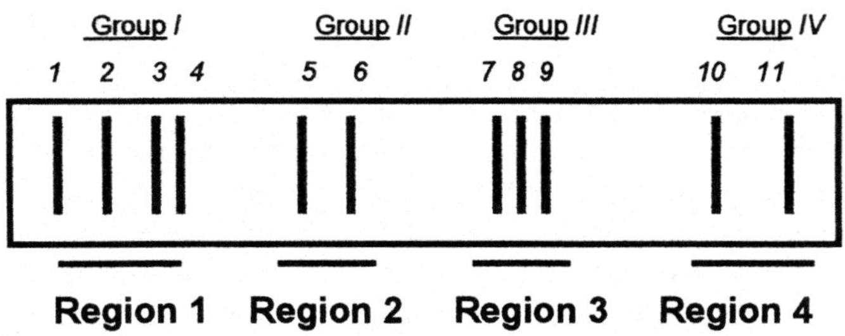

Fig. 1. A typical migration pattern displayed by group A rotavirus on a nondenaturing polyacrylamide gel. The 11 dsRNA segments migrate to one of four distinct regions on the gel and are grouped accordingly.

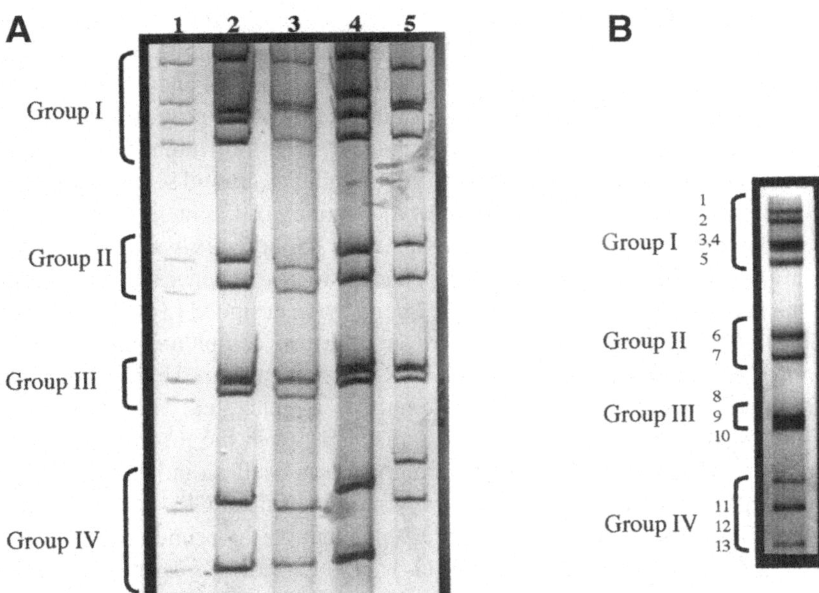

Fig. 2. Electropherotype patterns of rotavirus RNA generated by polyacrylamide gel electrophoresis. (A) Lanes 1–3 contain the typical "long" electropherotypes associated with group A and lanes 4–5 contain "short" electropherotypes. (B) A mixed electropherotype pattern isolated from a single rotavirus isolate.

polymerase chain reaction (PCR). Standard serological methods for virus identification often lack the required sensitivity to distinguish usefully between alternate isolates of a virus serotype. In addition, these strategies cannot detect newly evolved viral variants or identify the corresponding nucleotide mutations and amino acid substitutions that alter antigenic specificities. Molecular techniques including probe hybridization, restriction enzyme analysis, and in particular PCR-based assays have provided additional biochemical and serological information. These strategies have recognized technical advantages over conventional methods.

Nucleotide sequence analysis of each of the 11 dsRNA genomic segments from different rotavirus strains identified unique features relevant to the structure of rotavirus genes. These were conserved in all 11 dsRNA segments *(32)*. Significantly, the 5'- and 3'- ends of each genomic segment were found to be highly conserved among all strains analyzed. Each RNA segment had a single 5'-guanylate followed by a conserved sequence motif forming part of a 5'-noncoding region. Immediately downstream was located an open reading frame (ORF) coding for the specific protein product. The latter ORF was followed by another noncoding consensus sequence at the 3'-end. Using these sequence data, oligonucleotide primers were designed to be complementary to the negative ends of the VP4- and VP7-encoding RNA strands. These forward (VP7F; VP4F) and reverse (VP7R; VP4R) primer pairs (**Table 1**) were used in RT-PCR assays under defined reaction conditions to amplify the full-length (1062-bp) VP7 gene segment and a partial (867 bp) VP4 gene segment *(33,34)*. Comparative sequencing of the corresponding gene segments encoding VP4 and VP7 provided sufficient unique data to facilitate the design of molecular serotyping strategies.

Sequence analysis of VP7 gene segments from several different strains identified six discrete regions, designated A–F, with significant amino acid sequence divergence (**Fig. 3**). These regions were shown to be unique among different serotypes but highly conserved within any given serotype. Using each of these six variable regions as a blueprint for a distinct serotype, specific primers, whose sequences were complementary to the negative RNA strand of the VP7 gene, were designed *(34)*. These serotype-specific primers, along with a common (reverse) primer complementary to the 3'-end of the opposite strand, were included in a PCR reaction cocktail. Under defined conditions, PCR products of discrete segment lengths were generated. The variable regions under investigation are located at discrete distances from the distal end of the VP7 gene; thus the characteristic size of each amplicon is itself an indication of the viral serotype. **Figure 3** shows a schematic representation of this PCR-serotyping strategy. Corresponding sizes of each amplicon are dependent on the primer design *(35)*. Furthermore, the major human G-types (G1–G4) and other G-types can now be identified using this strategy. All major animal serotypes including G5, G6, G10, and G11 *(36)* can also be identified. This molecular strategy is sufficiently sensitive and has been extensively applied to several epidemiological studies to examine the geographical distribution of human rotavirus G-types *(14,17,22,37–40)*. A useful feature of this PCR typing protocol is its ability to detect mixed G-type infections. In this event, when more than one VP7 G-serotype is present, additional amplicons are detected after conventional agarose gel electrophoresis.

Table 1
Oligonucleotide Primers Used in PCR Amplification Methods

Primer	Primer Sequence (5'-3')
VP7F	5'-GGCTTTAAAAGAGAGAATTTCCGTCTGG-3'
VP7R	5'-GGTCACATCATACAATTCTAATCTAAG-3'
VP4F	5'-ATTTCGGACCATTTATAACC-3'
VP4R	5'-TGGCTTCGCCATTTTATAGACA-3'
aBT1	5'-CAAGTACTCAAATCAATGATG-3'
aCT2	5'-CAATGATATTAACACATTTTCTGTG-3'
aET3	5'-CGTTTGAAGAAGTTGCAACAG-3'
aDT4	5'-CGTTTCTGGTGAGGAGTTG-3'
aAT8	5'-GTCACACCATTTGTAAATTCG-3'
aFT9	5'-CTAGATGTAACTACAACTAC-3'
RVG9	5'-GGTCACATCATACAATTCT-3'
IT-1	5'-TCTACTTGGATAACGTGC-3'
2T-1	5'-CTATTGTTAGAGGTTAGAGTC-3'
3T-1	5'-TGTTGATTAGTTGGATTCAA-3'
4T-1	5'-TGAGACATGCAATTGGAC-3'

For complete serotype characterization of a viral strain, both the P-serotype (VP4) and G-seroytype (VP7) must be determined. Gentsch et al. ***(33)***, using a similar strategy, devised a PCR protocol that would facilitate the direct identification of rotaviral P-types. Comparative sequencing studies of VP4 genes from different strains revealed genetically distinct regions that were again used to design P-typing primers. These primers generated PCR products that were characteristic for the different P-types.

The potential genomic dynamics of the rotavirus necessitates continuous surveillance to monitor rotavirus infection and facilitate detection of novel strains. The combined use of electropherotyping and PCR serotyping protocols is useful in defining the epidemiology of rotavirus and identifying any novel strains in a population.

2. Materials

2.1. RNA Isolation Procedures for Rotavirus (see Note 1)

2.1.1. Inactivation of Endogenous Ribonucleases (see **Note 2**)

1. Diethyl-pyrocarbonate (DEPC).
2. RNase AWAY (Molecular Bioproducts, San Diego, CA).

2.1.2. RNA Extraction: SDS/Proteinase K Digestion–Phenol Chloroform Extraction

1. 10 mg/mL Proteinase K, prepared using DEPC water.
2. 10% (w/v) Sodium-dodecyl sulfate (SDS); prepared using DEPC water.
3. Chloroform.
4. Absolute ethanol.
5. 1:1 Phenol/chloroform (*see* **Note 3**).
6. Sterile 1.5-mL Eppendorf tubes.

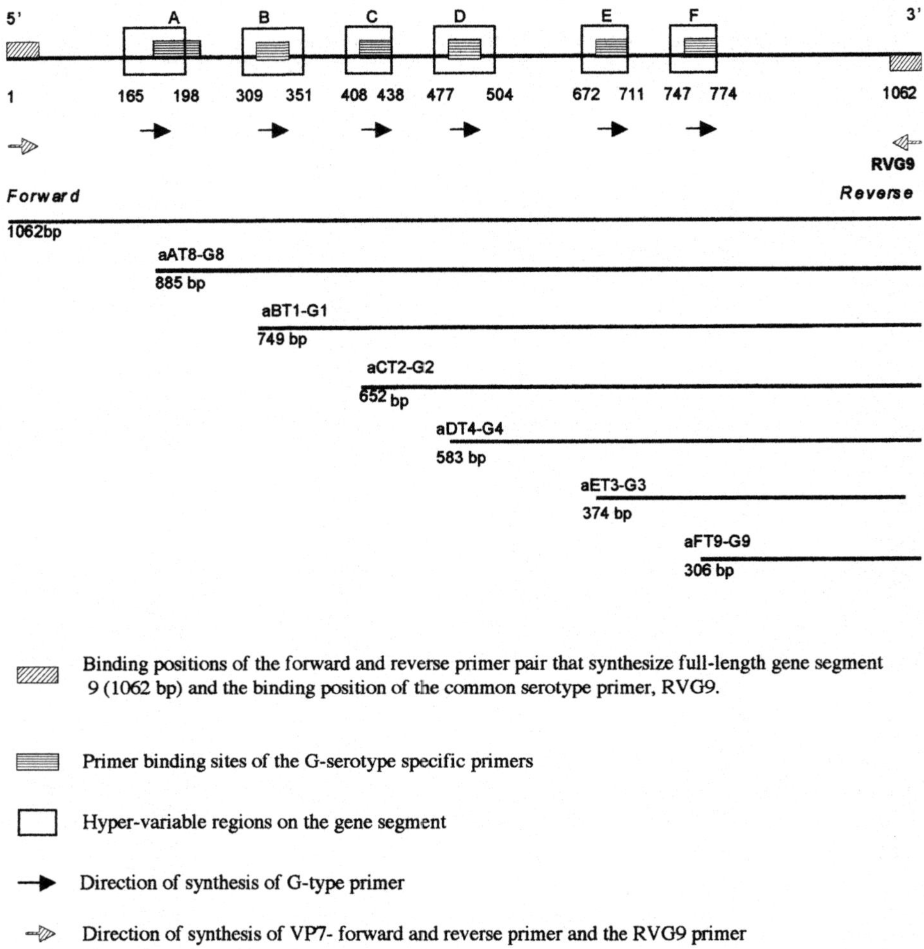

Fig. 3. Schematic representation of the seminested PCR assay that identifies the G-type of a rotavirus isolate. The locations of the variable regions on gene segment 9 are indicated. PCR primers identify distinct serotypes b y amplifying PCR products of defined length (*see* also **Table 1** for primer sequences).

2.1.3. Removal of Contaminating Genomic DNA (see ***Note 4****)*

1. 70% (v/v) Ethanol.
2. 3:1 Phenol/chloroform.
3. 3 *M* Sodium acetate.
4. DNase 1 (RNase free; Roche Diagnostics, East Sussex, UK).
5. 10X Tris buffer: 10 m*M* Tris-HCl, pH 8.3, 50 m*M* KCl, 5 m*M* $MgCl_2$.
6. Nuclease free water (Promega, Madison, WI).

2.2. Electropherotying of Rotavirus RNA Genomes

2.2.1. Polyacrylamide Gel Electrophoresis (PAGE)

1. 38% (w/v) Acrylamide/bisacrylamide mix: prepared by dissolving 74 g of acrylamide (molecular biology grade) and 2 g bis-acrylamide (molecular biology grade) in 200 mL DEPC-water. Store in a dark bottle at room temperature.
2. 1 *M* Tris-HCl, pH 6.8.
3. 1.5 *M* Tris-HCl, pH 8.8.
4. 10% (w/v) Ammonium persulfate (Sigma, Poole, UK).
5. 10% (w/v) SDS.
6. TEMED (*N,N,N',N'*-Tetramethylethylenediamine; Sigma).
7. Isobutanol.
8. 5X Tris-glycine buffer: 25 m*M* Tris-HCl, 250 m*M* glycine, pH 8.3, 0.1% (w/v) SDS.
9. Loading buffer: 62 m*M* Tris, 2% (w/v) SDS, 0.001% (w/v) bromophenol blue, 10% (v/v) glycerol.

2.2.2. Staining and Detection of Polyacrylamide Gels

1. *Buffer 1*: 10% (v/v) ethanol, 0.5% (v/v) glacial acetic acid.
2. 0.011 *M* Silver stain.
3. *Buffer 2*: 0.75 *M* NaOH containing 0.1 *M* formaldehyde and 0.0023 *M* sodium borohydride.
4. *Buffer 3*: 0.07 *M* sodium carbonate.

2.3. PCR Amplification Methods

1. 100 m*M* Stocks of each deoxyribonucleoside triphosphate (dNTP) dATP, dCTP, dGTP and dTTP (Promega).
2. dNTP working stock solutions containing 1.25 m*M* of each dNTP, prepared by diluting 2.5 μL of each dNTP in 190 μL of sterile water.
3. 10X PCR buffer: 100 m*M* Tris-HCl, pH 9.0, 500 m*M* KCl (Promega).
4. 25 m*M* $MgCl_2$ (Promega).
5. 5 U/mL *Taq* DNA Polymerase (Promega).
6. AMV (avian myeloblastosis virus)-reverse transcriptase (AMV-RT; Promega).
7. 5X AMV buffer: 250 m*M* Tris-HCl, pH 8.3, 250 m*M* KCl, 50 m*M* $MgCl_2$, 2.5 m*M* spermidine, 50 m*M* dithiotreitol (DTT).
8. 1 m*M* DTT (Sigma).
9. Dimethyl sulphoxide (DMSO).
10. Rnasin® ribonuclease inhibitor (Promega).
11. Sterile mineral oil (Sigma).
12. Sterile water.
13. Oligonucleotide primers (**Table 1**) were adjusted to the required concentrations. (All primers were obtained from Oswel Scientific and purified by high-performance liquid chromatography prior to use.)

2.4. Purification of PCR Products

1. PCR purification kit (QIAGEN, West Sussex, UK).

2.5. Conventional Agarose Gel Electrophoresis

1. Agarose (Promega).
2. 10X Tris-acetate EDTA (TAE): 40 m*M* Tris-HCl, pH 7.8, 40 m*M* glacial acetic acid, 2 m*M* EDTA.

3. 10X Gel loading buffer: 6.6 mL glycerol, 3.3 mL 10X TAE, 100 μL of 10% (w/v) bromophenol blue.
4. Ethidium bromide (10 mg/mL; Sigma).
5. DNA molecular weight markers: a range of DNA markers are commercially available (Roche Diagnostics) that have various fragment molecular weight ranges. In this study *three* DNA markers were used, including *DNA marker XIV*, which has 16 fragments that range in size from 100 to 1500 bp with an additional band at 2642 bp; *DNA marker III* providing 13 fragments with sizes ranging from 0.12 to 21.2 kbp; and *DNA marker V*, which contains 22 fragments ranging in sizes from 8 to 587 bp.

3. Methods

3.1. RNA Isolation Procedures for Rotavirus

3.1.1. Inactivation of Endogenous Ribonucleases

1. Prepare ribonuclease-free water by adjusting the volume of water required to contain 0.2% (w/v) DEPC.
2. Incubate the treated water at 37°C for 12 h and then autoclave at 121°C for 30 min.
3. Immerse glassware and nonsterile plasticware in a solution of 0.2% (v/v) DEPC and then autoclave at 121°C for 30 min.
4. Prepare buffers used in RNA procedures using DEPC-treated water. Buffers that cannot be autoclaved should be sterilized by filtration through a 0.2-μm filter.

3.1.2. RNA Extraction

1. Into a sterile 1.5-mL Eppendorf tube add the required volume of fecal filtrate (typically 400 μL). Adjust the volume to contain 1% (w/v) SDS and 100 μg/mL of proteinase K. Mix by pipeting and incubate at 37°C for 1 h.
2. To each reaction mixture add an equal volume of phenol/chloroform mixture (1:1) and vortex the sample for 30 s.
3. Centrifuge at 11,000*g* for 10 min, which separates the sample into two phases.
4. Collect the aqueous phase into a new sterile 1.5-mL Eppendorf tube and add an equal volume of chloroform.
5. Mix the sample by vortexing for 30 s followed by centrifugation at 11,000*g* for 10 min. Again collect the aqueous phase.
6. To precipitate the template 11 dsRNA segments, add 2 vol of cold absolute ethanol to the aqueous phase, and incubate the sample at –80°C overnight (12–18 h).
7. Recover the purified dsRNA by centrifugation (11,000*g* for 10 min) and then resuspend the pellet of RNA obtained in 100 μL nuclease-free water.

3.1.3. Removal of Contaminating Genomic DNA

1. Incubate the total RNA recovered (approx 100 μL) with 2 μL DNase 1 and 10.2 μL 10X Tris buffer (to give a final working solution of 1X) at 37°C for 30 min. This step allows sufficient time for any copurifying chromosomal DNA to be degraded.
2. To this solution add 80 μL of phenol/chloroform (3:1) reagent and vortex the sample for 1 min.
3. Centrifuge at 11,000*g* for 5 min and collect the aqueous phase as before into a sterile 1.5-mL Eppendorf tube.
4. Finally, precipitate the DNA-free purified dsRNA with 10 μL of 3 *M* sodium acetate and 300 μL of cold absolute ethanol.

5. Incubate the above overnight at –20°C and recover the dsRNA by centrifugation for 10 min at 10,000 rpm.
6. Wash the dsRNA pellet with 1 mL of cold, 70% (v/v) ethanol and resuspend the pellet in 100 μL DEPC-water.
7. Purified RNA can be analyzed by conventional agarose gel (1.5%) electrophoresis and rotaviral genomic segments visualized after staining with ethidium bromide (0.1 mg/mL). Store the dsRNA at –80°C.

3.2. Examination of the Electropherotype Patterns of Rotavirus Strains

3.2.1. Polyacrylamide Gel Electrophoresis

1. Vertical gel electrophoresis apparatus (Life Technologies, Paisley, UK).
2. Initially wash the glass plates in 1% (w/v) SDS, rinse in deionized water, and dry the plates completely.
3. Prepare 70 mL of a 10% (w/v) polyacrylamide gel by initially preparing the resolving gel (*see* **Note 5**).
4. Add (*in order*) 23.28 mL distilled water, 13.2 mL 38% (w/v) acrylamide mix, 12.5 mL 1.5 *M* Tris-HCl, pH 8.8, 1.5 mL 10% (w/v) SDS, 0.5 mL 10% (w/v) ammonium persulfate, and 0.02 mL TEMED.
5. Pour this mixture immediately into the gel cast leaving sufficient space for the stacking gel (approx 2 cm).
6. Overlay the acrylamide solution with isobutanol (*see* **Note 6**) and allow sufficient time for polymerization to occur (approx 30 min).
7. Pour off the overlay and wash the gel surface several times with deionized water.
8. Prepare 20 mL of the stacking gel (*see* **Note 5**) by adding (*in order*) 11.78 mL distilled water, 5.3 mL 38% (w/v) acrylamide mix, 2.5 mL 1 *M* Tris-HCl, pH 6.8, 0.2 mL 10% (w/v) SDS, 0.2 mL 10% (w/v) ammonium persulfate, and 0.02 mL TEMED. Mix the components and pour directly onto the resolving gel.
9. Insert a Teflon comb into the stacking gel and continue polymerization for a further 30 min at room temperature.
10. Remove the comb and wash the preformed wells with deionized water to remove any unpolymerized acrylamide.
11. Place the gel into the electrophoresis apparatus and add electrophoresis buffer to the top and bottom buffer reservoirs.
12. Prepare samples for electrophoresis by mixing 20 μL of sample RNA and 20 μl of gel loading dye (*see* **Note 7**).
13. Load all of this mixture directly onto the gel and perform electrophoresis for 16 h at 25 mA using 1X Tris-glycine, pH 8.3 buffer.

3.2.2. Staining and Detection of dsRNA Segments in Polyacrylamide Gels

1. Following electrophoresis, fix the gel for 30 min in a 10% (v/v) ethanol-acetic acid solution.
2. Silver stain with 0.011 *M* silver nitrate for 2 h then briefly (1 min) rinse in deionized water.
3. Transfer the gel to the developing solution (*buffer 2*) for 20 min, or until the RNA segments are clearly visible (not longer than 45 min).
4. Increase the intensity/visibility of the segments by immersing the gel in a solution of freshly prepared 0.07 *M* sodium carbonate for 10 min.
5. Visualize the electrophoretic patterns (electropherotypes) over a white light box.

3.3. PCR Methods

3.3.1. Reverse Transcriptase (RT)-Mediated PCR

1. Denature 1–5 µL of the dsRNA template (*see* **Note 8**) by heating to 97°C in the presence of 3.5 µL DMSO for 5 min (*see* **Note 9**). Place the reaction tubes immediately on ice.
2. To amplify full-length gene segment 9 (1062 bp) encoding VP7, assemble a PCR reaction cocktail containing the following: 5 µL 10X PCR buffer, 8 µL of the working dNTP mix (final concentration of each dATP, dCTP, dGTP, dTTP was 200 µM), 1.5 m*M* $MgCl_2$, 10 pmol of each primer (**Table 1**), 2.5 U *Taq* DNA polymerase, 4 U AMV-RT, 0.5 µL RNasin, and 1 m*M* DTT (which is added to the denatured template; *see* **Note 10**). The final reaction volume is 50 µL. Include an appropriate negative control containing all of the above except template dsRNA.
3. Centrifuge the tubes for 10 s and overlay the mixture with 100 µL of sterile mineral oil to prevent evaporation during thermal cycling.
4. Perform the RT step at 45°C for 30 min.
5. Then heat all reaction tubes to 70°C to inactivate the AMV-RT enzyme and cool on ice for 1 min. This step results in the production of a single-stranded complementary DNA (cDNA) copy of one of the double-stranded RNA molecules.
6. Amplify the cDNA products by PCR using the following cycling parameters: 94°C for 1 min, 48°C for 2 min, 72°C for 1 min (30 cycles) and finally 72°C for 7 min (1 cycle) to complete extension (*see* **Note 11**).
7. To amplify partial length gene segment 4 (867 bp), the same reaction conditions are required, with the substitution of 40 pmol of each VP4 forward and VP4 reverse primer (**Table 1**) to target gene segment 4 specifically.
8. Perform reverse transcription of the dsRNA template as previously described (**step 4** above). The reaction parameters for the VP4 primer pair are as follows; 94°C for 1 min, 53°C for 2 min, 72°C for 1 min (30 cycles), with a final extension at 72°C for 7 min (1 cycle).
9. Resolve RT-PCR products in a 1.5% (w/v) agarose gel, containing ethidium bromide (0.1 mg/mL) and visualize by UV illumination.

3.3.2. Purification of PCR Products

1. Purify RT-PCR products using a PCR purification kit according to the manufacturer's instructions (QIAGEN).

The RT-PCR products are subsequently used in seminested PCR assays to identify the G- and P-type of each isolate. In addition, the VP7 serotype could also be identified by direct DNA sequencing of the RT-PCR fragments followed by nucleotide comparison with sequenced strains in the databases using BLAST search tools (www.ncbi.nih.nlm.gov/BLAST).

3.3.3. Seminested PCR

1. Analyze PCR products from the RT-PCR reactions by seminested PCR to identify G- and/or P-types.
2. Identify the G-type of each isolate by adding sufficient template (*see* **Note 12**) to 5 µL of 10X PCR buffer (100 m*M* Tris-HCl, pH 9.0, 500 m*M* KCl), 8 µL of stock dNTP mix, 1 m*M* $MgCl_2$, 10 pmol of each G-serotype primer, and 10 pmol of the common reverse primer, RVG9 (**Tables 1** and **2** and **Fig. 3**).

Table 2
Characteristics of G-Typing Primers

Primer	Variable region (nucleotides)	Nucleotide position	Amplicon size (bp)	G-serotype
aBT1	B (309–351)	314–335	749	1
aCT2	C (408–438)	411–435	652	2
aET3	E (672–711)	689–709	374	3
aDT4	D (477–504)	480–498	583	4
aAT8	A (165–198)	178–198	885	8
aFT9	F (747–774)	757–776	306	9

Table 3
Characteristics of P-Typing Primers

Primer	Nucleotide position	Amplicon size (bp)	P-genotype	P-serotype
1T-1	339–356	345	[8]	1A
2T-1	474–494	483	[4]	1B
3T-I	259–278	267	[6]	2
4T-I	385–402	391	[9]	3

3. Add 1 U *Taq* DNA polymerase to a final volume of 50 mL.
4. Overlay all reaction mixtures with 100 mL of sterile mineral oil.
5. As thermal cycling conditions for amplification, use the following parameters: 97°C for 1 min, 46°C for 2 min, 72°C for 1 min (15 cycles), final extension at 72°C for 7 min (1 cycle).
6. Identification of the P-type isolates requires adding sufficient VP4 template to 5 µL 10X PCR buffer (100 m*M* Tris-HCl, pH 9.0, 500 m*M* KCl), 8 µL of stock dNTP mix, 1.5 m*M* $MgCl_2$, 2.5 U *Taq* DNA polymerase, 20 pmol of each P-typing primer, and 20 pmol of the VP4 common primer, Con2 (**Tables 1** and **3**).
7. The cycling parameters for the P-typing reaction are as follows; 97°C for 1 min, 50°C for 2 min, 72°C for 1 min (20 cycles), with a final extension at 72°C for 5 min (1 cycle).
8. Visualize all PCR products by agarose gel (1.5%) electrophoresis as previously described in **Subheading 3.3.1.**

4. Notes

1. The RNA isolation procedure described was used to isolate RNA directly from fecal matter. The method can also be applied to tissue culture samples, but the quality and quantity of RNA may not be as good. Alternative methods are also available.
2. Contamination with exogenous ribonucleases can result in the degradation of RNA; therefore proper pretreatment of reagents and experimental apparatus is a prerequisite. To reduce ribonuclease contamination of laboratory equipment, all surfaces and apparatus can be thoroughly wiped with RNase AWAY (Molecular Bioproducts, San Diego, CA).

Glassware and nonsterile plasticware can be immersed in a solution of 0.2% (v/v) DEPC and then autoclaved at 121°C for 30 min. Gloves should be worn at all times during procedures.
3. **Caution:** care should be taken when handling phenol/chloroform. As phenol is an acid, it can cause skin burns. This reagent should only be handled in a safety cabinet with extraction ventilation.
4. The presence of DNA in RNA samples could potentially interfere with subsequent analysis of the rotaviral genome and therefore it should be removed. DNA is degraded by treatment with a specific endonuclease enzyme, DNase 1. This enzyme digests DNA but does not affect the integrity of RNA.
5. Prepare sufficient gel volumes according to the PAGE apparatus used.
6. Isobutanol is added to prevent oxygen from diffusing into the gel, as this would inhibit polymerization.
7. The loading buffer described in **Subheading 2.2.1.** can be substituted for another appropriate loading buffer if preferred. The volume of sample that is loaded into the gel will depend on the size of the gel wells and thus can be adjusted accordingly.
8. Using excess RNA template can be detrimental to the RT-PCR assay, as it appears to inhibit the reaction. In general, small volumes of dsRNA template (1–5 µL) often produce better results.
9. Stronger binding forces exist between RNA-RNA compared with DNA-DNA and DNA-RNA; thus more stringent denaturing conditions are required to prevent strand reannealing. Prolonged denaturation (5 min) and the addition of nucleic acid destabilizers, such as DMSO, in the RT step are often recommended ***(34)***.
10. It is not essential but recommended to include RNasin in the assay to inactivate any contaminating ribonucleases that may be present in the reaction.
11. The thermocycler used in this study was a Pharmacia ATAQ controller (Pharmacia, Uppsala, Sweden).
12. The amount of cDNA template used in the serotyping PCR assays can be identified by performing initial template titrations. In general, 1–2 µL of cDNA template is sufficient for amplification. However, if the RT-PCR products are very weak, use more template.

References

1. Kapikian, A. Z. and Chanock, R. M. (1996) Rotaviruses. In: *Fields Virology* (Fields, B. N., et al, eds.). Philadelphia, pp. 1657–1708.
2. Cook, S. M., Glass, R. I., LeBaron, C. W., and Ho, M. S. (1990) Global seasonality of rotavirus infections. *Bull. WHO* **68,** 171–177.
3. Institute of Medicine (1986) Prospects for immunizing against rotavirus. In: *New Vaccine Development: Establishing Priorities. Diseases of Importance in Developing Countries*, vol. II. National Academy Press Washington, DC, pp. 308–318.
4. Ryan, M. J., Ramsay, M., and Brown, D. (1996) Hospital admissions attributable to rotavirus infection in England and Wales. *J. Infect. Dis.* **174(suppl 1),** S12–S18.
5. Woods, P. A., Gentsch, J. R., Gouvea, V., et al. (1982) Distribution of serotypes of human rotavirus in different populations. *J. Clin. Microbiol.* **30,** 781–785.
6. Gonzalez, S. A. L., Tomasini, L., Tortorici, M. A., and Affranchino, J. L. (1995) VP6 from porcine rotavirus strain CN86: amino acid sequence divergence with conservation of subgroup 11 specificity. *J. Gen Virol.* **76,** 221–224.

7. Fielding, P. A., Lambden, P. R., Caul, W. E. O., and Clarke I. N. (1994) Molecular characterisation of the outer capsid protein (VP4) gene from human group C rotavirus. *Virology* **204,** 442-446.
8. Arista, S., Vizzi, E., Ferraro, D., Cascio, A., and Di Stefano, R. (1997) Distribution of VP7 serotypes and VP4 genotypes among rotavirus strains recovered from Italian children with diarrhea. *Arch. Virol.* **142,** 2065–2071.
9. Gouvea, V., Santos, N., and do Carmo Timenetsky, M. (1994) Identification of porcine rotavirus G types by PCR. *J. Clin. Microbiol.* **32,** 1338–1340.
10. Gusmao, R. H., Mascarenhas, J. D. P., Gabbay, Y. B., et al. (1999) Rotavirus subgroups, G serotypes and electropherotypes in cases of nosocomial infantile diarrhoea in Belem, Brazil. *J. Trop. Pediatr.* **45,** 81–86.
11. Steele, A. D., Parker, S. P., Peenze, I., Pager, C.T., Taylor, M. B., and Cubitt, W.D. (1999) Comparative studies of human rotavirus serotype G8 strains recovered in South Africa and the United Kingdom. *J. Gen. Virol.* **80,** 3029–3034.
12. Gentsch, J. R., Woods, P. A., Ramachandran, M., et al. (1996) Review of G and P typing results from a global collection of rotavirus strains: Implications for vaccine development. *J. Infect. Dis.* **174(suppl 1),** S30–S36.
13. Isegawa, Y., Nakagomi, O., Nakagomi, T., and Ueda, S. (1992) A VP4 sequence conserved in human rotavirus strain AU-1 and feline rotavirus strain FRV-1. *J. Gen Virol.* **73,** 1936–1946.
14. Gault, E., Chikhi-Brachet, R., Delon, S., et al. (1999) Distribution of human rotavirus G types circulating in Paris, France, during the 1997-1998 epidemic: high prevalence of type G4. *J. Clin. Microbiol.* **37,** 2373–2375.
15. Ramachandran, M., Gentsch, J. R., Parashar, U. D., et al., and the National Rotavirus Strain Surveillance System Collaborating Laboratories. (1998) Detection and characterisation of novel rotavirus strains on the United States. *J. Clin. Microbiol.* **36,** 3223–3229.
16. Estes, M. (1996) Rotaviruses and their replication. In: *Fields Virology*, vol II (Fields, B. N., Knipe, D. M., and Howley, P. M., eds.). Lippincott-Raven, Philadelphia, PA, pp. 1625–1655.
17. O' Halloran, F., Lynch, M., Cryan, B., O' Shea, H., and Fanning, S. (2000) Molecular characterization of rotavirus in Ireland: detection of novel strains circulating in the population. *J. Clin. Microbiol.* **38,** 3370–3374.
18. Unicomb, L. E., Podder, G., Gentsch, J. R., et al. (1999) Evidence of high-frequency genomic reassortment of group A rotavirus strains in Bangladesh: emergence of type G9 in 1995. *J. Clin. Microbiol.* **37,** 1885–1891.
19. Santos, N., Lima, R. C. C., Pereira, C. F. A., and Gouvea, V. (1998) Detection of rotavirus types G8 and G10 among Brazilian children with diarrhea. *J. Clin. Microbiol.* **36,** 2727–2729.
20. Leite, J. P. G., Alfieri, A. A., Woods, P. A., Glass, R. I., and Gentsch, J. R. (1996) Rotavirus G and P types circulating in Brazil: characterisation by RT-PCR, probe hybridisation and sequence analysis. *Arch. Virol.* **141,** 2365–2374.
21. Burke, B., and Desselberger, U. (1996) Rotavirus pathogenicity. *Virology* **218,** 299–305.
22. Okada, J., Urasawa, T., Kobayashi, N., et al. (2000) New P serotype of group A human rotavirus closely related to that of a porcine rotavirus. *J. Med. Virol.* **60,** 63–69.
23. Sereno, M. M. and Gorziglia, M. I. (1994) The outer capsid protein VP4 of murine rotavirus strain Eb represents a tentative new P type. *Virology* **216,** 500–504.
24. Ramachandran, M.,. Das, B. K, Vij, A., et al. (1996) Unusual diversity of human rotavirus G and P genotypes in India. *J. Clin. Microbiol.* **34,** 436–439.

25. Bishop, R. F., Unicomb, L. E., and Barnes, G. L. (1991) Epidemiology of rotavirus serotypes in Melbourne, Australia, from 1973 to 1989. *J. Clin. Microbiol.* **29,** 862–868.
26. Sethi, . K., Olive, D. M., Strannegard, O. O., and Al-Nakib, W. (1988) Molecular epidemiology of human rotavirus infections based on genome segment variations in viral strains. *J. Med. Virol.* **26,** 249–259.
27. Lourenco, M. H., Nicolas, J. C., J. Cohen, J., Scherrer, R., and Bricout, F. (1981) Study of human rotavirus genome by electrophoresis: attempt of classification among strains isolated in France. *Ann. Virol.* **132,** 161–173.
28. Fukai, K., Sakai, T., and Kamata, H. (1998) Distribution of the G serotypes and P genotypes of bovine group A rotavirus isolated in Japan. *Aust. Vet. J.* **76,** 418–422.
29. Rodger, S. M. and Holmes, I. H. (1979) Comparison of the genomes of simian, bovine and human rotaviruses by gel electreophoresis and detection of genomic variation among bovine isolates. *J. Virol.* **30,** 839–846.
30. Palombo, E. A., Bugg, H. C., Masendycz, P. J., Coulson, B. S., Barnes, G. L., and Bishop, R. F. (1996) Multiple gene reassortants responsible for an outbreak of gastroenteritis in central and northern Australia. *J. Gen Virol.* **77,** 1223–1227.
31. Ramig, R. F. and Ward, R. L. (1999) Genomic segment reassortment in rotaviruses and other reoviridae. *Adv. Virus Res.* **39,** 163–207.
32. Patton, J. T. (1995) Structure and function of the rotavirus RNA-binding proteins. *J. Gen. Virol.* **76,** 2633–2644.
33. Gentsch, J. R., Glass, R. I., Woods, P. A., et al. (1992) Identification of group A rotavirus gene 4 types by polymerase chain reaction. *J. Clin. Microbiol.* **30,** 1365–1373.
34. Gouvea, V., Glass, R. I., Woods, P., et al. (1990) Polymerase chain reaction amplification and typing of rotavirus nucleic acid from stool specimens. *J. Clin. Microbiol.* **28,** 276–282.
35. Ushijima, H., Koike, H., Mukoyama, A., Hasegawa, A., Nishimura, S., and Gentsch, J. (1992) Detection and serotyping of rotavirus in stool specimens by using reverse transcription and polymerase chain reaction. *J. Med. Virol.* **38,** 292–297.
36. Gouvea, V., Santos, N., and Timenetsky, M. D. C. (1994) Identification of bovine and porcine rotavirus G types by PCR. *J. Clin. Microbiol.* **32,** 1338–1340.
37. Alfieri, A. A., Leite, J. P. G., Nakagomi, O., et al. (1996) Characterisation of human rotavirus genotype P[8]G5 from Brazil by probe hybridisation and sequence. *Arch. Virol.* **141,** 2353–2364.
38. Palombo, E. A., Clark, R., and Bishop, R. F. (2000) Characterisation of a "European-like" serotype G8 human rotavirus isolated in Australia. *J. Med. Virol.* **60,** 56–62.
39. Palombo, E. A., Masendycz, P. J., Bugg, H. C., Bogdanovic-Sakran, N., Branes, G. L., and Bishop, R. F. (2000) Emergence of G9 in Australia. *J. Clin. Microbiol.* **38,** 1305–1306.
40. Das B. K., Gentsch, J. R., Cicirello, H. G., et al. (1994) Characterization of rotavirus strains from newborns in New Delhi, India. *J. Clin. Microbiol.* **32,** 1820–1822.

9

Hepatitis A Virus

Molecular Detection and Typing

Glòria Sánchez, Cristina Villena, Albert Bosch, and Rosa M. Pintó

1. Introduction

Hepatitis A virus (HAV) infection is the leading cause of viral hepatitis throughout the world *(1)*. HAV infection is mainly propagated via the fecal-oral route *(2)*, and waterborne *(3)* and foodborne *(4–8)* outbreaks of the disease have been reported.

HAV, the prototype of the genus *Hepatovirus*, belongs to the family Picornaviridae. Its 7.5-kb single-stranded RNA genome bears different distinct regions: the 5' and 3' noncoding regions (NCR), the P1 region, which encodes the structural proteins VP1, VP2, VP3, and a putative VP4, and the P2 and P3 regions encoding nonstructural proteins associated with replication *(9)*. A single HAV serotype has been described, although seven genotypes have been defined *(9)*.

Since environmental samples usually contain low numbers of viral particles, sensitive methods such as molecular techniques based on nucleic acid amplification are required for their detection. However, even with the adoption of these techniques, the choice of the most adequate target is of relevant importance. The target region should be highly conserved, to increase the chance of detection, and should have an appropriate structure and length to allow sensitivity high enough for these kind of samples. As a target region, we have chosen a fragment of the 5'NCR flanked by highly conserved sequences that have been used for the primer design (forward primer from position 68 to position 85; reverse primer from position 222 to position 240 in the HM175 strain of HAV; GenBank accession number M14707) *(4)*. The internal part of this region, however, may present a certain degree of variation mainly owing to insertions and/or deletions, causing a variable size of the amplimer obtained, i.e., the wild-type HM175 strain gives a size of 174 bp whereas the cell-adapted pHM175 strain gives a size of 186 bp. For this reason it is extremely important to include a confirmative method such as Southern blot hybridization with an internal probe from a region not affected by the insertions/deletions *(4)*.

From: *Methods in Molecular Biology, vol. 268: Public Health Microbiology: Methods and Protocols*
Edited by: J. F. T. Spencer and A. L. Ragout de Spencer © Humana Press Inc., Totowa, NJ

When genotyping is the objective of the study, analysis of the sequence corresponding to the VP1X2A junction is the method of choice ***(10,11)***, allowing HAV isolates to be differentiated into seven genotypes ***(10)***. It should be pointed out that although the sequence employed aligns with the VP1X2A junction region used for genotyping purposes in other picornaviruses, such as poliovirus ***(10)***, in the case of HAV (position 3024–3191 of the HM175 strain) it actually represents a 2A sequence, since it contains only 1 codon of VP1 and 55 of 2A ***(12,13)***. This method consists of the amplification of a 360-bp fragment (from position 2949 to position 3308) that includes the previously mentioned genotyping region. The size of the amplimer (360 bp) induces a loss of detection sensitivity, which is partially overcome by the application of nested polymerase chain reaction (PCR) procedures. However, the use of nested PCR techniques may introduce important cross-contamination problems since the first amplimer product should be further manipulated into a second reaction. Consequently, there is a growing tendency to avoid nested reactions as much as possible in diagnostic laboratories. This is the reason why we have adopted amplification of a shorter fragment of the 5'NCR for the generic detection of HAV. HAV genotyping is subsequently performed on those samples that are clearly positive, by the 5'NCR method, in a single PCR reaction using the previously mentioned 360-bp fragment.

All nucleic acid polymerasing reactions are susceptible to inhibitors, and reverse transcriptase (RT) is especially sensitive to inhibitory substances, which may be found in water samples. In environmental studies, a water concentration step is frequently required to reduce the sampled volume to an amount able to be analyzed in the laboratory. Unfortunately, substances inhibitory to RT-PCR are concentrated along with the viruses. Preconditioning of the sample should then be performed prior to the molecular amplification. In the present work, the method of choice for this purpose has been lyophilization, which efficiently removes several volatile inhibitors and at the same time allows viral concentration ***(14)***.

2. Materials

2.1. Sample Preconditioning

1. Plastic containers.
2. Autoclaved (121°C for 45 min) molecular biology grade distilled water (conductance 2×10^{-6} ohm^{-1} cm^{-1}; Panreac, Barcelona, Spain; *see* **Notes 1** and **2**).
3. Freeze Drying Bench Top 3 from Virtis (Gardiner, NY) or similar.

2.2. RNA Extraction (see *Note 3*)

1. Washing buffer (L2): add 120 g of guanidinium isothiocyanate (Applichem, Darmstadt, Germany) to 100 mL of 0.1 *M* Tris-HCl, pH 6.4. Heat at 56°C to dissolve. Store in aliquots in the dark at room temperature. This solution is stable for 3 wk.
2. Lysis buffer (L6): add to 200 mL of L2 washing buffer 22 mL of 0.2 *M* EDTA, pH 8.0, and 2.44 mL Triton X-100. Store in aliquots in the dark at room temperature. This solution is stable for 3 wk.
3. Silica solution:
   a. Add 60 g of silicon dioxide (Sigma, St. Louis, MO) to 500 mL of autoclaved (121°C for 45 min) distilled H_2O in a 500 mL RNase-free bottle and keep it for 24 h at room temperature.

b. Aspirate 430 mL of the supernatant and resuspend the pellet in 500 mL of autoclaved (121°C for 45 min) distilled H_2O.
c. Allow to stand for 5 h at room temperature and aspirate 450 mL of the supernatant.
d. Add 600 µL of 8.7 *M* HCl to this supernatant and distribute it in aliquots of 0.5 mL before autoclaving at 121°C for 20 min.
e. Store in the dark at room temperature.
f. This solution is stable for 6 mo.

4. 70% Ethanol (*see* **Note 4**).
5. 100% Acetone.
6. TE buffer, pH 8: 10 m*M* Tris-HCl and 1 m*M* EDTA. Adjust the pH to 8.0. Aliquots are distributed in 1.5-mL tubes and autoclaved at 121°C for 45 min.
7. Vortex.
8. Microcentrifuge.
9. Heating block.
10. 1.5-mL Microcentrifuge tubes.
11. Tips with filter.
12. Gloves.

2.3. RT

1. Murine Moloney leukemia virus (M-MLV) reverse transcriptase RNase H Minus (Promega, Madison, WI) supplied with 5X RT buffer (250 m*M* Tris-HCl, pH 8.3, 375 m*M* KCl, 15 m*M* $MgCl_2$, 50 m*M* dithiothreitol).
2. Deoxynucleotide triphosphates (dNTPs) (Promega).
3. Reverse primers (HAV240: 5'-241GGAGAGCCCTGGAAGAAAGA222-3' and VP1-3285: 5'-3285AGTCACACCTCTCCAGGAAAACTT3308-3').
4. Autoclaved (121°C for 45 min) distilled H_2O.
5. Thermocycler AB 2700 (Applied Biosystems, Foster City, CA) or similar.
6. Micropipets and filter tips.
7. 0.2-mL PCR tubes.

2.4. PCR

1. Expand High Fidelity PCR enzyme (Roche, Mannheim, Germany) supplied with 10X Expand High Fidelity buffer and supplemented with 15 m*M* $MgCl_2$.
2. Deoxynucleotide triphosphates (dNTPs).
3. Reverse (HAV240: 5'-241GGAGAGCCCTGGAAGAAAGA222-3' and VP1-3285: 5'-3285AGTCACACCTCTCCAGGAAAACTT3308-3') and forward (HAV68: 5'-68TCACCGCCGTTTGCCTAG85-3' and VP1-2949: 5'-2949TATTTGTCTGTCA CAGAACAATCAG2973-3') primers.
4. Autoclaved (121°C for 45 min) distilled H_2O.
5. Thermocycler AB 2700 or similar.
6. Micropipets and filter tips.
7. 0.2-mL PCR tubes.

2.5. Electrophoresis

1. 10X TBE buffer: 0.9 *M* Tris-HCl, pH 8.3, 0.9 *M* boric acid, 0.02 *M* EDTA. Stable for long periods.
2. 10X TAE buffer: 0.4 *M* Tris-HCl, pH 8.3, 0.11 *M* acetic acid, 0.01M EDTA. Stable for long periods.

3. Seakem LE agarose (BioWhittaker, Walkersville, MD).
4. GelStar Nucleic Acid Gel Stain (BioWhittaker).
5. 6X loading buffer: 50% (v/v) glycerol, 0.25% (w/v) bromophenol blue (Sigma) in distilled water. Store at 4°C. Stable for long periods.
6. Prepare a working solution of the DNA molecular weight marker IX (72–1353 bp) (Roche) by mixing 10 μL of marker with 73.40 μL of TBE or TAE, and 16.6 μL of 6X loading buffer.
7. Horizontal gel tank, gel mold, gel combs (Bio-Rad, Hercules, CA) and power pack (Amersham Pharmacia Biotech Europe, Freiburg, Germany), or similar.
8. ImageMaster VDS gel imaging system analyzer (broad band UV 260–400 nm, peak at 312 nm; Amersham Pharmacia Biotech Europe) or similar.

2.6. Southern Blot

1. Positively charged nylon membrane (Roche).
2. Whatman 3MM paper.
3. Mini Trans-Blot Electrophoretic Transfer Cell (Bio-Rad) or similar.
4. TBE buffer.
5. 150 m*M* NaOH. Stable for long periods.
6. 20% sodium dodecyl sulfate (SDS). Stable for long periods.
7. 20X SSC buffer: 3 *M* NaCl, 300 m*M* sodium citrate, pH 7.0. Autoclave at 121°C for 20 min (*see* **Note 5**). Stable for long periods.
8. Digoxigenin (DIG)-labeled probe (5'-150TTAATTCCTGCAGGTTCAGG169-3').
9. Standard solution: 5X SSC buffer supplemented with 0.1% *N*-lauroylsarcosine, 0.02% SDS, and 1% blocking reagent (Roche). Heat to 60°C to dissolve the detergents and the blocking reagent. Dispense in 10-mL aliquots and store at –20°C.
10. Prehybridization solution: standard solution supplemented with 100 μg/mL salmon sperm DNA. To denature the salmon DNA, boil a 100X solution before being added to the standard solution (*see* **Note 6**).
11. Hybridization solution: DIG-labeled probe diluted in prehybridization solution.
12. Buffer I: 100 m*M* maleic acid, 150 m*M* NaCl, pH 7.5 (*see* **Note 7**). Autoclave at 121°C for 20 min. Stable for long periods.
13. Blocking solution: add 1 g of blocking reagent to 100 mL of buffer I. Autoclave at 121°C for 20 min, store at 4°C, and open it carefully in sterile conditions. Stable for 2 mo.
14. Anti-DIG alkaline phosphatase-labeled antibody (anti-DIG-AP) (0.75 U/μL; Roche).
15. Washing buffer: add 0.3 mL of Tween-20 to 100 mL of buffer I. Store at 4°C. Stable for long periods.
16. Phosphatase buffer: 100 m*M* Tris-HCl and 100 m*M* NaCl, pH 9.5. Store at 4°C. Stable for long periods.
17. 25 m*M* CSPD solution (Roche). Store in the dark.
18. Fixative solution: dilute 1:5 the fixative G350 (AGFA-Gevaert, Mortsel, Belgium). Store in the dark. Stable for several months.
19. Developer solution: dilute 1:5 the developer G150 (AGFA). Store in the dark. Stable for several months.
20. Film: AGFA curix RP2 100NIF 13 × 18.
21. Autoradiographic cassette (Gevamatic AGFA 18 × 24).
22. Plastic bags.
23. Trays.
24. Forceps.

25. Aluminum paper.
26. Plastic bag sealer.
27. UV lamp.
28. Water bath: Certomat (22–100 ± 0.5°C; Braun Biotech, Melsungen, Germany) or similar.
29. Vacuum oven.
30. Belly Dancer.
31. Dark room.

2.7. DNA Purification

2.7.1. Purification of an Amplimer From Single Band-Containing Solutions

1. High Pure PCR Product Purification Kit (Roche).
2. 96° Ethanol.
3. Standard tabletop centrifuge.
4. Microcentrifuge tubes.

2.7.2. Purification of an Amplimer From Mixed Band-Containing Solutions

1. Electrophoresis material.
2. Scalpel.
3. Heating block.
4. High Pure PCR Product Purification Kit (Roche).
5. 96° Ethanol.
6. Isopropanol.
7. Standard tabletop centrifuge.
8. Microcentrifuge tubes.

2.8. DNA Sequencing

1. ABI PRISM BigDye Terminator Cycle Sequencing ready Reaction Kit (Applied Biosystems).
2. Primer VP1-2949 (5'-2949TATTTGTCTGTCACAGAACAATCAG2973-3')
3. Autoclaved (121°C for 45 min) distilled H_2O.
4. 95° Ethanol ACS for analysis (Carlo Erba Reagenti, Italy).
5. 70% Ethanol.
6. Thermocycler AB 2700 (Applied Biosystems) or similar.
7. ABI PRISM 3700 DNA Analyzer (Applied Biosystems) or similar.
8. 0.2-mL Tubes.
9. Vortex.
10. Microcentrifuge.

3. Methods

3.1. Sample Preconditioning (see Note 8)

1. Collect 50 mL of sewage or wastewater in a plastic container.
2. Lyophilize for 48 h at –70°C temperature and 30–40 mtorr pressure.
3. Resuspend the solid pellet in 500 µL of water.

3.2. RNA Extraction Procedure Based on the Boom Method (15) (see Note 9)

1. Add 40 µL of silica solution to 900 µL of L6 buffer in a 1.5-mL microcentrifuge tube (*see* **Notes 10** and **11**).

2. Add 50 μL of sample.
3. Vortex the mixture for 5 s, keep it at room temperature for 10 min, and mix again for 5 s.
4. Centrifuge at 12,000*g* for 15 s and discard the supernatant (*see* **Notes 10** and **13**).
5. Wash the silica pellet twice with 900 μL of L2 buffer, twice with 900 μL of 70% ethanol, and once with 900 μL of acetone. Centrifuge for 15 s at 12,000*g* after each wash and discard the supernatants (*see* **Notes 10**, **12**, and **13**).
6. Place the tube, with the lid open, at 56°C in a dry heating block for 10 min.
7. Add 50 μL of TE buffer, vortex, and incubate for 10 min at 56°C with the lid closed (*see* **Notes 10** and **12**).
8. Vortex and centrifuge the tube at 12,000*g* for 2 min.
9. Recover the nucleic acid-containing supernatant in a new tube (*see* **Notes 13** and **14**) and store it at 4°C (*see* **Note 15**).

3.3. Two-Step RT-PCR for the Amplification of a Fragment of the 5′NCR and Confirmation by Southern Blot for the Generic HAV Detection (see Note 16)

3.3.1. RT

1. Add 10 μL of the extracted RNA to a 0.2-mL PCR tube.
2. Incubate the tube at 99°C for 5 min to denature the RNA.
3. Chill the tube on ice.
4. Prepare the RT mix (volume for one sample): 5.000 μL 5X RT buffer; 0.125 μL 100 μ*M* primer HAV 240; 2.500 μL 2 m*M* dNTPs; 0.040 μL M-MLV reverse trascriptase (200 U/μL); 7.300 μL Distilled H_2O; for a total of 15.000 μL.
5. Add the 15 μL of the RT mix to each tube containing denatured RNA.
6. Incubate the tubes at 45°C for 1 h.
7. Keep the tubes at 4°C.

3.3.2. PCR

1. Add 29.5 μL of distilled H_2O to a 0.2-mL PCR tube.
2. Prepare the PCR mix (volume for one sample): 5.00 μL 10X Expand High Fidelity-$MgCl_2$ buffer; 0.25 μL 100 μ*M* primer HAV 240; 0.25 μL 100 μ*M* primer HAV 68; 5.00 μL 2 m*M* dNTPs; 0.14 μL Expand High Fidelity PCR enzyme (3.5 U/μL); for a total of 10.50 μL.
3. Add the 10.5 μL of the PCR mix to each of the water-containing tubes and 10 μL of the RT product.
4. Transfer the samples to the thermocycler and run the following program: 4 min at 94°C; 40 cycles of 1 min at 94°C, 1 min at 55°C, 1 min 30 s at 72°C; and 10 min at 72°C.

3.3.3. Electrophoresis

1. Mix 10 μL of the PCR product with 2 μL of 6X loading buffer and dispense them in a well of a 1.5% TBE-agarose gel. In a parallel well dispense 5 μL of the working solution of the DNA molecular weight marker IX (Roche).
2. Connect the electric power at 70 V and run the samples for approx 2 h in TBE buffer.
3. Stain the DNA-containing agarose gel with a 1:10,000 dilution of GelStar in TBE buffer (*see* **Note 17**).
4. Analyze the DNA amplimers by observation of the stained bands in the gel analyzer using the UV light tray (*see* **Note 18**).

3.3.4. Southern Blot Hybridization

1. Transfer the gel to a glass Petri dish and submerge it in several volumes of 150 m*M* NaOH for 15 min to denature the double-stranded DNA. Keep the glass dish containing the gel on ice with constant shaking.
2. In the mean time, cut the nylon membrane (*see* **Note 19**) and rinse it in 0.5X TBE for 15 min. Wet fiber pads and filter papers with 0.5X TBE.
3. Wash and neutralize the gel by immersion in several volumes of 0.5X TBE. Keep the glass dish containing the gel on ice and shake constantly for 10 min.
4. Prepare the transfer cell as depicted in **Fig 1**. The gel should be in direct contact with the nylon membrane by the open end side of the well (*see* **Note 20**).
5. Close the cassette, and place it in the module.
6. Place 0.5X TBE buffer in the module.
7. Add the frozen Bio-Ice cooling unit.
8. Add a standard magnetic stir bar to maintain the buffer temperature and ion distribution in the tank.
9. Put on the lid, plug the cables into the power supply, and run for 1 h at 60 V.
10. Disassemble the blotting sandwich and remove the membrane.
11. UV-crosslink the wet membrane at 366 nm for 3 min.
12. Bake in a vacuum oven at 120°C for 30 min (*see* **Note 21**).
13. Place the membrane in a bag containing 20 mL of prehybridization solution per 100 cm^2 of membrane surface area.
14. Seal the bag and prehybridize overnight at 40°C.
15. Discard the prehybridization solution and add the hybridization solution containing the DIG-labeled probe at a 5 pmol/mL concentration and at the same volume/surface ratio.
16. Incubate for 2 h at 40°C.
17. Place the membrane in a glass dish and wash it briefly in several volumes of 2X SSC at room temperature with constant shaking.
18. Wash the membrane for 15 min in several volumes of 2X SSC-0.1% SDS at room temperature with constant shaking.
19. Incubate the membrane for 1 min in several volumes of buffer I at room temperature with constant shaking.
20. Block the membrane by gentle agitation in several volumes of blocking solution for 30 min at room temperature (*see* **Note 22**).
21. Dilute the anti-dig-AP 1:10,000 in blocking solution and mix.
22. Discard the blocking solution and incubate the membrane for 30 min in the antibody solution. Use a standard Petri dish for the incubation and the same volume/surface ratio as with the hybridization solution (*see* **Note 23**). Gentle agitation of the dish is required to ensure a constant and complete contact of the antigen-bound membrane and the antibody-containing solution.
23. Discard the antibody solution.
24. Wash the membrane twice (15 min per wash) in several volumes of washing buffer.
25. Equilibrate the membrane in several volumes of phosphatase buffer for two minutes.
26. Prepare the CSPD solution at a 1:100 in phosphatase buffer.
27. Discard the phosphatase buffer, avoiding, however, drying of the filter.
28. Transfer the membrane into a plastic bag with CSPD solution at the same volume/surface ratio as before. Remove any bubble. Protect the bag from the light.
29. Incubate the bag at room temperature for 5 min with constant shaking.

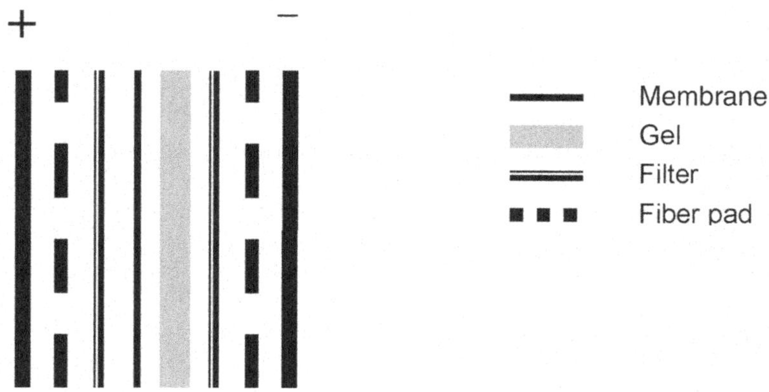

Fig. 1. Southern blot transfer cell.

30. Transfer the membrane to a new plastic bag (*see* **Note 24**).
31. Incubate for 15 min at 37°C (in the dark).
32. The chemilumininescent signal detection is performed by exposing the photographic film to the membrane-containing bag in the autoradiografic cassette (*see* **Note 25**).
33. Expose for 45 min (*see* **Note 26**).
34. Photographic development: transfer the film into several volumes of developer solution for 3 min. Rinse the film with water, transfer the film to several volumes of fixative solution for 3 min, and finally rinse the film again in water and let dry.

3.4. HAV Genotyping by Two-Step RT-PCR Amplification and Sequencing of a Fragment of the VP1X2A Junction Region

3.4.1. RT

1. Add 10 µL of the extracted RNA to a 0.2-mL PCR tube.
2. Incubate the tubes at 99°C for 5 min to denature the RNA.
3. Chill the tubes on ice.
4. Prepare the RT mix (volume for one sample): 5.000 µL 5X RT buffer; 0.125 µL 100 µM primer VP1-3285; 2.500 µL 2 m*M* dNTPs; 0.040 µL M-MLV reverse trascriptase (200 U/µL); distilled H_2O 7.300 µL; for a total of 15.000 µL
5. Add 15 µL of the RT-mix to each tube containing the denatured RNA.
6. Incubate the tubes at 45°C for 1 h.
7. Keep the tubes at 4°C.

3.4.2. PCR

1. Add 29.5 µL of H_2O in a new 0.2-mL PCR tube.
2. Prepare the PCR mix (*see* **Note 27**) (volume for one sample): 5.00 µL 10X PCR-$MgCl_2$ buffer; 0.25 µL 100 µ*M* primer VP1-2949; 0.25 µL 100 µ*M* primer VP1-3285; 5.00 µL 2 m*M* dNTPs; 0.14 µL Expand High Fidelity PCR enzyme (3.5 U/µL); for a total of 10.50 µL.
3. Add 10.5 µL of the PCR mix to each water-containing tube and 10 µL of RT product.
4. Transfer the samples to the thermocycler and run the following program: 4 min at 94°C; 40 cycles of 1 min at 94°C, 1 min at 55°C, 1 min 30 s at 72°C; and 10 min at 72C.

3.4.3. Electrophoresis

1. Mix 10 μL of the PCR product with 2 μL of 6X loading buffer and put them over a well of a 0.8% TBE-agarose gel. In a parallel well, dispense 5 μL of the working solution of the DNA molecular weight marker IX (Roche).
2. Connect the electric power up to 70 V and run the samples for approx 2 h in TBE buffer.
3. Stain the DNA-containing agarose gel with a 1/10,000 dilution of GelStar in TBE buffer (*see* **Note 17**).
4. Analyze the DNA amplimers by observation of the stained bands in the ImageMaster, using the UV light tray (*see* **Note 18**).

3.4.4. DNA Purification

3.4.4.1. Purification of an Amplimer From Single Band-Containing Solutions

1. Mix the remaining 40 μL of the PCR product with 200 μL of the Binding Buffer provided with the kit.
2. Transfer these 200 μL to the upper reservoir of a filter-tube inserted into a collection tube, both provided with the kit.
3. Centrifuge the combined tubes for 1 min at room temperature at 16,000*g* and discard the flowthrough solution.
4. Add 500 μL of wash buffer to the upper reservoir (1:5 dilution of the 5X wash buffer provided with the kit in 96°ethanol).
5. Centrifuge for 1 min at room temperature at 16,000*g* and discard the flowthrough solution.
6. Add 200 μL of wash buffer.
7. Centrifuge for 1 min at room temperature at 16,000*g* and discard the flowthrough solution.
8. Centrifuge for 5 s at room temperature at 16,000*g*.
9. Insert the filter-tube into a standard microcentrifuge tube.
10. Add 60 μL of the elution buffer (provided with the kit) to the filter-tube, and incubate for 3 min at room temperature.
11. Centrifuge the combined tubes for 1 min at room temperature at 16,000*g* and keep the DNA-containing flowthrough solution.
12. Estimate the quantity of DNA by comparison of its fluorescence intensity with that of the 360-bp band of the DNA Molecular Weight Marker IX (Roche; 1.425 ng/μL) run in parallel in an agarose gel and stained with GelStar.

3.4.4.2. Purification of an Amplimer From Mixed Band-Containing Solutions

1. Run an electrophoresis of the remaining 40 μL of PCR product, as previously described in **Subheading 3.4.3.** but using a 0.8% TAE-agarose gel instead of TBE-agarose. Prepare the agarose gel with a preparative comb.
2. Cut the desired band from the agarose gel with an ethanol-cleaned scalpel (*see* **Note 28**).
3. Weigh a 1.5-mL microcentrifuge tube, add the excised agarose fragment to the tube, and weigh it again to deduce the fragment weight.
4. Add binding buffer at a ratio of 300 μL/100 mg of agarose and mix vigorously.
5. Incubate at 56°C for a minimum of 10 min to dissolve the agarose. Mix vigorously every 2 min.
6. Add isopropanol at a ratio of 150 μL/100 mg of agarose and mix vigorously.
7. Transfer this solution (up to 700 μL) to the upper reservoir of a filter-tube inserted into a collection tube.

8. Centrifuge the combined tubes for 1 min at room temperature at 16,000*g* and discard the flowthrough solution. Repeat this step with any further volume exceeding 700 μL.
9. Proceed as in the purification of an amplimer from single band-containig solutions from (**Subheading 3.4.4.1.**) **step 4**.

3.4.5. DNA Sequencing

1. Prepare the sequencing mix: 4.00 μL Ready Reaction Mix (*see* **Note 29**); 0.64 μL 5 μ*M* primer VP1-2949 (*see* **Note 30**); x.xx μL 3-10 ng of purified DNA (*see* **Note 31**); x.xx μL distilled water (*see* **Note 31**) for a total of 10.00 μL.
2. Transfer the samples to the thermocycler and run the following program: 25 cycles of 10 s at 96°C; 5 s at 50°C; 4 min at 60°C.
3. Keep the reaction product at 4°C until further processing (*see* **Note 32**).
4. Prepare the precipitation mix: 10.00 μL sequencing product; 64.00 μL 95° ethanol; 26.00 μL distilled water; for a total 100.00 μL.
5. Mix vigorously by vortexing.
6. Allow to stand for 15 min at room temperature.
7. Centrifuge for 30 min at 16,000*g* and discard the supernatant.
8. Add 200 μL of 70% ethanol.
9. Centrifuge for 5 min at 16,000*g* and carefully discard all the supernatant.
10. Keep the tube open at room temperature until the pellet is completely dried.
11. Electrophorese the sequencing product on the ABI Prism Analyzer for 4.5 h according to the manufacturer's instructions.

4. Notes

1. Unless otherwise stated procedures described molecular biology grade distilled water should be used in all the procedures described.
2. Since the HAV genome is RNA and RNases are ubiquitous and highly resistant to heat inactivation, the plastic material (and whenever possible all solutions) should be autoclaved at 121°C for 45 min.
3. Preparation of these reagents is not necessary when RNA extraction kits such as the QIAamp Viral RNA Mini Kit (QIAGEN, Hilden Germany) are used. The advantages in using this kit are that it is less time-consuming and that the RNA recovery is similar or even better with certain kinds of samples such as shellfish. However, the price is comparatively expensive.
4. All ethanol solutions are prepared from 96° absolute ethanol.
5. The reason to autoclave (121°C for 20 min) most of the Southern blot reagents is not to prevent RNA degradation but rather to avoid bacterial contamination that could interfere with the hybridization reaction as well as immunological detection.
6. If the prehybridization solution is not used immediately, it should be kept on ice to avoid renaturation of the salmon DNA.
7. Since the initial pH of this buffer is around 1.5, the addition of NaOH beads is recommended to increase the pH abruptly to 7.0; at this point, adjust the pH with 1 *N* NaOH.
8. This step refers to a method for the removal of some potential inhibitors of the ulterior molecular techniques ***(14)***. At the same time it allows the viral particles present in the sample to be concentrated, since the initial volume of the sample is 50 mL and the final volume is 500 μL, which corresponds to a 100X concentration.
9. Owing to the crucial problem of cross-contamination in the PCR-based diagnostic procedures, physical separation between each step should be kept at maximum. The nucleic

acid extraction should be performed in a separate and specific room and in a vertical laminar flow hood (P2 level) since HAV is a human pathogen.

10. It is important to autoclave the microcentrifuge tubes with their lids open to avoid potential cross-contamination. For the same reason, use different pipets to add and remove reagents from the tubes, and change the filter tips at each step.
11. Completely resuspend the silica pellet from the stock.
12. Completely resuspend the silica pellet after each wash.
13. Manipulate the guanidinium isothiocyanate wastes very carefully, as well as all the reagents and materials in contact with it, and dispose of them in a container whose one-third part is 10 *N* NaOH.
14. Carefully eliminate all the silica particles. If the first supernatant has silica contamination, repeat the centrifugation step.
15. If further processing is delayed, keep the nucleic acid-containing supernatant frozen at –70°C.
16. Because of the crucial problem of cross-contamination in the PCR-based diagnostic procedures, physical separation between each step should be kept at maximum. The RT and PCR reaction mix should be prepared in a different room separately from the nucleic acid extraction room. The thermocycler should be located in a third separate facility, and, last but not least, the amplimer-containing tubes should be open and the gels run in a fourth separate room.
17. The DNA bands may also be stained with the traditional ethidium bromide. However, since environmental samples usually present low viral loads, the GelStar stain is suggested to increase the sensitivity (around 0.5–1 log higher).
18. The gel analyzer could be replaced by a UV transilluminator coupled to a Polaroid camera.
19. Avoid touching the nylon membrane with the fingers even when using gloves.
20. Use a tube to roll the air bubbles out gently.
21. The membrane can be used immediately for prehybridization or can be stored dry at 4°C for future use.
22. Longer blocking times may also be performed.
23. To reduce the background, it is preferable to incubate the membrane in Petri dishes rather than plastic bags. It is not useful to increase the incubation time since the background increases correspondingly.
24. To reduce the background, it is important to remove the CSPD reagent completely by changing the plastic bag container.
25. Since it is a photographic development, typical dark room conditions should be observed.
26. Longer exposition times may be used to increase the intensity of the bands. The same membrane may be used for several photographic expositions at different times.
27. Since the genotyping is based on a DNA sequence, it is important to use a DNA polymerase with proofreading activity such as the *Pwo* polymerase included in the Expand enzyme.
28. Avoid excising agarose by cutting the band as accurately as possible.
29. This mix contains all the reagents necessary for the sequencing reaction with the exception of the specific primer and template and should be maintained on ice during preparation of the complete sequencing mix.
30. Since HAV is not very variable, only one strand is sequenced. However, if the quality of the sequence is not good (double sequence, low intensity of peaks, and so on), the second strand is also analyzed by using the VP1-3285 primer. When mutations not described in any database are detected, the second strand is also analyzed to confirm it.

31. The volume necessary to set up a reaction with 3–10 ng of DNA. The remaining amount to a final volume of 10 μL is added as distilled water.
32. It is preferable to continue with the DNA precipitation on the same day.

References

1. Centers for Disease Control and Prevention (1995) *Hepatitis Surveillance Report No. 56.* Centers for Disease Control and Prevention, Atlanta, GA.
2. Mast, E. E. and Alter, M. J. (1993) Epidemiology of viral hepatitis: an overview. *Sem. Virol.* **4,** 273–283.
3. Bosch, A., Lucena, F., Díez, J. M., Gajardo, R., Blasi, M., and Jofre, J. (1991) Human enteric viruses and indicator microorganisms in a water supply associated with an outbreak of infectious hepatitis. *J. Am. Water Works Assoc.* **83,** 80–83.
4. Bosch, A., Sánchez, G., Le Guyader, F., Vanaclocha, H., Haugarreau, L., and Pintó, R.M. (2001) Human enteric viruses in coquina clams associated with a large hepatitis A outbreak. *Wat. Sci. Tech.* **43,** 61–65.
5. Halliday, M. L., Kang, L.-Y., Zhou, T.-K., et al. (1991) An epidemic of hepatitis A attributable to the ingestion of raw clams in Shanghai, China. *J. Infect. Dis.* **164,** 852–859.
6. Mele, A., Rastelli, M. G., Gill, O. N., et al. (1989) Recurrent epidemic hepatitis A associated with the consumption of raw shellfish, probably controlled through public health mesures. *Am. J. Epidemiol.* **130,** 540–546.
7. Reid, T. M. S. and Robinson, H. G. (1987) Frozen raspberries and hepatitis A. *Epidemiol. Infect.* **98,** 109–112.
8. Rosemblum, L. S., Mirkin, I. R., Allen, D. T., Safford, S., and Hadler, S. C. (1990) A multistate outbreak of hepatitis A traced to commercially distributed lettuce. *Am. J. Public Health* **80,** 1075–1080.
9. Hollinger, F. B. and Emerson, S. U. (2001) Hepatitis A virus. In: *Fields Virology*, 4 th ed., vol. 1 (Fields, B. N., Knipe. D. M., Howley, P. M., eds.). Lippincott-Williams & Wilkins, New York, pp. 799–840.
10. Robertson, B. H., Jansen, R. W., Khanna, B., et al. (1992) Genetic relatedness of hepatitis A virus strains recovered from different geographical regions. *J. Gen. Virol.* **73,** 1365–1377.
11. Taylor, M. B. (1997) Molecular epidemiology of South African strains of hepatitis A virus: 1982-1996. *J. Med. Virol.* **51,** 73–79.
12. Graff, J., Richards, O. C., Swiderek, K. M., et al. (1999) Hepatitis A virus capsid protein VP1 has a heterogeneous C terminus. *J. Virol.* **73,** 6015–6023.
13. Martin, A., Benichou, D., Chao, S. F., Cohen, L. M., and Lemon, S. M. (1999) Maturation of the hepatitis A virus capsid protein VP1 is not dependent on processing by the $3C^{pro}$ proteinase. *J. Virol.* **73,** 6220–6227.
14. Gajardo, R., Bouchriti, N., Pintó, R. M., and Bosch, A. (1995) Genotyping of rotaviruses isolated from sewage. *Appl. Environ. Microbiol.* **61,** 3460–3462.
15. Boom, R., Sol, C. J. A., Salimans, M. M. M., Jansen, C. L., Wertheim-van Dillen, P. M. E., and Van der Noordaa, J. (1990) Rapid and simple method for purification of nucleic acids. *J. Clin. Microbiol.* **28,** 495-503.

III

Fungi

10

Typing Fungal Isolates

Molecular Methods and Computerized Analysis

Aitor Rementeria, Ana Belén Vivanco, Ainara Cadaval, María Teresa Ruesga, Sonia Brena, José Pontón, Guillermo Quindós, and Javier Garaizar

1. Introduction

Infections caused by fungi (mycoses) are increasingly reported in many countries owing to greater life expectancy associated with an increase in quality of medical and surgical procedures, as well as the emergence of diseases or infections that affect the immune system such as AIDS. Nosocomial outbreaks of fungal infections are sometimes reported, and typing is then necessary to find the reservoirs, analyze the modes of transmission, study the antifungal susceptibility patterns, and investigate the susceptibility of the host ***(1,2)***.

In addition, the food industry is increasingly demanding typing methods that could help in selection of the best fungal strains, in order to incorporate them in the productive chains and augment the quality and security of food. This is the case for *Saccharomyces cerevisiae* in the wine industry ***(3,4)***: the selection and characterization of indigenous or autochthonous strains is an important objective for the production of high-quality certified wines.

Several genotyping methods are now widely used for strain delineation of medically or economically important microorganisms belonging to the kingdom Fungi ***(5)***. Most molecular typing methods are comparable to those already described for bacteria, although the peculiarities of their nucleic acids increase the number of available methods ***(6–8)***. Although typing procedures based on the analysis of nucleic acid sequences have been developed ***(9)***, most genotyping methods currently in use are electrophoretically based, and the procedures include the visual comparison of nucleic acid band profiles or their reading with the help of computerized software.

From: *Methods in Molecular Biology, vol. 268: Public Health Microbiology: Methods and Protocols*
Edited by: J. F. T. Spencer and A. L. Ragout de Spencer © Humana Press Inc., Totowa, NJ

Here we describe some of the most frequently used genotyping methods for fungi, based on polymerase chain reactions (PCR) ***(10–12)***, the isolation of chromosomal or mitochondrial DNA, and their restriction using endonuclease enzymes ***(1,8)***. The latter methods are exclusive for typing eukaryotic organisms and are based on the expected polymorphism obtained from the separation of large chromosomes using pulsed-field gel electrophoresis (PFGE) and the restriction of mitochondrial or chromosomal DNA. More sophisticated methods, such as those that combine endonuclease restriction with hybridization, are also available, although their use is less extensive and is limited mostly to research laboratories.

2. Materials

2.1. Randomly Amplified Polymorphic DNA Fingerprinting (RAPD-PCR)

1. Sabouraud dextrose agar with chloramphenicol (Pronadisa, Alcobendas, Madrid).
2. DNA extraction buffer: 2% v/v Triton X-100, 1% w/v sodium dodecyl sulfate (SDS), 100 m*M* NaCl, 10 m*M* Tris-HCl, pH 8.0, 1 m*M* Na_2 EDTA. Sterilize by autoclaving and store at 4°C.
3. Phenol/chloroform/isoamyl alcohol (25:24:1, v/v/v).
4. Glass beads. Sterilize by autoclaving.
5. Vortex.
6. 10X TE buffer: 10 m*M* Tris-HCl, pH 8.0, 1 m*M* Na_2EDTA. Sterilize by autoclaving and store at 4°C.
7. Microfuge (Sanyo).
8. Absolute ethanol (Merck).
9. RNAse buffer: 1X TE buffer. Add 1 µg/mL ribonuclease A (Sigma) prior to use.
10. 5 *M* NaCl.
11. Spectrophotometer (Kontron Instruments).
12. Ready-To-Go™ RAPD Analysis kit (Amersham Biosciences, Barcelona, Spain), providing room-temperature-stable beads containing buffer, PCR dNTPs, bovine serum albumin (BSA), AmpliTaq®, and Stoffel fragment and a set of PCR primers (**Table 1**).
13. Thermal cycler for PCR: Robocycler gradient 96 (Stratagene).
14. Tissue culture grade sterile water (Sigma).
15. Molecular grade agarose (Bio-Rad).
16. 10X TBE buffer: 890 m*M* Tris, 890 m*M* boric acid, 20 m*M* Na_2 EDTA. Sterilize by autoclaving and store at room temperature.
17. pGEM DNA marker (Promega).
18. Gel loading solution (Sigma).
19. Standard apparatus for the electrophoresis of agarose gels and power supply (Bio-Rad).

2.2. Mitochondrial DNA Restriction Analysis (mtREA)

1. Yeast extract peptone dextrose (YPD) broth (Sigma).
2. Spheroplast buffer: 1 *M* sorbitol, 0.1 *M* EDTA, pH 7.5. Sterilize by autoclaving and store at 4°C.
3. 2.5 mg/mL Lyticase solution (Sigma).
4. Microfuge (Sanyo).
5. Resuspension buffer: 50 m*M* Tris-HCl, 20 m*M* EDTA, pH 7.4.
6. 10% w/v SDS. Sterilize by autoclaving and store at room temperature.
7. 5 *M* Potassium acetate. Sterilize by autoclaving and store at 4°C.

Table 1
Primers and PCR Conditions for Performing RAPD of Several Species of Fungi

Fungi	Primers	Sequence	PCR conditions	
Saccharomyces cerevisiae	RAPD-Primer 2	5'-GTTTCGCTCC-3'	95°C 5 min	1 cycle
	RAPD-Primer 4	5'-AAGAGCCCGT-3'	95°C 1 min	45 cycles
	RAPD-Primer 6	5'-CCCGTCAGCA-3'	36°C 1 min	
			72°C 2 min	
Candida spp.	AB1-12	5'-CCTTGACGCA-3'	94°C 4 min	1 cycle
			94°C 1 min	40 cycles
			36°C 1 min	
			72°C 2 min	
			72°C 30 s	1 cycle
Scedosporium prolificans	UBC-701	5'-CCCACAACCC-3'	94°C 5 min	1 cycle
	AB1-08	5'-GTCCACACGG-3'	94°C 1 min	40 cycles
	AB1-11	5'-GTAGACCCGT-3'	36°C 1 min	
			72°C 2 min	
			72°C 2 min	1 cycle

8. Isopropanol.
9. 70% v/v Ethanol.
10. 10X TE buffer.
11. Spectrophotometer (Kontron Instruments).
12. Restriction endonucleases *Rsa*I (Sigma), *Hinf*I (Sigma), and *Alu*I (Bioline), and appropriate buffers (**Table 2**).
13. RNAse buffer: 1X TE buffer: add 1 µg/mL ribonuclease A (Sigma) prior to use.
14. 10X TBE buffer.
15. Molecular grade agarose (Bio-Rad).
16. Lambda DNA *Eco*RI/*Hind*III-digested marker (Sigma).
17. Gel loading solution (Sigma).
18. Standard apparatus for the electrophoresis of agarose gels and power supply (Bio-Rad).

2.3. PFGE Separation of Chromosomes (Karyotyping)

1. Sabouraud dextrose agar with chloramphenicol (Pronadisa).
2. TE buffer: 10 m*M* Tris-HCl, pH 8.0, 0.5 M EDTA. Sterilize by autoclaving and store at 4°C.
3. Zymolyase 20T (ICN).
4. Certified Low-Melt agarose (Bio-Rad).
5. β-Mercaptoethanol.
6. 0.5 *M* EDTA.
7. ESP buffer: 0.5 *M* EDTA, 1% w/v *N*-laurylsarcosine. Filter sterilize and store at 4°C. Add 2 mg/mL proteinase K (Bioline) prior to use.
8. PMSF solution: dissolve 17 mg phenylmethylsulfonyl fluoride (PMSF; Sigma) in 1 mL of isopropanol. **Caution:** PMSF is highly toxic by inhalation. Store at –20°C. If necessary, redissolve before using warming up to 56°C.

Table 2
Restriction Enzymes and Molecular Methods Recommended for Genotyping *Saccharomyces cerevisiae* and *Candida* spp.

Restriction enzyme	Recognition sequence	Molecular method	Fungi
*Rsa*I	5'-GT/AC-3'	mtREA	
*Hinf*I	5'-G/ANTC-3'	mtREA	*S. cerevisiae*
*Alu*I[a]	5'-AG/CT-3'	mtREA	
*Sfi*I	5'-GGCCNNNN/NGGCC-3'	PFGE rare cutting	*S. cerevisiae* *Candida* spp.

mtREA, mitochondrial DNA restriction analysis; PFGE, pulsed-field gel electrophoresis.

[a]*Alu*I gave the best discrimination analyzing strains of *S. cerevisiae* (data not shown).

9. 10X TE buffer.
10. 10X TBE buffer.
11. Pulsed-field certified agarose (Bio-Rad).
12. DNA Size Markers-Yeast Chromosomal (Bio-Rad).
13. CHEF DR-II pulsed-field electrophoresis system (Bio-Rad).

2.4. PFGE Using Rare Cutting Restriction Enzymes

1. Materials for performing PFGE from **Subheading 2.3., items 1–11**.
2. Rare cutting restriction endonuclease: *Sfi*I (Amersham Biosciences) and the appropriate buffer (**Table 2**).
3. Lambda ladder concatamers (Bio-Rad).
4. CHEF DR-II pulsed-field electrophoresis system (Bio-Rad).

2.5. Gels Processing and Analysis of Electrophoretical Patterns

1. 200 μg/mL Ethidium bromide stock solution (*see* **Note 1**).
2. Long-wave UV transilluminator for visualization of agarose gels (UltraLum).
3. MP4 Land camera (Polaroid).
4. Scanner ScanJet IICX (Hewlett Packard).
5. Software GelCompar 4.0 or superior (Applied Maths).
6. PC personal computer.
7. Laser printer.

3. Methods

3.1. RAPD-PCR

1. Grow the isolates on Sabouraud dextrose agar with chloramphenicol and incubate for 24–48 h at 28°C.
2. Collect the fungal cells in Eppendorf tubes and resuspend in 500 μL sterile distilled water. Spin in a microfuge 1700*g* for 5 min and decant supernatant. Wash the cells in 500 μL sterile water and spin once again.
3. Discard supernatant and resuspend pellet in 200 μL of DNA extraction buffer.

4. Add 200 μL of phenol/chloroform/isoamyl alcohol and approx 0.3 g of sterile glass beads. Mix vigorously in vortex during 30 s six times, placing the tubes in ice between each agitation.
5. Add 200 μL 1X TE buffer and mix by vortex. Centrifuge 5 min at 11,600*g* and transfer the supernatant to a new Eppendorf tube.
6. Add 200 μL of phenol/chloroform/isoamyl alcohol and spin once again until no white precipitate is visible at the aqueous-phenol interface.
7. Mix by inverting the final supernatant with 1 mL of cold absolute ethanol. Centrifuge at 11,600*g* for 10 min and discard supernatant.
8. Add to pellet 400 μL RNAse buffer and incubate for 1 h at 37°C.
9. Add 10 μL 5 *M* NaCl and 1 mL cold absolute ethanol. Mix by inverting and keep samples overnight at –20°C.
10. Spin 10 min at 11,600*g*, remove supernatant, and vacuum dry.
11. Rehydrate the DNA in 40 μL of 1X TE and keep at 4°C overnight. Measure the DNA concentration of the samples by spectrophotometry and prepare DNA stock solutions of 50 ng/μL. Keep at –20°C.
12. Carry out a PCR reaction with the commercial kit Ready-To-Go™. Mix the beads containing all reagents with 125 ng template DNA and 25 pmol primer in a 35 μL final volume (*see* **Note 2**).
13. Perform the PCR in the thermal cycler at the temperatures described in **Table 1**.
14. Prepare a 2% w/v molecular grade agarose gel in 1X TBE buffer, load the samples and DNA marker, and run the electrophoresis for 90 min at 100 V.
15. Dilute the stock solution of ethidium bromide to reach a concentration of 0.5 μg/mL. Incubate the gel in the ethidium bromide solution in the dark for approx 30–35 min. Wash the gel with distilled water and photograph the gel under 254 nm UV transillumination.

3.2. mtREA

1. Grow the isolates on 5 mL of YPD broth and incubate overnight at 28°C.
2. Spin at 2800*g* for 20 min. Resuspend the pellet in 500 μL of spheroplast buffer and transfer to an Eppendorf tube.
3. Add 20 μL of lyticase solution and place the samples at 37°C for 60 min (*see* **Note 3**).
4. After incubation, spin the samples for 5 min at 3400*g*. Discard supernatant and resuspend the pellet in 500 μL of resuspension buffer.
5. Add 50 μL 10% w/v SDS and incubate at 65°C for 30 min.
6. Immediately add 200 μL of 5 *M* potassium acetate, mix by inversion, and keep on ice for 30 min.
7. Spin at 11,600*g* for 10 min. Collect approx 600 μL of supernatant and transfer to new eppendorf tube.
8. Add the same volume of isopropanol and keep at room temperature for 10 min.
9. Spin the samples for 10 min at 11,600*g* and remove the supernatant.
10. Add 500 μL of 70% v/v ethanol and mix by inversion for 1–2 min. Spin at 11,600*g* for 10 min and discard the supernatant. Dry the pellet in a vacuum or at 37°C.
11. Rehydrate the DNA in 50 μL of 1X TE and keep at 4°C overnight. Measure the DNA concentration of the samples by spectrophotometry and store them at –20°C.
12. Carry out the restriction of 50 μg of the total DNA with 20 U *Rsa*I, 10 U *Hinf*I, or 10 U *Alu*I endonucleases with their respective buffers (**Table 2**) and 125 ng of ribonuclease A in a 50-μL final volume (*see* **Note 4**).

13. Prepare 1% w/v molecular grade agarose gel in 0.5X TBE buffer, mix samples and the molecular weight standard with loading buffer, and run an electrophoresis at constant 100 V for 95 min.
14. Stain the gel with ethidium bromide and take a photograph as described in **Subheading 3.1., step 15**.

3.3. PFGE Separation of Chromosomes (Karyotyping)

1. Grow the isolates on Sabouraud dextrose agar with chloramphenicol and incubate 24–48 h at 28°C.
2. Inoculate two loops of culture into 400 µL of YPD broth, measure the cell density, and adjust it to approx 1.5–2.0 using the McFarland scale.
3. Spin 400 µL of cell suspension at 2500*g* for 10 min.
4. Resuspend the pellet in 200 µL of TE buffer with 5 mg/mL of zymolyase 20T; solution should be prepared just before use.
5. Prepare by boiling a solution of 1.3% w/v low-melt agarose in distilled water and maintain it at 56°C in a water bath.
6. Mix 250 µL of warm agarose with the 200 µL of cell suspension and dispense into a block mold, allowing the blocks to solidify by placing them at 4°C for 30 min.
7. Incubate the blocks at 37°C for 24 h with 2 mL TE buffer and 120 µL β-mercaptoethanol.
8. Discard the buffer and wash the blocks with 3 mL distilled water for 15 min with gentle agitation at room temperature. Repeat this washing.
9. Discard the water and add 2 mL ESP buffer with 2 mg/mL of proteinase K added just before use. Incubate for 24 h at 50°C.
10. Remove the buffer carefully and wash the blocks with 3 mL distilled water for 15 min with gentle agitation at room temperature. Repeat this washing.
11. Discard the water and add 2 mL TE buffer and 30 µL of 100 m*M* PMSF solution for each sample. **Caution:** This step has to be performed in a fume hood and with the operator wearing gloves owing to the PMSF toxicity. Keep at room temperature for 30 min with shaking and repeat this step with fresh TE buffer/PMSF solution (*see* **Note 5**).
12. Decant the TE buffer and PMSF solution carefully, add 3 mL of distilled water, and place samples at room temperature for 15 min with gentle agitation. Repeat the washing with distilled water three times. Remove distilled water and add 3 mL of TE buffer. Keep the samples refrigerated.
13. Prepare 110 mL of 1.2% w/v pulsed-field certified agarose in 0.5X TBE buffer and cast the gel.
14. When the gel is hard, cut a small piece of the DNA block and cast the gel wells with it, making sure that they are placed flat against the leading edge of the well. Insert DNA Size Markers-Yeast Chromosomal agarose block as a molecular size marker. Seal the wells with warm agarose.
15. Transfer the gel together with 2 L of cold 0.5X TBE buffer to the PFGE system (CHEF-DRII) and leave for 15 min for equilibrium (*see* **Note 6**).
16. Program the pulse machine and run the electrophoresis, with a constant electrophoresis buffer temperature of 12°C. Voltages, times, and pulses are detailed in **Table 3** for different Fungi species.
17. Stain the gel with ethidium bromide and take a photograph as described above (**Subheading 3.1., step 15**).

Table 3
Electrophoretical Conditions for Karyotyping

Fungi	Electrophoretic conditions Pulse	Voltage (V)	Time (h)
S. cerevisiae	1. (70–70 s)–15 h 2. (120–120 s)–10 h	200	25
Candida spp.	(60–120 s)–24 h	230	24[a]

[a]Longer electrophoresis running time could increase the definition of chromosomal bands if necessary.

Table 4
Electrophoretical Conditions for PFGE Using Rare Cutting Restriction Enzymes for Fungi Genotyping

Fungi	Electrophoretical conditions Pulse	Voltage (V)	Time (h)
S. cerevisiae	1. (10–30 s)–22 h 2. (50–90 s)–7 h	200	29
Candida albicans	(5–35 s)–24 h	220	24

PFGE, pulsed-field gel electrophoresis

3.4. PFGE Using Rare Cutting Restriction Enzymes

1. Carry out **steps 1–12** as in the karyotyping protocol (**Subheading 3.3.**).
2. Cut a small piece of the DNA block and place it in an Eppendorf tube with 100 µL of 1X restriction enzyme buffer and incubate at room temperature for 1 h. Remove the buffer carefully with a pipet and replace with 50 µL final volume of fresh 1X restriction enzyme buffer with 20 U of *Sfi*I.
3. Cover the reaction volume with sterile mineral oil and incubate at 50°C for 16 h.
4. Prepare 110 mL of 1.2% w/v pulsed-field certified agarose in 0.5X TBE buffer and cast the gel.
5. When hard, cast the gel with the restricted DNA blocks, making sure that are placed flat against the leading edge of the well. Insert Lambda ladder concatamers agarose block as a molecular size marker. Seal the well with warm agarose.
6. Transfer the gel together with 2 L of cold 0.5X TBE buffer to the system (CHEF-DRII) and leave for 15 min for equilibrium (*see* **Note 6**).
7. Program the pulse machine and run the electrophoresis voltages, times, and pulses, with a constant electrophoresis buffer temperature of 12°C (**Table 4**).
8. Stain the gel with ethidium bromide and take a photograph as described above in **Subheading 3.1., step 15**.

3.5. Gels Processing and Analysis of Electrophoretical Patterns

1. Compare the banding patterns looking for band presence or absence visually or with the help of a computerized analyzer.
2. To perform the computerized analysis, images of gels are scanned and saved in a TIFF file and then analyzed by GelCompar version 4.0 or superior software (*see* **Note 7**).
3. After conversion and normalization of gels, the degrees of similarity of DNA fingerprints are determined by Dice or related coefficients.
4. Generate dendrograms using UPGMA (unweighted pair group method using arithmetic averages) algorithm.

4. Notes

1. Ethidium bromide is sensitive to light: use an opaque bottle to store the stock solution. Ethidium bromide is also carcinogenic. **Caution:** *always* wear gloves when handling materials containing ethidium bromide. Follow decontamination instructions using activated charcoal or incineration.
2. Several primers are available in the literature for RAPD procedures for fungal genotyping, and different levels of strain discrimination could be achieved depending on the species studied. Frequently, a combination of results obtained with several primers is necessary to increase their discriminatory resolution. RAPD has an inherent low interassay reproducibility owing to several technical factors, some of which could be overcome using kits that contain already mixed reagents, such as Ready-To-Go™ beads.
3. The fungal cell walls are decomposed using enzymes such as lyticase or zymolase that hydrolyze poly(β-1,3-glucose), producing spheroplasts in high osmotic solutions.
4. Restriction enzymes digest the eukaryotic chromosomes in small fragments; meanwhile the mitochondrial DNA is less sensitive, and its DNA is digested in larger fragments. Conventional nucleic acid electrophoresis will separate both classes of fragments in a resolvable way. Therefore, no tedious or complex procedures for specific isolation of mitochondrial DNA is then required. On the other hand, active growth of cells is required to obtain a higher proportion of mitochondrial DNA, although the corresponding increase in RNA should then be eliminated by addition of RNAse in order to visualize the DNA-restricted patterns.
5. PMSF is highly toxic when inhaled. Its use for proteinase K inactivation could be avoided, but more washing with distilled water is then required.
6. In some instances extensive DNA degradation is observed during electrophoresis. Addition of 75 μ*M* thiourea to the electrophoresis buffer could prevent such events.
7. GelCompar is licensed in the United States and several other countries to Bio-Rad. Other software such as Molecular Analyst runs on an Apple platform. The Bio Image Whole Band Analyzer runs on a Unix platform. We have compared and evaluated such software to determine whether results generated by these programs correlated adequately with visual interpretation of DNA patterns *(13)*. The software packages try to imitate the automatic band alignments that our eyes perform and accomplish it successfully in most instances when gels are of good quality. Computerized alignments of the internal control bands in heavily distorted gels are frequently performed inaccurately; therefore such gels should be eliminated from the comparison or the analysis repeated.

Acknowledgments

This work was supported by The University of Basque Country and project IE019, subproject DIAMOLFUN from the Basque Government. A. B. Vivanco and S. Brena were supported by predoctoral grants from The University of Basque Country and M. T. Ruesga by a predoctoral grant from the Ministerio de Educación y Ciencia, Spain.

References

1. Arif, S., Barkham, T., Power, E. G., and Howell, S. A. (1996) Techniques for investigation of an apparent outbreak of infections with *Candida glabrata*. *J. Clin. Microbiol.* **34,** 2205–2209.
2. Sutphin, J. E., Pfaller, M. A., Hollis, R. J., and Wagoner, M. D. (2002) Donor-to-host transmission of *Candida albicans* after corneal transplantation. *Am. J. Ophthalmol.* **134,** 120–121.
3. Clemons, K. V., Park, P., McCusker, J. H., McCullough, M. J., Davis, R. W., and Stevens, D. A. (1997) Application of DNA typing methods and genetic analysis to epidemiology and taxonomy of *Saccharomyces* isolates. *J. Clin. Microbiol.* **35,** 1822–1828.
4. Versavaud, A., Courcoux, P., Roulland, C., Dulau, L., and Hallet, J.-N. (1995) Genetic diversity and geographical distribution of wild *Saccharomyces cerevisiae* strains from the wine-producing area of Charentes, France. *Appl. Environ. Microbiol.* **61,** 3521–3528.
5. López-Ribot, J. L., McAtee, R. M., Kirkpatrick, W. R., Perea, S., and Patterson, T. F. (2000) Comparison of DNA-based typing methods to assess genetic diversity and relatedness among *Candida albicans* clinical isolates. *Rev. Iberoam. Micol.* **17,** 49–54.
6. Comi, G., Maifreni, M., Manzano, M., Lagazio, C., and Cocolin, L. (2000) Mitochondrial DNA restriction enzyme analysis and evaluation of enological characteristics of *Saccharomyces cerevisiae* strains isolated from grapes of the wine-producing area of Collio (Italy). *Int. J. Food Microbiol.* **58,** 117–121.
7. Querol, A., Barrio, E., and Ramón, D. (1992) A comparative study of different methods of yeast strain characterization. *System. Appl. Microbiol.* **15,** 439–446.
8. Gutiérrez, A. R., López, R., Santamaría, M. P., and Sevilla, M. J. (1997) Ecology of inoculated and spontaneous fermentations in Rioja (Spain) musts, examined by mitochondrial DNA restriction analysis. *Int. J. Food Microbiol.* **36,** 241–245.
9. Biswas, S. K., Yokoyama, K., Wang, L., Nishimura, K., and Miyaji, M. (2001) Typing of *Candida albicans* isolates by sequence analysis of the cytochrome b gene and differentiation from *Candida stellatoidea*. *J. Clin. Microbiol.* **39,** 1600–1603.
10. Lehmann, P. F., Lin, D., and Lasker, B. A. (1992) Genotypic identification and characterization of species and strains within the genus *Candida* by using random amplified polymorphic DNA. *J. Clin. Microbiol.* **30,** 3249–3254.
11. Alonso-Vargas, R., Garaizar, J., Pontón, J., and Quindós, G. (2000) Utility of random amplified polymorphic DNA in the discrimination between *Candida albicans* and *Candida dubliniensis*. *Rev. Iberoam. Micol.* **17,** 10–13.
12. San-Millán, R., Quindós, G., Garaizar, J., Salesa, R., Guarro, J., and Pontón, J. (1997) Characterization of *Scedosporium prolificans* clinical isolates by randomly amplified polymorphic DNA analysis. *J. Clin. Microbiol.* **35,** 2270–2274.
13. Rementeria, A., Gallego, L., Quindós, G., and Garaizar, J. (2001) Comparative evaluation of three commercial software packages for analysis of DNA polymorphism patterns. *Clin. Microbiol. Infect.* **7,** 331–336.

11

Fungal Isolation and Enumeration in Foods

Dante Javier Bueno, Julio Oscar Silva, and Guillermo Oliver

1. Introduction

Humans have now been growing and storing enough food for a long enough time that some rapidly evolving organisms, such as fungi, are moving into niches created by the exploitation of certain plants as food *(1)*.

Food is expected to be nutritious. The most important of the physicochemical conditions that affects fungal growth is related to the biological state of the food. Living foods, particularly fresh fruits, vegetables, and also grains and nuts before harvest, possess powerful defense mechanisms against microbial invasion. When the specific microorganisms overcome defense mechanisms, the spoilage of a living food starts. Other factors to consider are water activity, hydrogen ion concentration, temperature, gas tension, consistency, nutrient status, specific solute effect, and preservation *(1)*.

The consequences of mold contamination of foods are diverse *(2)*: unsightly appearance, chemical (removal or change of most of the constituents) and nutritional value changes, modification of organoleptic quality, difficulties in preservation, occupational hazards (mycoses, allergies), and toxicoses (mycotoxicoses).

It is possible to recognize a succession of three distinct mycoflora during the storage of cereals *(3)*, but they can also be mixed:

1. Field fungi growing and established before harvesting (*Alternaria*, *Fusarium*, *Helminthosporium*, *Cladosporium*).
2. Storage fungi taking over and dominanting in the silo (*Aspergillus* and *Penicillium*).
3. Advanced decay fungi (*Papulospora*, *Sordaria*, *Fusarium graminearum*, and members of the order Mucorales).

Some kind of foods are pelleted during the manufacturing process. This reduces fungal contamination, because the spores are relatively susceptible to heat; however, it does not eliminate all viable spores *(4,5)*. Even if fungal spores could be destroyed during this process, a quick contamination can occur later *(5)*. Therefore, Suárez *(6)* concluded that the pelleting process plays a minor role in fungal control.

From: *Methods in Molecular Biology, vol. 268: Public Health Microbiology: Methods and Protocols*
Edited by: J. F. T. Spencer and A. L. Ragout de Spencer © Humana Press Inc., Totowa, NJ

In this chapter, we describe the methods currently used to isolate and enumerate fungi from different foods. You can consult workshops from the International Commission on Food Mycology (ICFM) or books such as *Fungi and Food Spoilage* by J. I. Pitt and A. D. Hocking *(1)*, for more details.

2. Materials

2.1. Growth Media

1. Dichloran rose bengal chloramphenicol agar (DRBC): 10 g/L glucose, 5 g/L peptone, 1 g/L KH_2PO_4, 0.5 g/L $MgSO_4 7H_2O$, 15 g/L agar, 25 mg/L rose bengal (0.5 mL of a 5% w/v solution in water), 2 mg/L dichloran (1 mL of a 0.2% solution in ethanol), 100 mg/L chloramphenicol, pH 5.5–5.8. Sterilize by autoclaving at 121°C for 15 min. In the dark, the medium is stable for at least 1 mo at 1–4°C. Substances such as dichloran and rose bengal *(7)* inhibit the spread of mold colonies.
2. Potato dextrose agar (PDA): 400 g/L potatoes, 15 g/L glucose, 20 g/L agar, pH 6.0–6.2.
   a. Wash and cut the potatoes and then put them in 1000 mL of water.
   b. Steam or boil for 30–45 min.
   c. Filter the potato broth through cotton, and add water to recover the initial volume.
   d. Melt the agar in 500 mL of potato broth and add the glucose to the other 500 mL.
   e. Mix and sterilize by autoclaving at 121°C for 15 min and add antibiotic solution.
3. Czapek-Dox agar: 30g/L sucrose, 3 g/L $NaNO_3$, 0.5 g/L $MgSO_4$, 0.5 g/L KCl, 0.01 g/L $Fe_2(SO_4)_3$, 1g/L K_2HPO_4, 13 g/L agar; pH 7.3 ± 0.1. Sterilize by autoclaving at 121°C for 15 min and add antibiotic solution.
4. Malt extract agar (MEA): 20 g/L malt extract, 1 g/L peptone, 20 g/L glucose, 20 g/L agar, pH 5.6. Sterilize by autoclaving at 121°C for 15 min and add antibiotic solution.
5. Tap water agar (TWA): add 15 g/L agar to 1 L of tap water. Sterilize by autoclaving at 121°C for 15 min and add antibiotic solution. This can provide a substrate for growth and sporulation of plant pathogenic fungi such as *Fusarium, Drechslera, Bipolaris*, and some other dematicaeous Hyphomycetes.

2.2. Solutions

1. Stock solution of succinate/palmitate of chloramphenicol.
   a. Dissolve 1 g of succinate/palmitate of chloramphenicol in 5 mL of sterile distilled water or physiological solution.
   b. Transfer this solution to another recipient containing 95 mL of sterile distilled water or physiological solution. Use 1 mL of the stock solution for each 100 mL of steilized medium.
   c. Media containing chloramphenicol are easier to prepare, are not affected by autoclaving, and have greater long-term stability.
2. Peptone water solution (0.1%): dissolve 1 g of peptone in 1000 mL of distilled water and sterilize by autoclaving at 121°C for 15 min.

3. Methods

3.1 Direct Plating

Direct plating is the preferred method for detecting, enumerating, and isolating fungi from particulate foods such as grains and nuts. The results of this analysis are expressed as percentage infection of particles. This technique provides no direct indication of the extent of fungal invasion in individual particles. However, it is reasonable to assume that a high percentage of infection is correlated with extensive invasion in the particles *(1)*. For solid food, for which sampling involves cutting pieces of food, direct plating is essentially a qualitative technique *(8)*.

1. With surface desinfection (internal mycoflora): This process provides an effective measurement of inherent mycological quality. The process removes the inevitable surface contamination arising from dust and other sources.
   a. Immerse particles in a chlorine solution (0.4%) for 2 min. Stir, and then drain the chlorine (*see* **Note 1**).
   b. Rinse the particles twice with sterile distilled water.
   c. Leave the particles to dry in a laminar flow cabinet.
   d. Plate 6–20 particles per plate onto solidified agar depending on the particle size.
   e. Incubate plates upright for 5–7 d at 25–28°C (*see* **Notes 2** and **3**).
   f. Count the numbers of infected particles, and express results as a percentage (number of particles infected/number of total particles). Differential counting of various genera is often possible.
   g. Isolate different strains (*see* **Note 4**).
2. Without surface desinfection *(1,9)*: especially when surface contaminants become part of the downstream mycoflora (external mycoflora) or pelleted food for animals (total mycoflora, **Fig. 1**).
   a. Plate 6–20 particles per plate onto solidified agar, depending on particle size.
   b. Incubate plates upright for 5–7 d at 25–28°C (*see* **Notes 2** and **3**).
   c. Count the numbers of infected particles, and express the results as a percentage (number of particles infected/number of total particles). Differential counting of various genera is often possible.
   d. Isolate different strains (*see* **Note 4**).

3.2. Dilution Plating

Dilution plating is the appropriate method for mycological analysis of liquid or powdered foods. It is also suitable for grains intended for flour manufacture and in other situations in which total fungal contamination is relevant. It is usually possible to enumerate plates with up to 150 colonies, but if a high proportion of rapidly growing fungi are present, the maximum number will be lower. The minimum number may be 10–15 colonies. For yeast, enumeration is easier. In the absence of filamentous fungi, from 30 to 300 colonies per plate can be counted *(1)*.

1. Stomach or blending the materials for 2 or 1 min, respectively (*see* **Notes 5**, **6**, and **7**).
2. Make serial dilutions in peptone water until the ration is 1:10,000.
3. Inoculate 0.1 mL of the suspension of different dilutions onto solidified agar and spread it with a sterile bent glass rod (*see* **Note 8**).
4. Incubate plates upright for 5–7 d at 25–28°C (*see* **Notes 2** and **3**).
5. Count the numbers of viable spores, and express results as viable count per gram of sample (colony-forming unit/g).
6. Isolate different strains (*see* **Note 4**).

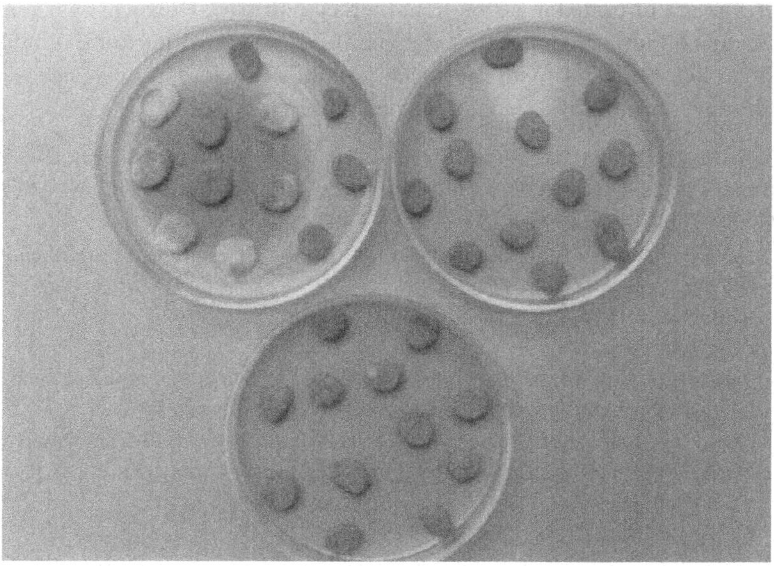

Fig. 1. Direct plating technique without surface desinfection of the kernel in pelleted food for animals.

4. Notes

1. For commodities such as peanuts or maize, in which high levels of *Aspergillus flavus* or *Penicillium* species may be present, Pitt and Hocking ***(1)*** recomend inmersing the particles in 70% ethanol for 2 min followed by 2 min in 0.4% clorine solution.
2. Some common fungi can shed large number of spores during handling, which in an inverted dish will be transferred to the lid. Reinversion of the Petri dish for inspection or removal of the lid may liberate spores into the air or onto solidified agar and cause important contamination problems.
3. In tropical regions, incubation at 30°C is recommended as a more realistic temperature for enumerating fungi from commodities stored at ambient temperatures. In cool regions such as Europe, 22°C has been recommended as the optimal incubation temperature ***(1)***.
4. Streaking techniques are commonly used for yeast isolation, similar to bacterial purification. For filamentous fungi, isolation depends on picking a small sample of hyphae or spores and placing this sample as a point inoculum on an appropriate medium.
5. Stomaching is recomended for dispersing and separating fungi from finely divided materials such as flour and spices, and for soft foods such as cheeses and meats.
6. Harder or particulate foods such as grains, nuts, or dried foods like dried vegetable should be soaked (30–60 min) before stomaching.
7. Blending is recomended for extremely hard particles.
8. Sterilize the rod by flaming it with ethanol before use.

References

1. Pitt, J. I. and Hocking, A. D. (1997) *Fungi and Food Spoilage*. Blackie Academic and Professional, London.
2. Moreau, C. 1979. Moulds, toxins and food. John Wiley & Sons, New York.
3. Christensen, C. M. and Kaufmann, H. H. (1965) Deterioration of stored grains by fungi. *Annu. Rev. Phytopathol.* **III,** 69–84.
4. Abarca, M. L., Bragulat, M. R., Castellá, G, and Cabañs, F. J. (1994) Mycoflora and aflatoxin-producing strains in animal mixed feeds. *J. Food Proteins* **57,** 248–256.
5. Wyatt, R. (1990) Importancia de los hongos en la salud aviar. *Avicultura Profesional* **8,** 48–50.
6. Suárez, O. (1999) Manejo de los granos en las fábricas de alimentos. *Industria Avícola* **10,** 18–21.
7. King, A. D., Hocking, A. D., and Pitt, J. I. (1979) Dichoran-rose bengal medium for enumeration and isolation of molds from foods. *Appl. Env. Microbiol.* **37,** 959–964.
8. Pitt J. I. (1989) Food mycology—an emerging discipline. *J. Appl. Bacteriol.* **67,** 1S–9S.
9. Bueno, D. J., Silva, J. O., and Oliver, G. (2001) Mycoflora in commercial pet foods. *J. Food Protection* **64,** 741–743.

12

Determination of Aflatoxins and Zearalenone in Different Culture Media

Dante Javier Bueno and Guillermo Oliver

1. Introduction

Some molds produce desirable changes in food, but most are merely esthetically undesirable. There has also been an increasing awareness that certain metabolic products of some molds commonly found on foods and feed are dangerous to humans and animals. These toxin substances, mycotoxins, are secondary metabolites produced by different fungi, especially *Aspergillus*, *Penicillium*, *Fusarium*, and, to a lesser degree, *Alternaria* *(1)*. The most important toxins for humans are aflatoxins, ochratoxin A, fumonisins, certain trichothecenes, and zearalenone *(2)*.

Aflatoxins are fungal metabolites produced by different *Aspergillus* species: *A. flavus*, *A. parasiticus*, and *A. nomius*. The most commonly encountered aflatoxins are B_1, B_2, G_1, G_2, M_1, and M_2, but aflatoxin B_1 (AFB_1; **Fig. 1**) is the most frequently found in contaminated samples, and aflatoxins B_2, G_1, and G_2 are generally not reported in the absence of AFB_1 *(3)*. The International Agency for Research on Cancer (IARC) considers that aflatoxins are carcinogenic (hepatocarcinogenic) to humans (group 1) and animals.

Zearalenone (ZEA; **Fig. 2**) is an estrogenic mycotoxin produced by several species of *Fusarium* (*F. acuminatum*, *F. culmorum*, *F. equiseti*, *F. graminearum*, *F. verticilliodes*, *F. oxysporum*, *F. poae*, *F. rosum*, *F. solani*, *F. semitectum*, and *F. sporotrichioides*) that primarily colonize different cereal grains *(4)*. Several reports were noted on the occurrence of ZEA along with various combinations of group B trichothecenes, fumonisins, aflatoxins, and ochratoxins. In most cases, the levels of ZEA were considered to be low; however, the toxicological significance is not known *(5)*. This toxin is not classifiable as to carcinogenicity in humans (group 3) by the IARC, but ZEA was implied in precocious sexual development in children in Puerto Rico and a breast enlargement in young boys in Italy *(6)*.

From: *Methods in Molecular Biology, vol. 268: Public Health Microbiology: Methods and Protocols*
Edited by: J. F. T. Spencer and A. L. Ragout de Spencer © Humana Press Inc., Totowa, NJ

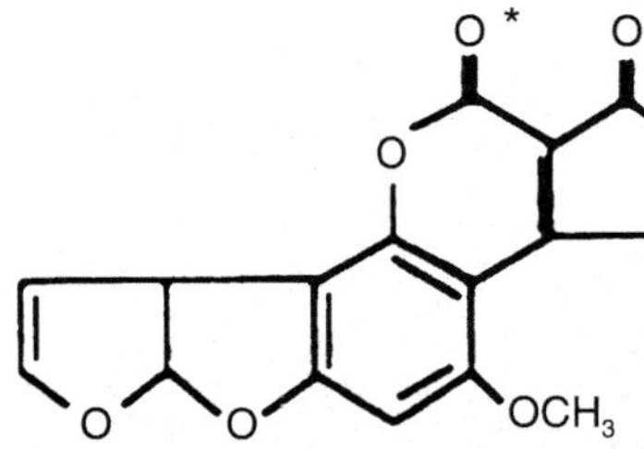

Fig. 1. Chemical structure of aflatoxin B_1.

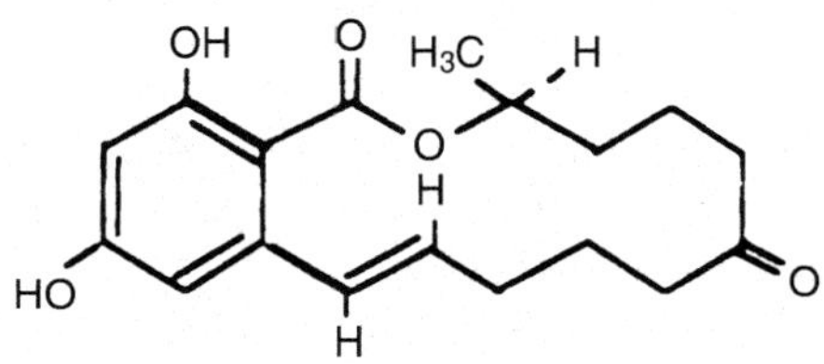

Fig. 2. Chemical structure of zearalenone.

The Association of Official Agricultural Chemists (AOAC) checks analytical methods through systematic interlaboratory studies to determine the performance characteristics for the intended analytical application; the method is then approved that has a good performance for each matrix *(7)*. In this chapter, we describe some methods to determine AFB_1 and ZEA in different culture media by high-performance liquid chromatography (HPLC) that are not included by the AOAC, but we usually use them in our laboratory to study the interaction of mycotoxin (or fungal mycotoxin producers) and lactic acid bacteria or adsorbents.

2. Materials

2.1. Growth Media

1. LAPTg broth *(8)*: 15 g/L peptone, 10 g/L triptone, 10 g/L yeast extract, 10 g/L glucose, 0.1% Tween-80, pH 6.5. Sterilize by autoclaving at 121℃ for 15 min.

2.2. Reagents

1. Standard solutions of AFB_1 or ZEA (Sigma, St. Louis, MO). **Caution:** Perform manipulations under a hood whenever possible, and take particular precautions, such as the use of the glove box, when toxins are in a dry form because of their electrostatic nature and the resulting tendency to disperse in working areas *(7)*.

2. Phosphate-buffered saline (PBS): 16.3 mL 0.5 *M* KH_2PO_4, 27.9 mL 0.5 *M* Na_2HPO_4, 955.8 mL demineralized water, pH 7.3.
3. Citrate buffer: 46.5 mL 0.1 *M* citric acid, 3.5 mL 0.1 *M* sodium citrate, 50 mL demineralized water, pH 3.0.
4. Solvents: dimethyl sulfoxide (DMSO) GC, chloroform (reagent grade), benzene (reagent grade), HPLC grade methanol, HPLC grade acetonitrile, HPLC grade water.

2.3. Apparatus

1. Liquid chromatograph: equipped with injection valve and 25–50 µL loop.
2. Liquid chromatographic column: Reverse phase C_{18} (5–10 µm) column, 250 (or 150) × 4.6 mm, with appropriate guard column.
3. Detector: UV detector with variable wavelength or fluorescence detector.

3. Methods

3.1. ZEA in PBS or Citrate Buffer

This is a simple technique; no extraction of ZEA from samples of PBS or citrate buffer is required. Neither of the buffers interferes with the elution of ZEA in the chromatogram.

1. Prepare stock solutions of ZEA in DMSO or methanol.
2. Evaporate the methanol with nitrogen or keep the DMSO.
3. Add PBS, pH 7.3, or citrate buffer, pH 3.0, to achieve the desired volume.
4. Inject directly into the HPLC apparatus (*see* **Note 1**).
5. Calculate the relative or absolute concentration of the toxin (*see* **Note 2**).

3.2. AFB_1 in PBS

As in **Subheading 3.1.**, no extraction of AFB_1 from samples of PBS is required. The PBS does not interfere with the elution of this toxin in the chromatogram.

1. Prepare stock solutions of AFB_1 in benzene/acetonitrile (97:3).
2. Evaporate the benzene/acetonitrile by heating in a water bath at 80°C for 10 min or with nitrogen (*see* **Note 3**).
3. Add PBS, pH 7.3, to achieve the desired volume.
4. Inject directly into the HPLC apparatus (*see* **Notes 4** and **5**).
5. Calculate the relative or absolute concentration of the toxin (*see* **Note 2**).

3.3. ZEA in LAPTg Broth

In some cases, it is necessary to take mycotoxin from a culture medium. Here we describe the method we use in our laboratory for LAPTg broth, which is a general medium used for the growth of lactic acid bacteria. This method is based on the very good solubility of ZEA in chloroform. It could be adapted to AFB_1 in other media like LTB broth ***(9)***, YES broth ***(10)***, or malt extract agar ***(11)***.

1. Prepare stock solutions of ZEA in DMSO or methanol.
2. Evaporate the methanol with nitrogen or keep the DMSO.
3. Add LAPTg broth, pH 6.5, to achieve the desire volume.
4. Extract ZEA with two portion of chloroform (3:1 v/v) (*see* **Note 6**).

5. Concentrate the chloroform extract to dryness under nitrogen (and by heating in a water bath at 60–65°C).
6. Dissolve in 1 mL of the mobile phase for ZEA.
7. Inject into the HPLC appartus (*see* **Note 1**).
8. Calculate the relative or absolute concentration of the toxin (*see* **Note 2**).

4. Notes

1. Water/acetonitrile/methanol (1.0:1.6:1.0, v/v/v) is used as the mobile phase with a flow rate of 1 mL/min. Detection is performed by fluorescence with excitation and emission wavelengths of 280 and 440 nm ***(12)***, respectively, or 254 nm with a UV detector. The assay is carried out at room temperature with an injection volume of 50 μL. The retention time is approx 7 min. Before the peak of the toxin, the solvent peak can be observed at 3 min.
2. Toxins are quantified by correlating peak areas of samples with those of standard curves in studies of mycotoxin production by fungi, or by the disappearance of mycotoxin.
3. Methanol at 50 μL of can be added after evaporating the benzene/acetonitrile; then make it to a desirable volume with PBS.
4. Water/acetonitrile/methanol (6:3:1, v/v/v) is used as the mobile phase with a flow rate of 1 mL/min. Detection is performed by fluorescence with excitation and emission wavelengths of 365 and 418 nm ***(13)***, respectively, or 365 nm with a UV detector. The assay is carried out at room temperature with an injection volume of 50 μL. The retention time is approx 11 min. Before the peak of the toxin, the solvent peak can be observed at 3 min.
5. An aliquot (200 μL) can be derivatized with trifluoroacetic acid/acetic acid/water (2:1:7) to increase the sensibility of the detection limit. The derivatization solution should be mixed with the aqueous aliquot and kept at 65°C for at least 8.5 min. The derivatized aflatoxin 50 μL is analyzed using an HPLC/fluorescence detection system with the same mobile phase details as above (precolumn derivatization).
6. The relation of the chloroform/aqueous phase (3:1; v/v) can be indicated for small volumes of aqueous phase (until 10 mL). A relation of 1:1 v/v or less could be fitted for bigger volumes.

References

1. Bueno, D. J., Salvano, M., Silva, J. O., González, S. N., and Oliver, G. (2001) Micotoxinas: diagnóstico y prevención en aves de corral. *Bol. Mico.* **16,** 23–36.
2. Pitt, J. I. (2000) Toxigenic fungi: which are important? *Med. Mycol.* **1,** 17–22.
3. Henry, S., Bosch, F. X., Bowers, J. C., Portier, C. J., Petersen, B. J., and Barraj, L. (1998) Aflatoxins. WHO Food Additives Series 40.
4. Pitt, J. I. and Hocking, A. (eds.) (1997) *Fungi and Food Spoilage.* Blackie Academic and Professional, London.
5. Trucksess M. W. (2001) Mycotoxins. *J. AOAC Int.* **84,** 202–212.
6. Pitt, J. I., Basflico, J. C., Abarca, M. L., and López, C. (2000) Mycotoxins and toxigenic fungi. *Med. Mycol.* **38 SI,** 41–46.
7. Scott P. M. (1995) *Natural Toxins. Official Methods of Analysis of AOAC International,* vol, II (Cunniff, P., ed.). AOAC, Arlington, VA, pp. 1–51.
8. Raibaud, P., Galpin, J. V., Ducluzeau, R., Mocquot, G., and Oliver, G. (1973) Le genre *Lactobacillus* dans le tube digestif du rat I.—Caractèes des souches homofermentaires isolées de rats holo et gnotoxéniques. *Ann. Microbiol.* **124,** 83–109.

9. Coallier-Ascah, J. and Idziak, E. S. (1985) Interaction between *Streptococcus lactis* and *Aspergillus flavus* on production of aflatoxin. *Appl. Environ. Microbiol.* **49,** 163–167.
10. Davis, N. D., Diener, U. L., and Eldridge, D. W. (1966) Production of aflatoxins B_1 and G_1 by *Aspergillus flavus* in a semisynthetic medium. *Appl. Microbiol.* **14,** 378–380.
11. Geisen, R. (1996) Multiplex polymerase chain reaction for the detection of potential aflatoxin and sterigmatocystin producing fungi. *System. Appl. Microbiol.* **19,** 388–392.
12. El-Nezami, H., Polychronaki, N., Salminen, S. and Mykkäen, H. (2002) Binding rather than metabolism may explain the interaction of two food-grade *Lactobacillus* strains with zearalenone and its derivative α-zearalenol. *Appl. Environ. Microbiol.* **68,** 3545–3549.
13. El-Nezami, H., Kankaanpaa, P., Salminen, S. and Ahokas, J. (1998). Ability of dairy strains of lactic acid bacteria to bind a common food carcinogen, aflatoxin B_1. *Food Chem. Toxicol.* **36,** 321–326.

IV

Other Pathogens

13

Intracellular Multiplication of *Legionella* Species and the Influence of Amoebae on Their Intracellular Growth in Human Monocytes

Mono Mac 6 Cells and Acanthamoeba castellanii *as Suitable In Vitro Models*

Birgid Neumeister

1. Introduction

1.1. Intracellular Multiplication of Legionellae Within Human Monocytes

Legionellae are important etiological agents of pneumonia. *Legionella pneumophila* (predominantly serogroup 1) is detected in most cases of legionellosis; other species only occasionally cause infections, predominantly in immunocompromized patients ***(1–4)***. Aquiferous technical systems are the primary source of infection (air-conditioning systems, refrigerators, showers, whirlpools, springs, taps, moisturizing equipment, medical nebulizers, and swimming pools). Legionellae are present in the water in these systems, within the amoebae, flagellates, and ciliates in which they replicate ***(2,5,6)***. After inhalation of contaminated aerosols, the bacteria multiply intracellularly within alveolar macrophages ***(7)***. The ability to multiply within monocytic host cells is usually considered to correspond to pathogenicity ***(8–13)***. The mechanisms of intracellular replication have been only partially characterized (recently summarized in ref. ***14***).

Analysis of the molecular pathogenesis of *Legionella* infection, both in the pathogen itself and in the host cell, is the subject of current research and may lead to new options in prophylaxis and treatment. We have established the human Mono Mac 6 cell line (MM6) ***(15)*** instead of the previously used histiocytic lymphoma cell line U 937 or the promyelocytic leukemia cell line HL-60 ***(16–18)*** to investigate the intracellular replication of legionellae and the molecular pathogenesis of *Legionella* infection within human monocytic host cells ***(19)***. MM6 cells represent a more mature mac-

From: *Methods in Molecular Biology, vol. 268: Public Health Microbiology: Methods and Protocols*
Edited by: J. F. T. Spencer and A. L. Ragout de Spencer © Humana Press Inc., Totowa, NJ

rophage-like cell line that expresses phenotypic and functional properties of mature monocytes and that does not need to be stimulated by phorbol esters or 1,25-dihydroxyvitamin D_3 ***(15,19)***. A good correlation between the prevalence of a given *Legionella* species and its intracellular multiplication in MM6 cells could be demonstrated.

In addition to *Legionella*, MM6 cells were found to support the intracellular growth of *Mycobacterium tuberculosis* ***(20)*** and *Chlamydia pneumoniae* ***(21)***, two other important bacterial agents involved in induction of pneumonia. Therefore, the MM6 model might be adaptable to investigations of the molecular pathogenesis of other intracellular bacteria that can replicate within human monocytes and induce disease.

1.2. Intracellular Multiplication of Legionellae Within Acanthamoeba castellanii *as an Environmental Host*

In the environment, legionellae survive by intracellular replication in protozoa. This intracellular replication seems to play a major role in bacterial ecology and pathogenesis. Some intracellular events following infection are common to both amoebae and monocytes. To analyze the intracellular multiplication of different *Legionella* species within a typical environmental host, axenic cultures of *A. castellanii* are preferentially used. We were able to show that only certain *Legionella* species are able to multiply in *A. castellanii*, possibly because of a specialized adaption to the amoebal host ***(19)***.

Apart from legionellae, amoebae in the environment have also been found to harbor bacteria of several other pathogenic genera, including *Burkholderia pickettii* ***(22)***, *Burkholderia cepacia* ***(23)***, and *Burkholderia pseudomallei* ***(24)***, *Chlamydia pneumoniae* ***(25)***, *Listeria* ***(26)***, *Mycobacterium avium* ***(27,28)***, *Pseudomonas aeruginosa* ***(29)***, *Vibrio cholerae* ***(30)***, and *E. coli* ***(31)***. Rickettsiae are also presumed to be endosymbionts of amoebae ***(32)***. As shown for *L. pneumophila*, *M. avium* is more virulent in an mouse model after intracellular growth within *A. castellanii* ***(27)***. Therefore, the *A. castellanii* model to study the behavior of *L. pneumophila* in an environmental host might also be useful to investigate the interaction between other bacterial pathogens and protozoa.

1.3. Enhancement of Bacterial Virulence by Protozoa

The virulence of legionellae for human alveolar macrophages is considered to be a consequence of their evolution as a parasite of amoebae. In 1980, Rowbotham ***(33)*** published the first report on intracellular multiplication of *L. pneumophila* within *Acanthamoeba* spp. and *Naegleria* spp. ***(33)***. Thereafter, several reports described replication of *Legionella* culture isolates from clinical samples within protozoa isolated from the presumed source of infection ***(34–41)***. Intracellular growth within protozoa enhances the ability of *L. pneumophila* to infect human monocytes ***(42)***, induces phenotypic modulation ***(43,44)***, and causes resistance to chemical disinfectants, biocides, and antibiotics ***(45,46)***. Inhalation of legionellae packaged in amoebae results in induction of more severe clinical cases of legionellosis ***(33,47)***.

Original methods appeared in **refs.** ***19*** and ***51***.

This speculation was supported by a recently published mouse model of coinhalation of *L. pneumophila* and *Hartmanella vermiformis*. Coinhalation with *H. vermiformis* significantly enhanced intrapulmonary growth of *L. pneumophila*, resulting in greater mortality in comparison with inhalation of legionellae alone ***(48)***. Intrapulmonary growth of mutant strains of *L. pneumophila* with reduced virulence for *H. vermiformis* but preserved virulence for monocytes was not significantly enhanced by coinhalation ***(49)***. Intrapulmonary growth of *L. pneumophila* was significantly greater in mice inoculated with *L. pneumophila*-infected *H. vermiformis* than in mice inoculated with an equivalent number of bacteria or coinoculated with *L. pneumophila* and uninfected *H. vermiformis* ***(50)***.

The mechanism of intrapulmonary growth enhancement of legionellae by amoebae remains to be determined. Amoeba-induced inhibition of proinflammatory cytokine production could be excluded, since coinhalation as well as inhalation of *L. pneumophila*-infected *H. vermiformis* induced significantly enhanced levels of interferon-γ and tumor necrosis factor-α in A/J mice similar to the levels induced during replicative *L. pneumophila* infection alone ***(48)***. Three possible explanations for the potentiating effect of coinhalation for *Legionella* infection were discussed: modification of the host response to *L. pneumophila* infection by amoebae, the function of amoebae as implanted host cells, or the enhanced virulence of amoeba-associated bacteria ***(48)***.

We established an in vitro coculture model of MM6 cells, *A. castellanii* and *Legionella* species of different intracellular growth rates within human monocytes, to analyze the underlying molecular and biochemical mechanisms ***(51)***. In this coculture model, reactants were separated using a cell culture chamber system that separates both cell types by a microporous polycarbonate membrane impervious to bacteria, amoebae, and human cells. Whereas *L. pneumophila* has shown a maximum 4-log multiplication within MM6 cells, which could not be further increased by coculture with *A. castellanii*, significantly enhanced replication of *L. gormanii*, *L. micdadei*, *L. steigerwaltii*, *L. longbeachae*, and *L. dumoffii* was noted after coculture with amoebae. This effect was seen only with uninfected but not with *Legionella*-infected amoebae. The supporting effect for intracellular multiplication in MM6 cells could be reproduced in part by addition of a cell-free coculture supernatant obtained from a coincubation experiment of uninfected *A. castellanii* and *Legionella*-infected MM6 cells, suggesting that amoeba-derived effector molecules are involved in this phenomenon. This coculture model allows investigations of molecular and biochemical mechanisms that are responsible for the enhancement of intracellular multiplication of legionellae in monocytic cells after interaction with amoebae.

2. Materials

2.1. Culture of Legionella *Bacteria*

1. BCYE-α agar (Oxoid). Store at 4°C; *pay attention to expiration date.*
2. *Legionella* species can be obtained from the American Type Culture Collection (ATCC, Rockville, MD). Process bacteria according to the instructions of ATCC.

2.2. Culture of Host Cells

2.2.1. Mono Mac 6 (MM6) Cells

1. MM6 cells can be obtained from the Department of Human and Animal Cell Cultures (Mascheroder Weg 1b, D-38124 Braunschweig, Germany, Phone: +49-531-2616-161; fax: +49-531-2616-150). For details of culture conditions, see the homepage of Prof. Ziegler-Heitbrock (www.monocytes.de).
2. 75-cm^3 vented culture flasks (Costar).
3. RPMI-1640 medium (Gibco). Store at 4°C.
4. Fetal calf serum (FCS; Myoclone super plus; Gibco). The serum has to be inactivated (30 min, 56°C). After inactivation, every batch of serum must be checked for cytotoxicity (by Trypan Blue exclusion) since some batches may be cytotoxic for MM6 cells. After inactivation and cytotoxicity testing, store serum at –20°C until use. Compatible batches of sera should be ordered in sufficient amounts (e.g., 20 bottles).
5. 2 m*M* L-glutamine (Gibco). Store at –20°C until use. After defrosting, store at 4°C.
6. 1% Nonessential amino acids (Gibco). Store at 4°C.
7. Oxalacetate, pyruvate, and bovine insulin (OPI; Sigma). Store at –20°C until use.
8. Columbia blood agar (Oxoid). Store at 4°C; *pay attention to expiration date.*
9. Mycoplasma plus™ PCR (Stratagene).

2.2.2. *Acanthamoeba castellanii*

1. *A. castellanii* can be obtained from the ATCC (cat. no. 30234).
2. 75-cm^3 Vented culture flasks (Costar).
3. Proteose pepton no. 3 (Difco).
4. Yeast extract (Difco).
5. Glucose (Sigma).
6. $MgSO_4$ (Sigma).
7. $CaCl_2$ (Sigma).
8. Sodium citrate dihydrate (Sigma).
9. $Fe(NH_4)_2(SO_4)_2 6H_2O$ (Sigma).
10. NaH_2PO_3 (Sigma).
11. K_2HPO_3 (Sigma).
12. Columbia blood agar (Oxoid). Store at 4°C; *pay attention to expiration date.*

2.3. Infection of Host Cells

1. 6-Well tissue culture plate (Greiner).
2. 24-Well tissue culture plate (Costar).
3. 2-mL Syringes and 27-gage needles.
4. Sterile Eppendorf tubes (capacity 2 mL).
5. Gentamicin (*pay attention to the different storage conditions according to the manufacturer's instructions*).

2.4. Coculture of MM6 and *A. castellanii*

1. Transwell® insert (Costar).
2. Columbia blood agar (Oxoid). Store at 4°C; *pay attention to expiration date.*
3. Millex®-GS 0.22-µm syringe-driven filter unit (Millipore).

3. Methods

3.1. Culture of Legionella *Bacteria*

Legionellae are fastidious, slow-growing bacteria. They are grown on BCYE-α agar at 35°C in 3% CO_2. After 3–5 d, small, round glistening colonies with a defined edge appear on the agar. The edge usually displays a pink or blue-green iridescence; the center of the colonies is usually grayish, with a characteristic speckled opalescence, the appearance of which resembles cut or ground glass. Later on, the centers of the colonies become creamy white and often lose the ground-glass appearance. At this time, large and small colonies are common and are therefore not evidence for contamination with other bacteria. Since legionellae do not grow on blood agar, this type of agar can be used for contamination control. For contamination of BCYE-α agar with molds (*see* **Note** 1).

3.2. Culture of Host Cells

3.2.1. Mono Mac 6 Cells

MM6 cells are cultured as replicative nonadherent monocytes under lipopolysaccharide-free conditions in 75-cm^3 vented culture flasks in 25 mL of RPMI-1640 medium supplemented with 10% fetal calf serum (Myoclone super plus), 2 m*M* L-glutamine, 1% nonessential amino acids, and OPI (containing 150 μg oxalacetate, 50 μg pyruvate, and 8.2 μg of bovine insulin per mL) at 37°C in 5% CO_2 (MM6 medium) and are diluted 1:3 twice a week in fresh medium ***(15)***. We recommend checking the cell culture regularly for sterility using Columbia blood agar and for *Mycoplasma* contamination using Mycoplasma plus™ polymerase chain reaction (PCR) according to the manufacturer's instructions (*see* **Notes 2** and **3**).

3.2.2. Acanthamoeba castellanii

A. castellanii are grown in 75-cm^3 vented culture flasks in 25 mL PYE broth (2% protease pepton no. 3, 0.1% yeast extract, 0.1 *M* glucose, 4 m*M* $MgSO_4$, 0.4 *M* $CaCl_2$, 0.1% sodium citrate dihydrate, 0.05 m*M* $Fe[NH_4]_2[SO_4]_2$, 2.5 m*M* NaH_2PO_3, and 2.5 m*M* K_2HPO_3, pH 6.5) at 35°C in 5% CO_2 ***(52)*** and are diluted 1:2 twice a week. Sterility testing should be performed as described for MM6 cells.

3.3. Infection of Host Cells

3.3.1. Mono Mac 6 Cells

1. Infection and incubation of MM6 cells after infection is performed in MM6 medium without FCS to exclude phagocytosis of bacteria via $F_c\gamma$ receptors.
2. Nonadherent MM6 cells are harvested by centrifugation at 400*g* for 10 min.
3. The pellet is washed twice in MM6 medium without FCS.
4. Legionellae are harvested from BCYE-α agar, suspended in MM6 medium without FCS, and adjusted to an OD_{578nm} of 0.2 (by a spectrophotometer) corresponding to a concentration of approx 3×10^8 legionellae/mL. For this concentration, bacteria can be centrifuged at 3000*g* for 10 min.
5. 2×10^7 MM6 cells are pelleted and resuspended with legionellae in a volume of 1.5 mL in a well of a 6-well tissue culture plate.

6. The bacteria-to-cell ratio can be selected from 100:1 to 0.1:1. It should be taken into account that legionellae are cytotoxic for monocytes at higher multiplicities of infection *(53)*. Therefore, lower bacteria-to-cell ratios are recommended.
7. Cocultures of bacteria and monocytes are then incubated at 35°C in 5% CO_2 for 2 h.
8. After this period, nonphagocytized bacteria are killed by addition of 4.5 mL MM6 medium without FCS containing 100 μg/mL gentamicin for 1 h at 35°C in 5% CO_2.
9. After three washes by centrifugation at 400*g* for 10 min, the cells are resuspended in 10 mL MM6 medium without FCS and distributed in 1-mL aliquots into the wells of a 24-well tissue culture plate, giving a concentration of 2×10^6 infected MM6 cells per well. This time point is defined as 0.
10. The cells are then incubated for an additional 72 h at 35°C in 5% CO_2.
11. Every 24 h, the contents of two wells are aspirated and pelleted by centrifugation at 400*g* for 10 min.
12. The supernatant is transferred into sterile tubes of approx 2 mL capacity.
13. 1 mL of sterile distilled water is added to the pellet, and final disruption of the cells is performed by aspirating the suspension through a 27-gage needle.
14. The supernatant and lysis fluid are pooled, and serial 10-fold dilutions are made.
15. 100 μL of each dilution is inoculated onto BCYE-α agar to determine the number of viable legionellae after multiplication in MM6 cells.
16. Colonies on the agar are counted on d 5 after incubation at 35°C in 5% CO_2.
17. For control, 1 mL of the original *Legionella* suspension in MM6 medium without FCS and without cells is incubated in a well of the tissue culture plate, and serial 10-fold dilutions are made at the same time points as indicated above.
18. Viability of MM6 cells is determined by Trypan Blue exclusion at the same time points as indicated above.
19. We recommend testing for contamination of the infected cell culture by subculture onto Columbia blood agar plates.

3.3.2. Acanthamoeba castellanii

Coculture of *A. castellanii* and bacteria is performed in an identical manner except that legionellae are suspended in PYE broth without glucose and the culture is incubated for 96 h.

3.4. Coculture of MM6 Cells and A. castellanii

1. Infected MM6 cells (*see* **Subheading 3.3.**) are resuspended in a mixture of 50% MM6 medium without FCS and 50% PYE broth without glucose. NaCl must be added to yield an NaCl concentration of 6.5 g/L (coculture medium) identical to the MM6 medium (*see* **Note 4**). This mixture was found to support growth of MM6 cells as well as *A. castellanii* in a manner comparable to the original cell culture media for these cells described in **Subheading 3.2**.
2. Infected MM6 cells are distributed in 1 mL aliquots into the wells of a 24-well tissue culture plate, giving a concentration of 1×10^6 infected MM6 cells per well.
3. 1×10^6 *A. castellanii* organisms are resuspended in coculture medium and are added using a Transwell® insert, which separates both cell types by a microporous polycarbonate membrane with a pore size of 0.1 μm, impervious to bacteria, amoebae, and MM6 cells.
4. Absence of legionellae from the upper chamber (*A. castellanii*) is determined by means of culture on BCYE-α agar, and sterility of the coculture is checked by culture on Columbia blood agar.

5. The coculture is incubated for 72 h.
6. Intracellular replication of bacteria within MM6 cells is detected as described above in **Subheading 3.3**.
7. Multiplication within MM6 cells in coculture with *A. castellanii* is compared with multiplication within MM6 cells in coculture medium without amoebae.
8. To demonstrate the presence of amoeba-derived effector molecule(s) in the coculture, a cell-free Transwell® supernatant is obtained by harvesting the content of a Transwell® insert (i.e., noninfected amoebae in coculture with infected MM6 cells, with both host cells separated by a microporous membrane), subsequent centrifugation (400*g* for 10 min), and filtration through a Millex®-GS 0.22-μm syringe-driven filter unit to obtain sterility.
9. 1×10^6 infected MM6 cells per well are then coincubated for 72 h with this cell-free supernatant. Intracellular multiplication of legionellae in MM6 cells is determined as described above (*see* **Subheading 3.3.**).
10. Multiplication within MM6 cells in coculture with the supernatant is compared with multiplication within MM6 cells in coculture medium.

4. Notes

1. BCYE-α agar is an ideal medium for growth of molds, especially *Aspergillus fumigatus*. Therefore, laboratories and equipment (especially incubators) should be free of molds.
2. Contamination of the cell culture is the main source of failure in using the in vitro models presented above. Common contaminants are fast-growing bacteria such as staphylococci and streptococci, especially when people having short-term or inadequate experience with cell cultures are working in the laboratory. Good staff training can prevent most contamination events. We recommend checking the cell cultures routinely (at least twice a week before dilution) as well as the in vitro infection experiments (parallel inoculation of lysis fluid onto BCYE-α agar and Columbia blood agar).

   A serious and often undetected problem is the contamination of cell cultures with *Mycoplasma* organisms. The origins are bovine serum, laboratory personnel, and mycoplasma-infected cell cultures; the latter spread the infection to other cell cultures within the same laboratory. It is important to keep in mind that *Mycoplasma* organisms, particularly in continuous cell lines such as MM6 cells or *A. castellanii*, grow slowly and do not destroy the host cells. However, the influence of *Mycoplasma* organisms on various parameters such as changes in metabolism, growth, viability, and morphology of the cells should not be underestimated. *Mycoplasma* organisms can seriously interfere with experiments. We therefore recommend checking the cell cultures at least quarterly for *Mycoplasma* contamination. A variety of techniques are available to detect *Mycoplasma* in cell cultures. The most reliable are enzyme immunoassays and PCR. We can recommend the Mycoplasma plus™ PCR (Stratagene). In case of *Mycoplasma* contamination, cell cultures should be discarded and the incubator should be disinfected. A new batch of cells should be cultured and used for in vitro infection. If no new batch is available, cell cultures contaminated with *Mycoplasma* can be treated with BM-Cyclin (Roche), an antibiotic combination of a pleuromutilin and a tetracycline derivative, without marked cytotoxic side effects. This treatment requires about 2–3 wk. In vitro infections should not be performed during this time since legionellae are sensitive to these antibiotics.
3. Intracellular multiplication of legionellae is a function of intact host cells. Monocytes or amoebae maintained under inadequate conditions may not support the intracellular growth of bacteria, although crude tests for viability (such as Trypan Blue exclusion) do not

indicate cell death. Therefore, cells should be diluted twice a week in fresh medium as described above. During washing procedures, cells should not be centrifuged faster than 400*g* to avoid disruption.

4. As described in **Subheading 3.**, the cell culture medium for coculture of MM6 cells and *A. castellanii* resembles a mixture of 50% MM6 medium without FCS and 50% PYE broth without glucose. This mixture is hypotonic, and it is necessary to add NaCl to achieve a salt concentration of 6.5 g/L identical to MM6 medium since hypotonic medium induces intracellular growth of non-replicative *Legionella* species owing to swelling of the host cells ***(51)***.

References

1. Brenner, D. J. (1987) Classification of the legionellae. *Semin. Respir. Infect.* **2,** 190–205.
2. Davis, G. S. and Winn, W. C. (1987) Legionnaire's disease: respiratory infection caused by *Legionella* bacteria. *Clin. Chest Med.* **8,** 419–439.
3. Fang, G. D., Yu, V. L., and Vickers, R. M. (1989) Disease due to the Legionellaceae (other than *Legionella pneumophila*). *Medicine* **68,** 116–132.
4. Nguyen, H. M., Stout, J. E., and Yu, V. L. (1991) Legionellosis. *Infect. Dis. Clin. North Am.* **5,** 561–584.
5. Hart, C. A. and Makin, T. (1991): Legionella in hospitals: a review. *J. Hosp. Infect.* **18(Suppl A),** 481–489.
6. Winn, W. C. (1988) Legionnaires disease: historical perspective. *Clin. Microbiol. Rev.* **1,** 60–81.
7. Horwitz, M. A. (1980) The Legionnaires' disease bacterium (*Legionella pneumophila*) multiplies intracellularly in human monocytes. *J. Clin. Invest.* **66,** 441–450.
8. Ciancotto, N. P., Eisenstein, B. I., Mody, C. H., Toews, G. B., and Engleberg, N. C. (1989) A *Legionella pneumophila* gene encoding a species-specific surface protein potentiates initiation of intracellular infection. *Infect. Immun.* **57,** 1255–1262.
9. Dowling, J. N., Saha, A. K., and Glew, R. H. (1992) Virulence factors of the family Legionellaceae. *Microbiol. Rev.* **56,** 32–60.
10. Horwitz, M. A. (1987) Characterization of avirulent mutant *Legionella pneumophila* that survive but do not multiply within human monocytes. *J. Exp. Med.* **166,** 1310–1328.
11. Pearlman, E., Jiwa, A. H., Engleberg, N. C., and Eisenstein, B. I. (1988) Growth of *Legionella pneumophila* in a human macrophage-like (U 937) cell line. *Microb. Pathogenesis* **5,** 87–95.
12. Yamamoto, Y., Klein, T. W., Newton, C. A., Widen, R., and Friedman, H. (1988) Growth of *Legionella pneumophila* in thioglycolate-elicited peritoneal macrophages from A/J mice. *Infect. Immun.* **56,** 70–375.
13. Yoshida, S. and Mizuguchi, Y (1986) Multiplication of *Legionella pneumophila* Philadelphia 1 in cultured peritoneal macrophages and its correlation to susceptibility of animals. *Can. J. Microbiol.* **32,** 438–442.
14. Swanson, M. S. and Hammer B. K. (2000) *Legionella pneumophila* pathogenesis: a fateful journey from amoebae to macrophages. *Annu. Rev. Microbiol.* **54,** 567–613.
15. Ziegler-Heitbrock, H. W. L., Thiel, E., Ftterer, A., Herzof, V., Wirtz, A., and Riethmller, G. (1988) Establishment of a human cell line (Mono Mac 6) with characteristics of mature monocytes. *Int. J. Cancer* **41,** 456–461.
16. Marra, A., Horwitz, M. A., and Shuman, H. A. (1990) The HL-60 model for the interaction of human macrophages with the Legionnaires' disease bacterium. *J. Immunol.* **144,** 2738–2744.

17. Pearlman, E., Jiwa, A. H., Engleberg, N. C., and Eisenstein, B. I. (1988) Growth of *Legionella pneumophila* in a human macrophage-like (U 937) cell line. *Microb. Pathogen.* **5,** 87–95.
18. Watanabe, M., Shimamoto, Y., Yoshida, S., et al. (1993) Intracellular multiplication of *Legionella pneumophila* in HL-60 cells differentiated by 1,25-dihydroxyvitamin D_3 and the effect of interferon γ. *J. Leukocyte Biol.* **54,** 40–46.
19. Neumeister, B., Schöiger, S., Faigle, M., Eichner, M., and Dietz, K. (1997) Multiplication of different *Legionella* species in Mono Mac 6 cells and in *Acanthamoeba castellani. Appl. Environ. Microbiol.* **63,** 1219–1224.
20. Barrow, E. L. W., Winchester, G. A., Staas, J. K., Quenelle, D. C., and Barrow, W. W. (1998) Use of microsphere technology for targeted delivery of rifampin to *Mycobacterium tuberculosis*-infected macrophages. *Antimicrob. Agents Chemother.* **42,** 2682–22689,
21. Heinemann, M., Susa, M., Simnacher, U., Marre, R., and Essig, A. (1996) Growth of *Chlamydia pneumoniae* induces cytokine production and expression of CD14 in a human monocytic cell line. *Infect. Immun.* **64,** 4872–4875.
22. Michel, R. and Haurder, B. (1997) Isolation of an *Acanthamoeba* strain with intracellular *Burkholderia pickettii* infection. *Zentralbl. Bakt.* **285,** 541–557.
23. Landers, P., Kerr, K. G., Rowbotham, T. J., et al. (2000) Survival and growth of *Burkholderia cepacia* within the free-living amoeba *Acanthamoeba polyphaga. Eur. J. Clin. Microbiol. Infect. Dis.* **19,** 121–123.
24. Inglis, T. J. J., Rigby, P., Robertson, T. A., Dutton, N. S., Henderson, M., and Chang, B. J. (2000) Interaction between *Burkholderia pseudomallei* and *Acanthamoeba* species results in coiling phagocytosis, endamebic bacterial survival, and escape. *Infect. Immun.* **68,** 1681–1686.
25. Essig, A., Heinemann, M., Simnacher, U., and Marre, R. (1997) Infection of *Acanthamoeba castellanii* by *Chlamydia pneumoniae. Appl. Environ. Microbiol.* **63,** 1396–1399.
26. Ly, T. M. C. and Mller, H. E. (1990) Ingested *Listeria monocytogenes* survive and multiply in protozoa. *J. Med. Microbiol.* **33,** 51–54.
27. Cirillo, J. D., Falkow, S., Tompkins, L. S., and Bermudez, L. E. (1997) Interaction of *Mycobacterium avium* with environmental amoebae enhances virulence. *Infect. Immun.* **65,** 3759–3767.
28. Steinert, M., Birkness, K., White, E., Fields, B., and Quinn, F. (1998) *Mycobacterium avium* bacilli grow saprozoically in coculture with *Acanthamoeba polyphaga* and survive within cyst walls. *Appl. Environ. Microbiol.* **64,** 2256–2261.
29. Michel, R., Burghardt, H., and Bergmann, H. (1995) *Acanthamoeba*, naturally intracellularly infected with *Pseudomonas aeruginosa*, after their isolation from a microbiologically contaminated drinking water system in a hospital. *Zentralbl. Hyg. Umweltmed.* **196,** 532–544.
30. Thom, S., Warhurst, D., and Drasar, B. S. (1992) Association of *Vibrio cholerae* with fresh water amoebae. *J. Med. Microbiol.* **36,** 303–306.
31. Barker, J., Humphrey, T. J., and Brown, M. W. (1999) Survival of *Escherichia coli* 0157 in a soil protozoan: implications for disease. *FEMS Microbiol. Lett.* **173,** 291–295.
32. Fritsche, T. R., Horn, M., Seyedirashti, S., Gautom, R. K., Schleifer, K. H., and Wagner, M. (1999) In situ detection of novel bacterial endosymbionts of *Acanthamoeba* spp. phylogenetically related to members of the order Rickettsiales. *Appl. Environ. Microbiol.* **65,** 206–212.
33. Rowbotham, T.J. (1980) Preliminary report on the pathogenicity of *Legionella pneumophila* for freshwater and soil amoebae. *J. Clin. Pathol.* **33,** 1179–1183.

34. Barbaree, J. M., Fields, B. S., Feeley, J. C., Gorman, G. W., and Martin, W. T. (1986) Isolation of protozoa from water associated with a legionellosis outbreak and demonstration of intracellular multiplication of *Legionella pneumophila*. *Appl. Environ. Microbiol.* **51,** 422–424.
35. Breiman, R.F., Fields, B.S., Sanden, G.N., Volmer, L.J., Meier, A., and Spika, J.S. (1990) Association of shower use with Legionnaires' disease. *JAMA* **263,** 2924–2926.
36. Fields, B. S., Sanden, G. N., Barbaree, J. M., et al. (1989) Intracellular multiplication of *Legionella pneumophila* in amoebae isolated from hospital hot water tanks. *Curr. Microbiol.* **16,** 131–137
37. Fields, B. S., Nerad, T. A., Sawyer, T. K., et al. (1990) Characterization of an axenic strain of *Hartmannella vermiformis* obtained from an investigation of nosocomial legionellosis. *J. Protozool.* **37,** 581–583.
38. Nahapetian, K., Challemel, O., Beurtin, D., Dubrou, S., Gounon, P., and Squinazi, F. (1991) The intracellular multiplication of *Legionella pneumophila* in protozoa from hospital plumbing systems. *Res. Microbiol.* **142,** 677–685.
39. Rowbotham, T. J. (1983) Isolation of *Legionella pneumophila* from clinical specimens via amoebae, and the interaction of those and other isolates with amoebae. *J. Clin. Pathol.* **36,** 978–986.
40. Sanden, G. N., Morrill, W. E., Fields, B. S., Breiman, R. F., and Barbaree, J. M. (1992) Incubation of water samples containing amoebae improves detection of legionellae by the culture method. *Appl. Environ. Microbiol.* **58,** 2001–2004.
41. Wadowsky, R. M., Butler, L. J., Cook, M. K., et al. (1988) Growth supporting activity for *Legionella pneumophila* in tap water cultures and implication of hartmannellid amoebae as growth factors. *Appl. Environ. Microbiol.* **54,** 2677–2682.
42. Cirillo, J. D., Falkow, S., and Tompkins, L. S. (1994) Growth of *Legionella pneumophila* in *Acanthamoeba castellanii* enhances invasion. *Infect. Immun.* **62,** 3254–3261.
43. Abu-Kwaik, Y., Fields, B. S., and Engleberg, N. C. (1994) Protein expression by the protozoan *Hartmanella vermiformis* upon contact with its bacterial parasite *Legionella pneumophila*. *Infect. Immun.* **62,** 1860–1866.
44. Barker, J., Lambert, P. A., and Brown, M. R. (1993) Influence of intra-amoebic and other growth conditions on the surface properties of *Legionella pneumophila*. *Infect. Immun.* **61,** 3503–3510.
45. Barker, J., Brown, M. R., Collier, P. J., Farrell, I., and Gilbert, P. (1992) Relationship between *Legionella pneumophila* and *Acanthamoeba polyphaga*: physiological status and susceptibility to chemical inactivation. *Appl. Environ. Microbiol.* **58,** 2420–2425.
46. Barker, J., Scaife, H., and Brown, R. W. (1995) Intraphagocytic growth induces an antibiotic resistant phenotype of *Legionella pneumophila*. *Antimicrob. Agents Chemother.* **39,** 2684–2688.
47. O'Brien, S. J. and Bhopal, R. S. (1993) Legionnaires' disease: the infective dose paradox. *Lancet* **342,** 5–6.
48. Brieland, J., McClain, M., Heath, L., et al. (1996) Coinoculation with *Hartmanella vermiformis* enhances replicative *Legionella pneumophila* infection in a murine model of Legionnaires' disease. *Infect. Immun.* **64,** 2449–2456.
49. Brieland, J., McClain, M., Legendre, M., and Engleberg, C. (1997) Intrapulmonary *Hartmanella vermiformis*: a potential niche for *Legionella pneumophila* replication in a murine model of legionellosis. *Infect. Immun.* **65,** 4892–4896.

50. Brieland, J. K., Fantone, J. C., Remick, D. G., Legendre, M., McClain, M., and Engleberg, C. (1997) The role of *Legionella pneumophila*-infected *Hartmanella vermiformis* as an infectious particle in a murine model of Legionnaires' disease. *Infect. Immun.* **65,** 5330–5333.
51. Neumeister, B., Reiff, G., Faigle, M., Dietz, K., and Lang, F. (2000) Influence of *Acanthamoeba castellanii* on intracellular growth of different *Legionella* species in human monocytes: establishment of an in-vitro coculture system. *Appl. Environ. Microbiol.* **66,** 914–919.
52. Moffat, J. F. and Tompkins, L. S. (1992) A quantitative model of intracellular growth of *Legionella pneumophila* in *Acanthamoeba castellanii. Infect. Immun.* **60,** 296–301.
53. Gao, L. Y. and Abu-Kwaik, Y. (1999) Apoptosis in macrophages and alveolar epithelial cells during early stages of infection by *Legionella pneumophila* and its role in cytopathogenicity. *Infect. Immun. 67,* 862–870.

14

Viability of Amoebae, Fungal Conidia, and Yeasts

Rapid Assessment by Flow Cytometry

Judith A. Noble-Wang, Shangtong Zhang, Daniel Price, and Donald G. Ahearn

1. Introduction

Conventional methods for the evaluation of antimicrobials and disinfecting solutions with microorganisms involve culture-based techniques, which are time-consuming and underestimate the number of viable organisms. Rapid detection and viability measurements of microorganisms in homogenous and heterogenous microbial populations have been greatly enhanced by recent advances in the use of fluorescent stains in flow cytometry (FCM) ***(1-5)***. FCM has been applied to enumerate, differentiate, and identify microorganisms, determine protein and DNA content of cells, analyze the physiological state of individual cells, and analyze the interaction of drugs, antibiotics, and antimicrobials with microbial cells ***(1–14)***. Four physiological states of cells can be distinguished by FCM: (1) reproductively viable, (2) metabolically active, (3) intact, and (4) permeabilized ***(15)***.

FCM permits a rapid and quantitative measurement of the optical characteristics of cells as they pass through, in a single file, a focused beam of light ***(4)***. As cells are carried within a fast-flowing fluid stream and through the focus of exciting light, three parameters are measured: forward angle light scatter, side angle light scatter, and fluorescence emitted by dyes that have specific interaction with intracellular components of individual cells. FCM data that are presented in histogram and dot plots can be generated to give information on a variety of properties of interest among cells in the population as a whole.

FCM offers major advantages in multiparameter data acquisition and multivariate data analysis, high-speed analysis, and cell-sorting capabilities. Disadvantages may be associated with the cost, which is usually over $100,000 (US) for a typical laser-

From: *Methods in Molecular Biology, vol. 268: Public Health Microbiology: Methods and Protocols*
Edited by: J. F. T. Spencer and A. L. Ragout de Spencer © Humana Press Inc., Totowa, NJ

based flow cytometer with just analyzing capabilities. Another disadvantage is that skilled personnel are usually required to operate these complex instruments so as to get optimum performance *(3)*. A schematic overview of flow cytometry is presented in **Fig. 1.**

FCM analyses with viability dyes, such as propidium idodide and oxonols, can assess cell viability after treatment with antimicrobial and disinfecting solutions. A breakdown in membrane integrity and a loss of transmembrane potential are indicators of cell death. Propidium iodide (absorbance: 535 nm; emission: 617 nm) has been widely used to indicate membrane integrity *(2,4,7,14)*. Propidium iodide stains nucleic acids but will not diffuse appreciably into intact cells; therefore, this dye preferentially stains dead cells that have permeable membranes *(4)*. Changes in membrane potential can be detected with some oxonol dyes such as bis-(1,3-dibutylbarbituric acid) trimethine oxonol (Ox). Oxonols are lipophilic anionic dyes, and their accumulation within cells is favored by a reduction in the magnitude of membrane potential that enables dye molecules to concentrate within the cell by association with intracellular macromolecules *(5)*.

2. Materials

2.1. Flow Cytometry Instrumentation and Analysis Software

1. FACSCalibur fluorescence-activated cell sorter system (Becton Dickinson, Heidelberg, Germany). Setting of analysis gates, data handling, with CELLQuest software (Becton Dickinson).
2. Illumination: 15-mW, 488-nm argon-ion laser.
3. FL1-H detector: 530/30 nm bandpass filter.
4. FL2-H detector: 585/42 nm bandpass filter.

2.2. Culture Medium

2.2.1. Amoeba: Proteose-Yeast Extract-Glucose Broth (PYG)

1. 20 g Proteose peptone (Beckon Dickinson, Franklin Lakes, NJ).
2. Yeast extract (Difco, Detroit, MI).
3. 900 mL Deionized H_2O.
4. Adjust pH to 6.5 with 1 *M* HCl and autoclave for 15 min at 121.5°C.
5. After autoclaving, add 50 mL of filter-sterilized 2 *M* glucose.
6. Add autoclaved and filter-sterilized stock salt solutions of the following: 10 mL 0.4 *M* $MgSO_47H\ _2O$, 8.0 mL 0.05 *M* $CaCl_2$, 10 mL sodium citrate ($Na_3C_6H_5O_72H\ _2O$), 10 mL 0.25 *M* $Na_2HPO_47H\ _2O$, 10 mL 0.25 *M* KH_2PO_4.

2.2.2. Fungi

1. Potato dextrose agar (PDA; Difco).
2. Sabouraud's dextrose agar (SAB).

2.2.3. Yeast

1. Bacto yeast nitrogen base (Difco) supplemented with 0.5% glucose (DYNB). Adjust pH to 5.5 with 1 *M* HCl.
2. Bacto Sabouraud's dextrose agar (SAB; Difco).

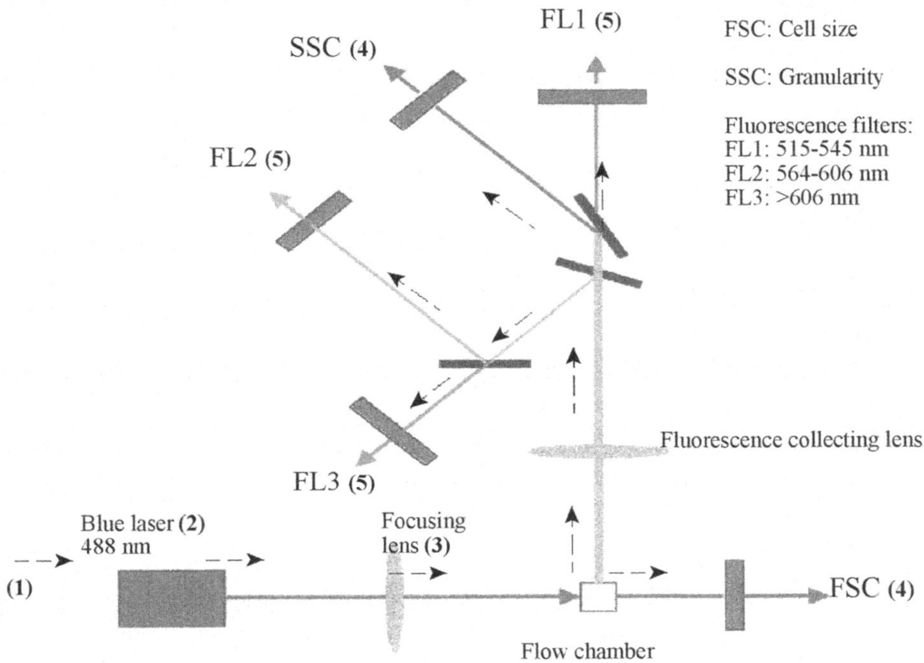

Fig. 1. The principle of the FACSCalibur™ Optical System ***(16)***. (1) A sample consisting of particles in suspension is injected into a nozzle, and the fluid containing the particles is hydrodynamically compressed to a thin stream, forcing the particles to flow in a single file. (2) The stream of cells intercepts the beam of the laser and absorbs the laser light. (3) As particles flow through the beam, light is scattered; scattered light is either deflected unmodified laser light or fluorescent light emitted from either a fluorochrome indigenous to the particle or one that has been attached to it. (4) Unmodified scattered laser light is measured at a low angle (FSC, forward angle light scatter) given a measurement of particle size. Scattered light can also be measured as SSC (side angle light scatter), which gives a measurement of granularity or cytoplasmic complexity. (5) The emitted light is captured by lenses and filters to measure selectively emitted light of particular wavelengths in a given detector. (6) Pulse signals sent by the detectors are converted into voltages and stored by computer software. Software integrates information from each detector for each particle.

2.3. Solutions

1. Phosphate-buffered saline (PBS; autoclaved and filter sterilized): 8 g NaCl, 0.2 g KCl, 1.44 g Na_2HPO_4, 0.24 g KH_2PO_4, 1000 mL deionized H_2O. Adjust pH to 7.4 with 1 *M* NaOH.
2. Spore dispersant: 0.9% NaCl containing 0.1% polysorbate-80 (NaCl-Tween-80) and 0.9% NaCl.
3. Dissolve heavy metals such as $AgNO_3$ and $HgCl_2$ (ACS reagent grade; Sigma, St. Louis, MO) in deionized distilled water (ddH_2O) to make 100 m*M* stock solutions.
4. Morpholine-ethanesulfonic acid buffer (10 m*M* MES; adjust pH to 5.5 with 1 *M* HCl).
5. Other antimicrobials or contact lens solution.

2.4. Viability Stains

1. 0.3% Methylene blue.
2. Propidium iodine (PI): 1 mg/mL in PBS (Molecular Probes, Eugene, OR) (*see* **Note 1**).

2.5. Supplies

1. Sterile pipets.
2. Micropipet tips.
3. 2.0-mL Cryogenic vials (Nalgene® Cryoware™).
4. Microcentrifuge tubes (1.7 mL capacity).
5. Polypropylene centrifuge tubes (15 and 50 mL capacity).

3. Methods

3.1. Evaluation of Trophozoites and Cysts of Acanthamoeba spp. Exposed to Antimicrobials (see ref. 2 and Note 2)

The following protocols describe FCM procedures for evaluation of amoebae, molds, and yeasts after exposure to antimicrobials. For each organism, a heat-killed suspension of cells is mixed with a viable cell population for establishment of FCM parameters (**Fig. 2**).

3.1.1. Culture Maintenance and Growth (see Note 3)

Cultures are maintained in the manner recommended by the curator of the appropriate culture collection. Maintain axenic strains of *A. castellanii* (ATCC 30234) and *A. polyphaga* (ATCC 30461) in 10 mL PYG medium at 25°C in 75-cm^2 tissue culture flasks or in 125 mL of PYG in 250-mL flasks.

3.1.2. Preparation of Acanthamoeba Inoculum

1. Preparation of trophozoite inoculum.
   a. Transfer amoebae (5 mL) to 125 mL of PYG medium in 250-mL flasks for 48 h.
   b. Incubate the amoebae at room temperature on a rotary shaker set at 125 rpm.
   c. Transfer approx 10–20 mL of 48-h trophozoites to 125 mL of PYG medium in 250-mL flasks for 24 h and incubate at room temperature on a rotary shaker set at 125 rpm.
   d. Harvest trophozoites in the exponential growth phase with centrifugation at 750*g* for 10 min and wash with PBS.
   e. Add 1–2 mL of PBS to pellet.
   f. Adjust to a concentration of 10^8 viable trophozoites/mL as determined from cell counts performed in a hemocytometer counting chamber.

Fig. 2. (*opposite page*) Determination of FCM analysis parameters for viability assessment with PI. PI viability stain absorbs laser light at 535 nm and emits fluorescent light at 617 nm; therefore, PI-stained cells demonstrate more FL2 than FL1 characteristics via the FL2 detector of the flow cytometer. FCM analysis parameters are selected with unstained viable and stained heat-killed populations of cells. The voltages are adjusted on the control panel for the FL1 and FL2 detectors so that viable populations produce histograms with a lower mean FL2 fluorescence (**A**) than heat-killed populations (**B**). Viable cells exclude PI and produce low fluorescence signals from the FL1 and FL2 detectors. Heat-killed populations, which absorb the stain and fluoresce red, produce a low fluorescent signal from the FL1 detector and a high fluorescent signal from the FL2 detector. A stained mixture of heat-killed and viable cells is applied to adjust the fluorescent compensation from the FL1 and FL2 detectors (**C**)

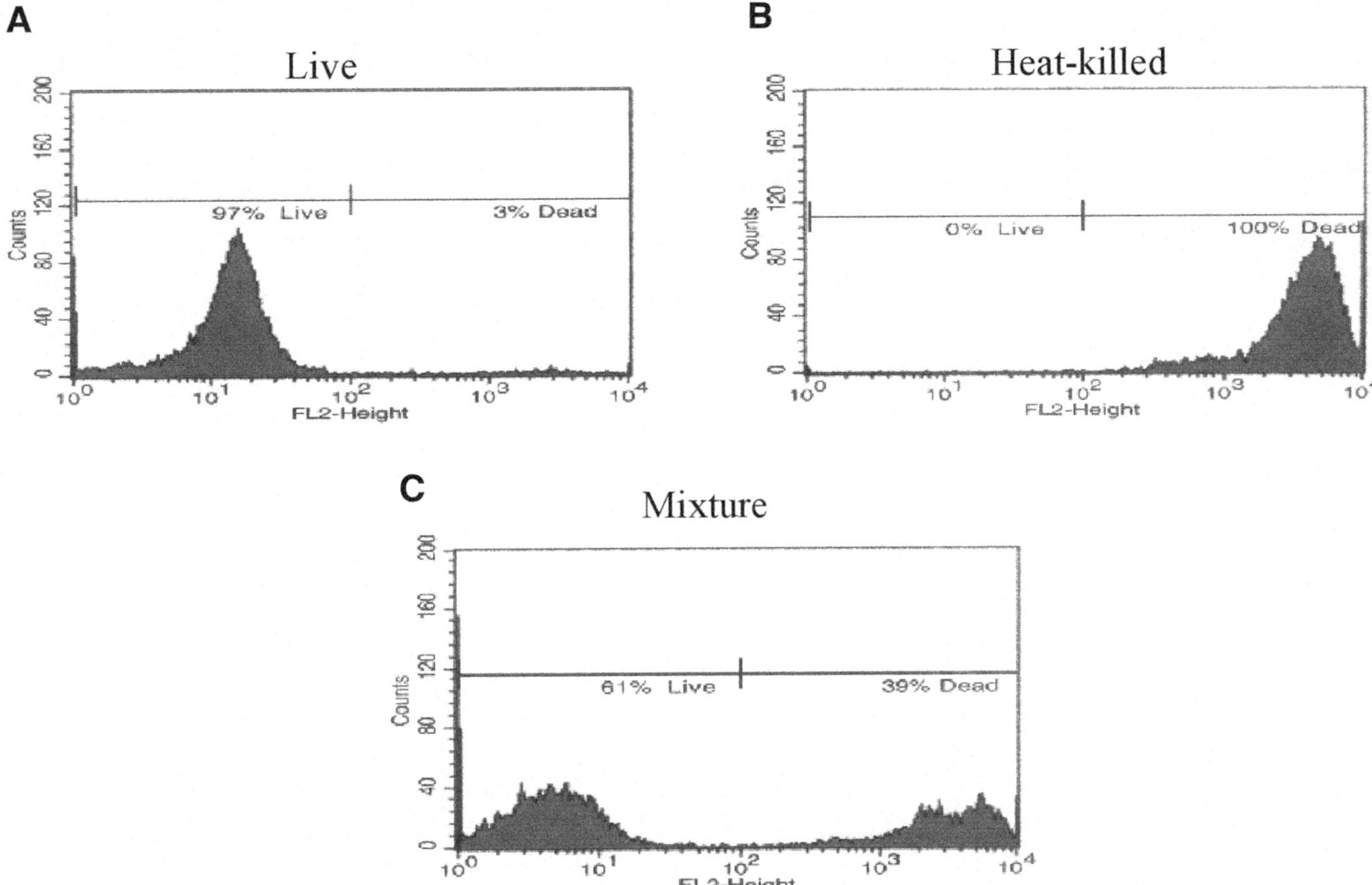
A
Live
97% Live
3% Dead
Counts
FL2-Height
B
Heat-killed
0% Live
100% Dead
Counts
FL2-Height
C
Mixture
61% Live
39% Dead
Counts
FL2-Height

2. Preparation of cyst inoculum.
   a. Prepare the cultures incubated in PYG for 3–6 wk at 25°C.
   b. Harvest cysts with centrifugation at 750*g* for 10 min and wash with PBS.
   c. Adjust suspension to a concentration of 10^8 viable cysts/mL as determined from cell counts performed in a hemocytometer counting chamber .
3. Stain 100 µL of amoeba suspension with 200 µL 0.3% methylene blue to determine viability. Nonviable amoebae stain blue, and viable amoebae exclude the stain.

3.1.3 Disinfection Test Procedure

1. Product samples to be tested should be representative of the product to be marketed. Aliquots should be taken directly from the final product container immediately prior to testing.
2. Prepare three to five cryogenic vials containing 990 µL of each disinfection solution.
3. Inoculate the sample tube of the product to be tested with 10 µL of amoeba suspension containing 10^8 organisms/mL to provide a final count of 1.0×10^6 viable amoebae/mL. The volume of inoculum should not exceed 1% of the sample volume. Ensure dispersion of the inoculum by adequate mixing with a single aspiration step from a 1-mL pipetor.
4. Inoculate cells into PBS controls to normalize the data obtained with the contact lens solutions.
5. Compare the amoeba suspensions in contact lens solutions for each test run with an inoculum suspended in PBS.
6. Store inoculated test solutions at room temperature for the manufacturer's specified disinfection time.

3.1.4. Viability Determination After Disinfection

1. Transfer suspensions of 1 mL containing treated trophozoites or cysts from the cryovials to 1.5-mL microcentrifuge tubes.
2. Harvest treated amoebae with centrifugation at 1000*g* for 3 min and suspend the pellets in 0.5 mL PBS.
3. Stain amoebae with 12.5 µL PI (1 mg/mL) for 1 min.
4. Determine percent viability based on the number of cells analyzed by flow cytometry (at least 10,000 amoebae). Determine the relative percent reduction by comparison with untreated amoeba suspensions in PBS. Analysis of stained cells should be performed within 30 min.

3.2. Evaluation of Fungal Conidia Exposed to Preservatives (see ref. 13)

3.2.1. Culture Maintenance and Growth

1. Maintain representative isolates of selected fungi (e.g., *Fusarium subglutinans*) on SAB slants.
2. Prepare working cultures on SAB or PDA plates.
3. Incubate cultures at 20–25°C for 7–14 d (time based on conidiogenesis).

3.2.2. Preparation of Fungal Conidia Inoculum

1. Harvest conidia from 7–14-d cultures with 0.9% NaCl containing 0.1% polysorbate-80 (NaCl-Tween-80).
2. Pour 10 mL of (NaCl-Tween-80) on the surface of fungal growth. Gently scrape the surface of fungal growth with a flame-sterilized loop to loosen conidia from hyphal structures.

3. Filter suspensions of conidia through glass wool. (This step removes most hyphal fragments.)
4. Harvest conidia with centrifugation and suspend pelleted conidia in 0.9% NaCl.
5. Adjust the suspensions for each species to give concentrations of about 10^7–10^8 conidia/mL.

3.2.3. Disinfection Test Procedure

1. Product samples to be tested should be representative of the product to be marketed.
2. Aliquots should be taken directly from the final product container immediately prior to testing.
3. Inoculate conidia of the selected fungi into full-strength preservative formulations and incubate at 20–25°C for 96 h. The preservative formulations should contain concentrations of conidia of 10^5–10^6/mL.

3.2.4. Viability Determination After Disinfection

1. Harvest the conidia from the preservative formulations with centrifugation and wash the pelleted conidia twice in 0.9% NaCl.
2. Stain conidia suspensions (1.0 mL) with 25 µL PI for 30 s prior to FCM analysis.

3.3. Evaluation of Yeasts Exposed to Heavy Metals (see *ref. 14*)

3.3.1. Culture Maintenance and Growth

1. Maintain stock cultures of *Candida albicans* or *Candida maltosa* (R-42) on SAB slants by transferring to fresh slants every 3–4 wk.
2. Grow working cultures in DYNB with agitation at 22°C for 18 h.

3.3.2. Preparation of Yeast Inoculum

1. Harvest cells from DYNB with centrifugation.
2. Wash cells twice with ddH_2O via centrifugation and suspend pellet in MES buffer to an optical density at 600 nm of 0.6 in a 1.0-cm light path (equivalent to 2.33×10^6 cells/mL of *C. maltosa* or 1.20×10^7 cells/mL of *C. albicans*).

3.3.3. Exposure of Cells to Heavy Metals

1. Add 20 µL of various concentrations of heavy metals (Ag[I] or Hg[II]) to 0.98 mL of cell suspension in microcentrifuge tubes and incubate tubes at 22°C for 1 h.
2. After exposure to heavy metals, harvest cells with centrifugation at 10,000*g* for 2 min.
3. Suspend cell pellet in 1.0 mL MES buffer for staining procedure.

3.3.4. Fluorescent Probe Staining Procedure

1. Add 20 µL of PI stock solution and incubate for 5 min at 22°C.

3.4. Flow Cytometric Analysis

1. Determine analysis parameter settings for determination of viability of cells.
2. Use control suspensions (heat killed and viable) to select analysis parameters on the flow cytometer. Heat-killed cells (90°C for 20 min) stained with PI (final concentration 25 mg/L) fluoresce red with a 585/42 bandpass filter. Viable cells exclude PI and are detected with a 530/30 bandpass filter. Sample parameter settings are given in **Table 1**.

Table 1
Sample Parameter Settings of Flow Cytometric Analysis

	Detector				Compensation	
Settings	FSC	SSC	FL1	FL2	FL1-%FL2	FL2-%FL1
Acanthamoeba spp.					1.8	50.8
Voltage	E-1	379	505	479–498		
Amp gain	3.61	1.00	1.00	1.00		
Mode	Linear	Linear	Log	Log		
Candida spp.					0	0
Voltage	E00	379–436	505–599	627–629		
Amp gain	1.71–2.88	1.00	1.00	1.00		
Mode	Linear	Linear	Log	Log		
Fusarium spp.					30.8	45.4
Voltage	E00	258	822	729		
Amp gain	1.00	1.00	1.00	1.00		
Mode	Log	Log	Log	Log		

FSC, Forward Scatter; SSC, Side Scatter; FL1, Green Fluorescence; FL2, orange-red fluorescence

4. Notes

1. PI is light-sensitive and toxic to cells. Protect cells from light after applying the PI viability stain. Incubate and perform FCM analysis within 30 min of application of PI, as the stain will kill cells with extended exposure.
2. The major advantage of the use of fluorophores and FCM over culture-based procedures for assessing viability is the rapid analysis and estimation of the efficacy of various antimicrobials on microorganisms. The disadvantages of FCM procedures include underestimation of lethality owing to the inability of viability fluorophores to discriminate between viable and less damaged cells (i.e., viable but nonrecoverable) after exposure to some antimicrobials *(3)*. In our FCM procedures, PI is capable of discriminating between viable intact cells and dead cells after treatment with most of the chosen antimicrobials; however, with certain antimicrobial exposures, lysis of cells resulted in an overestimated viability. Lysed particles are not stained with PI and therefore are estimated as viable; however, after careful examination of FSC (i.e., measurement of size) histogram plots we were able to discriminate between the lysed particles from intact cells and determine the discrepancy in the staining procedure.
3. It is generally recognized that the physiological state of the microorganism will determine its relative susceptibility to antimicrobials. The choice of media, particularly for the production of cysts of *Acanthamoeba*, will affect susceptibilities. Cysts induced from axenic cultures, particularly via increased concentration of $MgSO_4$, may show increased susceptibility to antimicrobials. Inclusion of a standard disinfection system as a control is advised.

Acknowledgment

Supported in part by NIH GM579450.

References

1. Avery, S. V., Harwood, J. L., and Lloyd, D. (1995) Quantification and characterization of phagocytosis in the soil amoeba *Acanthamoeba castellanii* by flow cytometry. *Appl. Environ. Microbiol.* **61,** 1124–1132.
2. Borazjani, R. N., May, L. L., Noble, J. A., Avery, S. V., and Ahearn, D. G. (2000) Flow cytometry for determination of the efficacy of contact lens disinfecting solutions against *Acanthamoeba* spp. *Appl. Environ. Microbiol.* **66,** 1057–1061.
3. Davey, H. A. and Kell, D. B. (1996) Flow cytometry and cell sorting of heterogeneous microbial populations: the importance of single-cell analyses. *Microbiol. Rev.* **60,** 641–696.
4. Deere, D., Shen, J., Vesey, G., Bell, P., Bissinger, P., and Veal, D. (1998) Flow cytometry and cell sorting for yeast viability assessment and cell selection. *Yeast* **14,** 147–160.
5. Epps, D. E., Wolfe, M. L., and Groppi, V. (1994) Characterization of the steady-state and dynamic fluorescence properties of the potential-sensitive dye *bis*-(1,3-dibutylbarbituric acid)trimethine oxonol (Dibac$_4$(3)) in model systems and cells. *Chem. Phys. Lipids* **69,** 137–150.
6. Green, L., Petersen, B., Steimel, L., Haeber, P., and Current, W. (1994) Rapid determination of antifungal activity by flow cytometry. *J. Clin. Microbiol.* **32,** 1088–1091.
7. Khunkitti, W., Avery, S. V., and Lloyd, D. (1997) Effects of biocides on *Acanthamoeba castellanii* as measured by flow cytometry and plaque assay. *J. Antimicrob. Chemother.* **40,** 227–233.
8. López-Amorós, R., Comas, J., and Vives-Rego, J. (1995) Flow cytometric assessment of *Escherichia coli* and *Salmonella typhimurium* starvation-survival in seawater using rhodamine 123, propidium iodide, and oxonol. *Appl. Environ. Microbiol.* **61,** 2521–2526.
9. Marie, D., Brussaard, C. P. D., Thyrhaug, R., Bratbak, G., and Vaulot, D. (1999) Enumeration of marine viruses in culture and natural samples by flow cytometry. *Appl. Environ. Microbiol.* **65,** 45–52.
10. Millard, P. J., Roth, B. L., Thi, H. P. T., Yue, S. T., and Haughland, P. (1997) Development of the FUN-1 family of fluorescent probes for vacuole labeling and viability testing of yeasts. *Appl. Environ. Microbiol.* **63,** 2897–2905.
11. Mller, S., Loffhagen, N., Bley, T., and Babel, W. (1996) Membrane-potential-related fluorescence intensity indicates bacterial injury. *Microbiol. Res.* **151,** 127–131.
12. Noble, J. A., Ahearn, D. G., Avery, S. V. and Crow Jr., S. A. (2002) Phagocytosis affects the biguanide sensitivity of *Acanthamoeba* spp. *Antimicrob. Agents Chemother.* **46,** 2064–2076.
13. Price, D., Drago, G., Noble, J., Simmons, R., Crow Jr., S. A., and Ahearn, D.,G. (2002) Rapid assessment of anti-mould efficacies of pressure-treated Southern pine. *J. Ind. Microbiol. Biotech.* **29,** 368–372.
14. Zhang, S. and Crow, Jr., S. (2001) Toxic effects of Ag(I) and Hg(II) on *Candida albicans* and *C. albicans*: a flow cytometric evaluation. *Appl. Environ. Microbiol.* **67,** 4030–4035.
15. Nebe-von Caron, G. and Badley, R. A. (1995) Viability assessment of bacteria in mixed populations using flow cytometry. *J. Microsc.* **179,** 55–66.
16. Becton Dickinson (1996) *FACSCalibur System User's Guide.* Becton Dickinson, Heidelberg, Germany, p. 23.

15

Detection and Differentiation of *Cryptosporidium* Oocysts in Water by PCR-RFLP

Lihua Xiao, Altaf A. Lal, and Jianlin Jiang

1. Introduction

Consumption of contaminated water has been implicated as a major source of *Cryptosporidium* infection in various outbreak investigations and case control studies. Surveys conducted in various regions of the United States demonstrated the presence of *Cryptosporidium* oocysts in 67–100% of wastewaters, 24–100% of surface waters, and 17–26.8% of drinking waters ***(1–4)***. The identity and human infective potential of these waterborne oocysts are not known, although it is likely that not all oocysts are from human-infecting *Cryptosporidium* species. Likewise, the source of the oocyst contamination is also not fully clear. Farm animals and human sewage discharge are generally considered to be the major sources of surface water contamination with *C. parvum* ***(5)***. Because *Cryptosporidium* infection is common in wildlife, it is conceivable that wildlife can also be a source for *Cryptosporidium* oocysts in waters ***(4)***. The presence of host-adapted *Cryptosporidium* spp. and genotypes makes it possible to develop molecular tools to assess the human infection potential and source of *Cryptosporidium* oocysts in water.

Currently, the identification of *Cryptosporidium* oocysts in environmental samples is largely made by the use of immunofluorescent assay (IFA) after concentration processes (Environmental Protection Agency [EPA] recommended information collection rule [ICR] method or method 1622/1623 or similar techniques). Because IFA detects oocysts from all *Cryptosporidium* parasites, the species distribution of *Cryptosporidium* parasites in environmental samples cannot be assessed. Although many surface water samples contain *Cryptosporidium* oocysts, it is unlikely that all these oocysts are from human-pathogenic species or genotypes, because only five genotypes of *Cryptosporidium* parasites (the *C. parvum* human and bovine genotypes, *C. meleagridis*, *C. canis*, and *C. felis*) are responsible for most human infections ***(6)***. Information on the source of *C. parvum* contamination is necessary for effective evalu-

From: *Methods in Molecular Biology, vol. 268: Public Health Microbiology: Methods and Protocols*
Edited by: J. F. T. Spencer and A. L. Ragout de Spencer © Humana Press Inc., Totowa, NJ

ation and selection of management practices for reducing *C. parvum* contamination of surface water and the risk of cryptosporidiosis. Thus, identification of oocysts to species and genotype levels is of public health importance.

Molecular techniques, especially polymerase chain reaction (PCR) and PCR-related methods, have been developed and used in the detection and differentiation of *Cryptosporidium* parasites for many years ***(22)***. Most of these techniques were designed for the analysis of fecal samples from humans and domestic animals. Earlier PCR methods ***(7–10)*** do not have the ability to differentiate species and thus can only be used for determination of the presence or absence of *Cryptosporidium* parasites. The primer sequences of these techniques (except for that described in ref. ***8***) are mostly based on undefined genomic sequences from *C. parvum* bovine isolates. These sequences tend to be more polymorphic than structural and housekeeping genes; therefore the primers based on them are unlikely to amplify DNA efficiently from *Cryptosporidium* spp. (such as *C. muris*, *C. baileyi*, *C. serpentis*, *C. canis*, and *C. felis*) and from *Cryptosporidium* genotypes (such as the fox, skunk, and opossum) that are more distant from the *C. parvum* bovine genotype.

Several PCR-restriction fragment length polymorphism (RFLP)-based tools have been developed for the detection and differentiation of *Cryptosporidium* parasites at the species level ***(11–17)***. All these techniques are based on the small subunit (SSU) rRNA gene. Unfortunately, primers for some of these techniques ***(11,13,14)*** used conserved sequences of eukaryotic organisms. Therefore, these primers also amplify DNA from organisms other than *Cryptosporidium* ***(18)***. A PCR-RFLP analysis of the internal transcribed spacers of the rRNA gene can also differentiate several *Cryptosporidium* parasites ***(19)***. Nucleotide sequencing-based approaches have also been developed for the differentiation of various *Cryptosporidium* spp. ***(20,21)***.

Various PCR based techniques have also been developed for the differentiation of *C. parvum* human and bovine genotypes ***(22)***. Both genotypes of *C. parvum* have been identified in humans, but the human genotype (genotype 1 or anthroponotic genotype) has been almost exclusively found in humans, whereas the bovine genotype (genotype 2 or zoonotic genotype) infects humans, ruminants, and some other animals. As a result, the human genotype was recently renamed *C. hominis*. Many of the genotyping tools used in these studies, however, cannot detect and differentiate other genotypes of *C. parvum* and non-*C. parvum Cryptosporidium* spp. This has limited the utility of these tools in the analysis of environmental samples. Even for clinical samples, their usefulness is compromised by the failure to detect *C. canis*, *C. meleagridis*, and *C. felis* in fecal samples from infected humans. Using newer molecular tools, we and others have found these parasites in AIDS patients as well as in children ***(6,23–26)***.

Numerous attempts have been made to apply PCR techniques in detecting *Cryptosporidium* oocysts in water samples ***(27–40)***. Most of these studies used water samples seeded with *Cryptosporidium* oocysts, and various successes were reported. One major obstacle is the presence of PCR inhibitors in waters, which are coextracted with DNA and inhibit PCR amplification of the target DNA. This has greatly reduced the sensitivity of PCR detection of oocysts in various water samples. The PCR inhibitors can be removed by an immunomagnetic separation (IMS) procedure ***(8)***, which is also a component of method 1622 or 1623 recommended for the detection of *Cryptosporidium* oocysts in water by the US EPA. Still, the sensitivity (5–10 oocysts in reagent-grade water) of the single-round PCR format adopted in most of these techniques has limited the usefulness of PCR in the analysis of water samples. This is because most water samples are likely to have fewer than five *Cryptosporidium* oocysts. The usefulness of PCR techniques, nevertheless, has been demonstrated by the detection of *Cryptosporidium* oocysts in water samples from the 1993 outbreak in Milwaukee, using a *Cryptosporidium* genus-specific PCR technique ***(8)***.

We have successfully used one such technique, the SSU rRNA-based nested PCR-RFLP method, in conjunction with IMS for the detection and differentiation of *Cryptosporidium* oocysts in storm water, raw surface water, and wastewater. It is the only technique that has been extensively used in the analysis of water samples and specimens from animals and humans ***(16,17,41,42)***. In one study, we analyzed 29 storm water samples collected from a stream that contributes to the New York City water supply system after storms and found 12 wildlife genotypes of *Cryptosporidium* parasites in 27 samples. Twelve of the 27 PCR-positive samples had multiple genotypes. Six of the genotypes were traced to sources (*C. baileyi* from birds, an unnamed species from snakes, two genotypes from opossums, according to data at the time of publication of the paper, a genotype from deer, and a genotype from muskrat by as shown by more recent data). The rest were wildlife genotypes that have never been found in humans or domestic animals, suggesting that wildlife was a major contributor to *Cryptosporidium* oocyst contamination in storm water in the area studied. This finding was consistent with the environmental setting (the catchment area was forested and isolated from agricultural activities) of the sampling site ***(41)***.

In another study we conducted, analysis of raw surface water samples collected from different locations (Maryland, Wisconsin, Illinois, Texas, Missouri, Kansas, Michigan, Virginia, and Iowa) in the United States with the SSU rRNA-based PCR-RFLP technique produced quite different results. A total of 49 surface water samples were analyzed, and 16 of the samples produced positive PCR amplification. Only four *Cryptosporidium* genotypes (*C. parvum* bovine genotype, *C. parvum* human genotype, *C. andersoni*, and *C. baileyi*) were found, all of which are genotypes commonly found in farm animals and/or humans, indicating that humans and farm animals are major sources of *Cryptosporidium* oocyst contamination in these waters. Similar results were also obtained from wastewater samples. Three *Cryptosporidium* genotypes (*C. parvum* bovine genotype, *C. andersoni*, and *C. muris*) were found in 7 of 34 grab samples (10 mL each) of raw wastewater collected from a treatment plant in Milwaukee. As expected, the diversity of *Cryptosporidium* parasites found in source waters and wastewaters was much lower than that in storm waters ***(42)***.

Promising results in the genotyping of *Cryptosporidium* parasites in water samples have also been generated in studies conducted by others. Using heat shock protein (HSP)70 sequence analysis of cell culture/PCR-amplified products, Di Giovanni et al. ***(29)*** have found six sequence types of *C. parvum* in raw surface water samples and filter backwash water samples. Comparison of these sequences with the HSP70 sequences we collected from various *Cryptosporidium* spp. and *C. parvum* genotypes indicates that these sequences were from three genotypes (bovine, mouse, and human genotypes) of *C. parvum* ***(21)***, suggesting that farm animals, rodents, and humans are responsible for *Cryptosporidium* oocyst contamination in these waters. Similarly, studies conducted in Northern Ireland and Japan have shown the presence of the bovine genotype of *C. parvum* in surface water ***(43,44)***. The molecular tools (HSP70 and TRAP-C2-based PCR) used in these studies, however, have probably limited the range of *Cryptosporidium* parasites found in water. The spectrum of *Cryptosporidium* oocysts found in surface water samples from Massachusetts using a PCR-sequencing analysis of the SSU rRNA gene is more similar to what we had found previously in surface water ***(12)***.

The method we are introducing here is a PCR-RFLP analysis of the SSU rRNA gene, for which extensive data are available on genetic heterogeneity ***(16,17)***. This method builds on existing procedures (filtration of water samples, elution and concentration of filtrates, IMS isolation of *Cryptosporidium* oocysts) used in the popular *Cryptosporidium* detection method 1622/1623 and involves DNA extraction, PCR, and restriction digestion. Because of the multicopy nature of the SSU rRNA gene (five copies per sporozoite in the bovine genotype of *C. parvum*) and the nested format of the PCR, this technique is one of the most sensitive PCR tools for the detection of *Cryptosporidium* oocysts. It amplifies and detects all known and unknown *Cryptosporidium* parasites and differentiates the most common species and genotypes. This method can supplement other detection methods such as 1622 and 1623 to provide information on the distribution of *Cryptosporidium* species and genotypes in water samples and the likely source of contaminations.

2. Materials

2.1. Supplies for IMS

1. Dynabeads® anti-*Cryptosporidium* kit (for the isolation of *Cryptosporidium* oocysts only; cat. no. 730.01 for 10 tests or cat. no. 730.11 for 50 tests), or Dynabeads® GC-Combo kit (for the isolation of both *Cryptosporidium* oocysts and *Giardia* cysts; cat. no. 730.02 for 10 tests or cat. no. 730.12 for 50 tests) (Dynal, Oslo, Norway).
2. Dynal Magnetic Particle Concentrators (Dynal MPC): Dynal MPC-S (cat. no. 120.20) and Dynal MPC-1 (product No. 120.01).

2.2. Supplies for DNA Extraction

QIAamp DNA mini kit: cat. no. 51304 (50 tests) or cat. no. 51306 (250 tests) (Qiagen, Valencia, CA).

2.3. Supplies for PCR-RFLP

1. Primary PCR primers.
   a. Forward (F1): 5'-TTCTAGAGCTAATACATGCG-3'.
   b. Reverse (R1): 5'-CCCATTTCCTTCGAAACAGGA-3'.
2. Secondary PCR primers.
   a. Forward (F2): 5'-GGAAGGGTTGTATTTATTAGATAAAG-3'.
   b. Reverse (R2): 5'-CTCATAAGGTGCTGAAGGAGTA-3'.

3. 10X PCR buffer with 15 m*M* Mg^{2+}, (cat. no. N808-0129, PE Applied Biosystems, Foster City, CA).
4. 100 m*M* dNTP (cat. no. U1240, Promega, Madison, WI). To make a 1.25 m*M* working solution, add 12.5 µL of each dNTP to 950 µL of distilled water. Store the working solution at –20°C before use.
5. Taq polymerase (cat. no. M2665, Promega).
6. 25 m*M* $MgCl_2$ (cat. no. A351F, Promega).
7. *Ssp*I (cat. no. R0132L, New England BioLabs, Beverly, MA).
8. *Vsp*I (cat. no. R6851, Promega).
9. *Dde*I (cat. no R0175L, New England BioLabs).

3. Methods

3.1. Immunomagnetic Separation of Cryptosporidium Oocysts From Water Pellets

1. Process 10 L of water samples through the filtration, elution, and concentration steps, following method 1622 or 1623 of the US EPA (*see* **Note 1**).
2. Wash the concentrated water pellets in 15-mL polypropylene tubes twice with distilled water by centrifugation at 1500*g* for 10 min.
3. Equilibrate in 10X Buffer A and 10X Buffer B from the Dynabeads kit to room temperature.
4. Add 1 mL of 10X Buffer A and 1 mL of 10X Buffer B into the 15-mL tube containing washed sample. Use only 0.5 mL of the water concentrate if the pellet is bigger than 0.5 mL in volume.
5. Resuspend the beads fully by vortexing the vial for 10 s, and add 100 µL of Dynabeads to the 15-mL tube.
6. Add distilled water to give a final volume of 10 mL.
7. Rotate at 15–20 rpm for 1 h at room temperature.
8. Prepare the 1X dilution of Buffer A. One milliliter of 1X Buffer A will be required for each sample.
9. At the end of incubation, capture the Dynabeads in the 15-mL tube using MPC-1, and decant the solution in the tube
10. Resuspend the Dynabeads with 1 mL of 1X Buffer A, and transfer the suspension into 1.5-mL microfuge tube.
11. Capture the Dynabeads in a microfuge tube using MPC-S, and decant the solution in the tube. The Dynabeads with bound *Cryptosporidium* oocysts will be used in DNA extraction (*see* **Note 2**).

3.2. DNA Extraction Using the QIAamp DNA Mini Kit

1. Add 180 µL of Buffer ATL to a 1.5-mL microfuge tube containing IMS-isolated *Cryptosporidium* oocysts, and vortex for 30 s.
2. Freeze-thaw five times at –70°C (or on dry ice) and 56°C.
3. Add 20 µL of proteinase K to the tube, vortex for 10 s, and incubate at 56°C overnight.
4. Add 200 µL of Buffer AL to the sample, vortex, and incubate the tube at 70°C for 10 min.
5. Centrifuge at full speed to precipitate the undigested pellet.
6. Transfer the supernatant into a new 1.5-mL tube.
7. Add 200 µL of ethanol to the sample and vortex for 15 s.

8. Carefully transfer the mixture to a QIAamp spin column without wetting the rim, and centrifuge the column at 6000*g* for 1 min.
9. Place the spin column in a clean 2-mL collection tube, and discard the tube containing the filtrate.
10. Add 500 μL of Buffer AW1 without wetting the rim, and centrifuge at 6000*g* for 1 min.
11. Place the spin column in a clean 2-mL collection tube, and discard the tube containing the filtrate.
12. Add 500 μL of Buffer AW2 without wetting the rim, and centrifuge at full speed for 3 min.
13. Place the spin column into a clean 1.5-mL microfuge tube, and discard the tube containing the filtrate.
14. Add 100 μL of Buffer AE, and incubate the tube at room temperature for 1 min.
15. Centrifuge the tube at 6000*g* for 1 min.
16. Save the filtrate containing DNA and the store the extraction at –20°C.

3.3. PCR-RFLP Analysis of the SSU rRNA Gene

1. 1. Primary PCR.
   a. Preparation of master mixture. For each PCR reaction, prepare the following (*see* **Note 3**): 10 μL 10X Perkin-Elmer PCR buffer, 16 μL dNTP (1.25 m*M*), 2.5 μL F1 primer (40 ng/μL), 2.5 μL R1 primer (40 ng/μL), 6 μL $MgCl_2$ (25 m*M*), 4 μL Bovine serum albumin (10 mg/mL), 57.5 μL Distilled water, and 0.5 μL Taq polymerase, for a total of 99 μL.
   b. Add 99 μL of the master mixture to each PCR tube.
   c. Add 1 μL of DNA sample to each tube.
   d. Run the following PCR program: 94°C, 3 min; 35 cycles of 94°C for 45 min, 55°C for 45 min, and 72°C for 1 min; 72°C for 7 min; and 4°C soaking.
2. Secondary PCR.
   a. A. Preparation of master mixture. For each PCR reaction, prepare the following: 10 μL 10X Perkin-Elmer PCR buffer, 16 μL dNTP (1.25 m*M*), 5 μL F2 primer (40 ng/μL), 5 μL R2 primer (40 ng/μL), 6 μL $MgCl_2$ (25 m*M*), 55.5 μL distilled water, and 0.5 μL Taq polymerase for a total of 98 μL.
   b. Add 98 μL of the master mixture to each PCR tube.
   c. Add 2 μL of the primary PCR reaction to each tube.
   d. Run the following PCR program: 94°C, 3 min; 35 cycles of: 94°C for 45min, 58°C for 45min, and 72°C for 1 min; 72°C for 7 min; and 4°C soaking.
   e. Run electrophoresis on a 1.5% agarose gel with 20 μL of the PCR product.

3.4. RFLP

1. Prepare master mixture using the formula shown in **Table 1**, which is for one restriction digestion reaction (*see* **Note 4**).
2. Transfer 30 μL of master mixture to each tube, add 10 μL of secondary PCR reaction to the tube, and mix well.
3. Incubate in a 37°C waterbath for 2 h or overnight.
4. Run electrophoresis on a 2% argrose gel with the entire 40 μL from the restriction digestion reaction, using procedures standardized in the laboratory.
5. Identify *Cryptosporidium* species and genotypes based on RFLP banding patterns (*see* **Note 5**).

Table 1
Formula for Master Mixture Preparation

Restriction enzyme	Buffer	Water (μL)	Enzyme (μL)
*Ssp*I	4 μL of New England BioLabs Buffer SspI	22	4
*Vsp*I	4 μL of Promega Buffer D	24	2
*Dde*I	4 μL of New England BioLabs Buffer 3	24	2

4. Notes

1. Protocols for EPA methods 1622 and 1623 can be downloaded from the EPA website (http://www.epa.gov/waterscience/methods/1622.pdf and http://www.epa.gov/waterscience/methods/1623.pdf). Even though filtration of 10-L water samples is recommended, larger volumes of finished water can and should be filtered and used in the analysis. For raw wastewater, filtration with the standard Envirocheck capsule filters can be problematic. We normally process the pellets from 50 mL of grab samples of raw wastewater directly for IMS without filtration, after they are washed twice by centrifugation at 1500*g* for 10 min.
2. IMS-oocyst pellets can be stored at –20°C before they are used in DNA extraction. No detachment of Dynabeads from oocysts is needed prior to the DNA extraction.
3. The magnesium concentration used in both primary and secondary PCR is 3 m*M*, which is higher than normal PCR. Even though concentrations lower than 3 m*M* generally do not work well for the SSU rRNA-based PCR, it is recommended that the magnesium concentration should be optimized in each laboratory prior to sample analysis.
4. We generally do *Ssp*I and *Vsp*I restriction digestions to differentiate common *Cryptosporidium* species and genotypes. *Dde*I digestion is performed to differentiate *C. andersoni* from *C. muris* and is done only when results of *Ssp*I digestion have shown the RFLP pattern of *C. andersoni*/*C. muris*. In most areas, *C. andersoni* is found much more frequently in water than *C. muris*.
5. *Ssp*I and *Vsp*I RFLP patterns for some common *Cryptosporidium* species and genotypes are shown in **Table 2** and **Fig. 1**, and the *Dde*I RFLP patterns for *C. andersoni* and *C. muris* are shown in **Fig. 2**. Occasionally, owing to the genetic heterogeneity between parasites of the same genotype and the multicopy nature of the SSU rRNA gene, RFLP banding patterns deviations from the characteristic patterns can occur. New and unusual *Cryptosporidium* parasites are also sometimes seen in water samples. A few genotypes cannot be differentiated from each other by RFLP analysis. These require DNA sequencing of the secondary PCR product using the secondary forward and reverse primers for confirmation. The most polymorphic region of the SSU rRNA sequences of known *Cryptosporidium* parasites are shown in **Fig. 3**. Likewise, isolates within each genotype can differ somewhat in SSU rRNA sequences owing to the presence of heterogeneous copies of the gene and intragenotypic variartions. Thus minor differences (<5 bp) in the SSU rRNA sequences generally do not warrant new genotype designation.

Table 2
Restriction Fragment Length Polymorphism in the SSU rRNA Gene of Common *Cryptosporidium* spp. and Genotypes

Species	PCR fragment	Ssp I digestion[a]	Vsp I digestion[a]
C. muris/C. andersoni	833	385, 448	102, 731
C. serpentis	831	370, 414	102, 729
C. baileyi	826	254, 572	102/104, 620
C. felis	864	390, 426	102/104, 182, 476
C. meleagridis	833	108, 254, 449	102/104, 171, 456
C. wrairi	834	109, 254, 449	102/104, 628
C. saurophilum	834	109, 255, 418	102/104, 628
C. canis	829	105, 254, 417	94/102, 633
Cryptosporidium ferret genotype	837	111, 254, 449	102/104, 174, 457
Cryptosporidium pig genotype	838	365, 453	102/104, 632
Cryptosporidium marsupial genotype	837	109, 254, 441[b]	102/104, 631
C. parvum human genotype	837	111, 254, 449	70, 102/104, 561
C. parvum bovine genotype A gene	834	108, 254, 449	102/104, 628
C. parvum bovine genotype B gene	831	119, 254, 449	102/104, 625
C. parvum mouse genotype	838	112, 254, 449	102/104, 175, 457

[a]Lengths are in base pairs; only sizes of visible bands are shown.
[b]An additional upper band (about 583 bp) from the heterogeneous copy of the gene is usually present.

Fig. 1. (*opposite page*) Differentiation of common *Cryptosporidium* species and genotypes by a nested PCR-RFLP procedure based on the SSU rRNA gene. Lane 1, *C. muris* or *C. andersoni*; lane 2, *C. serpentis*; lane 3, *C. baileyi*; lane 4, *C. felis*; lane 5, *C. meleagridis*; lane 6, *C. wrairi*; lane 7, *Cryptosporidium* pig genotype; lane 8, *C. canis*; lane 9, *C. saurophilum*; lane 10, *Cryptosporidium* ferret genotype; lane 11, *Cryptosporidium* marsupial genotype; lane 12, *C. parvum* mouse genotype; lane 13, *C. parvum* bovine genotype; and lane 14, *C. parvum* human genotype. The upper panel shows *Ssp*I digestion products, and the lower panel shows *Vsp*I digestion products. Molecular markers are 100-bp ladders.

Fig. 2. (*opposite page*) Differentiation of *C. andersoni* and *C. muris* by RFLP analysis of SSU rRNA gene PCR products using *Dde*I. Lanes 1 and 2, *C. andersoni*; lanes 3 and 4, *C. muris*; and lane 5, *C. andersoni* and *C. parvum* human genotype. The upper panel shows *Ssp*I digestion products, and the lower panel shows *Dde*I digestion products. The top band in lane 5 of the *Ssp*I products was owing to partial digestion. Molecular markers are 100-bp ladders.

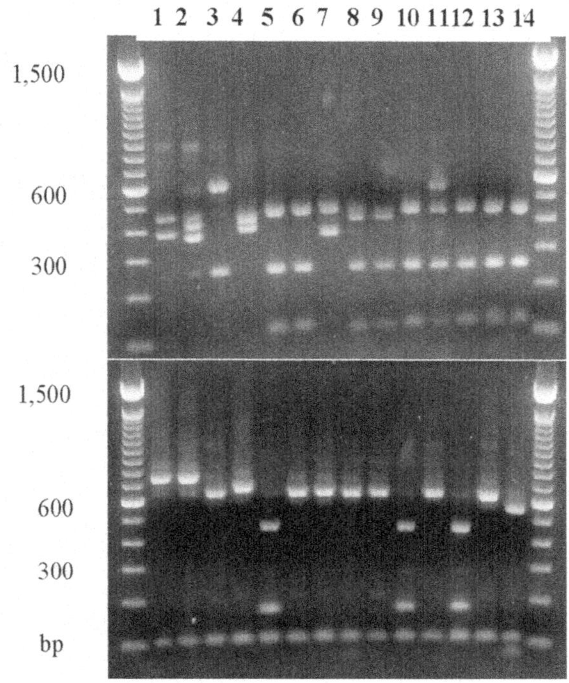
1 2 3 4 5 6 7 8 9 10 11 12 13 14
1,500
600
300
1,500
600
300
bp

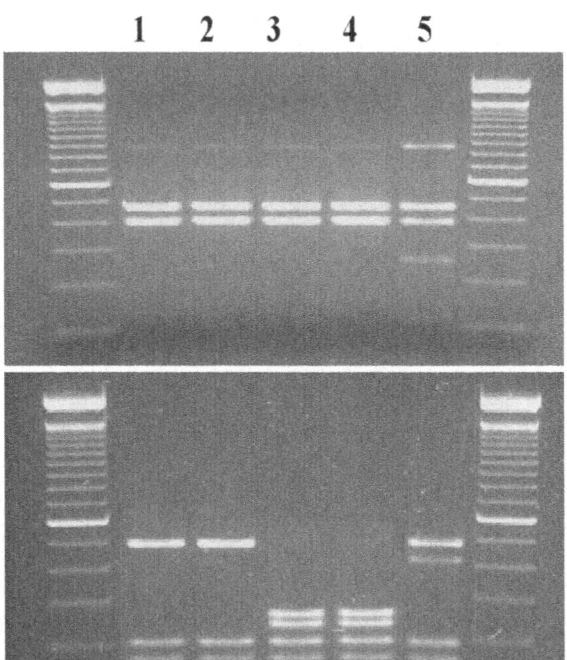
1 2 3 4 5

```
Human           TTGGATTTCTGTTA-ATAATTTA----TATAAAATATTTTG------ATGAATATT----TATATAATATTAACATAATT
Rabbit          ..............-........----.....T........------..AG.....----....................
Bovine          ..............-........----..............------.........----....................
Mouse           ..............-........----.............A------..T......----....................
Ferret          ..............-........----..............------..T......----....................
C. wrairi       ..............-........----....T.........------.A-......----....................
C. meleagridis  ..............-........----....T.......-.------..T......----....................
Pig             ..............-........----....T......---------T.T......----....................
Marsupial       ..............-........----....TT...C...T------TAAGG.G..----....................
Opossum I       ..............-........----....G....C...T------T.AGG.G..----....................
Coyote          ..............-........----....T.......-A--------AC.....----....................
C. canis        ..............-........----....T.......-A--------AC.....----....................
Bear            ..............-........----....T........A--------AT.....----....................
Deer mouse      ............C.-...G...T---G...GT....C..CA--------ACG....----C....G..G..G........
Opossum II      ..............-.......TTATG...GC.....A.A.TTAAAA..ATG....GTTG...G................
Fox             ..............-........----.....T..T..---------.ACG....-----....................
Skunk           ..............-........----....T........A------..T......----....................
C. felis        ..............-...CC...----....T........TTTT---T.A......A--A...G...G............
C. saurophilum  ..............-........----....T......AC.----------G....----....................
Deer            ..AAT.........-..-T....----....C....C.AC.----------G....----....................
Cattle          ..AATC........-..-T....----....T.....CAC.----------.....----....................
Goose           ...........C..-..-T...G---C....C....CCAC.----------G....----...G...A....G......C
C.baileyi       ..............-..-.C...----....C....CCAC.----------G....----.......C............
Snake           ..............-..-TC.CG----....TT...C.AC.----------G...G----.....G.C............
C. andersoni    ............GT........----.-.T.T..CCAA.----------G..A---TAT....T..C.....CC..
C. muris        ............GT....C..----.-.T.T..C.AA.----------G..A---TAT....T..C.....CC..
C. galli        ............GC..C..-..----..-TT.TC.CCAA.----------G..A----TA.........C.....CC.C
C. serpentis    ............GT..T-....----.-.T.T...AA.----------G..A.---A.T........C.....CC..
Tortoise        ............GT..G...AT----A.-TTG.....AA.----------....----TAT....TC..C.....CC..
```

Fig. 3. Sequence diversity among *Cryptosporidium* species and genotypes in the polymorphic region of the SSU rRNA gene. Dots denote sequence identity to the *C. parvum* human genotype (top sequence) and dashes denote deletions. Human, *C. parvum* human genotype; rabbit, *C. parvum* rabbit genotype; bovine, *C. parvum* bovine genotype; mouse, *C. parvum* mouse genotype; ferret, *Cryptosporidium* ferret genotype; pig, *Cryptosporidium* pig genotype; marsupial, *Cryptosporidium* marsupial genotype; opossum I, *Cryptosporidium* opossum genotype I

```
Human              CAT-ATTACTATTTTT-------------TTTTTTA----------------------------GTATATGAAATTTTAC
Rabbit             ...-............-------------....--.----------------------------................
Bovine             ...-........A-----------------.A....----------------------------................
Mouse              ...-........AA..A------------.......----------------------------................
Ferret             ...-.......AA...-------------..G...G----------------------------................
C. wrairi          ...-........A.A.-------------..--...----------------------------................
C. meleagridis     ...-.......AA...A------------..----.----------------------------................
Pig                ...-........AA..-------------...A...----------------------------................
Marsupial          ...-........A..---------------......----------------------------................
Opossum I          ...-........--.---------------......----------------------------................
Coyote             ...-........------------------.-.A..----------------------------................
C. canis           ...-........------------------.-.A..----------------------------..........C.....
Bear               ...-........------------------AAC...----------------------------................
Deer mouse         .G.-......--------------------......----------------------------.....C..........
Opossum II         ...-........A.GGTGGATTGGTGAAG.......CTTTTTTCCAGTCACACCGGGAATTATG................
Fox                ...-.........A..-------------A..-...----------------------------........G.......
Skunk              ...-........A..A--------------......----------------------------................
C. felis           ...-...TT..AGAC.GA-----------A......GTTTTG---------------------ATA..............
C. saurophilum     ...-......T.A-----------------......G--------------------------A................
Deer               ...-......---------------------.....----------------------------..........C.....
Cattle             ...-......---------------------.....----------------------------..........C.....
Goose              .GC-.....CTCGC--------------------------------------------------....GC.G........
C.baileyi          ..C-......--------------------.A...A---------------------------A....G.....C.....
Snake              ...-......--------------------...A..----------------------------................
C. andersoni       .C.-....TAT.--------------------C.A.AT--------------------------A....G..........
C. muris           .C.-....TAT..-------------------C.A.AT--------------------------A....G....C.....
C. galli           .C.T....TATC----------------------..AT--------------------------A..G.G.G........
C. serpentis       .C.-....TAT.--------------------....AT--------------------------A....G..G.......
Tortoise           .C.-.....A..---------------------...TT--------------------------.....G..........
```

Fig. 3. (*continued*) (related to the *Cryptosporidium* marsupial genotype); coyote, *C. canis* coyote genotype; bear, *Cryptosporidium* bear genotype; deer mouse, *Cryptosporidium* deer mouse genotype; opossum II, *Cryptosporidium* opossum genotype II; fox, an unnamed *Cryptosporidium* sp. in foxes; deer, an unnamed *Cryptosporidium* sp. in deer; cattle, an unnamed *Cryptosporidium* sp. in cattle (rare in prevalence); goose, an unnamed *Cryptosporidium* sp. in geese; snake, an unnamed intestinal *Cryptosporidium* sp. in snakes; and tortoise, an unnamed gastric *Cryptosporidium* sp. in tortoises.

References

1. LeChevallier, M. W., Norton, W. D., and Lee, R. G. (1991) *Giardia* and *Cryptosporidium* spp. in filtered drinking water supplies. *Appl. Environ. Microbiol.* **57,** 2617–2621.
2. LeChevallier, M. W., Norton, W. D., and Lee, R. G. (1991) Occurrence of *Giardia* and *Cryptosporidium* spp. in surface water supplies. *Appl. Environ. Microbiol.* **57,** 2610–2616.
3. Madore, M. S., Rose, J. B., Gerba, C. P., Arrowood, M. J., and Sterling, C. R. (1987) Occurrence of *Cryptosporidium* oocysts in sewage effluents and selected surface waters. *J. Parasitol.* **73,** 702–705.
4. Rose, J. B. (1997) Environmental ecology of *Cryptosporidium* and public health implications. *Annu. Rev. Public Health* **18,** 135–161.
5. Meinhardt, P. L., Casemore, D. P., and Miller, K. B. (1996) Epidemiologic aspects of human cryptosporidiosis and the role of waterborne transmission. *Epidemiol. Rev.* **18,** 118–136.
6. Xiao, L., Bern, C., Limor, J., et al. (2001) Identification of 5 types of *Cryptosporidium* parasites in children in Lima, Peru. *J. Infect. Dis.* **183,** 492–497.
7. Chrisp, C. E. and LeGendre, M. (1994) Similarities and differences between DNA of *Cryptosporidium parvum* and *C. wrairi* detected by the polymerase chain reaction. *Folia Parasitol.* **41,** 97–100.
8. Johnson, D. W., Pieniazek, N. J., Griffin, D. W., Misener, L., and Rose, J. B. (1995) Development of a PCR protocol for sensitive detection of *Cryptosporidium* oocysts in water samples. *Appl. Environ. Microbiol.* **61,** 3849–3855.
9. Laxer, M. A., Timblin, B. K., and Patel, R. J. (1991) DNA sequences for the specific detection of *Cryptosporidium parvum* by the polymerase chain reaction. *Am. J. Trop. Med. Hyg.* **45,** 688–694.
10. Webster, K. A., Pow, J. D., Giles, M., Catchpole, J., and Woodward, M. J. (1993) Detection of *Cryptosporidium parvum* using a specific polymerase chain reaction. *Vet. Parasitol.* **50,** 35–44.
11. Awad-el-Kariem, F. M., Warhurst, D. C., and McDonald, V. (1994) Detection and species identification of *Cryptosporidium* oocysts using a system based on PCR and endonuclease restriction. *Parasitology* **109,** 19–22.
12. Jellison, K. L., Hemond, H. F., and Schauer, D. B. (2002) Sources and species of Cryptosporidium oocysts in the Wachusett reservoir watershed. *Appl. Environ. Microbiol.* **68,** 569–575.
13. Kimbell, L. M., Miller, D. L., Chavez, W., and Altman, N. (1999) Molecular analysis of the 18S rRNA gene of *Cryptosporidium* serpentis in a wild-caught corn snake (*Elaphe guttata guttata*) and a five-species restriction fragment length polymorphism-based assay that can additionally discern *C. parvum* from *C. wrairi*. *Appl. Environ. Microbiol.* **65,** 5345–5349.
14. Leng, X., Mosier, D. A., and Oberst, R. D. (1996) Differentiation of *Cryptosporidium parvum*, *C. muris*, and *C. baileyi* by PCR-RFLP analysis of the 18S rRNA gene. *Vet. Parasitol.* **62,** 1–7.
15. Sturbaum, G. D., Reed, C., Hoover, P. J., Jost, B. H., Marshall, M. M., and Sterling, C. R. (2001) Species-specific, nested PCR-restriction fragment length polymorphism detection of single Cryptosporidium parvum oocysts. *Appl. Environ. Microbiol.* **67,** 2665–2668.
16. Xiao, L. H., Escalante, L., Yang, C. F., et al. (1999) Phylogenetic analysis of *Cryptosporidium* parasites based on the small-subunit rRNA gene locus. *Appl. Environ. Microbiol.* **65,** 1578–1583.

17. Xiao, L. H., Morgan, U. M., Limor, J., et al. (1999) Genetic diversity within *Cryptosporidium parvum* and related *Cryptosporidium* species. *Appl. Environ. Microbiol.* **65,** 3386–3391.
18. Sulaiman, I. M., Xiao, L. H., and Lal, A. A. (1999) Evaluation of *Cryptosporidium parvum* genotyping techniques. *Appl. Environ. Microbiol.* **65,** 4431–4435.
19. Morgan, U. M., Deplazes, P., Forbes, D. A., Spano, F., Hertzberg, H., Sargent, K. D., Elliot, A., and Thompson, R. C. A. (1999) Sequence and PCR-RFLP analysis of the internal transcribed spacers of the rDNA repeat unit in isolates of *Cryptosporidium* from different hosts. *Parasitology* **118,** 49–58.
20. Morgan, U. M., Monis, P. T., Fayer, R., Deplazes, P., and Thompson, R. C. A. (1999) Phylogenetic relationships among isolates of *Cryptosporidium*: evidence for several new species. *J. Parasitol.* **85,** 1126–1133.
21. Sulaiman, I. M., Morgan, U. M., Thompson, R. C., Lal, A. A., and Xiao, L. (2000) Phylogenetic relationships of *Cryptosporidium* parasites based on the 70-kilodalton heat shock protein (HSP70) gene. *Appl. Environ. Microbiol.* **66,** 2385–2391.
22. Fayer, R., Morgan, U., and Upton, S. J. (2000) Epidemiology of *Cryptosporidium*: transmission, detection and identification. *Int. J. Parasitol.* 30, 1305–1322.
23. Morgan, U., Weber, R., Xiao, L., et al. (2000) Molecular characterization of *Cryptosporidium* isolates obtained from human immunodeficiency virus-infected individuals living in Switzerland, Kenya, and the United States. *J. Clin. Microbiol.* 38, 1180–1183.
24. Pedraza-Diaz, S., Amar, C., Iversen, A. M., Stanley, P. J., and McLauchlin, J. (2001) Unusual *Cryptosporidium* species recovered from human faeces: first description of *Cryptosporidium felis* and *Cryptosporidium* ‘dog type’ from patients in England. *J. Med. Microbiol.* 50, 293–296.
25. Pedraza-Diaz, S., Amar, C., and McLauchlin, J. (2000) The identification and characterisation of an unusual genotype of *Cryptosporidium* from human faeces as *Cryptosporidium meleagridis. FEMS Microbiol. Lett. 189,* 189–194.
26. Pieniazek, N. J., Bornay-Llinares, F. J., Slemenda, S. B., et al. (1999) New *Cryptosporidium* genotypes in HIV-infected persons. *Emerg. Infect. Dis.* **5,** 444–449.
27. Chung, E., Aldom, J. E., Carreno, R. A., et al. (1999) PCR-based quantitation of *Cryptosporidium parvum* in municipal water samples. *J. Microbiol. Methods* **38,** 119–130.
28. Chung, E., Aldom, J. E., Chagla, A. H., et al. (1998) Detection of *Cryptosporidium parvum* oocysts in municipal water samples by the polymerase chain reaction. *J. Microbiol. Methods* **33,** 171–180.
29. Di Giovanni, G. D., Hashemi, F. H., Shaw, N. J., Abrams, F. A., LeChevallier, M. W., and Abbaszadegan, M. (1999) Detection of infectious *Cryptosporidium parvum* oocysts in surface and filter backwash water samples by immunomagnetic separation and integrated cell culture-PCR. *Appl. Environ. Microbiol.* **65,** 3427–3432.
30. Gibbons, C. L., Rigi, F. M., and Awad-El-Kariem, F. M. (1998) Detection of *Cryptosporidium parvum* and *C. muris* oocysts in spiked backwash water using three PCR-based protocols. *Protist* **149,** 127–134.
31. Kaucner, C. and Stinear, T. (1998) Sensitive and rapid detection of viable Giardia cysts and *Cryptosporidium parvum* oocysts in large-volume water samples with wound fiberglass cartridge filters and reverse transcription-PCR. *Appl. Environ. Microbiol.* 64, 1743–1749.
32. Kostrzynska, M., Sankey, M., Haack, E., et al. (1999) Three sample preparation protocols for polymerase chain reaction based detection of *Cryptosporidium parvum* in environmental samples. *J. Microbiol. Methods* 35, 65–71.

33. Mayer, C. L. and Palmer, C. J. (1996) Evaluation of PCR, nested PCR, and fluorescent antibodies for detection of *Giardia* and *Cryptosporidium* species in wastewater. *Appl. Environ. Microbiol.* **62,** 2081–2085.
34. Rochelle, P. A., De Leon, R., Johnson, A., Stewart, M. H., and Wolfe, R. L. (1999) Evaluation of immunomagnetic separation for recovery of infectious *Cryptosporidium parvum* oocysts from environmental samples. *Appl. Environ. Microbiol.* **65,** 841–845.
35. Rochelle, P. A., De Leon, R., Stewart, M. H., and Wolfe, R. L. (1997) Comparison of primers and optimization of PCR conditions for detection of *Cryptosporidium parvum* and *Giardia lamblia* in water. *Appl. Environ. Microbiol.* **63,** 106–114.
36. Rochelle, P. A., Ferguson, D. M., Handojo, T. J., De Leon, R., Stewart, M. H., and Wolfe, R. L. (1997) An assay combining cell culture with reverse transcriptase PCR to detect and determine the infectivity of waterborne *Cryptosporidium parvum. Appl. Environ. Microbiol.* 63, 2029–2037.
37. Sluter, S. D., Tzipori, S., and Widmer, G. (1997) Parameters affecting polymerase chain reaction detection of waterborne *Cryptosporidium parvum* oocysts. *Appl. Microbiol. Biotechnol.* 48, 325–330.
38. Stinear, T., Matusan, A., Hines, K., and Sandery, M. (1996) Detection of a single viable *Cryptosporidium parvum* oocyst in environmental water concentrates by reverse transcription-PCR. *Appl. Environ. Microbiol.* **62,** 3385–3390.
39. Toze, S. (1999) PCR and the detection of microbial pathogens in water and wastewater. *Water Res.* **33,** 3545–3556.
40. Wiedenmann, A., Kruger, P., and Botzenhart, K. (1998) PCR detection of *Cryptosporidium parvum* in environmental samples—a review of published protocols and current developments. *J. Indust. Microbiol. Biotechnol.* **21,** 150–166.
41. Xiao, L., Alderisio, K., Limor, J., Royer, M., and Lal, A. A. (2000) Identification of species and sources of *Cryptosporidium* oocysts in storm waters with a small-subunit rRNA-based diagnostic and genotyping tool. *Appl. Environ. Microbiol.* 66, 5492–5498.
42. Xiao, L., Singh, A., Limor, J., Graczyk, T. K., Gradus, S., and Lal, A. (2001) Molecular characterization of *Cryptosporidium* oocysts in samples of raw surface water and wastewater. *Appl. Environ. Microbiol.* **67,** 1097–1101.
43. Lowery, C. J., Moore, J. E., Millar, B. C., et al. (2001) Occurrence and molecular genotyping of *Cryptosporidium* spp. in surface waters in Northern Ireland. *J. Appl. Microbiol.* **91,** 774–779.
44. Ono, K., Tsuji, H., Rai, S. K., et al. (2001) Contamination of River Water by *Cryptosporidium parvum* oocysts in Western Japan. *Appl. Environ. Microbiol.* **67,** 3832–3836.

16

Genotyping of *Cryptosporidium parvum* With Microsatellite Markers

Giovanni Widmer, Xiaochuan Feng, and Sultan Tanrıverdi

1. Introduction

Recent outbreaks of cryptosporidiosis caused by *Cryptosporidium parvum* in the United States and other countries ***(1,2)***, as well as the emergence of cryptosporidiosis as a frequent cause of morbidity and mortality in immunodeficient individuals ***(3)***, have raised the interest of the research community in this parasite. The genus *Cryptosporidium*, phylum Apicomplexa, comprises an undefined number of species, of which only *C. parvum* is of public health concern. Cryptosporidiosis is contracted through the ingestion of oocysts, the stage of the parasite produced in large numbers by infected hosts. Because the oocysts are small, typically about 5 μm in diameter, and lack species-specific morphological features, there is a need for molecular markers to distinguish between human-infectious *C. parvum* and other species that do not (or only infrequently) cause disease in humans. Genetic characterization of *Cryptosporidium* oocysts using restriction fragment length or sequence polymorphism has revealed host-associated genotypes, that are often referred to as species ***(4–6)***. In addition, *C. parvum* was found to include two genotypes, designated type 1 and type 2 (*see* **Note 1**). Type 1 is almost exclusively found in humans, whereas type 2 infects humans and various mammalian hosts ***(7–9)***. The frequent occurrence of *Cryptosporidium* oocysts in untreated surface water and the potential for contamination of drinking water ***(10,11)*** have emphasized the need for molecular markers to track the source of oocysts within a watershed or water distribution system, and to discriminate between oocysts infectious to humans and nonpathogenic species. Genetic markers are also needed to study the taxonomy of *Cryptosporidium*.

Several laboratories have identified microsatellites in the genome of *C. parvum* and have investigated the level of polymorphism at these loci ***(12–15)***. For instance, 10 alleles of marker 5B12 have been found to date among *C. parvum* isolates from various geographical and host origins. Multilocus haplotypes based on such markers are suitable for discriminating individual isolates (*see* **Note 2**) of *C. parvum*.

From: *Methods in Molecular Biology, vol. 268: Public Health Microbiology: Methods and Protocols*
Edited by: J. F. T. Spencer and A. L. Ragout de Spencer © Humana Press Inc., Totowa, NJ

In an attempt to develop rapid and cost-effective methods for typing isolates of *C. parvum*, we have pursued two methods, a traditional polymerase chain reaction (PCR) method followed by gel electrophoresis, and real-time PCR using SYBR Green I melting curve analysis for allele identification.

2. Materials

1. Nycodenz (*N,N'*-bis[2,3 dihydroxypropyl]acetamido-2,4,6-tri-iodo-isophthalamide; Histodenz™Sigma, St. Louis, MO).
2. Centrifuge tubes, ultraclear, for swing bucket rotor.
3. Immunomagnetic separation kit (Crypto-Scan®IMS, ImmuCell, Portland, ME) or Dynabeads anti-Cryptosporidium (Dynal, Lake Success, NY), or Aureon Crypto Kit, (Aureon Biosystems, Vienna, Austria).
4. Ethyl ether.
5. STE buffer: 100 m*M* NaCl, 10 m*M* Tris-HCl, pH 8.0, 1 m*M* EDTA.
6. Geneclean kit for DNA isolation (Qbiogene, Carlsbad, CA).
7. Zirconia/silica or glass beads, 0.5 mm (Biospec Products, Bartlesville, OK).
8. Mini-Beadbeater (Biospec Products).
9. 10X PCR buffer: 40 m*M* Tris-HCl, pH 8.3, 200 m*M* KCl, 0.004% gelatin.
10. 100 m*M* Deoxynucleoside triphosphates (Promega, Madison, WI).
11. 25 m*M* $MgCl_2$ solution.
12. Thermal cycler (MJ Research, Watertown, MA).
13. LightCycler (Roche, Diagnostics, Mannheim, Germany).
14. Polynucleotide kinase.
15. $[\gamma^{32}P]$ATP.
16. TBE buffer: 0.09 *M* Tris-borate, 1 m*M* EDTA, pH 8.0.
17. Spreadex Polymer NAB (Elchrom Scientific, Cham, Switzerland).
18. High Pure PCR template preparation kit (Roche Diagnostics).

3. Methods

First we describe several methods for extracting DNA from *Cryptosporidium* oocysts for PCR analysis. These methods are suitable for extracting DNA directly from fecal samples or from environmental samples containing oocysts. Two sample preparation methods are discussed: (1) direct extraction of fecal/environmental DNA; and (2) isolation of oocysts followed by DNA extraction. Direct extraction of DNA is faster, but we frequently prefer to purify the oocysts first, as a portion of these oocysts can subsequently be used for additional studies or animal propagation. Extracting DNA from purified oocysts also reduces the probability of contamination of the DNA with PCR inhibitors. Second, two PCR-based methods for analyzing microsatellite polymorphisms in *Cryptosporidium* are discussed: (1) traditional PCR, and (2) real-time PCR.

3.1. Extraction of DNA

3.1.1. Extraction of Fecal and Environmental DNA

3.1.1.1. ProteinaseK/SDS Method

DNA extracted from feces or water pellets containing *Cryptosporidium* oocysts is often suitable for PCR detection and genotyping. Since PCR inhibitors may copurify with environmental DNA ***(16,17)***, some investigators prefer to purify the oocysts first and then extract DNA from the oocysts. Using purified or partially purified oocysts may also be an advantage when working with a sample containing low concentrations of oocysts. Methods for purifying oocysts are discussed below in **Subheading 3.1.2.**

1. For the proteinase K/sodium dodecyl sulfate (SDS) extraction method, a 100–500 μL volume of feces or water pellet is homogenized in STE buffer until a slurry is obtained. The volume of buffer added to the sample depends on the consistency; typically an equal volume of sample and buffer are used.
2. Following homogenization, proteinase K is added at a concentration of 200 μg/mL and SDS to 0.1%.
3. The samples are then incubated overnight at 45°C and the DNA recovered by absorption to a silica matrix (GeneClean).

3.1.1.2. Beadbeater Method

1. One small volume of feces or water pellet is diluted with 2 vol of 10 m*M* Tris-HCl, pH 8.0, 25 m*M* EDTA, and 2% SDS in a 2-mL screw-cap microcentrifuge tube.
2. To 1 mL of this slurry, 200 μL of glass beads (0.5 mm diameter) and an equal volume of phenol/chloroform are added and agitated for 2 min at 5000 rpm in a beadbeater.
3. Samples are spun in a microcentrifuge to separate the phenol and precipitate the beads.
4. The aqueous phase containing the DNA is transferred to a new microcentrifuge leaving behind the pellet containing the beads.
5. NaCl is added to the sample from a 5 *M* solution to obtain a final concentration of 0.7 *M* and the solution is extracted with an equal volume of chloroform.
6. Final purification of the DNA is performed with GeneClean.

3.1.1.3. High Pure Extraction Method

1. A portion of 0.5 g of feces is homogenized in 2–5 mL of physiological (0.9%) saline and filtered through gauze.
2. The filtrate is centrifuged at 4000*g* for 10 min, and the precipitate containing the oocysts is resuspended by vortexing in 200 μL sterile distilled water.
3. This suspension is then subjected to three cycles of 37/–80°C freeze-thawing.
4. The lysate is then mixed with 200 μL of binding buffer and 40 μL proteinase K, both provided with the High Pure PCR template preparation kit, and incubated for 15 min at 72°C.
5. Isopropanol (100 μL) is then added and the mixture applied to the High Pure filter tube.
6. The tubes are spun at 6000*g* for 1 min in a microcentrifuge.
7. After centrifugation, 500 μL of inhibitor removal buffer (from the kit) is added and the centrifugation repeated.
8. The same volume of washing buffer (from the kit) is added to the filter tube and the centrifugation step repeated. This wash cycle is performed twice.
9. One volume of 200 μL elution buffer (from the kit) heated to 70°C is added to the filter tube, and the tubes are centrifuged at 6000*g* for 1 min.
10. A 1–10 μL portion of the eluate containing the DNA is then added directly to the PCR reaction.

3.1.2. Oocyst Purification

3.1.2.1. Oocyst Purification on Density Gradients

Several density media and gradients have been described for floating or sedimenting oocysts from fecal samples. These methods are used individually or in combination with other purification steps until the required level of oocyst purification is achieved.

To our knowledge, no systematic comparison of various oocyst purification methods has been published, and each investigator needs to experiment with different methods before deciding on the one that is most appropriate and meets the requirements of the experiment. Even if a comparative study of different purification protocols were performed, it is likely that the results achieved in one laboratory would not directly apply to another, since the sample matrix has a major impact on oocyst recovery. What may give excellent results with sludge or a diarrheic sample may not be satisfactory with solid fecal samples.

1. In our laboratory, after much experimenting with different gradients, we favor for most fecal samples an initial floatation on a high-density solution followed by sedimentation on an isopycnic gradient or by IMS.
2. We have experimented with two density media for floating oocysts from fecal slurries or sludge samples: 4 *M* NaCl and 36% w/v Nycodenz. Recoveries with these media are in the 30–50% range. Sodium chloride works well with most samples and is cheap. Nycodenz is more expensive but has the advantage of low osmolarity even when dissolved at high concentration ***(18)***, and it appears to float the oocysts better than salt, potentially improving oocyst recovery (**Fig. 1**).
3. Whichever floatation solution is used, a volume of fecal slurry is mixed with 2 vol of saturated (6.1 *M*) NaCl and centrifuged at approx 2000*g* for 15 min.
4. Floatation on Nycodenz is achieved by mixing 1 vol of fecal slurry with 10 vol of 40% Nycodenz, followed by centrifugation under the same conditions.
5. In both methods, but particularly with Nycodenz, the oocysts concentrate in the uppermost layer and can be recovered by aspiration.
6. Before proceeding to the next purification step, the density of the oocyst suspension recovered from the floats is reduced by diluting the oocyst suspension with 2–3 vol of water and precipitation in a microcentrifuge (~3000*g*).

Oocysts partially purified by floatation can be further purified by sedimentation on a continuous or step gradient. Several rate-zonal or isopycnic gradients have been used, including the popular Sheather's sugar flotation method ***(19)***. Cesium chloride, Percoll, and Nycodenz have also been used ***(20,21)*** (*see* **Note 3**). In our laboratory, we favor sedimentation on a 15/30% w/v Nycodenz step gradient.

1. Depending on the diameter of the centrifugation tube, a 1–5 mL of a 30% Nycodenz solution is used as a cushion.
2. The volume of high and low density fraction will depend on the diameter of the tube. The minimum volume to obtain a good separation of the interphase, where the oocysts will collect, from the pellet at the bottom, and the supernatant at the top, should be used. For instance, for a 13-mm-diameter tube, as used in a SW41 swing bucket rotor (Beckman, Fullerton, CA), 2–3 mL of 30% solution will be sufficient to form a cushion.
3. Onto this cushion, 5–10 mL of 15% solution is slowly layered, taking care not to mix the two phases.
4. Finally, the semipurified oocyst suspension is layered onto the 15% solution.
5. The tubes are then centrifuged in a swing bucket rotor at 10,000–15,000*g* for 30 min to 1 hour. To avoid the collapse of thin-walled centrifuge tubes during centrifugation, water or mineral oil can be added to increase the sample volume, although in our experience this is not necessary when working at these relative low *g*-forces.

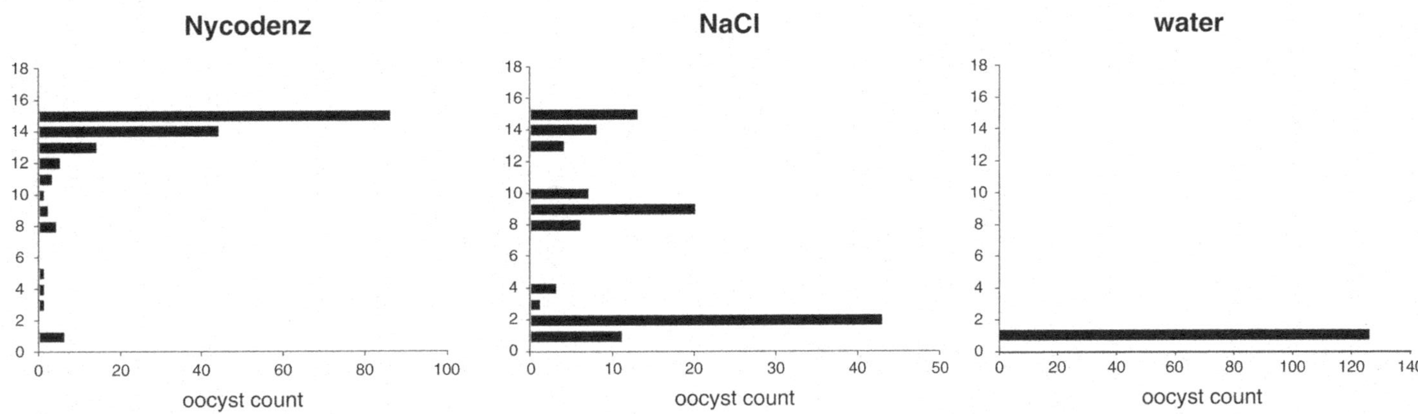

Fig. 1. Comparison of oocyst floatation on NaCl and Nycodenz. Oocysts (2.75×10^5) were mixed with 3 mL 40% Nycodenz or 3 mL saturated NaCl and spun at 2000*g* for 15 min approximately. Fractions of 200 μL were collected starting with the top layer, and the oocysts were counted using a hemocytometer. A mock floatation in water was performed in parallel.

6. For most molecular applications, and certainly for PCR analysis, oocysts obtained using a flotation step followed by gradient centrifugation are sufficiently clean, and DNA can be readily extracted using the proteinase K/SDS or the High Pure method described above.
7. Before proceeding with extraction of DNA from oocysts, oocysts are collected from the interphase between the high- and low-density phase using aspiration, diluted with a minimum of 1 vol of water, and precipitated by centrifugation. We have not observed any inhibitory effect to PCR from residual Nycodenz.

3.1.2.2. Oocyst Purification by Immunomagnetic Capture

Immunomagnetic capture of oocysts can be used as a quick alternative to gradient methods, or in combination with floatation or sedimentation. The choice of which method or combination of methods to use will depend on several variables, but primary considerations are sample volume, sample consistency, and the desired purity of the final oocyst preparation. Large sample volumes with low oocyst concentrations will require an initial concentration, and possibly a floatation before the oocysts can be IMS purified. The limitations of IMS are cost and sample volume (maximum 8 mL). Cost of IMS is an important consideration if large numbers of samples need to be processed. We typically use IMS when purifying oocysts from "difficult samples," i.e., samples with large amounts of solids and low concentrations of oocysts, and when high yields and/or high purity is needed. We have used immunomagnetic beads from three different manufacturers (Dynal, ImmuCell, and Aureon Biosystems) and have observed no obvious differences in recovery, although we have not systematically compared the kits. The recovery of non-parvum oocysts with these different products remains to be investigated, and researchers wishing to recover *Cryptosporidium* oocysts from other species should be aware that these products were developed for the purification of *C. parvum*. When working with human or bovine fecal samples, fat is often a problem, as it interferes with IMS.

1. To remove fat, fecal samples are first extracted with ether by adding an equal volume of 1% Tween-80, followed by vigorous shaking with a 40% vol of ether.
2. The ether and water phases are separated by centrifugation for 10 min at 4000*g* and the ether and fatty interphase removed by aspiration (*see* **Note 4**).
3. The pellet containing the oocysts can then be washed to remove residual ether and processed directly by IMS.
4. The instructions provided with each IMS kit are similar and sufficiently detailed to perform the procedure.
5. Leighton glass tubes can be purchased from the kit manufacturers, but glass tubes with similar dimension and a flat surface to provide a good contact with the magnets work equally well (*see* **Note 5**).

3.2. PCR Analysis of Cryptosporidium *Microsatellites*

Restriction fragment length polymorphism (RFLP) has been used for identifying *Cryptosporidium* species ***(22)*** and for subspecies identification ***(9,23)***. In contrast, microsatellite and DNA sequence polymorphism has revealed the presence of genetic heterogeneity within *C. parvum* types. These markers are useful for epidemiological studies and source tracking and have recently been applied to the study of genetic recombination ***(24)***.

3.2.1. Conventional PCR Analysis of C. parvum Microsatellites

1. PCR amplifications are performed in 25-μL volumes containing 1X PCR buffer, 2 m*M* $MgCl_2$, 0.2 m*M* dNTPs, and 10 pMol of each forward and reverse primer ***(13)***. If the amplification products are to be radioactively labeled, 2 pMol of 5' ^{32}[P]-labeled reverse primer is used instead of unlabelled reverse primer.
2. Primers are end-labeled with polynucleotide kinase and [γ-^{32}P]ATP.
3. The PCR is run in a conventional thermal cycler using the following temperature parameters: 36 cycles of 15 s at 94°C, 25 s at the respective primer annealing temperature, and 40 s at 72°C. We use a MiniCycler or a PTC-100 thermal cycler from MJ Research.
4. For gel analysis, radiolabeled PCR products (unpurified) are mixed with an equal volume of loading buffer (10 m*M* NaOH, 95% formamide, 0.05% bromophenol blue, 0.05% xylene cyanol), heat-denatured for 1–2 min at 95–100°C, and fractionated on a 8% polyacrylamide, 8 M urea sequencing gel at 2000 V/36 W. The electrophoresis buffer is 0.5X TBE in the top chamber and 1X TBE in the bottom chamber.
5. Following electrophoresis, the gel is dried at 80°C under vacuum and the PCR products visualized by autoradiography.

Because of the inconvenience of working with radioactivity, an alternative method to visualize PCR amplicons was developed. In this method, amplicons are electrophoresed on 10% native polyacrylamide supplemented with 10% Spreadex Polymer NAB and visualized by immersing the gel in an ethidium bromide solution (**Fig. 2**). Although this method may not resolve bands differing in size by a single base pair, it resolves size differences as small as 2 bp.

3.2.2. Real-Time PCR Analysis

Real-time PCR combined with melting curve analysis (MCA) is currently being evaluated as an alternative method for genotyping *C. parvum* using microsatellite polymorphisms. Because MCA eliminates the need for electrophoretic analysis, this approach is much quicker than traditional gel analysis of PCR products. In our laboratory we have tested several PCR protocols targeted at microsatellites using a LightCycler real-time PCR system. Although this method is still being evaluated, it appears that repeats containing alternating G/C and A/T are amplified more efficiently than those consisting only AT repeats (*see* **Note 6**).

Microsatellite Cp492 ***(24,25)*** consists of a variable number of CAC repeats. To date, we have identified 6 Cp492 alleles in *C. parvum* and a different allele in *C. meleagridis*. Various primers flanking the repeat were synthesized such that nested amplifications can be performed. Different primer combinations are also useful in case samples fail to amplify owing to the presence of unknown nucleotide polymorphisms in the region flanking the repeat. Because Cp492 is located within an open reading frame, positioning the PCR primers such that their 3' end does not fall on the third codon position reduces the possibility of primer failures.

1. PCR amplification in a LightCycler is performed in 20-μL vol containing 1X LC Fast Start DNA Master Mix for SYBR Green I, 4 m*M* $MgCl_2$, 10 pMol of each forward and reverse primer, and 1–10 μL DNA template.
2. The PCR premix and DNA template are added to the LightCycler capillaries, and the capillaries capped and centrifuged using the appropriate centrifuge adapters (Roche Diagnostics).

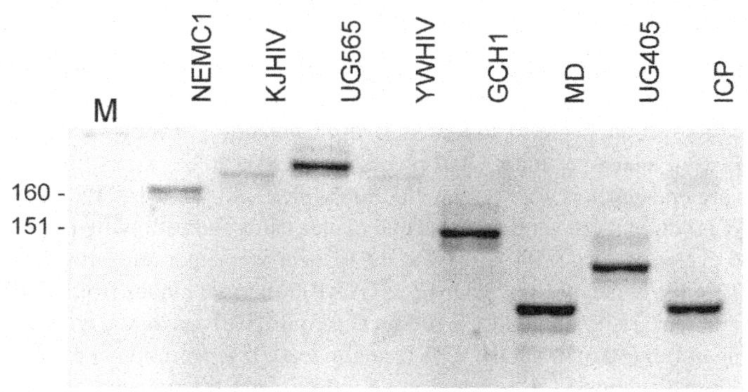

Fig. 2. Fractionation of Cp492 microsatellite locus on a 10% acrylamide gel supplemented with 10% Spreadex. *C. parvum* isolates NEMC1 ***(26)***, KJHIV, UG565, and YWHIV are type 1; GCH1 ***(27)***, MD ***(28)***, UG405 ***(13)***, and ICP ***(29)*** are type 2.

3. A typical amplification is performed using the following temperature conditions: one cycle of 10 min at 95°C, 40 cycles of 0–1 s at 95°C, 4–5 s at the respective annealing temperature, and 7–22 s (depending on the length of template) at 72°C.
4. Temperature transitions of 20°C/s seem to be adequate for most amplifications, but certain sequences seem to amplify better under different conditions. Optimal amplification conditions, including the concentration of $MgCl_2$, need to be determined empirically (*see* **Note 7**).
5. PCR products are typed using MCA analysis (**Fig. 3**) by raising the temperature from 45 to 95°C at a rate of 0.05°C/s.
6. As with traditional PCR methods, a positive control amplification using a standard template and a negative control are included with each reaction.
7. When using a new PCR protocol, it is desirable to confirm by gel analysis that an amplicon of the expected size has been amplified. However, electrophoresis is not needed once the range of melting temperatures for a specific marker has been determined.

4. Notes

1. Several designations for *C. parvum* types are used in the literature. Here we will adhere to the type 1/type 2 definition adopted in 1999 at a meeting entitled "*Cryptosporidium* spp. Systematics and Waterborne Challenges in Public Health," sponsored by the EPA Office of Water, Crystal City, Virginia April 29–30.
2. The term *isolate* is used to describe a population of parasites originating from a host. Isolate may originate from naturally infected animals or humans or may be artificially maintained in the laboratory by serial propagation.
3. Isopycnic gradients of CsCl have also been used by other investigators, but we have experienced difficulty extracting DNA from oocysts purified on CsCl.
4. For extracting the ether supernatant and the fatty interphase, a Pasteur pipet connected to a source of vacuum works well. If clogging of the pipet occurs while one is aspirating the interphase, the opening of the pipet can be enlarged by breaking off the tip of a clean

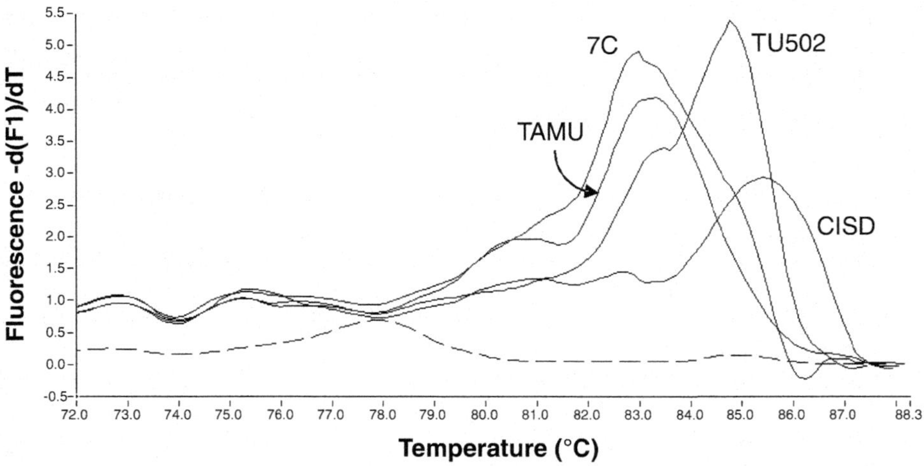

Fig. 3. SYBR Green I melting curve analysis of Cp492 amplicons from one *C. parvum* type 1 isolate (TU502) and three type 2 isolates (TAMU, 7C, CISD). Because the differences in the melting temperature are determined by the length of the amplicon and by the base composition, melting temperatures do not always correlate with amplicon length or gel mobility. In the example shown, isolate TU502 has a lower melting temperature than isolate CISD in spite of the fact that TU502 has the highest number of repeats. F1, SYBR Green fluorescence measured at 530 nm. The negative value of the first derivative of F1 over the temperature is plotted on the *y*-axis.

pipet and repeating the aspiration. As with all procedures involving ether, buildup of pressure should be prevented by periodically venting the tubes while performing the extraction.

5. Because of the relatively high cost of the immunomagnetic products, it is often possible to reduce the volume of immunomagnetic beads and adjust the buffer volume proportionally. Unless we expect to extract a large number of oocysts, say more than 10^7, the volume of beads and buffers can be reduced to one-half (50-µL bead suspension) or one-third (30-µL bead suspension) of the recommended volumes. For routine genetic analyses, this method will recover sufficient oocysts for extracting DNA and performing multiple PCR assays.
6. Several modifications to the standard amplification program were attempted, but the amplification curves obtained with (AT) microsatellites remained flat in comparison with those amplified from nonrepeat sequences or repeats containing C or G residues. This observation indicates suboptimal amplification. In addition, melting curves obtained from (AT) microsatellites frequently display multiple peaks. We assume that this is caused by partial melting of the AT-rich region at a lower temperature than that of the sequences flanking the repeat. If this assumption is confirmed, (AT) repeats may not be suitable for real-time PCR typing. Additional microsatellites containing repeats of G and/or C residues are being evaluated as alternative markers.
7. The general applicability of SYBR Green I real-time PCR for allele discrimination is still being investigated. Preliminary results indicate that AT microsatellites are not efficiently amplified. Repeats containing G and C nucleotides appear to amplify better, but rela-

tively large differences in repeat length (6 bp or more) may only result in relatively small differences in melting temperature. This observation contrasts with those from real-time PCR methods based on internal fluorescent probes, which can generate measurable and reproducible melting profiles on the basis of single-nucleotide polymorphisms ***(30,31)***.

References

1. Barwick, R. S., Levy, D. A., Craun, G. F., Beach, M. J., and Calderon R. L. (2000) Surveillance for waterborne-disease outbreaks—United States, 1997–1998. *MMWR* **49,** 1–35.
2. McLauchlin. J., Amar, C., Pedraza-Diaz, S., and Nichols, G. L. (2000) Molecular epidemiological analysis of *Cryptosporidium* spp. in the United Kingdom: results of genotyping *Cryptosporidium* spp. in 1,705 fecal samples from humans and 105 fecal samples from livestock animals. *J. Clin. Microbiol.* **38,** 3984–3990.
3. Hunter, P. R. and Nichols, G. (2002) Epidemiology and clinical features of *Cryptosporidium* infection in immunocompromised patients. *Clin. Microbiol. Rev.* **15,** 145–154.
4. Morgan, U. M., Sargent, K. D., Elliot, A., and Thompson, R. C. (1998) *Cryptosporidium* in cats—additional evidence for *C. felis. Vet. J.* **156,** 159–161.
5. Lindsay, D. S., Upton, S. J., Owens, D. S., Morgan, U. M., Mead, J. R., and Blagburn, B. L. (2000) *Cryptosporidium andersoni* n. sp. (Apicomplexa: Cryptosporiidae) from cattle, *Bos taurus. J. Eukaryot. Microbiol.* **47,** 91–95.
6. Fayer, R., Trout, J. M., Xiao, L., Morgan, U. M., Lai, A. A., and Dubey, J. P. (2001) *Cryptosporidium canis* n. sp. from domestic dogs. *J. Parasitol.* **87,** 1415–1422.
7. Awad-El-Kariem, F.M., Robinson, H.A., Petry, F., McDonald, V., Evans, D., and Casemore, D. (1998) Differentiation between human and animal isolates of *Cryptosporidium parvum* using molecular and biological markers. *Parasitol. Res.* **84,** 297–301.
8. Peng, M. M., Xiao, L., Freeman, A. R., et al. (1997) Genetic polymorphism among *Cryptosporidium parvum* isolates: evidence of two distinct human transmission cycles. *Emerg. Infect. Dis.* **3,** 567–573.
9. Widmer, G. (1998) Genetic heterogeneity and PCR detection of *Cryptosporidium parvum. Adv. Parasitol.* **40,** 223–239.
10. LeChevallier, M. W., Norton, W. D., and Lee, R. G. (1991) Occurrence of *Giardia* and *Cryptosporidium* spp. in surface water supplies. *Appl. Environ. Microbiol.* **57,** 2610–2616.
11. Smith, H. V. and Rose J. B. (1998) Waterborne cryptosporidiosis: current status. *Parasitol. Today* **14,** 14–22.
12. Cacciò S., Homan, W., Camilli, R., Traldi, G., Kortbeek, T., and Pozio, E. (2000) A microsatellite marker reveals population heterogeneity within human and animal genotypes of *Cryptosporidium parvum. Parasitology* **120,** 237–244.
13. Feng, X., Rich, S. M., Akiyoshi, D., et al. (2000) Extensive polymorphism in *Cryptosporidium parvum* identified by multilocus microsatellite analysis. *Appl. Environ. Microbiol.* **66,** 3344–3349.
14. Strong, W. B., Gut, J., and Nelson, R. G. (2000) Cloning and sequence analysis of a highly polymorphic *Cryptosporidium parvum* gene encoding a 60-kilodalton glycoprotein and characterization of its 15- and 45-kilodalton zoite surface antigen products. *Infect. Immun.* **68,** 4117–4134.
15. Leav, B. A., Mackay, M. R., Anyanwu, A., et al. (2002) Analysis of sequence diversity at the highly polymorphic Cpgp40/15 locus among *Cryptosporidium* isolates from human immunodeficiency virus-infected children in South Africa. Infect. Immun. **70,** 3881–3890.

16. Trevors, J. T. and van Elsas, J. D. (1995) Introduction to nucleic acids in the environment: methods and applications. In: *Nucleic Acids in the Environment. Methods and Applications* (Trevors, J. T. and van Elsas, J. D., eds.). Springer-Verlag, Berlin, pp. 1–7.
17. Sluter, S. D., Tzipori, S., and Widmer, G. (1997) Parameters affecting polymerase chain reaction detection of waterborne *Cryptosporidium parvum* oocysts. *Appl. Microbiol. Biotechnol.* **48,** 325–330.
18. Rickwood, D. (1987) *Centrifugation. A Practical Approach.* IRL, Washington, DC, p. 29.
19. Anderson, B. C. (1981) Patterns of shedding of cryptosporidial oocysts in Idaho calves. *J. Am. Vet. Med. Assoc.* **178,** 982–984.
20. Current, W. L. (1990) Techniques and laboratory maintenance of Cryptosporidium. In: *Cryptosporidiosis of Man and Animals* (Dubey, J. P., Speer, C. A., and Fayer, R., eds.). CRC, Boca Raton, FL, pp. 31–49.
21. Widmer, G., Tzipori, S., Fichtenbaum, C. J., and Griffiths, J. K. (1998) Genotypic and phenotypic characterization of *Cryptosporidium parvum* isolates from people with AIDS. *J. Infect. Dis.* **178,** 834–840.
22. Xiao, L., Alderisio, K., Limor, J., Royer, M., and Lal, A. A. (2000) Identification of species and sources of *Cryptosporidium* oocysts in storm waters with a small-subunit rRNA-based diagnostic and genotyping tool. *Appl. Environ. Microbiol.* **66,** 5492–5498.
23. Sulaiman, I. M., Xiao, L., and Lal, A. A. (1999) Evaluation of *Cryptosporidium parvum* genotyping techniques. *Appl. Environ. Microbiol.* **65,** 4431–4435.
24. Feng, X., Rich S. M., Tzipori S., and Widmer G. (2002) Experimental evidence for genetic recombination in the opportunistic pathogen *Cryptosporidium parvum. Mol. Biochem. Parasitol.* **119,** 55–62.
25. Strong W. B. and Nelson, R. G. (2000) Preliminary profile of the *Cryptosporidium parvum* genome: an expressed sequence tag and genome survey sequence analysis. *Mol. Biochem. Parasitol.* **107,** 1–32.
26. Widmer, G., Akiyoshi, D., Buckholt, M. A., et al. (2000) Animal propagation and genomic survey of a genotype 1 isolate of *Cryptosporidium parvum. Mol. Biochem. Parasitol.* **108,** 187–197.
27. Tzipori, S., Rand, W., Griffiths, J., Widmer, G., and Crabb, J. (1994) Evaluation of an animal model system for cryptosporidiosis: therapeutic efficacy of paromomycin and hyperimmune bovine colostrum-immunoglobulin. *Clin. Diagn. Lab. Immunol.* **1,** 450–463.
28. Okhuysen, P. C., Rich, S. M., Chappell, C. L., et al. (2002) Infectivity of a *Cryptosporidium parvum* isolate of cervine origin for healthy adults and interferon-gamma knockout mice. *J. Infect. Dis.* **185,** 1320–1325.
29. Carraway, M., Tzipori, S., and Widmer, G. (1996) Identification of genetic heterogeneity in the *Cryptosporidium parvum* ribosomal repeat. *Appl. Environ. Microbiol.* **62,** 712–716.
30. Limor, J. R., Lal, A. A., and Xiao, L. (2002) Detection and differentiation of *Cryptosporidium* parasites that are pathogenic for humans by real-time PCR. *J. Clin. Microbiol.* **40,** 2335–2338.
31. Tanrıverdi, S., Tanyeli, A., Baslamisli, F., et al. (2002) Detection and genotyping of oocysts of *Cryptosporidium parvum* by real-time PCR and melting curve analysis. *J. Clin. Microbiol.* **40,** 3237–3244.

17

Immunomagnetic Separation of Pathogenic Organisms From Environmental Matrices

Gary P. Yakub and Kathleen L. Stadterman-Knauer

1. Introduction

One of the most difficult challenges in the analysis of environmental samples is to separate the organism of interest from a sample that is high in background debris. Immunomagnetic separation (IMS) is one technique that has been developed to accomplish this in a rapid and reliable assay.

Immunomagnetic separation (or biomagnetic separation) involves a superparamagnetic, monodispersed, polystyrene microsphere that is coated with a specific ligand. When added to a heterogeneous target suspension, the microspheres bind to the desired target. Using a powerful magnet, the microsphere-target complex is then removed from the suspension *(1)*. Many different targets of interest can be isolated with this technique, including fungal/bacterial cells or spores, protozoan parasites, cellular and subcellular material, proteins, and nucleic acid products. This wide range of application makes IMS one of the most versatile techniques available for the purification of target products from heterogeneous sample matrices.

1.1. Molecular Separation

Molecular separations involve the binding of a superparamagnetic microsphere to a nucleic acid product such as a segment of cDNA or tRNA. This technique is useful for the purification of nucleic acid products prior to downstream applications such as PCR and hybridization *(2–6)*.

1.2. Clinical Separation

Clinical separations involve the isolation of a target from a sample of human tissue, body fluid, or waste product. Some of the targets that have been isolated include proteins *(7,8)*, specific cell types such as cancer cells *(9,10)*, peripheral blood mononuclear cells such as monocytes, T-cells, B-cells, and granulocytes *(11–13)*, and a variety of pathogenic organisms such as *Shigella* spp. *(14)*, *Escherichia coli* 0157 strains *(15)*, *Mycobacterium* spp. *(16,17)*, and *Bordetella pertussis (18)*.

From: *Methods in Molecular Biology, vol. 268: Public Health Microbiology: Methods and Protocols*
Edited by: J. F. T. Spencer and A. L. Ragout de Spencer © Humana Press Inc., Totowa, NJ

1.3. Environmental Separation

Environmental separations involve the isolation of a target organism from an environmental matrix such as soil, surface or ground water, raw and treated wastewaters, animal tissue, fluid, feces, and food or food byproducts. In the environmental field, IMS is most often used to isolate organisms that are pathogenic to humans. Organisms such as *Cryptosporidium parvum* ***(19–21)***, *E. coli* O157:H7 ***(22–27)*** *Mycobacterium* ***(30,31)***, *Helicobacter pylori* ***(32,33)***, *Vibrio* spp. ***(34,35)***, *Microsporidia* ***(36)***, *Yersinia enterocolitica* ***(37)***, *Bacillus* spp. ***(38,39)***, *Coxiella* ***(40)***, and hepatitus A virus ***(41)*** have all been successfully isolated with the IMS technique. Additionally, fungal plant pathogens ***(42)***, equine pathogens ***(43)***, and sulfate-reducing bacteria of importance to the offshore oil industry ***(44)*** have also been isolated from environmental matrices by the IMS procedure.

The basic IMS procedure is fairly straightforward. A sample or sample concentrate is obtained. One or more buffers are added to optimize the separation environment. The proper immunomagnetic beads are added and the sample is mixed for a period of time to allow the beads to associate with the target organism. A powerful magnet is used to separate the bead-target complex from the remaining debris. The beads are then dissociated from the target organisms and removed from the suspension. What remains are the purified target organisms. The specific methodology can be somewhat variable owing to the inherent variability of the different target organisms, and an exhaustive presentation of each individual organism's technique would be beyond the scope of this volume. A representative environmental IMS methodology will be presented in detail to allow the reader to become familiar with the basic techniques. This methodology can then be easily adapted to other target organisms of interest.

1.4. IMS of Cryptosporidium parvum

As a model, the remaining sections of this chapter will deal with the immunomagnetic separation of the human intestinal parasite, *C. parvum*, from environmental matrices. *C. parvum* is a coccidian protozoan pathogen of the mammalian digestive tract. It infects a variety of domestic animals and is highly prevalent in cattle. It forms an environmentally stable oocyst, 4–6 μm in diameter. The main infection mode is the fecal-oral route. Oocysts can enter the potable water supply through fecal contamination ***(45)***. The resulting disease in humans, cryptosporidiosis, causes gastrointestinal upset, profuse watery diarrhea, cramps, and nausea. This disease is self-limiting in healthy immunocompetent hosts but can be fatal to immunosuppressed individuals as well as the very old and the very young ***(46)***. The infectious dose for 50% of immunocompetent individuals (ID_{50}) is 132 to as few as 30 oocysts ***(47)***. For the immunosuppressed, the infectious dose has been estimated to be as low as one oocyst ***(48)***. The mortality rates of *C. parvum* among these individuals vary from 52 to 68% ***(49)***. The largest documented outbreak of cryptosporidiosis occurred in the spring of 1993 in Milwaukee. Oocysts passed through the drinking water treatment plant and infected the general population. An estimated 403,000 people were affected, and several people died as a result of the disease ***(50)***. Because of the risk to the immunocompromised and the potential for oocysts to pass through drinking water treatment systems,

Cryptosporidium has become one of the most important contaminants found in drinking water today ***(49)***.

Current methods to enumerate oocysts from environmental and drinking waters include IMS. The United States Environmental Protection Agency (EPA) method 1623 ***(51)*** for water samples includes an initial concentration step using a cartridge filter. IMS is then used to isolate and purify the oocysts from the sample concentrate, followed by an immunofluorescent microscopic assay. Researchers have reported average oocyst recoveries in deionized water ranging from 68 to 83% using the Dynal kit described below ***(52)***. At turbidity levels up to 500 NTU, oocyst recoveries were similar to the recoveries obtained in deionized water. At a turbidity of 5000 NTU, oocyst recoveries greater than 35% were obtained ***(52)***. Other researchers have reported average recoveries of oocysts ranging from 62 to 100% in seeded environmental water concentrates with turbidities ranging from 210 NTU to 11,480 NTU ***(53)***.

2. Materials

1. Flat-sided sample tubes, 16 × 125 mm Leighton type with 60 × 10 mm flat-sided magnetic capture area (Dynal, Lake Success, NY, L10, cat. no. 740.03 or equivalent) (*see* **Note 1**).
2. Sample mixer: Dynal cat. no. 947.01 or equivalent (*see* **Note 2**).
3. Magnetic particle concentrator for 10 mL and test tubes (Dynal MPC-1®, cat. no. 120.01 or equivalent).
4. Magnetic particle concentrator for microcentrifuge tubes (Dynal MPC-M®, cat. no. 120.09 or equivalent (*see* **Note 3**).
5. Merifluor Direct Immunofluorescence Assay for *Cryptosporidium/Giardia* (Meridian Diagnostics, Cincinnati, OH, cat. no. 250050).
6. Dynabeads® anti-*Cryptosporidium* kit (Dynal cat. no. 730.01 or equivalent) (*see* **Notes 4–6**). Each kit contains ***(54)***:
   a. Anti-*Cryptosporidium* immunomagnetic beads, coated with purified antibodies against *Cryptosporidium* that are covalently bonded to the bead surface. The beads are supplied as a suspension in phosphate-buffered saline (PBS), pH 7.4, with 0.1% bovine serum albumin (BSA).
   b. 10X SL buffer A (clear, colorless solution).
   c. 10X SL buffer B (magenta solution).
7. Hydrochloric acid, 0.1 *N*, ACS grade or equivalent.
8. Sodium hydroxide, 1.0 *N*, ACS grade or equivalent.

3. Methods *(51)*

1. Begin with a water sample or water sample concentrate of approx 10 mL contained in a Leighton tube (*see* **Notes 7–10**).
2. Prepare a 1X dilution of SL buffer A (clear, colorless solution) in reagent water from the 10X solution provided with the Dynal kit. A volume of approx 1.5 mL 1X SL buffer A will be needed for each sample analyzed.
3. To each Leighton tube containing a sample, add 1 mL of the 10X SL buffer A (not the diluted 1X SL buffer A).
4. To each Leighton tube from **step 3**, add 1 mL. 10X SL buffer B (magenta solution).

5. To each Leighton tube from **step 4**, add 100 µL of the anti-*Cryptosporidium* bead suspension (*see* **Note 11**). Tighten the screw cap on the Leighton tube.
6. Affix the sample tube to the Dynal sample mixer and rotate at approx 18 rpm for 1 h at room temperature.
7. After rotating 1 h, remove the sample tube from the mixer and place it in the MPC-1 concentrator with the flat side toward the magnet.
8. Gently rock the tube through 90°, tilting cap end and base end up and down in turn, for 2 min with approx one tilt/s (*see* **Notes 12–15**).
9. After tilting, rapidly remove the cap and pour the liquid into a suitable container. Do not remove the Leighton tube from the MPC-1 concentrator during this step (*see* **Notes 16** and **17**).
10. Return the Leighton tube to the upright position and remove the tube from the MPC-1 concentrator. Resuspend the sample in 1 mL of the 1X SL buffer A (prepared in **step 2**). Mix gently to resuspend all material (*see* **Note 18**). *Do not vortex.*
11. Quantitatively transfer all liquid to a labeled 1.5-mL microcentrifuge tube (*see* **Note 19**).
12. Place the microcentrifuge tube into the MPC-M concentrator with the magnetic strip in place.
13. Gently rock the MPC-M concentrator through 180° by hand for 1 min with approx one rock/s (*see* **Note 20**).
14. Immediately aspirate the remaining liquid from the microcentrifuge tube and cap (*see* **Notes 21** and **22**). If you are processing more than one sample, conduct three 90° rolling actions between the processing of each tube. Do not remove the tubes from the MPC-M concentrator during this step (*see* **Note 23**).
15. Remove the magnetic strip from the MPC-M concentrator (*see* **Note 24**).
16. Add 50 µL of 0.1 *N* hydrochloric acid, cap the microcentrifuge tube, and vortex vigorously for 10–15 s (*see* **Notes 25** and **26**).
17. Place the tube in the MPC-M concentrator without the magnetic strip and allow to stand in a vertical position for at least 10 min at room temperature.
18. Vortex vigorously for 5–10 s (*see* **Note 26**).
19. Replace the magnetic strip in the MPC-M concentrator and allow to stand undisturbed for approx 10 s.
20. Prepare a well slide from the Merifluor kit by adding 5 µL of 1.0 *N* sodium hydroxide to each sample well (*see* **Note 27**). Prepare one well for each processed Leighton tube.
21. Without removing the microcentrifuge tube from the MPC-M concentrator, transfer all the liquid from the tube and cap onto the prepared well slide (*see* **Notes 28** and **29**).
22. Allow the slide to sit undisturbed overnight to air dry (*see* **Note 30**). Proceed with an appropriate immunofluorescent assay to identify the *Cryptosporidium* oocysts (*see* **Note 31**).

4. Notes

1. Our laboratory utilizes Dynal (Oslo, Norway) as our sole supplier for all IMS reagents and equipment. Dynal equipment is an integral part of US EPA method 1623 ***(51)*** and has been shown by other researchers to provide the best recoveries for Cryptosporidium ***(52,53)***.
2. The sample mixer holds up to twelve 10-mL Leighton tubes. If you plan to process batches of more than 12 samples frequently, it would be advisable to purchase two mixers.
3. It is advisable to purchase one MPC-1 and one MPC-M concentrator for each analyst who will be working at the same time, to minimize sample processing time.

4. If you are testing for both *Cryptosporidium* and *Giardia*, Dynal makes a combination kit for the simultaneous recovery of both organisms (cat. no. 730.02).
5. Store the Dynal kits at 4–8°C and use prior to the expiration date listed on the carton.
6. One test kit contains enough reagent for 10 tests.
7. Because some organisms are found rather infrequently in nature, a general procedure must be developed to concentrate a large water sample down to a 10-mL. volume. For *Cryptosporidium* oocysts, much research is currently being conducted to determine which methods of sample concentration are best suited to a particular matrix. Processes such as centrifugation, density gradient floatation, and microfiltration are currently being investigated for utility in specific environmental matrices. For samples in which bacteria are the targets, a pre-enrichment step may be necessary.
8. It is recommended that the water concentrate packed pellet volume not exceed a 0.5 mL vol. If this is the case, vortex sample for 2 min, subdivide into two or more Leighton tubes, and bring the volume of each tube back to 10 mL. Pool all results for the final answer.
9. Water concentrates high in some chemical or physical compounds may exhibit an inhibitory effect on the IMS protocol. Two compounds known to cause interference are turbidity ***(52)*** and dissolved iron ***(55)***. Analyzing quality control recoveries of the intended target organisms can help to characterize the matrix in terms of interference as well as to monitor the separation process.
10. Allow all Dynal kit reagents to come to room temperature before use.
11. Be sure to vortex the bead suspension for a minimum of 10 s to resuspend the beads. After vortexing, invert the tube and inspect for signs of beads still sitting on the bottom of the tube. (Beads appear as reddish brown grains in the *Cryptosporidium* kit.) Vortex as long as needed to resuspend all the beads.
12. It is possible during the tilting process for the Leighton tube to fall out of the MPC-1 concentrator. Be sure to support the tube during the tilting process. Do not rely on the clip to hold the tube in place.
13. Use a mechanical countdown timer to time all steps accurately in the procedure.
14. Ensure that the tilting procedure is not interrupted during this step to avoid binding of magnetizable material in the sample.
15. The 90° vertical agitation used with the Dynal magnetic particle concentrator ensures that particulate matter and sample debris remain suspended in the solution in the Leighton tube while the bead-oocyst complexes are captured by the magnet ***(53)***.
16. Because of the possible presence of unrecovered pathogens, pour the liquid into an autoclavable container and autoclave prior to disposal.
17. Researchers have reported that 3.1–4.6% of the oocysts were not captured in the initial separation ***(53)***.
18. During mixing, be sure to wash all the material off the flat side of the tube. This is best accomplished by adding the 1X SL buffer A near the top of the tube with a micropipet and allowing it to wash down the inside square surface.
19. Transfer the bead suspension either by direct pouring or the use of a micropipet. There will not be much room to add rinsate to the microcentrifuge tube, so restrict rinsing of the Leighton tube to three 200-µL vol of reagent water. Do not overfill the microcentrifuge tube.
20. At the end of this step, the beads should appear as a distinct brown dot at the back of the tube.
21. For this step, we recommend using a disposable glass Pastuer pipet and bulb. The long thin tip is less likely to disturb the bead pellet. Lower the tip into the solution along the

front of the tube wall, gently aspirate out the solution, and remove the tip following the front wall of the microcentrifuge tube.

22. Dispose of this liquid into the same container used in **step 9** (*see* **Note 15**).
23. Researchers have reported that 0.6–3% of the oocysts remained in the microcentrifuge tube ***(53)***.
24. This is the beginning of the dissociation step, whereby the superparamagnetic beads are removed from the target oocysts. The procedure outlined here is followed by slide fixing and microscopic examination of the captured oocysts. If the captured oocysts are to be subjected to viability studies or further molecular examination (i.e., PCR), other researchers have outlined dissociation protocols that will not interfere with these downstream activities ***(53,56)***.
25. Researchers have reported that 2–8% of the oocysts remain attached to the paramagnetic beads following the acid dissociation step ***(53)***.
26. Do not vortex so vigorously as to force the suspension into the underside of the microcentrifuge tube cap.
27. The Merifluor kit contains positive and negative control suspensions. Place a drop of each suspension onto separate wells of a treated well slide and process with the samples.
28. A micropipetor is adequate for this step. We use a 10–100-μL adjustable micropipetor set for approx 80 μL in order to ensure that all the liquid is collected. The transfer may be accomplished in several passes.
29. Place the solution from the microcentrifuge tube directly into the bead of sodium hydroxide on the slide well. This will ensure adequate acid neutralization.
30. We cover the slides with an inverted plastic bowl to prevent disturbance or contamination.
31. Our laboratory currently utilizes the Merifluor kit listed in **Subheading 2.** (Materials), following the procedure outlined in Section 14 of USEPA Method 1623 ***(51)***.

References

1. Dynal A. S. (1998) *Dynabeads® Biomagnetic Applications in Cellular Immunology*. Dynal, Oslo.
2. Jakobsen, K. S., Haugen, M., Sæbøe-Larssen, S., Hollung, K., Espelund, M., and Hornes, E. (1994) Direct mRNA isolation using magnetic oligo (dT) beads: a protocol for all types of cell cultures, animal and plant tissues. In: *Advances in Biomagnetic Separation* (Uhlén, M., Hornes, E., and Olsvik, O., eds.). Eaton, Natick, MA, pp. 61–72.
3. Korn, B., Sedlacek, Z., and Poustka, A. (1994) Isolation of transcribed and conserved sequences from large genomic regions by magnetic capture. In: *Advances in Biomagnetic Separation* (Uhlén, M., Hornes, E., and Olsvik, O., eds.). Eaton, Natick, MA, pp. 79–90.
4. Tagle, D. A., Swaroop, M., Elmer, L., et al. (1994) Magnetic bead capture of cDNAs: a startegy for isolating expressed sequences encoded within large genomic segments. In: *Advances in Biomagnetic Separation* (Uhlén, M., Hornes, E., and Olsvik, O., eds.). Eaton, Natick, MA, pp. 91–106.
5. Mörl, M., Dorner, M., and Pääbo, S. (1994) Direct purification of tRNAs using oligonucleotides coupled to magnetic beads. In: *Advances in Biomagnetic Separation* (Uhlén, M., Hornes, E., and Olsvik, O., eds.). Eaton, Natick, MA, pp. 107–112.
6. Mizutani, S., Asada, M., Wada, H., Yamada, A., and Kodama, C. (1994) Magnetic separation in molecular studies of human leukemia. In: *Advances in Biomagnetic Separation* (Uhlén, M., Hornes, E., and Olsvik, O., eds.). Eaton, Natick, MA, pp. 127–134.
7. Lawn, S. D., Roberts, B. D., Griffin, G., Folks, T., and Butera, S. (2000) Cellular compartments of human immunodeficiency virus type 1 replication in vivo: determination by pres-

ence of virion-associated host proteins and impact of opportunistic infection. *J. Virol.* **74,** 139–194.

8. Rossomando, E. F. and White, L. (1994) *In situ* immunomagnetic capture of proteins from body fluids. In: *Advances in Biomagnetic Separation* (Uhlén, M., Hornes, E., and Olsvik, O., eds.). Eaton, Natick, MA, pp. 187–194.
9. Zhong, X. Y., Kaul, S., and Lin, Y. S. (2000) Sensitive detection of micrometastases in bone marrow from patients with breast cancer using immunomagnetic isolation of tumor cells in combination with reverse transcriptase/polymerase chain reaction for cytokeratin-19. *J. Cancer Res. Clin. Oncol.* **126,** 212–218.
10. Calenda, V., Bolmont, C., and Chermann, J.-C. (1994) Immunomagnetic bone marrow cells separation: a rapid and efficient negative/positive cell selection method. In: *Advances in Biomagnetic Separation* (Uhlén, M., Hornes, E., and Olsvik, O., eds.). Eaton, Natick, MA, pp. 205–209.
11. Wang, J. E., Jørgensen, P. F., Almlöf, M., et al. (2000) Peptidoglycan and lipoteichoic acid from Staphylococcus aureus induce tumor necrosis factor alpha, interleukin 6 (IL-6), and IL-10 production in both T cells and monocytes in human whole blood model. *Infect. Immun.* **68,** 3965–3970.
12. Bergquam, E., Avery, N., Shiigi, S., Axthelm, M., and Wong, S. (1999) Rhesus rhadinovirus establishes a latent infection in B lymphocytes in vivo. *J. Virol.* **73,** 7874–7876.
13. Sleasman, J. W., Leon, B., Aleixo, L., Rojas, M., and Goodenow, M. (1997). Immunomagnetic selection of purified monocyte and lymphocyte populations from peripheral blood mononuclear cells following cryopreservation. *Clin. Diagn. Lab. Immunol.* **4,** 653–658.
14. Islam, D. and Lindberg, A. (1992) Detection of *Shigella dysenteriae* type 1 and *Shigella flexneri* in feces by immunomagnetic isolation and polymerase chain reaction. *J. Clin. Microbiol.* **30,** 2801–2806.
15. Karch, H., Janetzki-Mittmann, C., Aleksic, S., and Datz, M. (1996) Isolation of enterohemorrhagic *Escherichia coli* 0157 strains from patients with hemolytic-uremic syndrome by using immunomagnetic separation, DNA-based methods, and direct culture. *J. Clin. Microbiol.* **34,** 516–519.
16. Li, Z., Bai, G. H., Fordham von Reyn, C., et al. (1996). Rapid detection of *Mycobacterium avium* in stool samples from AIDS patients by immunomagnetic PCR. *J. Clin Microbiol.* **34,** 1903–1907.
17. Mazurek, G. H., Reddy, V., Murphy, D., and Ansari, T. (1996) Detection of *Mycobacterium tuberculosis* in cerebrospinal fluid following immunomagnetic enrichment. *J. Clin. Microbiol.* **34,** 450–453.
18. Stark, M., Reizenstein, E., Uhlén, M., and Lundeberg, J. (1996) Immunomagnetic separation and solid-phase detection of *Bordetella pertussis. J. Clin. Microbiol.* **34,** 778–784.
19. Periera, M. G. C., Atwill, E., and Jones, T. (1999) Comparison of sensitivity of immunofluorescent microscopy to that of a combination of immunofluorescent microscopy and immunomagnetic separation for detection of *Cryptosporidium parvum* oocysts in adult bovine feces. *Appl. Environ. Microbiol.* **65,** 3236–3239.
20. Deng, M. Q. and Cliver, D. O. (1998) Differentiation of *Cryptosporidium parvum* isolates by a simplified randomly amplified polymorphic DNA technique. *Appl. Environ. Microbiol.* **64,** 1954–1957.
21. Deng, M. Q., Cliver, D.O., and Mariam, T. W. (1997) Immunomagnetic capture PCR to detect viable *Cryptosporidium parvum* oocysts from environmental samples. *Appl. Environ. Microbiol.* **63,** 3134–3138.

22. Oberst, R. D., Hays, M. P., Bohra, L. K., et al. (1998) PCR based DNA amplification and presumptive detection of *Escherichia coli* O157:H7 with an internal fluorogenic probe and the 5' nuclease (TaqMan) assay. *Appl. Environ. Microbiol.* **64,** 3389–3396.
23. Chapman, P. A., Cerdan Malo, A. T., Siddons, C. A., and Harkin, M. (1997) Use of commercial enzyme immunoassays and immunomagnetic separation systems for detecting *Escherichia coli* 0157 in bovine fecal samples. *Appl. Environ. Microbiol.* **63,** 2549–2553.
24. Tomoyasu, T. (1998) Improvement of the immunomagnetic separation method selective for *Escherichia coli* 0157 strains. *Appl. Environ. Microbiol.* **64,** 376–382.
25. Tsai, W. L., Miller, C. E., and Richter, E. R. (2000) Determination of the sensitivity of a rapid *Escherichia coli* O157:H7 assay for testing 375-gram composite samples. *Appl. Environ. Microbiol.* **66,** 4149–4151.
26. Pyle, B. H., Broadaway, S. C., and McFeters, G. A. (1999) Sensitive detection of *Escherichia coli* O157:H7 in food and water by immunomagnetic separation and solid-phase laser cytometry. *Appl. Environ. Microbiol.* **65,** 1966–1972.
27. Yu, H. and Bruno, J. G. (1996) Immunomagnetic-electrochemiluminescent detection of *Escherichia coli* O157 and *Salmonella typhimurium* in foods and environmental water samples. *Appl. Environ. Microbiol.* **62,** 587–592.
28. Hanai, K., Satake, M., Nakanishi, H., and Venkateswaran, K. (1997) Comparison of commercially available kits with standard methods for detection of *Salmonella* strains in foods. *Appl. Environ. Microbiol.* **63,** 775–778.
29. Favrin, S. J., Jassim, S. A., and Griffiths, M. W. (2001) Development and optimization of a novel immunomagnetic separation-bacteriophage assay for detection of *Salmonella enterica* serovar enteritidis in broth. *Appl. Environ. Microbiol.* **67,** 217–224.
30. Roberts, B. and Hirst, R. (1997) Immunomagnetic separation and PCR for detection of *Mycobacterium ulcerans. J. Clin. Microbiol.* **35,** 2709–2711.
31. Grant, I. R., Ball, H. J., and Rowe, M. T. (1998) Isolation of *Mycobacterium paratuberculosis* from milk by immunomagnetic separation. *Appl. Environ. Microbiol.* **64,** 3153–3158.
32. Osaki, T., Taguchi, H., Yamaguchi, H., and Kamiya, S. (1998) Detection of *Helicobacter pylori* in fecal samples of gnotobiotic mice infected with *H. pylori* by an immunomagnetic-bead separation technique. *J. Clin. Microbiol.* **36,** 321–323.
33. Enroth, H. and Engstrand, L. (1995) Immunomagnetic separation and PCR detection of *Helicobacter pylori* in water and stool specimens. *J. Clin. Microbiol.* **33,** 2162–2165.
34. Tomoyasu, T. (1992) Development of the immunomagnetic enrichment method selective for *Vibrio parahaemolyticus* serotype K and its application to food poisoning study. *Appl. Environ. Microbiol.* **58,** 2679–2682.
35. Vuddhakul, V., Chowdhury, A., Laohaprertthisan, V., et al. (2000) Isolation of a pandemic O3:K6 clone of a vibrio parahaemolyticus strain from environmental and clinical sources in Thailand. *Appl. Environ. Microbiol.* **66,** 2685–2689.
36. Dowd, S. E., Gerba, C. P., Enriquez, F. J., and Pepper, I. L. (1998) PCR amplification and species determination of microsporidia in formalin-fixed feces after immunomagnetic separation. *Appl. Environ. Microbiol.* **64,** 333–336.
37. Kapperud, G., Vardund, T., Skjerve, E., Hornes, E., and Michaelsen, T. E. (1993) Detection of pathogenic *Yersinia enterocolitica* in foods and water by immunomagnetic separation, nested polymerase chain reactions, and colorimetric detection of amplified DNA. *Appl. Environ. Microbiol.* **59,** 2938–2944.
38. Blake, M. R. and Weimer, B. C. (1997) Immunomagnetic detection of *Bacillus stearothermophilus* spores in food and environmental samples. *Appl. Environ. Microbiol.* 63: 1643–1646.

39. Bruno, J. G. and Yu, H. (1996) Immunomagnetic-electrochemiluminescent detection of *Bacillus anthracis* spores in soil matrices. *Appl. Environ. Microbiol.* **62,** 3474–3476.
40. Muramatsu, Y., Yanase, T., Okabayashi, T., Ueno, H., and Morita, C. (1997) Detection of *Coxiella burnetii* in cow's milk by PCR-enzyme-linked immunosorbent assay combined with a novel sample preparation method. *Appl. Environ. Microbiol.* **63,** 2142–2146.
41. Jothikumar, N., Cliver, D. O., and Mariam, T. W. (1998) Immunomagnetic capture PCR for rapid concentration and detection of hepatitus a virus from environmental samples. *Appl. Environ. Microbiol.* **64,** 504–508.
42. Hutchison, K. A., Perfect, S. E., O'Connell, R. J., and Green, J. R. (2000) Immunomagnetic purification of *Colletotrichum lindemuthianum* appressoria. *Appl. Environ. Microbiol.* **66,** 3464–3467.
43. Biswas, B., Vemulapalli, R., and Dutta, S. K. (1994) Detection of *Ehrlichia risticii* from feces of infected horses by immunomagnetic separation and PCR. *J. Clin. Microbiol.* **32,** 2147–2151.
44. Christensen, B., Torsvik, T., and Lien, T. (1992) Immunomagnetically captured thermophilic sulfate-reducing bacteria from North Sea oil field waters. *Appl. Environ. Microbiol.* **58,** 1244–1248.
45. Fayer, R., Speer, C. A., and Dubey, J. P. (1997) The general biology of *Cryptosporidium.* In: *Cryptosporidium and Cryptosporidiosis* (Fayer, R. ed.). CRC, Boca Raton, FL, pp. 1–42.
46. Current, W. L., Reese, N. C., Erns, J. V. D. Bailey, W. S., Heyman, M. B., and Weistein, W. M. (1983) Human cryptosporidiosis in immunocompetant and immunodeficient persons: studies of an outbreak and experimental transmission. *N. Engl. J. Med.* **308,** 1252–1257.
47. Dupont, H. L., Chappel, C. L., Sterling, C. R., Okhuysen, P. C., Rose, J. B., and Jakubowski, F. (1995) The infectivity of *Cryptosporidium parvum* in healthy volunteers. *N. Engl. J. Med.* **332,** 855–859.
48. Casemore, D. P., Wright, S. E., and Coop, R. L. (1997) Cryptosporidiosis—human and animal epidemiology. In: *Cryptosporidium and Cryptosporidiosis* (Fayer, R., ed.). CRC, Boca Raton, FL, pp. 65–92.
49. Rose, J. B. (1997) Environmental ecology of *Cryptosporidium* and public health implications. *Ann. Rev. Public Health 18,* 135–161.
50. Mackenzie, W., Neil, M., Hoxie, N., et al. (1994) A massive outbreak in Milwaukee of Cryptosporidium infection transmitted through the public water supply. *N. Engl. J. Med.* **331,** 161–167.
51. United States Environmental Protection Agency (1999) *Method 1623:* Cryptosporidium *and* Giardia *in Water by Filtration/IMS/FA.* EPA821-R-99-006. EPA, Washington, DC.
52. Bukhari, Z., McCuin, R. M., Fricker, C. R, and Clancy, J. L. (1998) Immunomagnetic separation of Cryptosporidium parvum from source water samples of various turbidities. *Appl. Environ. Microbiol.* **64,** 4495–4499.
53. Rochelle, P. A., De Leon, R., Johnson, A., Stewart, M. H., and Wolfe, R. L. (1999) Evaluation of immunomagnetic separation for recovery of infectious Cryptosporidium parvum oocysts from environmental samples. *Appl. Environ. Microbiol.* **65,** 841–845.
54. Dynal. (2000) GC Combo Kit Product Literature. Revision 06. Dynal, Oslo, Norway.
55. Yakub, G. P. and Stadterman-Knauer, K. L. (2000) Evaluation of immunomagnetic separation for recovery of *Cryptosporidium parvum* and *Giardia duodenalis* from high-iron matrices. *Appl. Environ. Microbiol.* **66,** 3628–3631.
56. Di Giovanni, G. D., Hashemi, F. H., Shaw, N. J., Abrams, F. A., LeChevallier, M. W., and Abbaszadegan, M. (1999) Detection of infectious *Cryptosporidium parvum* oocysts in surface and filter backwash water samples by immunomagnetic separation and integrated cell culture-PCR. *Appl. Environ. Microbiol.* **65,** 3427–3432.

18

Detection of *Erysipelothrix rhusiopathiae* in Clinical and Environmental Samples

Silvana G. Fidalgo and Thomas V. Riley

1. Introduction

Erysipelothrix rhusiopathiae is pathogenic for both animals and humans, causing erysipelas in swine and erysipeloid in humans *(1)*. In swine, disease may be either acute or chronic, resulting in the development of arthritis and endocarditis *(2)*. In Japan, erysipelas remains an animal hygiene problem causing great economic loss as infected swine are disused *(3)*. Human infection closely resembles that seen in swine, with both acute and chronic forms also. The most common presentation is erysipeloid, a localized cutaneous infection *(4)*. In Western Australia, an erysipeloid-like infection referred to as "crayfish poisoning" occurs in lobster fishermen and handlers *(5)*. A second type of presentation is a generalized cutaneous form involving lesions that progress from the initial site of infection or appear in remote areas *(6)*. The third and most serious form of disease is a septicemia that is almost always linked to endocarditis *(4)*. The mortality rate in *Erysipelothrix* endocarditis is still high (38%) *(7)* and can be explained by the use of vancomycin (to which *Erysipelothrix* spp. are inherently resistant) as empirical therapy. Therefore, it is critical to have an early diagnosis of *E. rhusiopathiae* infection *(8)*.

Unfortunately, several problems exist with the diagnosis of *E. rhusiopathiae* infections by conventional cultural procedures, and these infections are often incorrectly diagnosed. First, because of their very small colony size and slow growth rates, it is difficult to isolate *E. rhusiopathiae* from heavily contaminated specimens *(4)*. Various selective media have been described to improve the isolation of *E. rhusiopathiae* from contaminated specimens; however, not all contaminants are inhibited. The development of two polymerase chain reaction (PCR) methods has created an opportunity to greatly improve the efficiency with which these organisms are detected and identified. Makino et al. *(9)* designed a PCR method that amplifies a 407-bp DNA fragment derived from the 16S rRNA coding sequence. The primers in this method are specific for the genus *Erysipelothrix* and do not differentiate between the species *(9)*. A second set of

From: *Methods in Molecular Biology, vol. 268: Public Health Microbiology: Methods and Protocols*
Edited by: J. F. T. Spencer and A. L. Ragout de Spencer © Humana Press Inc., Totowa, NJ

primers designed by Shimoji et al. ***(10)*** amplifies a 937-bp DNA fragment which is derived from a sequence associated with virulence of *E. rhusiopathiae*. These primers are specific for *E. rhusiopathiae* only. Shimoji et al. ***(10)*** also utilized a selective enrichment medium based on tryptic soy broth containing ethidium bromide and sodium azide. More recently, a third PCR method, which differentiates between four different species of *Erysipelothrix*, has been publisheded ***(11)***.

We recently undertook a survey of seafood to evaluate various methodologies for detection of *Erysipelothrix* spp. ***(12)***. The tryptic soy broth used by Shimoji et al. ***(10)***, with a 48-h incubation followed by a double-round PCR, was the best method. With some modifications to the two PCR methods designed by Makino et al. ***(9)*** and Shimoji et al. ***(10)***, and with the use of selective tryptic soy broth for enrichment, this chapter presents a method for the rapid detection of *E. rhusiopathiae* in both clinical and environmental samples.

2. Materials

2.1. Preparation of Sample

1. Trypticase soy broth (TSB), pH 7.6 (Becton Dickinson, Cockeysville, MD) supplemented and made selective with: 0.3% Tris-HCl, 0.1% Tween-80, 0.03% sodium azide, and 5 μg/mL crystal violet (TSB/S).
2. McCartney bottles.
3. 1.5-mL Microcentrifuge tubes.

2.2. Preparation of PCR Master Mixes and Amplification

1. 0.2-mL microtubes (certified DNase/RNase free).
2. Openers for 0.2-mL microtubes.
3. Microtube racks.
4. Sterile diethylpyrocarbonate (DEPC)-treated water: Add 400 μL of DEPC to 400 mL deionized water that has been filtered through a MilliQ filter system. Shake thoroughly, incubate at 37°C overnight, and then autoclave at 121°C for 60 min.
5. Thermal cycler (Perkin Elmer 9600 suggested).
6. 10X PCR buffer II: 100 m*M* Tris-HCl, pH 8.3, and 500 m*M* KCl (Applied Biosystems).
7. 25 m*M* $MgCl_2$ (Appled Biosystems).
8. 5 U/μL Ampli*Taq* Gold™ DNA polymerase (Appled Biosystems).
9. Deoxyribonucleoside triphosphate (dNTP) mix (25 m*M* each of dATP, dCTP, dGTP, and dTTP; Fisher Biotec).
10. Primer sets: *see* **Table 1** for sequences (Gibco BRL, Paisley, UK); for use make up 10-μ*M* stocks in DEPC water and then autoclave at 121°C for 60 min.

2.3. Analysis of PCR Products

1. Electrophoresis gel tank, tray, and comb.
2. Electrophoresis power supply capable of 200 V.
3. DNA grade agarose (DNase and RNase free).
4. Ethidium bromide (10 mg/mL): add 1 g ethidium bromide to 100 mL of distilled water. Stir in magnetic stirrer for several hours, wrap the container in aluminium foil, and store at 2–8°C.

Table1
Primers Used for PCR Detection of *Erysipelothrix* spp. and *E. rhusiopathiae*

Primer[a]	Nucleotide sequence 5'→3'	Length of product (bp)	Detection
MOIO1[b]	AGATGCCATAGAAACTGGTA	407	*Erysipelothrix* spp.
MOIO2[b]	CTGTATCCGCCATAACTA		
ER1[c]	CGATTATATTCTTAGCACGCACGCAACG	937	*E. rhusiopathiae*
ER2[c]	TGCTTGTGTTGTGATTTCTTGACG		

[a]Primer sequences for MOIO1 and -2 and ER1 and -2 from refs. ***6*** and ***8***, respectively.
[b]GenBank/EMBL accession no. M23728.
[c]GenBank/EMBL accession no. D64177.

5. 50X TAE buffer: add 121 g of Tris base, 23.6 mL glacial acetic acid, 50 mL 0.5 *M* EDTA, pH 8.0. Dissolve and make up to 500 mL with deionized water.
6. Electrophoresis buffer: add 40 mL 50X TAE buffer and 100 µL ethidium bromide stock (10 mg/mL) to 1960 mL deionized water. Store at 2–8°C.
7. 6X Loading dye: add 1 g bromophenol blue and 160 g sucrose to 400 mL deionized water. Mix and store at 2–8°C.
8. TE buffer, pH 8.0: 10 m*M* Tris-HCl, pH 8.0, 1 m*M* EDTA pH 8.0.
9. DNA Molecular Weight V III marker (Boehringer, Mannheim, Germany): For use, dilute 1:10 with TE buffer.
10. UV transilluminator.
11. Polaroid camera.

3. Methods (*see* Notes 1 and 2)

3.1. Preparation of PCR Master Mixes

Volumes given are based on 16 and 20 µL per reaction mixture for the first- and second-round mix tubes, respectively. Thaw out the following reagents and keep them on ice at all times: 10X PCR buffer II, 25 m*M* $MgCl_2$, 5 U/µL Ampli*Taq* Gold DNA polymerase, 100m*M* dNTP mix, and 10 µ*M* stocks of primers (MOIO1 and -2 and ER1 and -2).

3.2. Erysipelothrix *spp. Master Mixes*

1. Using aseptic techniques, add the following volumes of each reaction component to a sterile McCartney bottle to make 100 first-round mix tubes: 1134 µL of DEPC water, 200 µL of 10X PCR buffer II, 160 µL of 25 m*M* $MgCl_2$, 16 µL of dNTP pool, 10 µL of 5U/µL Ampli*Taq* Gold DNA polymerase, 40 µL of 10 µ*M* MOIO1 and -2 primers.
2. To make 100 second-round mix tubes, add the following volumes: 1527 µL of DEPC water, 204 µL of 10X PCR buffer II, 163.2 µL of 25 m*M* $MgCl_2$, 16.3 µL of dNTP pool, 10 µL of 5 U/µL Ampli*Taq* Gold DNA polymerase, 40.8 µL of 10 µ*M* MOIO1 and -2 primers.
3. Use the vortex mixer and gently but thoroughly mix the contents.
4. Place rows of sterile 0.2 mL microtubes into a microtube rack with the lids open.

5. Dispense 16 and 20 µL of the reaction mixtures into the microtubes for the first- and second-round mix tubes, respectively.
6. Color code the tips of the microtubes for the first and second rounds, and make sure the lids are placed on firmly.
7. Place the rack with the microtubes in the –70°C freezer for 1 h.
8. Once the reaction mixtures are frozen, they can be removed from the rack and stored in boxes in the –70°C freezer until later required.

3.3. E. rhusiopathiae *Master Mixes*

1. Add the following volumes to a McCartney bottle for 100 first-round mix tubes: 1174 µL of DEPC water, 200 µL of 10X PCR buffer II, 120 µL of 25 m*M* MgC12, 16 µL of dNTP pool, 10 µL of 5U/µL Ampli*Taq* Gold DNA polymerase, 40 µL of 10 µ*M* MOIO1 and -2 primers.
2. For 100 second-round mix tubes, add the following volumes to another McCartney bottle: 1565.7 µL of DEPC water, 204 µL of 10X PCR buffer II, 122.4 µL of 25 m*M* $MgCl_2$, 16.3 µL of dNTP pool, 10 µL of 5U/µL Ampli*Taq* Gold DNA polymerase, 40.8 µL of 10 µ*M* MOIO1 and -2 primers.
3. Follow the same procedure as for *Erysipelothrix* master mixes from **step 3, Subheading 3.2.**

3.4. *Preparation of Samples* (see *Note 4*)

1. Dispense 10 mL sterile TSB/S into a McCartney bottle and inoculate the organism to be tested.
2. Use the vortex mixer to mix the contents thoroughly.
3. Incubate the broth for 48 h at 37°C.
4. Following the 48-h incubation, vortex the broth culture vigurously.
5. With a sterile plastic transfer pipet, remove 1.5 mL of the culture and transfer it to a sterile microcentrifuge tube.
6. Centrifuge the culture at 10,000*g* for 3 min to pellet the suspension.
7. Remove the supernatant and add 1.5 mL DEPC water to the pellet.
8. Wash the pellet by vortexing it into suspension.
9. Centrifuge the suspension once again at 10,000*g* for 3 min.
10. Remove the supernatant and resuspend the pellet in 100 µL DEPC water.

3.5. *Preparation of DNA Template*

1. Boil the 100-µL suspensions at 100°C for 15 min.
2. Pellet the debris at 10,000*g* for 3 min.
3. Transfer the supernatant containing the DNA to a sterile microcentrifuge tube.
4. Store the microcentrifuge tube containing the DNA extract in the –70°C freezer until required.

3.6. *PCR Amplification* (see *Note 5*)

Inoculation of the mix tubes should be conducted in a laminar flow cabinet if possible, or if not, in a separate area or laboratory from where the mix tubes were prepared. Plastic sleeves should be worn while working in the cabinet. While inoculating the second-round mix tubes from the first, do not open them by hand; use the openers at all times.

3.6.1. First-Round Amplification

1. Thaw out the DNA extract from first-round *Erysipelothrix* and *E. rhusiopathiae* mix tubes, and then place them on ice.
2. Label mix tubes according to samples tested.
3. In a laminar flow cabinet, vortex the DNA extract and add 4 μL to the corresponding *Erysipelothrix* and *E. rhusiopathiae* mix tubes.
4. Place the lid firmly on the microtubes and gently mix the contents (*see* **Note 6**).
5. Place the *Erysipelothrix* mix tubes in the thermal cycler and the *E. rhusiopathiae* mix tubes on ice.
6. For the *Erysipelothrix* mix tubes amplify the reaction by using the following cycle parameters: 94°C for 15 min and then 45 cycles of denaturation at 94°C for 30 s, annealing at 50°C for 30 s, and extension at 72°C for 45 s, followed by an additional extension step at 72°C for 7 min and cooling to 4°C.
7. Once the amplification of the *Erysipelothrix* mix tubes is completed, place them in the fridge.
8. Gently mix the contents of the *E. rhusiopathiae* mix tubes and place them in the same thermal cycler.
9. For the *E. rhusiopathiae* mix tubes, amplify the reaction by using the following cycle parameters: heating at 94°C for 5 min and then 30 cycles of denaturation at 94°C for 1 min, annealing at 63°C for 30 s, and extension at 72°C for 1 min, followed by an additional extension step at 72°C for 7 min and cooling to 4°C.
10. Following amplification, also place the *E. rhusiopathiae* mix tubes in the fridge.

3.6.2. Double (Second-Round) Amplification

1. Place the second-round mix tubes on a rack in line with the first-round mix tubes.
2. Place all the equipment that is required in the cabinet and then put on plastic sleeves.
3. Use a microtube opener to open the first-round mix tube.
4. Remove 0.4 μL from the first-round mix tube.
5. Use a clean opener to open the second-round mix tube.
6. Inoculate the 0.4 μL from the first round into the second-round mix tube.
7. Remove the second-round mix tubes from the block, gently mix the contents, and place them in the thermal cycler.
8. Then amplify the second-round mix tubes using the same cycle parameters as for the first round for both *Erysipelothrix* and *E. rhusiopathiae* mix tubes.
9. Remove the samples from the thermal cycler and store them at 4°C.

3.7. Analysis of PCR Products

1. Prepare a 2% agarose gel (w/v) by adding 1.75 g of agarose to 70 mL deionized water, plus 1.4 mL 50X TAE buffer and 3.5 μL ethidium bromide (10 mg/mL).
2. Once the gel is set, place it in the tank with the electrophoresis buffer just covering the surface of the gel.
3. Add 4 μL of the loading dye into each sample, gently resuspend, and load 12 μL of the samples into the well. Also load the molecular weight markers.
4. When loading is complete, place the lid on the tank and plug into the power pack.
5. Run the gel at a constant voltage of 100–120 V until the dye front approaches the edge of the gel.
6. Switch off the power pack and remove the gel tray.

7. Put on the UV eye shield.
8. Place the gel on the UV transilluminator and switch on.
9. Visualize and photograph the products (*see* **Note 7**).

4. Notes

1. Each section of the methods should be performed either in separate laboratories or in different areas to avoid contamination. Preparation of the master mixes should be carried out in a specimen-free laboratory. Inoculation of the first- and second-round mix tubes should be performed in separate laminar flow cabinets to prevent cross-contamination. Analysis of the PCR products should also be performed in a separate area or laboratory.
2. Gloves should be worn at all times to avoid contamination with DNAses and RNAses.
3. When preparing the master mixes, overestimating the final required volume by one or two reactions to allow for volume losses and pipeting errors is recommended. Also, make a minimum batch of 25 reaction mixtures for the first and second rounds. Large batch numbers are recommended, as very small volumes of the primers are used.
4. It is recommended that the following be used as positive and negative controls. *E. rhusiopathiae* ATCC 19414 can be used as the positive control for both PCR methods, and *E. tonsillarum* ATCC 43339 can be used as the negative control for the *E. rhusiopathiae*-specific PCR.
5. A DEPC water negative control should be used for both PCR methods.
6. Make sure the lids on the centrifuge and microcentrifuge tubes are placed on firmly to prevent evaporation of the suspension during boiling or PCR amplification.
7. Samples should be tested twice on separate days to check for reproducibility. If discrepant results are obtained, then the analysis should be repeated. If the third PCR is positive, then the overall results are considered positive. If the third PCR is negative, then the overall result is considered negative.

Acknowledgments

This work was supported by the Western Australian Fisheries Industry Council and a grant from the Fisheries Industry Research and Development Corporation.

References

1. Wood, R. L. and Shuman, R. D. (1975) Swine erysipelas. In: *Diseases of Swine* (Scholl, E. and Leman, A. D., eds.). Iowa State University Press, Ames, IA, pp. 565–620.
2. Takahashi, T., Sawada, T., Takagi, M., Seto, K., Kanzaki, M., and Maruyama, T. (1984) Serotypes of *Erysipelothrix rhusiopathiae* strains isolated from slaughter pigs affected with chronic erysipelas. *Jpn. J. Vet. Sci.* **46,** 149–153.
3. Takahashi, T. (1996) Swine erysipelas and *Erysipelothrix. Jpn. J. Bacteriol.* **51,** 707–715.
4. Reboli, A. C. and Farrar, W. E. (1992) The genus *Erysipelothrix*. In: *The Prokaryotes, A Handbook on the Biology of Bacteria: Ecophysiology, Isolation, Identification and Applications* (Balows, A., Truper, H. G., Dworkin, M., Harder, W., and Schleifer, K., eds.). Springer-Verlag, New York, pp. 1629–1642.
5. Brooke, C. J., McLaughlin, V., Mee, B. J., and Riley, T. V. (1999) An investigation into "crayfish poisoning" in Western Australia. *Med. J. Aust.* **170,** 288.
6. Klauder, J. V. (1944) *Erysipelothrix rhusiopathiae* infection in swine and in human beings: comparative study of the cutaneous lesions. *Arch. Dermatol. Syphilol.* **50,** 151–159.

7. Gorby, G. L. and Peacock, J. E., Jr. (1988) *Erysipelothrix rhusiopathiae* endocarditis: microbiologic, epidemiologic, and clinical features of an occupational disease. *Rev. Infect. Dis.* **10,** 317–325.
8. Grieco, M. H. and Sheldon, C. (1970) *Erysipelothrix rhusiopathiae. Ann. NY Acad. Sci.* **174,** 523–532.
9. Makino, S., Okada, Y., Maruyama, T., et al. (1994) Direct and rapid detection of *Erysipelothrix rhusiopathiae* DNA in animals by PCR. *J. Clin. Microbiol.* **32,** 1526–1531.
10. Shimoji, Y., Mori, Y., Hyakutake, K., Sekizaki, T., and Yokomizo, Y. (1998) Use of enrichment broth cultivation-PCR combination assay for rapid diagnosis of swine erysipelas. *J. Clin. Microbiol.* **36,** 86–89.
11. Takeshi, K., Makino, S., Ikeda, T., et al. (1999) Direct and rapid detection of Erysipelothrix spp. DNAs prepared from bacterial strains and animal tissues. *J. Clin. Microbiol.* **37,** 4093–4098.
12. Fidalgo, S. G., Wang, Q., and Riley, T. V. (2000) Comparison of methods for detection of *Erysipelothrix* spp. and their distribution in some Australasian seafoods. *Appl. Environ. Microbiol.* **66,** 2066–2070.

19

Interaction Between Lactic Acid Bacteria and Gastrointestinal Nematodes of Caprine Origin

Diana Draksler, María Cecilia Monferran, and Silvia González

1. Introduction

To compare the level of parasitism with gastrointestinal nematodes in sheep and goats, several studies have been conducted *(1–3)*. They have generally shown that goats were more infected than sheep, as they exhibited higher worm burdens and egg excretion. This difference between two host species has been attributed not only to a difference in feeding behavior, but also to a lesser ability of goats to develop resistance to trichostrongylate infection *(4)* (In kids and lambs the greatest damage is observed from weaning until 1 yr of age; mature mothers, before and after parturition and during suckling, are affected *[5]*).

In the last few decades, the most common tool (and frequently the only one) used for controlling internal parasites in livestock was the anthelmintic drugs. The application of anthelmintic treatments; must be accompanied by epidemic data and determinations supporting the appropriate timing and frequency of animal treatment. This view has not always been respected. In our country, because of a decrease in price, the anthelmintic drugs were used indiscriminately, causing the resistance we see today *(6,7)*. This serious problem, added to the objective of producing organic foods without drug residuals, calls for better use of the antiparasitic drugs and for the developmentment of alternative methods that are ecologically viable and without risks for human health *(8)*.

Little information is available on the interactions between bacteria and intestinal nematodes of caprine origin. Some reports note ovicidal activity of different strains of *Bacillus thuringiensis* on the eggs of zooparasitic nematodes *(9,10)*. Recent work found inhibitory actions of lactic bacteria on gastrointestinal nematodes (both of caprine origin). In the present chapter we describe the methods used for the determination of interactions between lactic acid bacteria and nematodes.

From: *Methods in Molecular Biology, vol. 268: Public Health Microbiology: Methods and Protocols*
Edited by: J. F. T. Spencer and A. L. Ragout de Spencer © Humana Press Inc., Totowa, NJ

2. Materials

2.1. Bacterial Strains

1. LAPTg medium *(11)* for growth of microorganisms: 1.5 g/L meat peptone, 1.0 g/L yeast extract, 1.0 g/L triptone, 1.0 g/L glucose, 0.1% Tween-80.
2. Sterile physiological solution.

2.2. Nematodes

2.2.1. Collection of Fecal Samples

1. Latex gloves.
2. Very clean glass or plastic recipients (sterility is not necessary), with covers, to be labeled.
3. Recipient with ice for transporting the samples under refrigeration (at 4°C).

2.2.2. Microscopic Examination of Previous Enrichment

1. Fecal samples.
2. 25 mL Precipitation glasses.
3. Glass rods.
4. Strainer with wire mesh.
5. Cover slips and slides.
6. Lugol solution.
7. NaCl saturated solution.

2.2.3. Fecal Nematode Egg Counts

1. Fecal samples.
2. McMaster recount camera.
3. Mortar with pestle (for pounding).
4. Strainer with wire mesh.
5. Pasteur pipets.
6. 100 mL NaCl-saturated solution.

2.2.4. Coprocultures

1. Fecal samples.
2. 5- and 10-cm Petri dishes.
3. Mortar with pestle (for pounding).
4. Pasteur pipets.
5. Distilled or mineral water (without chlorine).
6. Inert material (5-mm Telgopor pearls).
7. 1-mL Bacterial suspensions.
8. Assay tubes.

2.2.5. Third-Stage Larvae (L3) Counts

1. Camera to count the larvae.
2. Pasteur pipets.

3. Methods

3.1. Bacterial Strains

1. Activate the microorganisms (at 37°C for 14–18 h) by three consecutive passages in 5 mL LAPTg broth.

2. Centrifuge the cultures were at 3000 rpm for 10 min.
3. Discard the supernants and resuspend the precipitates in 5 mL of sterile physiological solution (SPS).
4. Wash the cells twice in 5 mL of SPS and finally resuspend in SPS up to 1 mL.

3.2. Nematodes

3.2.1. Collection of Fecal Samples (see **Note 1**)

1. With gloves on, take 6–10 g of fecal matter per animal by anal ring stimulation, avoiding contamination with free life nematodes or other elements that can lead to erroneous diagnosis.
2. When you have many animals, screening can take place in 8–10% of the same species.
3. Place the samples in the recipient and label (*see* **Note 1**)
4. Send to the laboratory under refrigerated conditions.

3.2.2. Microscopic Examination of Previous Enrichment: Willis' Technique (12) (see **Note 2**)

1. NaCl-satured solution: Dissolve by agitation and heat 360 g NaCl in 1000 mL of distilled water (*see* **Note 3**).
2. Place in the precipitation glass a sample of fecal matter about the size of a chickpea (or a teaspoonful) (*see* **Note 4**).
3. Fill the fourth part of the precipitation flask with saturated saline solution and mix with a glass rod until homogenization.
4. Strain to another glass, rinsing the first one with the saturated saline solution.
5. Fill the precipitation glass with the saline solution until the border forms a meniscus.
6. Place a clean slide on the liquid surface for 15 min. Then quickly invert the slide, add a drop of Lugol, put a cover on, and observe microscopically (*see* **Note 5**).

3.2.3. Fecal Nematode Egg Counts: Modified McMaster Method (13)

1. Homogenize in a mortar 5 g of fecal matter with the saturated solution, filter (through the strainer with wire mesh) in a precipitation glass, and add more saturated solution up to a volume of 100 mL (dilution 1:20).
2. Stir gently to avoid exaggerated formation of air bubbles and fill camera's cells with a Pasteur pipet.
3. Place on a microscope plate and after 2 or 3 min to begin the count using a magnification of 60–70×. The nematode eggs are added to the interior surface of the camera's cell, in the same plane that eventually bubbles.
4. Eggs from one, two, or four cells can be counted. If the eggs contained in one cell are counted, multiply by 40; if the eggs contained in two cells are counted, multiply by 20; and if eggs from four cells are counted, multiply by 10 (*see* **Note 5**).
5. Correction factor according to fecal consistency (for sheep and goats):

Normal	× 1
Mash not formed	× 2
Liquid thick	× 3
Liquid diarrhea	× 4

3.2.4. Coprocultures: Modified Roberts and O'Sullivan Technique (see **Note 4**)

1. Place 3 g of fecal matter in the mortar and crush.
2. Place the crushed feces in a 5-cm Petri dish and add the bacterial suspension under evaluation. Add some Telgopor pearls and chlorine-free water until a consistent mass is achieved.
3. Cover the Petri dish and incubate at 25–28°C during 10 d.
4. For the control coprocultures, proceed in the same way but without adding the bacterial suspension.
5. To collect the infecting larvae (L3), fill a Petri dish with chlorine-free water up to the edge; use another 10-cm Petri dish for a cover. Pour quickly to avoid spilling the water.
6. Place 5 mL of chlorine-free water in a 10-cm Petri dish and leave for 3–4 h.
7. Collect the water in the Petri dish containing the larvae and place them in an assay tube. Keep under refrigeration for 2–3 h and then remove the supernatants, placing 3–4 mL in the assay tube (*see* **Notes 5 and 6**).

3.2.5. Third-Stage Larvae (L3) Counts

1. After 3–4 h the larvae in the Petri dish will displace to the water of the dish. Using a Pasteur pipet, place a small volume of water containing larvae in the camera to count the number of larvae.
2. During microscopic observation, count the dead L3 larvae first. Then kill the remaining larvae by heat (placing the camera near the burner), and count the total L3 larvae.
3. Calculate viable total L3 larvae from the difference between total L3 and dead L3 larvae.
4. Continue in the same way for each coproculture (*see* **Note 5**).

3.3. Interpretation of Results

1. Recovery index of infectant larvae (L3). This index is considered to be the relationship between the number of L3 larvae obtained, and the egg counts by gram of feces (eggs per gram [epg]) multiplied by 3. Significant differences between the recovery indices obtained from control and test coprocultures are calculated, to establish the possible action of probiotic bacteria on parasitic eggs or first larval stages.
2. Dead L3 larvae number. Compare control and test coprocultures for dead L3 larvae numbers. A statistically significant difference between them indicates a larvicidal action from the bacteria studied (*see* **Note 5**).

4. Notes

1. Make notes of (1) the date of sample collection; (2) the time (mainly if the sample doesn't take chemical preserving because some helminthic eggs evolve quickly, which can make later identification difficult); (3) the animal species; (4) identification of the animal (for individual samples) or number of animals (for a flock); (5) sample consistency; (6) presence of mature parasites or blood, or any other fact that you consider important to remember.
2. This procedure allows separation of cysts, eggs, and other parasitic structures by employing a liquid of a higher specific weight that those of the parasites. Then parasites are recovered from the surface, and organic matter remains at the bottom of the recipient.
3. NaCl-saturated solution is the most utilized, but others hypersatured solutions such as magnesium sulphate (density 1.3 g/L), or zinc sulfate (density 1.2–1.25 g/L) can be used.

4. The fecal samples collected for coprocultures do not contain preserving solutions. The use of preserving solutions is only for to obtaining from fecal eggs different larval nematodes that are so similar they cannot be identified. On the other hand, observation of larvae obtained, allows systematic classification of different nematode parasites.
5. All laboratory material must be washed with water and hypochlorite sodium after use.
6. If it is not possible to identify all the larvae the same day, the assay tube can be maintained under refrigeration (12–14°C) for up to 4 mo.

References

1. Le Jambre, L. F. and Royal, W. M. (1976) A comparison of worm burdens in grazing Merino sheep and Angora goats. *Aust. Vet. J.* **52,** 181–183.
2. Pomroy, W. E., Lambert, M. G., and Betteridge, K. (1986) Comparison of faecal strongylate egg counts of goats and sheep on the same pasture. *N.Z. Vet. J.* **37,** 23–26.
3. Jallow, O. A., McGregor, B. A., Anderson, N., and Holmes, J. H. G. (1994) Intake of trichostrongylid larvae by goats and sheep grazing together. *Aust. Vet. J.* **71,** 361–364.
4. Hoste, H. and Chartier, C. (1998) Response to challenge infection with Haemonchus contortus and Trichostronylus colubriformis in dairy goats. Consequences on milk production. *Vet. Parasitol.* **74,** 43–54.
5. Romero, J. R. (2000) Selection of sheep with lower necessity of anthelmintic treatment. III Congreso Argentino de Parasitología, November, 2000, Mar del Plata, Argentina.
6. Eddi, C., Caracostantologo, J., Peña, M., et al. (1996) The prevalence of anthelmintic resistance in nematode parasites in sheep in southern Latin America: Argentina. *Vet. Parasitol.* **62,** 189–197.
7. Fiel, C. A., Saumell, C. A., Steffan, P. E., Rodríguez, E. M., and Salaberry, G. (2000) Resistencia de los nematodes trichostrongylideos—*Cooperia y Trichostrongylus*—a tratamientos con avermectinas en bovinos de la Pampa Húmeda, Argentina. *Rev. Med. Vet.* **81,** 310–315.
8. Saumell, C. A. (2000) No chemical methods for the control of helminths in veterinary practice. III Congreso Argentino de Parasitología, November, 2000, Mar del Plata, Argentina.
9. Bone, L. W., Bottjer, K. P., and Gill, S. S. (1985) *Trichostronylus colubriformis*: egg lethality due to *Bacillus thuringiensis* crystal toxin. *Exp. Parasitol.* **60,** 314–322.
10. Bottjer, K. P., Bone, L. W., and Gill, S. S. (1985) Nematoda: susceptibility of the egg to *Bacillus thuringiensis* toxins. *Exp. Parasitol.* **60,** 239–244.
11. Raibaud, P., Caulet, M., Galpin, J. V., and Mocquot, G. (1961) Studies on the bacterial flora of the alimentary tract of pigs. II. Streptococci: selective enumeration and differentiation of the dominant group. *J. Appl . Bacteriol.* **24,** 285–291.
12. Willis, H. H. (1921) A simple levitation method for the detection of hookworm ova. *Med. J. Aust.* **8,** 375.
13. Raynaud, J. P. (1970) Etude de l'efficacité d'une technique de coproscopie quantitative pour le diagnostic de routine et le controle des infestations parasitaires des bovins, ovins, equins et porcins. *Ann. Parasitol. Hum. Comp.* **45,** 321–342.

20

Molecular Detection of Genes Responsible for Cyanobacterial Toxin Production in the Genera *Microcystis*, *Nodularia*, and *Cylindrospermopsis*

Brendan P. Burns, Martin L. Saker, Michelle C. Moffitt, and Brett A. Neilan

1. Introduction

Cyanobacteria are ubiquitous in the freshwater environment. Their success as a group in a wide range of aquatic habitats has been attributed to their unique physiological characteristics and their high adaptive ability over a wide range of environmental conditions. They are capable of reaching very high biomass levels, often dominating the other aquatic biota, and under some circumstances can accumulate near the water surface, producing scums. Such cyanobacterial "blooms" are of particular concern in reservoirs used to supply potable water. Dense aggregations of cyanobacterial cells may block water filters, and many species produce compounds that affect the taste and odor of water supplies. Of greatest concern, however, is the potential of many bloom-forming cyanobacteria to produce a wide range of toxic substances. These natural compounds, known as cyanotoxins, are chemically diverse and are usually either neuro- or hepatotoxic in pathology.

1.1. Chemical and Toxicological Diversity of Microcystins, Nodularin, and Cylindrospermopsin

Among the most common hepatotoxins encountered in the freshwater environment are the cyclic heptapeptide toxins (containing seven peptide-linked amino acids) known as microcystins (MCYSTs). Over 60 variants of MCYSTs have been found to date, most with median lethal dose (LD_{50}) values (ip mouse bioassay) from 50 to 500 μg/kg and molecular weights ranging from 800 to 1100 *(1)*. The general structure of MCYST is given in **Fig. 1A**. Most of the structural variants of MCYST are formed by the substitution of L-amino acids at positions 2 or 4, or by the demethylation of amino acids at positions 3 and/or 7 (**Fig. 1A**) *(1)*. The MCYSTs are produced by several

From: *Methods in Molecular Biology, vol. 268: Public Health Microbiology: Methods and Protocols*
Edited by: J. F. T. Spencer and A. L. Ragout de Spencer © Humana Press Inc., Totowa, NJ

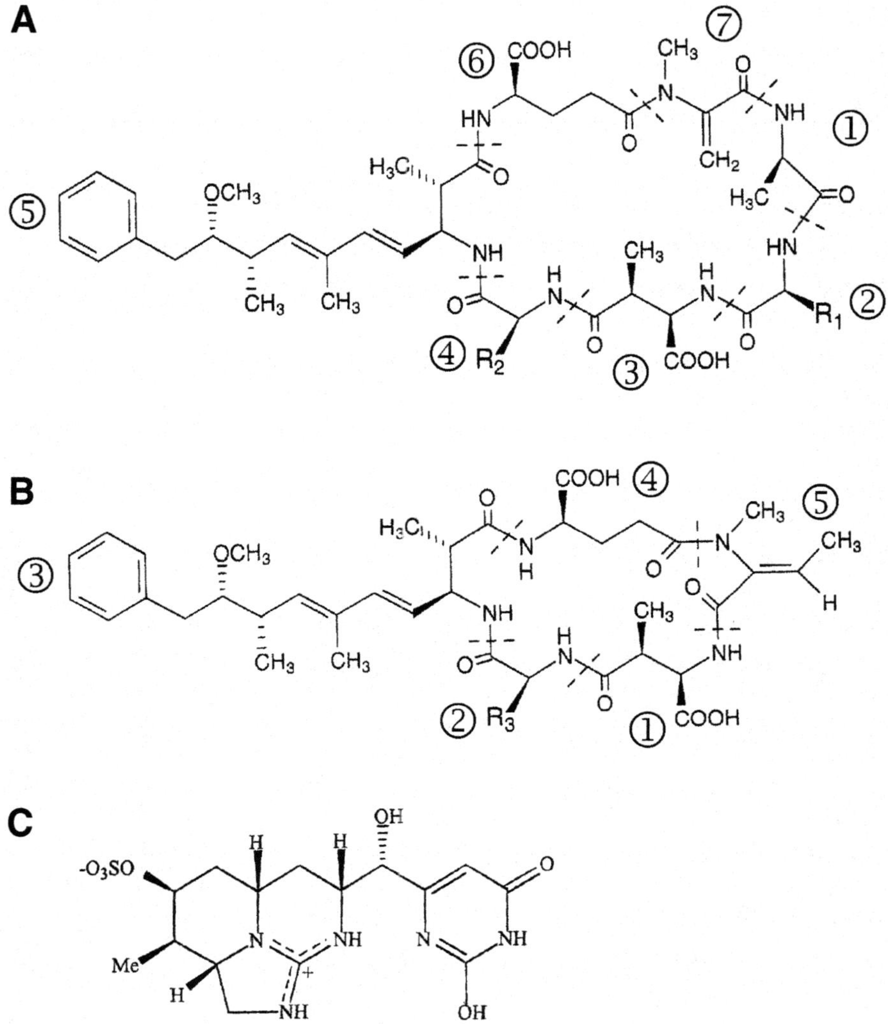

Fig. 1. General structures of microcystin (**A**), nodularin (**B**), and cylindrospermopsin (**C**).

genera of cyanobacteria including *Microcystis*, *Anabaena*, *Planktothrix*, *Nostoc*, *Hepalosiphon*, and *Anabaenopsis* ***(1)*** and have caused death in humans and other animals ***(2–5)***. The MCYSTs have also been shown to be potent tumor promotors in rats and have been linked to liver cancer in humans ***(6,7)***.

A compound with similar chemical structure and also similar toxicological properties has been identified from the brackish water cyanobacterium *Nodularia*. The compound termed nodularin is a pentapeptide (containing five peptide-linked amino acids).

To date, only two toxic variants of nodularin have been reported, one with a demethylated amino acid at position 1 and the other with a demethylated amino acid at position 3 (**Fig. 1B**) *(1)*. Nodularin has been implicated in the death of domestic animals *(8)*.

Another cyanotoxin of increasing concern throughout the world is cylindrospermopsin (**Fig. 1C**). This hepatotoxic compound is a cyclic guanidine alkaloid with a molecular weight of 415 and LD_{50} (ip mouse bioassay) of 200 μg/kg. Only one toxic structural variant of cylindrospermopsin (named 7-epicylindrospermopsin) has been reported *(9)*. This compound differs from cylindrospermopsin by the lack of the 7-hydroxy function and also by differences in the uracil nucleus. Cylindrospermopsin has been implicated in an outbreak of human gastroenteritis *(10,11)* and mortality in cattle *(12)* and has been shown to display carcinogenic activity *(13)*. Cylindrospermopsin is produced by several genera of cyanobacteria including *Anabaena*, *Aphanizomenon*, *Cylindrospermopsis*, *Umezakia*, and *Raphidiopsis* *(14–16)*.

1.2. Variation in Toxicity of Cyanobacterial Isolates at the Subspecies Level

The production of MCYST, nodularin, and cylindrospermopsin by potentially toxic strains of cyanobacteria is complicated by the fact that toxin production not only differs spatially and temporally within a bloom population, but also between morphologically indistinct strains within a single population *(1)*. Until very recently evaluation of strain toxigenicity was not possible since microscopic examination of a bloom sample does not provide information on toxigenicity or the toxic bloom-forming potential of a particular strain present. Recent developments, using molecular techniques, have resulted in the identification of the genes responsible for the production of some of the most commonly occurring freshwater cyanotoxins, namely, microcystins *(17,18)*, nodularin *(19)*, and cylindrospermopsin *(16)*. These methods, based on polymerase chain reaction (PCR), are of rapidly increasing use in the identification of potentially toxic cyanobacterial bloom populations and will in the future be important tools for the management of water quality.

In this chapter we detail methods for cyanobacterial DNA extraction followed by PCR methods used for the molecular detection of microcystin, nodularin, and cylindrospermopsin toxigenicity from laboratory-cultured cyanobacterial cells. It should be noted that the following techniques are also applicable to natural bloom samples (consisting of multiple genotypes) and can also be adapted for the use of single cyanobacterial colonies/trichomes/filaments as the DNA template for PCR.

2. Materials

Caution: All cyanobacterial culture material should be treated with care because of the potential risk of serious adverse acute and long-term effects resulting from inhalation or skin exposure. Wear gloves at all times, and when dealing with lyophilized material, manipulation should be carried out under ventilation.

For the following procedures, all chemicals should be of analytical grade quality.

Table 1
Characteristics of Oligonucleotide Primers Used

Designation	Target gene	Sequence	Annealing temp. (°C)
PKSM4	Cylindrospermopsin (PKS)	5'-GAAGCTCTGGAATCCGGTAA	55
PKSM5		5'-AATCCTTACGGGATCCGGTGC	
CPSF	Cylindrospermopsin (PS)	5'-AGTATATGTTGCGGGACTCG	55
CPSRC		5'-CCCGCCAAGACAGAAGG	
MSF	Microcystin	5'-ATCCAGCAGTTGAGCAAGC	60
MSR		5'-TGCAGATAACTCCGCAGTTG	
NPF	Nodularin	5'-TATTTTGTGGTGGAGAAGCACCTA	49
NPR		5'-GGAACTATCTGATAATTAGAC	
27F	16S rRNA	5'-AGAGTTTGATCCTGGCTCAG	50
809R		5'-GCTTCGGCACGGCTCGGGTCGATA	

PKS, polyketide synthetase; PS, peptide synthetase.

2.1. Cyanobacterial DNA Extraction

1. TNE buffer: 50 m*M* Tris-HCl, pH 8.0, 5 m*M* EDTA, pH 8.0, and 50 m*M* NaCl.
2. 50 mg/mL Lysozyme.
3. 10 mg/mL Proteinase K.
4. 10% Sodium dodecyl sulfate.
5. Phenol/chloroform/isoamylalcohol (25:24:1).
6. 4 *M* Ammonium acetate.
7. Isopropanol.
8. 70% Ethanol.
9. Agarose and gel electrophoresis equipment.

2.2. Polymerase Chain Reaction

1. GeneAmp® PCR System 2400 (Perkin Elmer).
2. PCR reaction buffer: 67 m*M* Tris-HCl, pH 7.6, 16 m*M* $(NH_4)SO_4$, 0.45% Triton X-100, 0.2% gelatin.
3. 2.5 m*M* $MgCl_2$.
4. 200 μ*M* Deoxyribonucleotide triphosphates.
5. Forward/reverse primers (**Table 1**).
6. *Taq* DNA polymerase.
7. Sterile deionized H_2O.
8. Agarose and gel electrophoresis equipment.

3. Methods

3.1. Cyanobacterial DNA Extraction

Total genomic DNA can be extracted from lyophilized (or fresh) cyanobacteria using a modification of a technique for purification of DNA from Gram-negative bacteria ***(20)***.

1. Suspend 5–10 mg of lyophilized cyanobacterial cells (or fresh culture equivalent) in 500 µL of TNE buffer. Vortex for 10 s and then add 10 µL of lysozyme (50 mg/mL). Incubate the solution at 55°C for 30 min.
2. Add 10 µL proteinase K (10 mg/mL) and 20 µL of 10% sodium dodecyl sulfate. Incubate at 55°C for 10 min, or until the solution has cleared (complete cell lysis; *see* **Note 1**).
3. Chill the solution on ice for 10 min and extract twice with an equal volume of phenol/chloroform/isoamylalcohol (25:24:1). Vortex the solution for 10 s and centrifuge at 12,000*g* for 6 min.
4. Add the supernatant to an equal volume of 4 *M* ammonium acetate.
5. Precipitate total genomic DNA by addition of 2 vol of isopropanol followed by centrifugation (12,000*g*) for 12 min at room temperature. Remove the supernatant, taking care not to make contact with the DNA pellet, and air-dry. Then resuspend the genomic DNA in 40 µL of sterile H_2O.
6. Test the purity and concentration of the DNA by gel electrophoresis using a 1.5% gel and following standard gel electrophoresis protocols ***(21)***. Maintain the DNA (which can be diluted to the appropriate concentration with sterile deionized H_2O) under refrigeration for the PCR analyses described below.

3.2. Polymerase Chain Extraction

The following is the standard PCR protocol used for cyanobacterial toxin analyses, with the only variations being the specific primers used and their respective annealing temperatures, which are listed in **Table 1**. Small-volume PCR is performed for rapid and highly specific generation of DNA molecules.

1. PCR reactions are performed in a 20-µL reaction volume, and thermal cycling is conducted using a GeneAmp PCR System 2400. The reaction mix consists of reaction buffer (67 m*M* Tris-HCl, 16 m*M* $(NH_4)_2SO_4$, 0.45% Triton X-100, 0.2 % gelatin), 2.5 m*M* $MgCl_2$, 200 µ*M* deoxyribonucleotide triphosphates, ~1 ng chromosomal DNA template (*see* **Subheading 3.1.**), 1 pmol forward and reverse primers and 0.5 U *Taq* DNA polymerase. The final reaction volume is adjusted to 20 µL with sterile deionized H_2O.
2. Initial denaturation of template DNA is achieved by incubation at 95°C for 2 min, and standard reactions are then subjected to 30 cycles of 95°C for 1.5 min, primer annealing between 49 and 60°C (**Table 1**), and extension at 72°C for 50 s. A further 7-min extension is carried out at the end of the 30 cycles for each PCR.
3. As a positive control, cyanobacterial-specific 16S rRNA primers are used for each sample amplifying the region 27–809 (*E. coli* designation) (*see* **Notes 2–5**).

3.3. Indentificaiton of Toxin Genes and Interpretation of Results

3.3.1. Microcystis and Microcystins

The recent identification of the locus responsible for microcystin synthesis in *Microcystis aeruginosa* ***(17,18)*** has shown that microcystin synthetase is a member of the non-ribosomal peptide synthetase family ***(17,22)***. Microcystin was the first cyanobacterial peptide for which a non-ribosomal synthesis by the thio-template mechanism was discovered ***(23)***. Insertional mutagenesis of one of these putative microcystin synthetase genes from the *Microcystis* strain PCC 7806 led to the complete loss of all microcystin variants, thus providing evidence for a function of this gene in microcystin biosynthesis ***(17)***. Subsequently, the entire microcystin synthetase

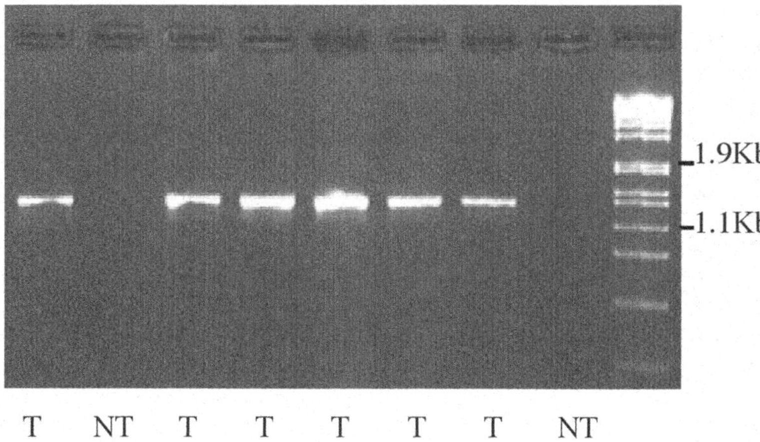

Fig. 2. PCR amplification of the *N*-methyl transferase region of microcystin synthetase from an environmental bloom ***(22)***. Samples that proved toxic (T) or nontoxic (NT) by the phosphatase inhibition assay are indicated.

(*mcy*) gene cluster of the strain PCC 7806 has been sequenced. It comprises 10 genes, encoding peptide synthetases, polyketide synthases, and tailoring functions. Two large bidirectionally transcribed operons (A–C and D–J) span a region of 55 kb. Further knockout mutations have confirmed the involvement of other *mcy* genes in microcystin biosynthesis ***(15)*** (see **Fig. 2** for sample results).

3.3.2. Nodularia *and Nodularins*

Nodularin is a peptide thought to be produced via a large multienzyme complex made up of peptide synthetase and polyketide synthase enzymes, similar to the biosynthesis of microcystin. Recent studies by Moffit and Neilan ***(19)*** have identified peptide-synthetase (PS) and polyketide synthetase (PKS)-like gene regions in *Nodularia.* The corresponding PCR using this toxic *Nodularia*-specific primer identified all nodularin-producing *N. spumigena* strains within the toxic bloom-forming cluster and was unable to amplify nontoxic *Nodularia* strains and toxic strains from other genera (**Fig. 3**).

Specific primers have also been designed to detect the presence of nodularin synthetase genes in environmental samples, hence directly detecting the presence of the nodularin biosynthetic pathway ***(19)***.

3.3.3. Cylindrospermopsis *and Cylindrospermopsin*

Biochemical studies and structural analysis have suggested a hybrid PS-PKS system for the synthesis of cylindrospermopsin by *C. raciborskii* ***(22)***. Schembri et al. ***(16)*** screened 17 toxic and nontoxic cyanobacterial strains for the presence of PKS and PS genes by hybridization of specifically designed radiolabeled probes. The

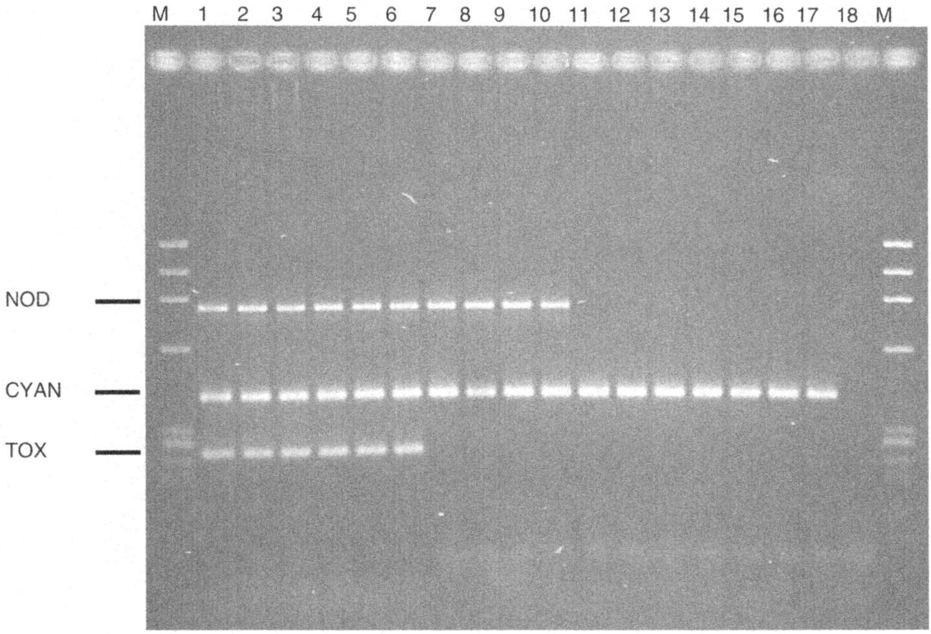

Fig. 3. Analysis of 16S rRNA gene regions using cyanobacterial- and *Nodularia*-specific oligonucleotide primers (**Table 1**). The cyanobacterial specific 16S rDNA PCR amplified a 400-bp fragment from all strains analyzed (CYAN). The *Nodularia* genus-specific 16S rDNA PCR amplified a 780-bp fragment from all *Nodularia* strains analyzed (NOD). The toxic *N. spumigena*-specific 16S rDNA PCR amplified a 200-bp fragment from all toxic planktic bloom-forming *Nodularia* (TOX).

hybridization studies revealed that the PKS and PS genes were found only in the CYL-producing strains examined, *C. raciborskii* and *Anabaena bergi*. No PKS or PS genes were detected in non-CYL-producing strains, and the PKS and PS genes appeared to be linked in that either both were detected or both were absent in any one strain.

DNA sequence analysis and subsequent amino acid sequence determination of the PKS and PS gene fragments from *C. raciborskii* revealed a high identity to genes encoding similar functions in other bacteria. The similarity in sequence and function of the genes investigated by Schembri et al. ***(16)*** to other PKS and PS genes, combined with the finding that these genes are found only in strains that produced CYL, suggests a hybrid PS-PKS system in the production of CYL.

Using the PCR protocol described in **Subheading 3.2.** and the primers in **Table 1**, the PCR resulted in the amplification of a 650-bp fragment of the PKS gene and an approx 500-bp fragment of the PS gene (**Fig. 4**). The observation that the PKS and PS genes are either both present or absent in strains examined suggests that they are physically linked.

1 2 3 4 5 6 7 8 9 10

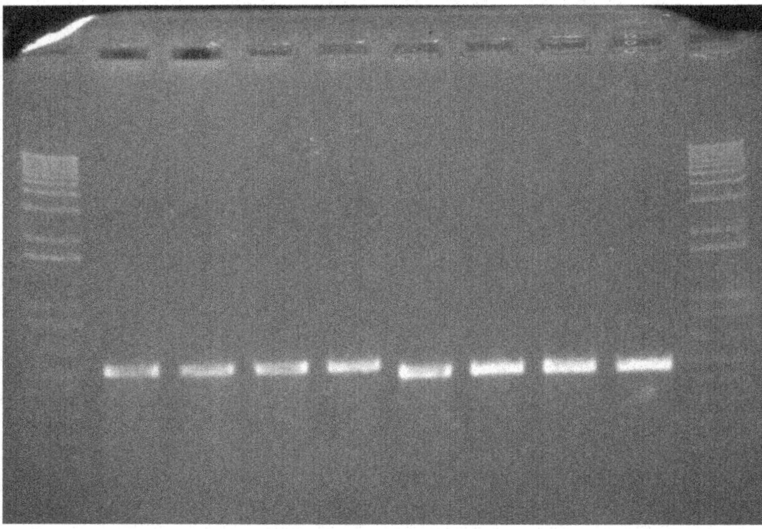

Fig. 4. PCR amplification of regions of cylindrospermopsin from representative strains. Lanes 2–5, amplification of PKS using primers PKSM4 and PKSM5; lanes 6–9, amplification of PS using primers CPSF and CPSRC. Lanes 2 and 6, *C. raciborskii* AWT205; lanes 3 and 7, *C. raciborskii* SDC; lanes 4 and 8, *C. raciborskii* GOON; lanes 5 and 9, *Aphanizomenon ovalisporum*; lanes 1 and 10, molecular weight markers.

4. Notes

1. **Steps 1** and **2** can be repeated if cyanobacterial material has not undergone complete cell lysis. The incubation in **step 2** can be extended for several hours or left overnight to facilitate cell lysis.
2. As these PCR protocols can also be used on nonaxenic environmental samples, in addition to the 16S rRNA control control, any PCR of samples that results in a negative result for the toxin gene can be repeated with the addition of cyanobacterial DNA that is known to contain the respective toxin gene. The "spiking" of these samples ensures that there are no inhibitors present in these samples that could result in the production of false-negative results.
3. If these spiked samples are still negative, the addition of a reagent that binds inhibitors such as humic acids (e.g., GeneReleaser®; BioVentures) to the final DNA suspension is recommended.
4. As with optimization of most PCRs, if nonspecific products are obtained in the reaction, one or more of the following factors can be changed to minimize the occurrence of these products: increase annealing temperature, or decrease one or more of the following: annealing time, extension time, or primer, DNA template, or *Taq* polymerase concentration.
5. If the PCR product is very weak, a bump-up PCR can be performed, whereby the PCR product is diluted (1:50) and then used as the template in a repeat of the same PCR. The number of cycles is normally reduced to 18–20.

Acknowledgments

This work was supported by the Australian Research Council and the CRC for Water Quality and Treatment.

References

1. Sivonen, K. and Jones, G.(1999) Cyanobacterial toxins. In: *Toxic Cyanobacteria in Water* (Chorus, I. and Bartram, J. eds.). E & FN Spon, London, pp. 41–91.
2. Kuiper-Goodman, T., Falconer, I., and Fitzgerald, J. (1999) Human health aspects. In: *Toxic Cyanobacteria in Water* (Chorus, I. and Bartram, J., eds.). E & FN Spon, London, pp. 113–141.
3. Carmichael, W. W. (1994) The toxins of cyanobacteria. *Sci. Am.* **270**,78–86.
4. Carmichael, W. W. (1992) Cyanobacterial secondary metabolites—the cyanotoxins. *J. Appl. Bacteriol.* **72,** 445–459.
5. Jochimsen, E. M., Carmichael, W. W., An., J., et al. (1998) Liver failure and death following exposure to microcystin toxins at a dialysis centre in Brazil. *N. Engl. J. Med.* **338,** 873–878.
6. Falconer, I. R. and Humpage, A. R. (1996) Tumour promotion by cyanobacterial toxins. *Phycologia* **35,** 74–79.
7. Falconer, I. R. (1999) An overview of problems caused by toxic blue-green algae occurring in water supplies. *Environ. Toxicol.* **14,** 5–12.
8. Francis, G. (1878) Poisonous Australian lake. *Nature* **18,** 11–12.
9. Banker, R., Teltsch, B., Sukenik, A., and Carmeli, S. (2000) 7-Epicylindrospermopsin, a toxic minor metabolite of the cyanobacterium *Aphanizomenon ovalisporum* from Lake Kinneret, Israel. *J. Nat. Prod.* **63,** 387–389.
10. Bourke, A. T. C., Hawes, R. B., Neilson, A., and Stallman, N. D. (1983) An outbreak of hepatoenteritis (the Palm Island mystery disease) possibly caused by algal intoxication. *Toxicon* **3,** 45–48.
11. Byth, S. (1980) Palm Island mystery disease. *Med. J. Aust.* **2,** 40–42.
12. Saker, M. L., Thomas, A. D., and Norton, J. H. (1999) Cattle mortality attributed to the toxic cyanobacterium *Cylindrospermopsis raciborskii* in an outback region of north Queensland. *Environ. Toxicol.* **14,** 179–183.
13. Humpage, A. R., Fenech, M., Thomas, P., and Falconer, I. R. (2000) Micronucleus induction and chromosome loss in transformed human white cells indicate clastogenic and aneugenic action of the cyanobacterial toxin, cylindrospermopsin. *Mutat. Res.* **472,** 155–161.
14. Banker, R. S., Carmeli, O., Hadas, B., Teltsch, R., Porat, R., and Sukenik, A. (1997) Identification of cylindrospermopsin in *Aphanizomenon ovalisporum* (Cyanophyceae) isolated from Lake Kinneret, Israel. *J. Phycol.* **35,** 613–616.
15. Harada, K., Ohtani, I., Iwamoto, K., et al. (1994) Isolation of cylindrospermopsin from a cyanobacterium *Umezakia natans* and its screening method. *Toxicon* **32,** 73–84.
16. Schembri, M. A., Neilan, B. A., and Saint, C. P. (2001) Identification of genes implicated in toxin production in the cyanobacterium *Cylindrospermopsis raciborskii. Environ. Toxicol.* **16,** 413–421.
17. Dittmann, E., Neilan, B. A, Erhard, M., von Döhren, H., and Börner, T. (1997) Insertional mutagenesis of a peptide synthetase gene that is responsible for hepatotoxin production in the cyanobacterium *Microcystis aeruginosa* PCC 7806. *Mol. Microbiol.* **26,** 779–787.

18. Tillett, D., Dittmann, E., Erhard, M., von Döhren, H., Börner, T., and Neilan, B.A. (2000) Structural organization of microcystin biosynthesis in *Microcystis aeruginosa* PCC7806: an integrated peptide-polyketide synthetase system. *Chem. Biol.* **7,** 753–764.
19. Moffit, M. C. and Neilan, B. A. (2001) On the presence of peptide synthetase and polyketide synthetase genes in the cyanobacterial genus *Nodularia. FEMS Microbiol. Lett.* **196,** 207–214.
20. Neilan, B. A. (1995) Identification and phylogenetic analysis of toxigenic cyanobacteria by multiplex randomly amplified polymorphic DNA PCR. *Appl. Environ. Microbiol.* **61,** 2286–2291.
21. Sambrook, J., Fritsch, E. F., and Maniatis, T. (1989) *Molecular Cloning: A Laboratory Manual.* Cold Spring Harbor Laboratory Press, Cold Spring Harbor, NY.
22. Tillett, D., Parker, D. L., and Neilan, B.A. (2001) Detection of toxigenicity by a probe for the microcystin synthetase A gene (mcyA) of the cyanobacterial genus *Microcystis*, comparison of toxicities with 16S rRNA and phycocyanin operon (phycocyanin intergenic spacer) phylogenies. *Appl. Environ. Microbiol.* **67,** 2810–2818.
23. Arment, A. R. and Carmichael, W. W. (1996) Evidence that microcystin is a thiotemplate product. *J. Phycol.* **32,** 591–597.
24. Burgoyne, D. L., Hemscheidt, T. K., Moore, R. E., and Runnegar, M. (2000) Biosynthesis of cylindrospermopsin. *J. Org. Chem.* **65,** 152–156.

V

Bacteriocins and Other Inhibitors

21

Purification of Antilisterial Bacteriocins

Jean-Marc Berjeaud and Yves Cenatiempo

1. Introduction

In recent years, numerous contamination outbreaks, involving various pathogens (i.e., *Listeria* and *Salmonella*), have increased concern over food preservation. Research efforts have focused on the discovery of new molecules targeting such foodborne pathogens and therefore able to inhibit and or kill them. Lactic acid bacteria (LAB) extensively used in fermented foods for thousands of years not only improve their flavor and texture but also inhibit pathogenic and spoilage microorganisms. LAB inhibitory activity is primarily owing to pH decrease and competition for substrates. Antagonistic activity of LAB also depends on secreted antimicrobial compounds with a poor selectivity, such as metabolic compounds (i.e., hydrogen peroxide, acetoin, and others) or more specific ones like bacteriocins. The latter are proteinaceous compounds, ribosomally synthesized and subsequently secreted by Gram-positive as well as Gram-negative bacteria. Their antimicrobial activity is generally restricted to strains phylogenetically related to the producers.

A classification of bacteriocins produced by LAB was first proposed by Klaenhammer in 1993 *(1)* and was modified by Nes et al. in 1996 *(2)*; class I and class II bacteriocins are the most abundant and thoroughly studied *(3,4)*. Bacteriocins from both classes exhibit antilisterial activity. Class I bacteriocins, namely, lantibiotics, have been widely studied, and among them, nisin is used in many countries as a preservative in food products *(5)*. These bacteriocins are characterized by the presence, in their primary structure, of post-translationally modified amino acid residues (i.e., lanthionine and methylanthionine) that are formed *(3)*. Class II bacteriocins, containing three subclasses, consist of small peptides that do not bear any modified amino acid residue. The most studied subclass corresponds to class IIa, also termed anti-*Listeria* bacteriocins *(4)*. These peptides share strong structural homologies in their N-terminal domain (**Table 1**), with the presence of one disulfide bond and a net positive charge. Their C-terminal domain is more variable but appears quite hydrophobic.

From: *Methods in Molecular Biology, vol. 268: Public Health Microbiology: Methods and Protocols*
Edited by: J. F. T. Spencer and A. L. Ragout de Spencer © Humana Press Inc., Totowa, NJ

Moreover, some of these bacteriocins, namely, sakacin G *(6)*, pediocin PA-1 *(7)*, enterocin A *(8)*, coagulin *(9)*, and divercin V41 *(10)*, are characterized by the presence of a second disulfide bond in the C-terminal region (**Table 1**).

According to studies on pediocin PA-1 *(11)* and mesentericin Y105 *(12)*, these peptides act by pore formation in the cytoplasmic membrane of target strains and consecutive dissipation of proton motive force. Recently, the presence of a mannose-specific phosphotransferase, as a putative receptor, in the sensitivity of *Listeria* to leucocin A *(13)* or *Enterococcus (14)* and *Listeria (15)* to mesentericin Y105 was demonstrated. Analysis of the genetic determinants of several class IIa bacteriocins revealed that genes involved in the production and transport of the bacteriocin and immunity are organized in one or two operon-like structures. Most of them are located on a plasmid and possess at least two genes that encode proteins homologous to ABC transporters and accessory proteins, probably involved in the transport of class IIa bacteriocins *(7,16,17)*. One noticeable exception is divercin V41, whose operon is located on a bacterial chromosome and that seems to be devoid of any accessory protein gene *(10)*.

The number of such peptidic compounds may be limited since many of the newly detected strains express previously described bacteriocins *(4)*. Consequently, detection and characterization of still unknown antagonistic peptides is becoming difficult. One of the more time-consuming steps in such studies consists of the purification of antagonistic compounds. In a previous work *(18)*, we described an efficient purification method of mesentericin Y105, our class IIa bacteriocin model. We applied this rapid and efficient method, with some improvements, first to other known class IIa bacteriocins from their producer strain cultures and second to any culture supernatant showing antilisterial activity. Since all class IIa bacteriocins displayed a positive charge and a hydrophobic C-terminal part (**Table 1**), we thus used cation exchange and reverse phase chromatography to purify these peptides. The method developed consists of a straightforward four-step process. After removal of bacterial cells and heating of the culture supernatant, a cation exchange chromatography, eluted with increasing NaCl concentrations, was performed, followed by solid-phase extraction on a C18 reverse phase and finally reverse-phase high-performance liquid chromatography (HPLC) on a C8 column.

2. Materials

1. Culture medium (MRS broth, DIFCO), sterilized for 12 min at 110°C and stored at 4°C for up to 1 mo (*see* **Note 1**).
2. Elution solutions for cation exchange chromatography: acidified H_2O, acidified 0.1 *M* NaCl, and acidified 0.5 *M* NaCl (*see* **Note 2**). The cation exchange phase chosen consisted of carboxy-methyl cellulose (*see* **Note 3**). After usage, the column was filled with 2% sodium azide and stored at room temperature.
3. For solid phase extraction, prior to sample deposit, each cartridge (C18 Sep-pak plus, Waters) was washed with 10 mL of ethanol (99%), 5 mL of acetonitrile (HPLC grade), and H_2O (milliQ). Sample washing and elution were realized with 5 mL of 20 m*M* ammonium acetate (*see* **Note 4**) (final concentration) containing increasing quantities of acetonitrile (HPLC grade). Acetonitrile-containing buffers were stored at room temperature for up to 2 mo.

Table 1
Sequence Alignment of Class IIa Bacteriocins

Bacteriocin[a]	Residue number[b]: 1 — 10 — 20 — 30 — 40	Refs
Sakacin G	KYYGNGVSCNSHGCSVNWGQAWTCGVNHLANGGHGVC	*6*
Mesentericin Y105	KYYGNGVHCTKSGCSVNWGEAASAGIHRLANGGNGFW	*23*
Leucocin A	KYYGNGVHCTKSGCSVNWGEAFSAGVHRLANGGNGFW	*24*
Leucocin C	KNYGNGVHCTKKGCSVDWGYAATNIANNSVMNGLTG	*25*
Sakacin P	KYYGNGVHCGKHSCTVDWGTAIGNIGNNAAANWATGWNAGG	*26*
Mundticin	KYYGNGVSCNKKGCSVDWGKAIGIIGNNSAANLATGGAAGWSK	*27*
Sakacin A	ARSYGNGVYCNNKKCWVNRGEATQSIIGGMISGWASGLAGM	*28*
Piscicocin V1a	KYYGNGVSCNKNGCTVDWSKAIGIIGNNAAANLTTGGAAGWNKG	*29*
Bacteriocin BM1	AISYGNGVYCNKEKCWVNKAENKQAITGIVIGGWASSLAGMGH	*30*
Bacteriocin B2	VNYGNGVSCSKTKCSVNWGQAFQERYTAGINSFVSGVASGAGSIGRRP	*30*
Bavaricin MN	TKYYGNGVYCNSKKCWVDWGQAAGGIGQTVVxGWLGGAIPGK	*31*
Enterocin P	ATRSYGNGVYCNNSKCWVNWGEAKENIAGIVISGWASGLAGMGH	*32*
Bacteriocin 31	ATYYGNGLYCNKQKCWVDWNKASREIGKIIVNGWVQHGPWAPR	*33*
Bifidocin B	KYYGNGVTCGLHDCRVDRGKATCGIINNGGMWGDIG	*34*
Pediocin PA-1	KYYGNGVTCGKHSCSVDWGKATTCIINNGAMAWATGGHQGNHKC	*7*
Coagulin	KYYGNGVTCGKHSCSVDWGKATTCIINNGAMAWATGGHQGTHKC	*9*
Enterocin A	TTHSGKYYGNGVYCTKNKCTVDWAKATTCIAGMSIGGFLGGAIPGKC	*8*
Divercin V41	TKYYGNGVYCNSKKCWVDWGQASGCIGQTVVGGWLGGAIPGKC	*10*
Consensus[c]	YYGNGV C C V WG A I	

[a]Leucocin A is identical to leucocin B ***(35)***, piscicocin V1a is identical to piscicolin 126 ***(36)***, sakacin A is identical to curvacin A ***(37)***, carnobacteriocin BM1 is identical to piscicocin V1b ***(29)***, and pediocin PA-1 is identical to pediocin AcH ***(38)***, and pediocin SJ-1 ***(39)***. Bavaracin A ***(40)*** is probably identical to sakacin P.

[b]Residue numbering is according to the sequence of sakacin G.

[c]The consensus sequence includes residues conserved by at least 75%. The residues conserved by more than 90% are underlined.

4. Reverse phase HPLC.
   a. Eluent A: H_2O (milliQ)/Trifluoroacetic acid (TFA) 0.1%.
   b. Eluent B: 80% Acetonitrile (HPLC grade)/20% H_2O/0.07% TFA (*see* **Note 5**). Separation can be conducted on any analytical C8 HPLC column (*see* **Note 6**).
5. HPLC can be conducted on any apparatus able to generate a binary gradient fitted with a UV detector. Elutions were monitored at 220 nm.

3. Methods (*see* Note 7)

1. Cell-free supernatant from an overnight culture (100 mL; *see* **Note 8**) of bacteriocin producer strain was obtained by centrifugation at 6000*g* for 15 min and heating, to inactivate extracellular proteases, at 70°C for 20 min (*see* **Note 9**).
2. After cooling, the supernatant was diluted with 1 vol of acidified water. Diluted supernatant (200 mL) was loaded, at a flow rate of 4 mL/min, on a column filled with weak cation exchange carboxy-methyl cellulose equilibrated with water (*see* **Note 10**). After washing

successively with water (100 mL) and 0.1 *M* NaCl (150 mL) (*see* **Note 11**), anti-*Listeria* bacteriocin was eluted with 0.5 *M* NaCl (200 mL). The pH of all the solutions was maintained below 6 with 1 *M* HCl.

3. Active fraction was applied to a solid-phase extraction C18 cartridge (Sep-pak plus, Waters) equilibrated with water (*see* **Note 12**). After washing successively with 5 mL of 0, 10, 20, and 30% (*see* **Note 13**) acetonitrile containing 20 m*M* ammonium acetate solutions, bacteriocin was eluted from the cartridge with 10 mL of 80% acetonitrile (*see* **Note 14**) containing 20 m*M* ammonium acetate prior to lyophilization (*see* **Note 15**).
4. The extract was solubilized with 1 mL of 50% acetonitrile aqueous solution and injected on a C8 reverse-phase HPLC analytical column (*see* **Note 16**). Elution was monitored for absorbance at 220 nm (*see* **Note 17**). Separation was carried out using a water/acetonitrile/trifluoroacetic acid 0.1% (v/v) solvent system. After an initial 5-min wash with 20% of eluent B (*see* **Subheading 2.**, Materials), elution was achieved in 30 min at a flow rate of 0.8 mL/min with a 10-min linear gradient from 20 to 40% of eluent B and then 20 min from 40 to 52% of eluent B. More hydrophobic compounds were then removed from the column first, using a linear gradient from 50 to 100% of eluent B and second by a 5-min washing at 100% of eluent B (*see* **Note 18**). Chromatograms obtained from various bacteriocin-producing bacterial strains are presented in **Fig. 1**. In most cases, bacteriocin peaks were well defined and isolated from other compounds in the injected fraction and consequently were easy to collect.

With the method described in the present work, very pure preparations of class IIa bacteriocins were obtained repeatedly with high reproducibility. This method appeared to be simpler and faster than other purification protocols described so far and particularly well adapted to fast identification of antilisterial bacteriocin. Recently, this method was successfully applied to the purification of bacteriocin produced by lactic acid bacteria from an industrial collection ***(19)***. Of the five bacteriocins produced by eight active detected strains, four were easily identified as the previously known pediocin PA-1, sakacin A, sakacin P, and enterocin A (**Table 1**). The other peptide, a newly discovered bacteriocin named Sakacin G, was purified and further characterized ***(6)***. To date, all class IIa bacteriocins produced by the lactic acid bacteria tested in our laboratory (underlined in **Table 1**) were purified without any noticable problem. Moreover, we showed earlier ***(18)*** that a non-antilisterial class II bacteriocin, mesentericin B105 ***(20)*** (also called mesenterocin 52B ***[21]***), which is active toward *Leuconostoc*, copurified with mesentericin Y105 using the method described, suggesting that it was applicable to bacteriocins other than class IIa. More interestingly, nisin A as well as nisin Z, produced by *Lactococcus lactis* strains, were also purified according to the currently described method (unpublished results) (*see* **Note 19**). Nisin A and Z are lantibiotics active toward *Listeria*, that are structurally differents from class IIa bacteriocins. In conclusion, the purification process developed in the laboratory, originally intended for antilisterial class IIA bacteriocins, can be extended to any type of anti-listerial bacteriocin and could even be a more widely applicable method.

4. Notes

1. Culture media for LAB are classically sterilized at 120°C for 20 min. However, it appears that the extent of hydrophobic and colored cumbersome contaminants depends on the duration and temperature of sterilization. Selected conditions correspond to a low extent of contaminants with good sterilization to prevent contaminations.

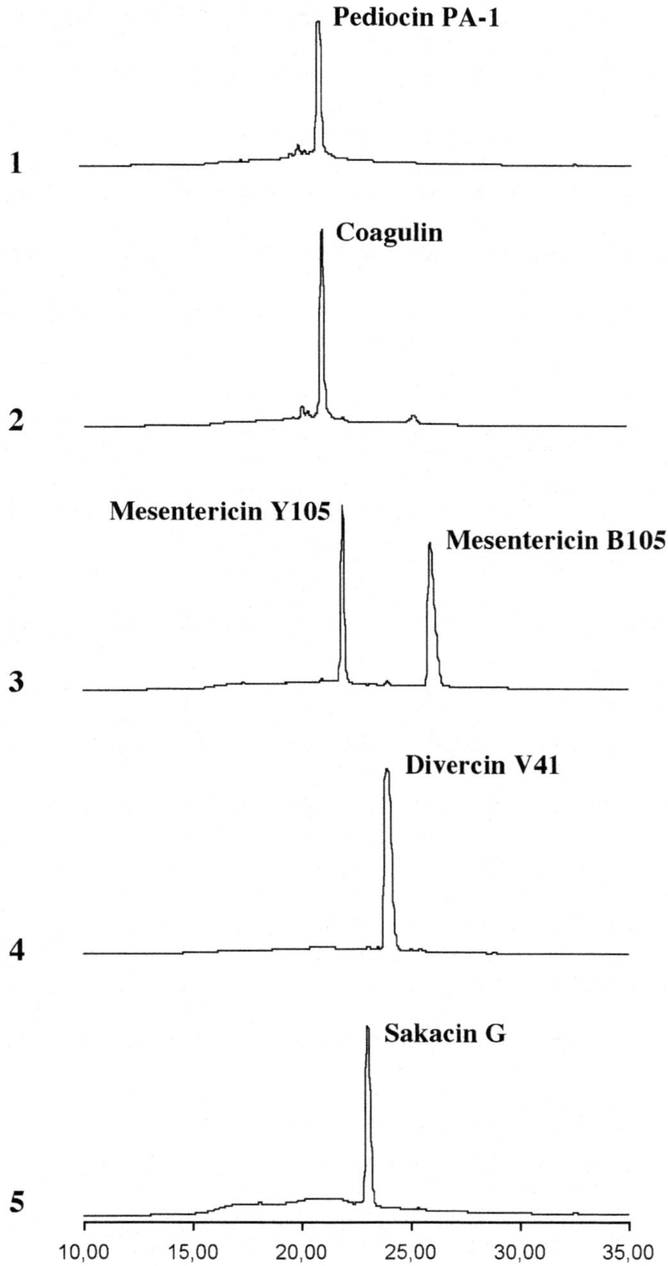

Fig. 1. HPLC elution profile recorded at 220 nm of active fractions after cation exchange chromatography and solid phase extraction from culture supernatants of: 1, *Pediococcus acidilactici* P1521; 2, *Bacillus coagulans* 14; 3, *Leuconostoc mesenteroides* Y105; 4, *Carnobacterium divergens* V41; and 5, *Lactobacillus sakei* 2512.

2. All aqueous solutions were made with fresh milliQ (Waters) water and their pH was adjusted to 5 with 1 *M* HCl. They are stored at room temperature for up to 1 mo.
3. Another low-force cation exchange phase could be used.
4. Bacteriocin resolution appeared to be better with ammonium acetate as a counterion than TFA.
5. Protein sequencing grade TFA was used. We noted that a combination of acetonitrile and TFA absorbs light at 220 nm, so the TFA ratio in eluent B was adjusted to give a horizontal baseline in chromatograms.
6. Kromasil phase was preferentially used because of its resistance and low cost.
7. When this method is applied for the first time on a particular bacterial strain, we recommend testing all the fractions for their anti-*Listeria* activity, in order to, eventually, adjust volumes of phase and eluents used. Bacteriocin assay could be performed using a well diffusion method ***(22)***.
8. Phase quantities and chromatography column lengths were designed for mesentericin Y105 quantities produced by 100 mL of culture supernatant from *Leuconostoc mesenteroides* Y105. They might be adjusted for purification from higher volumes or from bacterial strains producing higher quantities of bacteriocins, such as divercin V41 ***(22)***.
9. This step is not absolutely necessary and can be omitted in case of "emergency." It greatly depends on the bacterial strain cultivated and its ability to secrete proteases.
10. Previously, this step was replaced by an ammonium sulfate precipitation. Briefly, 36 g of ammonium sulfate were added to 100 mL of culture supernatant, in a centrifugation bottle, and then gently stirred overnight at 4°C. The protein fraction, containing bacteriocin, was obtained after a 45-min centrifugation at 10,000*g* and 4°C. The pellet was resuspended with 10 mL before solid phase extraction. This step was time- and material-consuming. Indeed, this protein-enriched fraction contained too much hydrophobic compounds for one solid phase extraction cartridge, suggesting the use of two cartridges in tandem. Moreover, colored contaminants were very difficult to eliminate.
11. The culture supernatant displayed a brown color and, after loading on the column, the solid phase appeared yellow. The 0.1 *M* NaCl washing must be conducted until the phase appears totally white again.
12. Loading, washing, and elution of the cartridge were made manually. The "flow" of this chromatographic step is very critical. Ideally, it must be maintained at around 2 drops/s.
13. Brown contaminants, were removed during all these washings. A washing solution of 5 mL/cartridge is sufficient.
14. Bacteriocins were generally eluted at 40% acetonitrile. However, for the most hydrophobic peptides, residual activity is detected in an 80% sample, with no additional contaminants, convincing us to elute them with 10 mL of such a solution.
15. Lyophilization permits concentration of samples without degradation.
16. If only a rapid identification of bacteriocins produced is desired, a complete purification is almost not necessary. Ionspray mass spectrometry analysis of the concentrated 80% solid phase extraction fraction is sufficient in many cases.
17. Peptide elution is generally monitored at a wavelength of 214 nm, corresponding to the absorbance of peptidic bond, or 280 nm, associated with aromatic amino acids, especially tryptophan. For bacteriocins, we choose 220 nm, which has a better sensitivity than 280 nm and less "noise" than 214 nm.
18. A simpler linear gradient could give good separations but, after multiple attempts made in order to purify mesentericin Y105, the protocol presented gave the best results. The first rapid step, until 40% of eluent B, allows us to eliminate hydrophilic compounds, and the slower second step permits us to obtain a good resolution around the bacteriocins.

19. The chromatogram obtained from the purification of nisin A and Z showed more contaminants than those obtained from class IIa bacteriocins. We supposed that this was owing to the culture medium; the M17 broth used for the *Lactococcus lactic*-producing nisin A culture contained more byproducts than the MRS broth used for cultivating LAB-producing class IIa bacteriocins.

References

1. Klaenhammer, T. R. (1993) Genetics of bacteriocins produced by lactic acid bacteria. *FEMS Microbiol. Rev.* **12,** 39–85.
2. Nes, I. F., Diep, D. B., Havarstein, L. S., Brurberg, M. B., Eijsink, V., and Holo, H. (1996) Biosynthesis of bacteriocins in lactic acid bacteria. *Antonie Van Leeuwenhoek* **70,** 113–128.
3. McAuliffe, O., Ross, R. P., and Hill, C. (2001) Lantibiotics: structure, biosynthesis and mode of action. *FEMS Microbiol. Rev.* **25,** 285–308.
4. Ennahar, S., Sashihara, T., Sonomoto, K., and Ishizaki, A. (2000) Class IIa bacteriocins: biosynthesis, structure and activity. *FEMS Microbiol. Rev.* **24,** 85–106.
5. Schillinger, U., Geisen, R., and Holzapfel, W. H. (1996) Potential of antagonistic microorganisms and bacteriocins for the biological preservation of food. *Trends Food Sci. Technol.* **7,** 158–164.
6. Simon, L., Fremaux, C., Cenatiempo, Y., and Berjeaud, J. M. (2002) Sakacin G, a new type of antilisterial bacteriocin. *Appl. Environ. Microbiol.* **68,** 6416–6420.
7. Marugg, J. D., Gonzalez, C. F., Kunka, B. S., et al. (1992) Cloning, expression, and nucleotide sequence of genes involved in production of pediocin PA-1, and bacteriocin from *Pediococcus acidilactici* PAC1.0. *Appl. Environ. Microbiol.* **58,** 2360–2367.
8. Aymerich, T., Holo, H., Havarstein, L. S., Hugas, M., Garriga, M., and Nes, I. F. (1996) Biochemical and genetic characterization of enterocin A from *Enterococcus faecium*, a new antilisterial bacteriocin in the pediocin family of bacteriocins. *Appl. Environ. Microbiol.* **62,** 1676–1682.
9. Le Marrec, C., Hyronimus, B., Bressollier, P., Verneuil, B., and Urdaci, M. C. (2000) Biochemical and genetic characterization of coagulin, a new antilisterial bacteriocin in the pediocin family of bacteriocins, produced by *Bacillus coagulans* I(4). *Appl. Environ. Microbiol.* **66,** 5213–5220.
10. Metivier, A., Pilet, M. F., Dousset, X., et al. (1998) Divercin V41, a new bacteriocin with two disulphide bonds produced by *Carnobacterium divergens* V41: primary structure and genomic organization. *Microbiology* **144,** 2837–2844.
11. Bruno, M. E. C. and Montville, T. J. (1993) Common mechanistic action of bacteriocins bacteria from lactic acid bacteria. *Appl. Environ. Microbiol.* **59,** 3003–3010.
12. Maftah, A., Renault, D., Vignoles, C., et al. (1993) Membrane permeabilization of *Listeria monocytogenes* and mitochondria by the bacteriocin mesentericin Y105. *J. Bacteriol.* **175,** 3232–3235.
13. Ramnath, M., Beukes, M., Tamura, K. and Hastings, J.W. (2000) Absence of a putative mannose-specific phosphotransferase system enzyme IIAB component in a leucocin A-resistant strain of *Listeria monocytogenes*, as shown by two-dimensional sodium dodecyl sulfate-polyacrylamide gel electrophoresis. *Appl. Environ. Microbiol.* **66,** 3098–101.
14. Hechard, Y., Pelletier, C., Cenatiempo, Y., and Frere, J. (2001) Analysis of sigma(54)-dependent genes in *Enterococcus faecalis*: a mannose PTS permease (EII(Man)) is involved in sensitivity to a bacteriocin, mesentericin Y105. *Microbiology* **147,** 1575–11580.

15. Dalet, K., Cenatiempo, Y., Cossart, P., The European *Listeria* Genome Consortium, and Héchard, Y. (2001) A sigma(54)-dependent PTS permease of the mannose family is responsible for sensitivity of *Listeria monocytogenes* to mesentericin Y105. *Microbiology* **147 (Pt. 12),** 3263–3269.
16. Axelsson, L. and Holck, A. (1995) The genes involved in production of and immunity to sakacin A, a bacteriocin from *Lactobacillus sake* Lb706. *J. Bacteriol.* **177,** 2125–2137.
17. Huhne, K., Axelsson, L., Holck, A., and Krockel, L. (1996) Analysis of the sakacin P gene cluster from *Lactobacillus sake* Lb674 and its expression in sakacin-negative *Lb. sake* strains. *Microbiology* **142,** 1437–1448.
18. Biet, F., Berjeaud, J. M., Worobo, R. W., Cenatiempo, Y., and Fremaux, C. (1998) Heterologous expression of the bacteriocin mesentericin Y105 using the dedicated transport system and the general secretion pathway. *Microbiology* **144,** 2845–2854.
19. Simon, L., Fremaux, C., Cenatiempo, Y., and Berjeaud, J. M. (2001) Luminescent method for the detection of antibacterial activities. *Appl. Microbiol. Biotechnol.*, in press (OnLine D.O.I. 10.1007/s00253-001-0833-3).
20. Héchard, Y., Berjeaud, J.M. and Cenatiempo, Y. (1999) Characterization of the *mesB* gene and expression of bacteriocins by *Leuconostoc mesenteroides* Y105. *Curr. Microbiol.* **39,** 265–269.
21. Revol-Junelles, A. M., Mathis, R., Krier, F., Fleury, Y., Delfour, A. and Lefebvre, G. (1996) *Leuconostoc mesenteroides* subsp. *mesenteroides* FR52 synthesizes two distinct bacteriocins. *Lett. Appl. Microbiol.* **23,** 120–124.
22. Guyonnet, D., Fremaux, C., Cenatiempo, Y., and Berjeaud, J.M. (2000) Method for rapid purification of class IIa bacteriocins and comparison of their activities. *Appl. Environ. Microbiol.* **66,** 1744–1748.
23. Fremaux, C., Hechard, Y., and Cenatiempo, Y. (1995) Mesentericin Y105 gene clusters in *Leuconostoc mesenteroides* Y105. *Microbiology* **141,** 1637–1645.
24. Hastings, J. W., Sailer, M., Johnson, K., Roy, K. L., Vederas, J. C., and Stiles, M.E. (1991) Characterization of leucocin A-UAL 187 and cloning of the bacteriocin gene from *Leuconostoc gelidum. J. Bacteriol.* **173,** 7491–500.
25. Papathanasopoulos, M. A., Dykes, G. A., Revol-Junelles, A. M., Delfour, A., von Holy, A., and Hastings, J. W. (1998) Sequence and structural relationships of leucocins A-, B- and C-TA33a from *Leuconostoc mesenteroides* TA33a. *Microbiology* **144,** 1343–1348.
26. Tichaczek, P. S., Vogel, R. F., and Hammes, W. P. (1994) Cloning and sequencing of sakP encoding sakacin P, the bacteriocin produced by *Lactobacillus sake* LTH 673. *Microbiology* **140,** 361–367.
27. Bennik, M. H., Vanloo, B., Brasseur, R., Gorris, L. G., and Smid, E. J. (1998) A novel bacteriocin with a YGNGV motif from vegetable-associated *Enterococcus mundtii*: full characterization and interaction with target organisms. *Biochim. Biophys. Acta* **1373,** 47–58.
28. Axelsson, L., Holck, A., Birkeland, S. E., Aukrust, T., and Blom, H. (1993) Cloning and nucleotide sequence of a gene from *Lactobacillus sake* Lb706 necessary for sakacin A production and immunity. *Appl. Environ. Microbiol.* **59,** 2868–2875.
29. Bhugaloo-Vial, P., Dousset, X., Metivier, A., et al. (1996) Purification and amino acid sequences of piscicocins V1a and V1b, two class IIa bacteriocins secreted by *Carnobacterium piscicola* V1 that display significantly different levels of specific inhibitory activity. *Appl. Environ. Microbiol.* **62,** 4410–4416.
30. Quadri, L. E., Sailer, M., Roy, K. L., Vederas, J. C., and Stiles, M. E. (1994) Chemical and genetic characterization of bacteriocins produced by *Carnobacterium piscicola* LV17B. *J. Biol. Chem.* **269,** 12204–12211.

31. Kaiser, A. L. and Montville, T. J. (1996) Purification of the bacteriocin bavaricin MN and characterization of its mode of action against *Listeria monocytogenes* Scott A cells and lipid vesicles. *Appl. Environ. Microbiol.* **62,** 4529–4535.
32. Cintas, L. M., Casaus, P., Havarstein, L. S., Hernandez, P. E., and Nes, I. F. (1997) Biochemical and genetic characterization of enterocin P, a novel sec-dependent bacteriocin from *Enterococcus faecium* P13 with a broad antimicrobial spectrum. *Appl. Environ. Microbiol.* **63,** 4321–4330.
33. Tomita, H., Fujimoto, S., Tanimoto, K., and Ike, Y. (1996) Cloning and genetic organization of the bacteriocin 31 determinant encoded on the *Enterococcus faecalis* pheromone-responsive conjugative plasmid pYI17. *J. Bacteriol.* **178,** 3585–3593.
34. Yildirim, Z., Winters, D. K., and Johnson, M. G. (1999) Purification, amino acid sequence and mode of action of bifidocin B produced by *Bifidobacterium bifidum* NCFB 1454. *J. Appl. Microbiol.* **86,** 45–54.
35. Felix, J. V., Papathanasopoulos, M. A., Smith, A. A., von Holy, A., and Hastings, J.W. (1994) Characterization of leucocin B-Ta11a: a bacteriocin from *Leuconostoc carnosum* Ta11a isolated from meat. *Curr. Microbiol.* **29,** 207–212.
36. Jack, R. W., Wan, J., Gordon, J., et al. (1996) Characterization of the chemical and antimicrobial properties of piscicolin 126, a bacteriocin produced by *Carnobacterium piscicola* JG126. *Appl. Environ. Microbiol.* **62,** 2897–2903.
37. Tichaczek, P. S., Vogel, R. F., and Hammes, W. P. (1993) Cloning and sequencing of curA encoding curvacin A, the bacteriocin produced by *Lactobacillus curvatus* LTH1174. *Arch. Microbiol.* **160,** 279–283.
38. Motlagh, A., Bukhtiyarova, M., and Ray, B. (1994) Complete nucleotide sequence of pSMB 74, a plasmid encoding the production of pediocin AcH in *Pediococcus acidilactici. Lett. Appl. Microbiol.* **18,** 305–312.
39. Schved, F., Lalazar, A., Lindner, P., and Juven, B. J. (1994) Interaction of the bacteriocin produced by *Pediococcus acidilactici* SJ-1 with the cell envelope of *Lactobacillus* spp. *Lett. Appl. Microbiol.* **19,** 281–283.
40. Larsen, A. G., Vogensen, F. K., and Josephsen, J. (1993) Antimicrobial activity of lactic acid bacteria isolated from sour doughs: purification and characterization of bavaricin A, a bacteriocin produced by *Lactobacillus bavaricus* MI401. *J. Appl. Bacteriol.* **75,** 113–122.

22

The Hazard Analysis and Critical Control Point System in Food Safety

Anavella Gaitan Herrera

1. Introduction

The Hazard Analysis and Critical Control Point (HACCP) system is a preventive method of ensuring food safety. Its objectives are the identification of consumer safety hazards that can occur in the production line and the establishment of a control process to guarantee a safer product for the consumer; it is based on the identification of potential hazards to food safety and on measures aimed at preventing these hazards. HACCP is the system of choice in the management of food safety ***(1–3)***. The principles of HACCP are applicable to all phases of food production ***(4)***, including basic husbandry practices, food preparation and handling, food processing, food service, distribution systems, and consumer handling and use ***(5–13)***. The HACCP system is involved in every aspect of food safety production (according to the UN Food and Agriculture Organization [FAO] and the International Commission on Microbiological Specifications for Foods [ICMSF]) ***(14–18)***.

The most basic concept underlying the HACCP system is that of prevention rather than inspection. The control of processes and conditions comprises the critical control point (CCP) element (*see* **Note 1**). HACCP is simply a methodical, flexible, and systematic application of the appropriate science and technology for planning, controlling, and documenting the safe production of foods ***(6–11,19,20)*** (*see* **Note 2**).

The successful application of HACCP requires the full commitment and involvement of management and the workforce, using a multidisciplinary approach that should include, as appropriate, expertise in agronomy, veterinary health, microbiology, public health, food technology, environmental health ***(21)***, chemistry, engineering, and so on according to the particular situation ***(22,23)***. Application of the HACCP system is compatible with the implementation of total quality management (TQM) systems such as the ISO 9000 series ***(15–17)***.

From: *Methods in Molecular Biology, vol. 268: Public Health Microbiology: Methods and Protocols*
Edited by: J. F. T. Spencer and A. L. Ragout de Spencer © Humana Press Inc., Totowa, NJ

As stated previously, international food commerce is regulated by the World Trade Organization (WTO), which ensures that all economic relations involving foods are regulated by the norms, guidelines, and recommendations of the *Codex Alimentarius* Commission, Office of International Epizootics (OIE) and of the International Phytosanitary Protection Convention (IPPC) ***(24,25)*** (*see* **Note 3**). Thus, food-exporting countries may require additional resources to enhance the ability of their food industries to meet the requirements. Adequate steps should be taken to facilitate food trade, such as training of personnel, technology transfer, and strengthening of the national food control system ***(15,16)***.

The HACCP system covers all types of potential risks or hazards to food safety, whether biological, chemical, or physical, occurring naturally in the food or environment or caused by errors in food processing ***(26,27)***. HACCP is not a zero-risk system, but it is designed to minimize the risk of food safety hazards. The growing acceptance of the HACCP system worldwide by industry, governments, and consumers makes it likely that it will be the most universal tool for guaranteeing food safety in the twentieth century ***(28–34)***.

1.1. History (1–3,24)

Theories of quality management were regarded as a major factor in increasing the quality of Japanese products in the 1950s. Dr. W. E. Deming and others were active in developing TQM systems.

In the 1960s, the Pillsbury Company, the US Army, and the US National Aeronautics and Space Administration (NASA) developed a program for the production of safe foods for the US space program. NASA had two principal safety issues: the potential problems with food particles in the space capsule under conditions of zero gravity and the protection of food from all pathogens and biological toxins.

The US Army's Natick Laboratories were able to predict *what* might be wrong (a hazard), as well as *how* and *where* the problem could occur in the food production process. Thus it was possible to select points at which measurements and/or observations could be made; these points were called critical control points.

The Pillsbury Company introduced and adopted the HACCP system to guarantee the greatest safety possible while reducing end-product inspection and testing. Pillsbury presented the HACCP system to the public for the first time in 1971 at a food safety conference in the United States. In 1973, the Pillsbury Company published *Food Safety Through the Hazard Analysis and Critical Control Point System*, the first document detailing the HACCP technique. It later served as the basis for US Food and Drug Administration (FDA) inspector training programs ***(35,36)***.

The US National Academy of Sciences recommended in 1985 that the HACCP approach be adopted in food processing establishments to ensure food safety. In 1988, ICMSF published a book recommending the HACCP system as a basis of quality control with regard to hygiene and microbiology, and the International Association of Milk, Food, and Environmental Sanitarians (IAMFES) has recommended the broad application of HACCP to food safety ***(16)***.

The *Guidelines for the Application of Hazard Analysis and Critical Control Point (HACCP) System* of the *Codex Alimentarius* has become the point of reference for international food safety requirements. The *Codex Alimentarius* Commission adopted the *Guidelines* (ALINORM 93/13[a], Appendix II) at its 20th session, held in Geneva, in 1993. The revised *Recommended International Code of Practice—General Principles of Food Hygiene* (CAC/RCP 1-1969, Rev. 3 [1997]) was adopted by the *Codex Alimentarius* Commission during its 22nd session in June of 1997 ***(15–17,37,38)***.

1.2. Foodborne Diseases

A foodborne disease outbreak is defined by the US Centers for Disease Control and Prevention as an incident in which two or more persons experience a similar illness after ingestion of a common food and epidemiological analysis implicates the food as the source of the illness.

The foods most frequently involved are foods of animal origin. In 48% of the outbreaks between 1973 and 1987 in the United States in which the vehicle was identified, beef, chicken, eggs, pork, finfish, shellfish, turkey, or dairy products were involved. For a foodborne illness to occur, the pathogen or its toxin(s) must be present in the food. The mere presence of a pathogen, however, does not necessarily result in foodborne disease ***(26,27,39)***.

Foodborne illnesses are generally classified as:

1. *Foodborne infection*: a disease that results from ingesting food containing living harmful microorganisms, such as *Salmonella*, *Shigella*, hepatitis A virus, and *Trichinella spiralis* ***(40)***.
2. *Foodborne intoxication*: toxins or poisons from bacteria or mold growth are present in the ingested food ***(26)***.
3. *Foodborne toxin infection*: results from eating a food containing a large amount of disease-causing microorganisms that are capable of producing or discharging toxin once they are ingested, e.g., *Vibrio cholerae* and *Clostridium perfringens*, respectively ***(26,39)***.

1.3. Definitions

Control (noun)	To state wherein correct procedures are being followed and criteria are being met.
Control (verb)	To take all necessary actions to ensure and maintain compliance with criteria established in the HACCP plan.
Control measure	Any action and activity that can be used to prevent or eliminate a food safety hazard or reduce it to an acceptable level.
Corrective action	Any action to be taken when the results of monitoring at the CCPs indicates a loss of control.
Critical control point (CCP)	A step at which control can be applied and is essential to prevent or eliminate a food safety hazard or reduce it to an acceptable level.
Critical limit	A criterion that separates acceptability from unacceptability.
Deviation	Failure to meet a critical limit.
Flow diagram	A systematic representation of the sequence of steps or operations used in the production or manufacture of a particular food item.
HACCP	A system that identifies, evaluates, and controls hazards that are significant for food safety.

HACCP plan	A document prepared in accordance with the principles of the HACCP system to ensure control of hazards that are significant for food safety in the segment of the food chain under consideration.
Hazard	A biological, chemical, or physical agent in, or condition of, food with the potential to cause an averse health effect.
Hazard analysis	The process of collecting and evaluating information on hazards and conditions leading to their presence to decide which are significant for food safety and therefore should be addressed in the HACCP plan.
Monitor	The act of conducting a planned sequence of observations or mea surements of control parameters to assess whether a CCP is under control.
Step	A point, procedure, operation, or stage in the food chain including raw materials, from primary production to final consumption.
Validation	Obtaining evidence that the elements of the HACCP plan are effective.
Verification	The application of methods, procedures, tests, and other evaluations, in addition to monitoring to determine compliance with the HACCP plan (*Hazard Analysis and Critical Control Point (HACCP) System and Guidelines for its Application* (Annex to CAC/RCP 1-1969, Rev 3 [1997]), are reproduced in **Subheading 1.4.** ***(1–3,16,37,41)***.

1.4. Principles of the HACCP System (1–4)

The HACCP system consists of the following seven principles:

1. Conduct a hazard analysis. Identify the potential hazard(s) associated with food production at all stages, from primary production, processing, manufacture, and distribution until the point of consumption. Assess the likelihood of occurrence of the hazard(s) and identify the measures for their control.
2. Determine the CCPs. Determine the points, procedures, or operational steps that can be controlled to eliminate the hazard(s) or minimize its (their) likelihood of occurrence ***(42)***. A "step" means any stage in food production and/or manufacture including the receipt and/or production of raw materials, harvesting, transport, formulation, processing, storage, and so forth (*see* **Note 4**).
3. Establish critical limit(s). Establish critical limit(s) that must be met to ensure the CCPs are under control.
4. Establish a system to monitor control of the CCPs. Establish a system to monitor control of the CCPs by scheduled testing or observations.
5. Establish the corrective action to be taken when monitoring indicates that a particular CCP is not under control.
6. Establish procedures for verification to confirm that the HACCP system is working effectively.
7. Establish documentation concerning all procedures and records appropriate to these principles and their application.

1.5. Application of the HACCP Principles

The application of HACCP principles consists of the following tasks as identified in the Logic Sequence for Application of HACCP ***(4,42a,43)***:

1. Assemble HACCP team.
2. Describe product.

3. Identify intended use.
4. Construct flow diagram.
5. On-site verification of flow diagram.
6. List all potential hazards associated with each step, conduct a hazard analysis, and consider any measures to control identified hazards (*see* principle 1, **Subheading 1.4.**).
7. Determine CCPs (*see* principle 2).
8. Establish critical limits for each CCP (*see* principle 3).
9. Establish a monitoring system for each CCP (*see* principle 4).
10. Establish corrective actions (*see* principle 5).
11. Establish verification procedures (*see* principle 6).
12. Establish documentation and record keeping (*see* principle 7).

2. Materials, Equipment, and Personnel

2.1. The HACCP Team

It is very important to have full commitment to the HACCP initiative from management at all levels. Both the company and the personnel involved in the development of the HACCP plan must be totally committed to its implementation ***(22,44)***. The first task is to assemble a team having the knowledge and expertise to develop an HACCP plan. The team should be multidisciplinary and could include plant personnel from production/sanitation, quality assurance, laboratory, engineering, and inspection. The team should also include personnel who are directly involved in daily processing activities, as they are more familiar with the specific variability and limitations of the operations. It may be necessary to hire an expert in public health risks associated with the product or process.

2.1.1. Team Composition

Those who will be involved in hazard identification.
Those who will be involved in determination of CCPs.
Those who will monitor CCPs.
Those who will verify operations at CCPs.
Those who will examine samples and perform verification procedures.

2.1.2. Required Knowledge

Technology and equipment used on the processing lines.
Practical aspects of the food operations.
The flow and technology of the process.
Applied aspects of food microbiology.
HACCP principles and techniques.

2.1.3. Scope Issues

Limit the study to a specific product and process.
Define the type(s) of hazards to be included (e.g., biological, chemical, physical).
Define the part of the food chain to be studied.

2.1.4. Coordinator Duties

Ensure that the composition of the team meets the needs of the study.
Suggest changes to the team if necessary.
Coordinate the team's work.

Ensure that the agreed established plan is followed.
Share the work and responsibilities.
Ensure that a systematic approach is used.
Ensure that the scope of the study is met.
Chair meetings so that team members can freely express their ideas.
Represent the team before management.
Provide management with an estimate of the time, money, and labor required for the study.

2.1.5. Training Requirements

It is essential that the team members be trained in the *Codex* General Principles of Food Hygiene and the guidelines for the application of the HACCP system to ensure that the team will work together with a common focus and use the same approach and terminology.

2.2. Resources

The necessary resources for the HACCP study include the following:

Time for team meetings and administration.
Costs of initial training.
Necessary documents.
Access to analytical laboratories.
Access to information sources to answer questions raised by the team (e.g., universities, public and private research authorities, government and public authorities, scientific and technical literature, databases).

3. Methods

3.1. Describe Product and Identify Intended Use

Each food product should be described completely including all ingredients/processing methods/packaging materials/other used in the formulation of the product, to assist in the identification of all possible associated hazards. In brief, the product description should include the name, ingredients and composition, potential to support microbial growth (water activity [A_w], pH, and so on), brief details of the process and technology used in production, appropriate packaging, and intended use, including target population (*see* **Note 5**).

This information will be essential, particularly for microbiological hazards, because the product's composition needs to be assessed in relation to the ability of different pathogens to grow. The product to which the HACCP plan applies should be described on Forms 1 and 2 (**Tables 1** and **2**).

3.1.1. Formulation of Product

What raw materials or ingredients are used? Are microorganisms of concern likely to be present in or on these materials, and if so what are they? If food additives or preservatives are used, are they used at acceptable levels, and at those levels do they accomplish their technical objective? Will the pH of the product prevent microbial growth or inactivate particular pathogens? Will the A_w of the product prevent microbial growth? What is the oxidation/reduction potential (*E*h) of the product?

Table 1
Form 1: Product Description

1. Product name(s)	Canned mushrooms
2. Important product characteristics of end product (e.g., A_w, pH, and so on)	pH 4.8–6.5 (low acid) $A_w > 0.85$ (high moisture)
3. How the product is to be used	Normally heated before serving (casseroles, garnishes, and so on) or sometimesserved un heated (salads, appetizers, and so on)
4. Packaging	Hermetically sealed metal container
5. Shelf life	Two years plus, at normal retail shelf temperatures
6. Where the product will be sold	Retail, institutions, and food service. Could be consumed by high-risk groups (infirm, immunocompromised, elderly)
7. Labeling instructions	None required to ensure product safety
8. Special distribution control	No physical damage, excess humidity, or temperture extremes

DATE:__________________ **APPROVED BY:**______________________________

Table 2
Form 2: Product Ingredients and Incoming Material

Product Name(s): Canned mushrooms

Raw material	Packaging material	Dry ingredients
Mushrooms (domestic, white)	Cans Ends	Salt Ascorbic acid Citric acid

Other

Water (municipal)

DATE:__________________ **APPROVED BY:**______________________________

3.1.2. Processing and Preparation Checklist

Can a contaminant reach the product during preparation, processing, or storage? Will microorganisms or toxic substances of concern be inactivated during cooking, reheating, or other processing? Could any microorganisms or toxins of concern contaminate food after it has been heated? Would more severe processing be acceptable or desirable? Is the processing based on scientific data? How does the package or container affect survival and/or growth of microorganisms? How much time is taken for each step of processing, preparation, storage and display? What are the conditions of distribution?

3.1.3. Form 1: Product Description

1. Product name (common name) or group of product names (The grouping of like products is acceptable as long as all hazards are addressed.)
2. Important end-product characteristics: properties or characteristics of the food under review that are required to ensure its safety (e.g., A_w, pH/preservatives).
3. How the product is to be used (i.e., ready-to-eat/further processing required, heated prior to consumption).
4. Type of package, including packaging material and packaging conditions (e.g., modified atmosphere).
5. Shelf life, including storage temperature and humidity if applicable.
6. Where the product will be sold (e.g., retail, institutions, further processing).
7. Labeling instructions (e.g., handling and usage instructions).
8. Special distribution control (e.g., shipping conditions).

3.2. Construct Flow Diagram and On-Site Confirmation of Flow Diagram

It is easier to identify routes of potential contamination, to suggest methods of control, and to discuss these among the HACCP team if there is a flow diagram (**Table 3**). The review of the flow of raw materials from the point at which they enter the plant, through processing to departure, is the feature that makes HACCP a specific and important tool for the identification and control of potential hazards (**Table 4**).

Data may include but are not restricted to the following:

All ingredients and packaging used (biological, chemical, physical data).
Sequence of all process operations (including raw material addition).
Time/temperature history of all raw materials and intermediate and final products, including the potential for delay.
Flow conditions for liquids and solids.
Product recycle/rework loops.
Equipment design features.

3.3. List All Potential Hazards Associated With Each Step, Conduct Hazard Analysis, and Consider Any Measures to Control Identified Hazards

The *Codex Hazard Analysis and Critical Control Point (HACCP) System and Guidelines for its Application* [Annex to CAC/RCP 1-1969, Rev. 3 (1997)] defines a hazard as "A biological, chemical or physical agent in, or condition of, food with the potential to cause an adverse health effect." The hazard analysis is necessary to identify for the HACCP plan which hazards are of such a nature that their elimination or reduction to acceptable levels is essential to the production of a safe food.

The contamination of a food with a very low level of these microbes can still cause illness if the food is subsequently mishandled (**Table 5**).

3.3.1. Hazards

3.3.1.1. Biological Hazards

Foodborne biological hazards include microbiological organisms such as bacteria, viruses, fungi, and parasites. They are commonly associated with humans and with raw products entering the food establishment (**Table 6**).

Table 3
Flow Diagram

Product Name(s): Canned mushrooms

Mushrooms (raw)	Empty cans/ends	Dry ingredients	Water (municipal)
1. Receiving	2. Receiving	3. Receiving	4. Intaking
5. Storing	6. Storing	7. Storing	
8. Dumping/washing	9. Inspecting/depalletizing	10. Dumping	
11. Blanching	12. Conveying	13. Mixing	
14. Conveying/inspecting	15. Washing		
16. Slicing/dicing	17. Brine injecting		
18. Foreign object removing	19. Filling		
	20. Weighing		
	21. Water filling		
	22. Head-spacing		
	23. End feeding/closing/ inspecting		24. Chlorinating
	25. Thermal processing		
	26. Cooling		
	27. Conveying/drying		
	28. Labeling/storing		
	29. Shipping		

DATE:_______________ **APPROVED BY:**_______________________________________

Table 4
Plant Schematic/Floor Plan

Product Name(s): Canned mushrooms

The diagram should show the product flow and employee traffic patterns in each individual plant to identify and eliminate crosscontamination potentials.

DATE:_____________ **APPROVED BY:**_______________________________________

Bacteria, spore-forming: *Clostridium botulinum*, *Clostridium perfringens*, *Bacillus cereus*.
Bacteria, non-spore-forming: *Brucella abortis*, *Brucella suis Campylobacter* spp, Pathogenic *Escherichia coli* (*E. coli* O157:1-17, EHEC, EIEC, ETEC, EPEC), *Listeria monocytogenes*, *Salmonella* spp. (*S. typhimurium*, *S. enteriditis*), *Shigella* (*S. dysenteriae*), *Staphylococcus aureus*, *Streptococcus pyogenes*, *Vibrio cholerae*, *Vibrio parahaemolyticus*, *Vibrio vulnificus*, *Yersinia enterocolitica*.
Viruses: hepatitis A and E, Norwalk virus group, rotavirus.
Protozoa and parasites: *Cryptosporidium parvum*, *Diphyllobothrium latum*, *Entamoeba histolytica*, *Giardia lamblia*, *Ascaris lumbricoides*, *Taenia solium*, *Taenia saginata*, *Trichinella spiralis*.

Table 5
Clinical Response of Adults to Varying Challenge Doses for Enteric Pathogens

Organism	Challenge dose (cells)
Shigella dysenteriae	10^1–10^4
Shigella flexneri	10^2–10^9
Vibrio cholerae	10^3–10^9
Salmonella Typhi	10^4–10^9
Salmonella species (non-typhi)	10^5–10^{10}
Escherichia coli (pathogenic types)	10^6–10^{10}
Clostridium perfringens	10^8–10^9
Yersinia enterocolitica	10^9

From ref. ***44a***.

3.3.1.2. Chemical Hazards

Naturally occurring chemicals.
- Allergens.
- Mycotoxins (e.g., aflatoxin).
- Scombrotoxin (histamine).
- Ciguatoxin.
- Mushroom toxins (**Table 7**).

Shellfish toxins
- Paralytic shellfish poisoning (PSP).
- Diarrheic shellfish poisoning (DSP).
- Neurotoxic shellfish poisoning (NSP).
- Amnesic shellfish poisoning (ASP).
- Pyrrolizidine alkaloids.
- Phytohemagglutinin.n

Added chemicals
- Polychlorinated biphenyls (PCBs).
- Agricultural chemicals: pesticides, fertilizers, antibiotics, and growth hormones.
- Prohibited substances: direct and indirect.
- Toxic elements and compounds: lead, zinc, cadmium, mercury, arsenic, and cianide.

Food additives
- Vitamins and minerals.
- Contaminants: lubricants, cleaners, sanitizers, coatings, paints, refrigerants, water or steam treatment chemicals, and pest control chemicals.

From packaging materials
- Plasticizers
- Vinyl chloride.
- Printing/coding inks.
- Adhesives.
- Lead.
- Tin.

3.3.1.3. Physical Hazards

Illness and injury can result from hard foreign objects in food. These physical hazards can result from contamination and/or poor practices at many points in the food chain from harvest to consumer, including those within the food establishment (**Tables 8 and 9**).

3.3.2. How to Conduct a Hazard Analysis

3.3.2.1. Review Incoming Material

For each incoming material (ingredient or packaging material), write B, C, or P directly (**Table 10**) to indicate the potential of a biological, chemical, or physical hazard. Be specific when describing the hazards. For example, instead of writing "bacteria in incoming ingredient," write "*C. botulinum* in incoming mushroom."

To facilitate the identification of potential hazards, answer the following questions for each incoming material:

- Could pathogenic microorganisms, toxins, chemicals, or physical objects possibly be present on/in this material?
- Are any returned or reworked products used as ingredients? If yes, is there a hazard linked to that practice?
- Are preservatives or additives used in the formulation to kill microorganisms or inhibit their growth or extend shelf life?
- Are any ingredients hazardous if used in excessive amounts? (For example, nitrites could be a chemical hazard if used excessively.)
- Could any ingredients, if used in amounts lower than recommended or if omitted altogether, result in a hazard because of microbial vegetative or sporulated cell outgrowth?
- Do the amount and type of acid ingredients and the resulting pH of the final product affect growth or survival of microorganisms?
- Do the moisture content and the water activity (A_w) of the final product affect microbial growth? Do they affect the survival of pathogens (parasites, bacteria, fungi)?
- Should adequate refrigeration be maintained for products during transit or in holding?

3.3.2.2. Evaluate Processing Operations for Hazards

Identify all realistic potential hazards related to each processing operation, the product flow, and the employee traffic pattern.

- Assign a number to each processing step on the process flow diagram (**Table 11**) horizontally from receiving to shipping.
- Examine each step on the process flow diagram and determine whether a hazard (biological, chemical, or physical) exists for that operation
- Write B for biological, C for chemical, and P for physical beside each operation where such a hazard has been identified (**Table 11**).
- Review the plant schematic and employee traffic pattern on Form 4 in the same manner.

To help in determining whether a hazard exists, the following questions should be answered for each processing step:

- Could contaminants reach the product during this processing operation? (Consider personnel hygiene, contaminated equipment or material, cross-contamination from raw materials, leaking valves or plates, dead ends [niches], splashing, and so on.)

Table 6
Hazard Identification: Biological Hazards

Product Name(s): Canned mushrooms
List all biological hazards related to ingredients, incoming material, processing, product flow, and so on

Identified biological hazards	Controlled at
Ingredients/materials	
Mushrooms	
Could contain *C. botulinum* or other pathogenic organisms, yeasts, or molds	
Dry ingredients	
Could contain bacterial spores; could contain rodent excrement	
Water	
Could contain coliform or spore-forming bacteria or other microorganisms	
Empty cans/ends	
Could arrive with serious internal double seam or body plate defects, which could result in leakage causing postprocess contamination; could arrive with serious external double seam, body plate, lacquer/coating defects or damage which could result in leakage causing postprocess contamination	
Process Steps	
5. Refrigerated mushroom storing	
Improper storage temperature and humidity could result in increase of bacterial load	
6. Can/end storing	
Physical damage could result in serious double seam defects, which could result in postprocess contamination with pathogenic bacteria	
Could be contaminated with rodent excrement	
7. Dry ingredient storing	
Could be contaminated with rodent excrement	
9. Can depalletizing/inspection	
Incorrect cans, physical damage or serious visible defects could result in leakage and postprocess contamination with pathogenic bacteria	
11. Mushroom blanching	
Improper cleaning of the blancher could result in the growth of thermophilic bacteria in mushrooms	
Inadequate blanching could result in insufficient removal of gases, which could cause stress on double seams and perforations and lead to postprocess contamination with pathogenic bacteria	
Excessive blanching could result in textural changes to the mushrooms, which could result in inadequate thermal processing	
12. Can conveying	
Physical damage could result in the formation of defective double seams, which could lead to postprocess contamination with pathogenic bacteria	
20. Weighing	
Overfilled cans not properly rejected for overweight could be underprocessed	
21. Water filling	
Inadequate temperature could result in low initial temperature and subsequent underprocessing	

(*continued*)

Table 6 (continued)

Identified biological hazards	Controlled at
22. Headspacing	
Insufficient headspace could result in excessive internal pressure during processing causing distorted seams and leakage contamination	
23. End feeding/closing/inspecting	
Ends with damaged curls or other serious defects could result in leakage and contamination with pathogenic bacteria	
Improperly formed double seams could result in leakage and contamination with pathogenic bacteria	
25. Thermal processing	
Nonvalidated process or vent schedule could result in underprocessing and survival of pathogenic bacteria	
Improper flow patterns in processing area could result in heat-processed cans being contaminated with unclean water from unprocessed baskets of cans	
Improper flow design in processing area could result in retort baskets missing the retort, allowing growth of pathogenic bacteria	
Excessive time lapse between closing and retorting could result in excessive buildup of bacteria, some of which could survive the thermal process	
Lack of adherence to time, temperature, and other critical factors of the scheduled process or vent schedule could result in inadequate heat treatment, allowing the survival of pathogenic bacteria	
26. Cooling	
Insufficient chlorinated cooling water could result in contamination of product during contraction of cans	
Excess chlorine in cooling water could result in corrosion and subsequent leakage and contamination of product	
Insufficient contact time between the chlorine and water could result in contamination of product during contraction of the cans	
Insufficient or excessive cooling could result in thermophilic spoilage or postprocess contamination because of leakage of corroded cans	
27. Conveying/drying	
Contaminated water from wet and unclean postprocess equipment could contaminate product	
28. Labeling/storing	
Physical damage to cans could result in leakage and contamination of product	
High temperatures could result in growth of thermophilic bacteria	
29. Shipping	
Physical damage to cans could result in leakage and contamination of product	

DATE: _______________ **APPROVED BY:** _______________________________

Table 7
Hazard Identification: Chemical Hazards

Product Name(s): Canned mushrooms
List all chemical hazards related to ingredients, incoming material, processing, product flow, and so on.

Identified chemical hazards	Controlled at
Ingredients/Materials	
Mushrooms	
Could contain pesticide residues	
Could contain heat-stable staphylococcal enterotoxin from improper handling	
Water	
Could be contaminated with dissolved heavy metals or toxic substances	
Empty cans/ends	
Cans/ends could be contaminated with greases/oils or cleaning chemicals	
Process Steps[a]	
6. Can/end storing	
Cans/ends could become contaminated with nonfood chemicals as a result of improper storage	
7. Dry ingredient storing	
Food ingredients could become contaminated with nonfood chemicals if improperly stored	
11. Mushroom blanching	
Cleaning-chemical residues could contaminate the mushrooms	
If live steam is used, boiler water additives could carry over and contaminate the product	
14, 16, 19, 23. Mushroom conveying, mushroom slicing/dicing, filling, end feeding/closing	
Cleaning-chemical residues or lubricants could contaminate the mushrooms	

DATE:______________ **APPROVED BY:** ____________________________

[a]Numbers are from flow diagram in **Table 11**.

- Could any microorganisms of concern multiply during this processing operation to the point where they constitute a hazard? (consider temperature, time).

3.3.2.3. Observe Actual Operating Practices

- Observe the operation long enough to be confident that it comprises the usual process or practices.
- Observe the employees (e.g., could raw or contaminated product cross-contaminate workers' hands, gloves, or equipment used for finished or post-process product?).
- Observe hygienic practices and note the hazards.
- Analyze whether there is a kill step (process that destroys all microorganisms) during the process (if so, attention should be focused on potential crosscontamination after this processing operation).

Table 8
Examples of Physical Hazards

Material	Injury potential	Sources
Glass	Cuts, bleeding; may require surgery to find or remove	Bottles, jars, light fixtures, utensils, gauge covers, and so on.
Wood	Cuts, infection, choking; may require surgery to remove	Field sources, pallets, boxes, building materials
Stones	Choking, broken teeth	Fields, buildings
Metal	Cuts, infection; may require surgery to remove	Machinery, fields, wire, employees
Insulation	Choking; long-term if asbestos	Building materials
Bone	Choking	Improper processing
Plastic	Choking, cuts, infection; may require surgery to remove	Packaging, pallets, equipment
Personal effects	Choking, cuts, broken teeth; may require surgery to remove	Employees

3.3.2.4. Take Measurements

- Measure product temperatures, considering heat processing and cooling or chilling operations: take measurements at the coldest point of the product when heat processing is evaluated and at the warmest point of the product when cooling or chilling is evaluated (frequently at the center of the largest piece).
- Measure time/temperature for cooking, pasteurizing, canning cooling (rates), storing, thawing, reconstituting, and so on.
- Measure the dimension of the containers used to hold foods being cooled and the depth of the food mass.
- Measure pressure, headspace, venting procedure, adequacy of container closure, initial temperatures, and any other factors critical to the successful delivery of a scheduled process.
- Measure the pH of the product during processing and also of the finished product, measuring pH at room temperature whenever possible.
- Measure A_w of the product, running duplicate samples whenever possible (because of variations) and remembering to make corrections for ambient temperatures, as necessary.

3.3.2.5. Analyze the Measurements

- Plot time/temperature measurements using a computer or on graph paper.
- Interpret controlled data vs optimal growth temperatures of microorganisms and temperature ranges at which they can multiply.
- Estimate and evaluate probable cooling rates; interpret cooling rates and compare the measured temperatures with temperature ranges within which bacteria of concern multiply rapidly vs temperature at which growth begins, slows, and ceases ***(44b,44c)***; determine whether covers are used on containers to cool down foods (which may delay cooling but may also prevent contamination); if containers are stacked against each other in a manner affecting cooling or heating time, evaluate the impact.
- Compare A_w and pH values with ranges at which pathogens multiply or are eliminated.
- Evaluate the shelf stability of the product.

Table 9

Hazard Identification: Physical Hazards

Product Name(s): Canned mushrooms
List all physical hazards related to ingredients, incoming material, processing, product flow, and so on.

Identified physical hazards	Controlled at
Ingredients/Materials	
Mushrooms	
Could be contaminated with harmful extraneous materials, e.g., glass, metal, plastic, wood	
Empty cans	
Could contain metal fragments, and so on	
Dry ingredients	
Could be contaminated with harmful extraneous materials	
Ingredients/Materials[a]	
1. Mushroom receiving	
Inadequate protection against harmful extraneous material could result in contamination of mushrooms	
2. Can/end receiving	
Inadequate protection against harmful extraneous material could result in contamination of cans and ends	
3. Dry ingredient receiving	
Inadequate protection against harmful extraneous material could result in contamination of ingredients	
5. Mushroom storing	
Inadequate protection against harmful extraneous material could result in contamination of raw mushrooms	
6. Can/end storing	
Inadequate protection against harmful extraneous material could result in contamination	
7. Dry ingredient storing	
Inadequate protection against harmful extraneous material could result in contamination of food ingredients	
9. Can inspection/depalletizing	
Empty cans coming from storage could contain harmful extraneous material, which could result in contamination of food product	
12. Can conveying	
Inappropriate design and protection against harmful extraneous material could result in contamination of food product	
14. Mushroom convening/inspection	
Inappropriate design and protection against harmful extraneous material could result in contamination of the mushrooms	
16. Mushrooms slicing/dicing	
Product could become contaminated with metal fragments from plant machinery	

(continued)

Table 9 (continued)

Identified physical hazards	Controlled at
18. Foreign object removal Inadequate monitoring of foreign object removal could allow foreign objects to contaminate the product	
19. Filling Filled cans of mushrooms could become contaminated with metal fragments from the filling equipment	

DATE:__________ **APPROVED BY:** ____________________

[a]Numbers are from the flow diagram in **Table 11**.

Table 10
Product Ingredients and Incoming Material

Product Name(s): Canned mushrooms

Raw material	Packaging material	Dry ingredients
Mushrooms (domestic, white) B, C, P	Cans B, C, P Ends B, C Citric acid	Salt B, C Ascorbic acid B, C B, C

Other
Water (municipal) B, C

DATE:__________ **APPROVED BY:** ____________________

B, biological; C, chemical; P, physical.

3.3.3. Control Measures

Control measures are any actions and activities that can be used to prevent or eliminate a food safety hazard or reduce it to an acceptable level.

3.3.3.1. Controlling Biological Hazards

Biological hazards can be controlled by limiting, removing, or altering the growth kinetics that microorganisms need to survive, grow, and reproduce. They can be destroyed, eliminated, or controlled by thermal processing (heating or cooking), freezing, or drying.

Table 11
Flow Diagram

Product Name(s): Canned mushrooms

Mushroom (raw)	Empty cans/ends	Dry ingredients	Water (municipal)
1. Receiving **P**	2. Receiving **P**	3. Receiving **P**	4. Intaking
5. Storing **BP**	6. Storing **BCP**	7. Storing **BCP**	
8. Dumping/washing	9. Inspecting/depalletizing **BP**	10. Dumping	
11. Blanching **BC**	12. Conveying **BP**	13. Mixing	
14. Conveying/inspecting CP	15. Washing		
16. Slicing/dicing **CP**	17. Brine injecting		
18. Foreign object removing	19. Filling **CP**		
	20. Weighing **B**		
	21. Water filling **B**		
	22. Head-spacing **B**		
	23. End feeding/closing/ inspecting **BC**		24. Chlorinating
	25. Thermal processing **B**		
	26. Cooling **B**		
	27. Conveying/drying **B**		
	28. Labeling/storing **B**		
	29. Shipping B		

DATE:____________ **APPROVED BY:** ________________________

B, biological; C, chemical; P, physical.

3.3.3.1.1. For Bacteria, Control Measures Include:

- Temperature/time control (proper control of refrigeration and storage time, for example, minimizes the proliferation of microorganisms).
- Heating and cooking (thermal processing) for an adequate time and at an adequate temperature to eliminate microorganisms or reduce them to acceptable levels.
- Cooling and freezing.
- Fermentation and/or pH control (for example, lactic acid-producing bacteria in yoghurt inhibit the growth of other microorganisms that do not tolerate the acidic conditions and competition).
- Addition of salt or other preservatives, which at acceptable levels may inhibit growth of microorganisms.
- Drying, which may use enough heat to kill microorganisms or may remove enough water from the food to prevent certain microorganisms from growing even when drying is conducted at lower temperatures.
- Packaging conditions (vacuum packaging, for example, can be used to inhibit microorganisms that require air to grow).

- Source control, i.e., control of the presence and level of microorganisms by obtaining ingredients from suppliers who can demonstrate adequate controls over the ingredients (e.g., suppliers that follow an HACCP program).
- Cleaning and sanitizing, which can eliminate or reduce the levels of microbiological contamination.
- Personal and hygienic practices, which can reduce the levels of microbiological contamination.

3.3.3.1.2. For Viruses, Control Measures Include:

- Thermal processing—heating or cooking methods such as steaming, frying, or baking—which may destroy many but not all viruses. (The type of virus determines the appropriate controls.)
- Personal hygienic practices, including the exclusion of workers affected by certain viral diseases, e.g., hepatitis.

3.3.3.1.3. For Parasites (Worms and Protozoa), Control Measures Include:

- Dietary control (infection from *Trichinella spiralis* in pork, for example, has decreased as a result of better control of the pigs' diet and environment)—a method not always practical, however, for all species of animals used for food. (The diet and environment of wild fish, for example, cannot be controlled.)
- Heating, drying, or freezing.
- Salting or brining.
- Visual examination, which can be used in some foods to detect parasites (e.g., a procedure called *candling* can be used for certain fish).
- Good personal hygiene practices by food handlers, proper disposal of human feces, and proper sewage treatment.

3.3.3.2. Controlling Chemical Hazards

- Source control, i.e., specifications for raw materials and ingredients and vendor certification that harmful chemicals or levels are not present.
- Processing control, i.e., formulation control and the proper use and control of food additives and their levels.
- Proper segregation of nonfood chemicals during storage and handling.
- Control of incidental contamination from chemicals (e.g., greases, lubricants, water and steam treatment chemicals, paints).
- Labeling control, i.e., ascertaining that the finished product is accurately labeled with ingredients and known allergens.

3.3.3.3. Controlling Physical Hazards

- Source control (i.e., specifications for raw materials and ingredients and vendor certification that unacceptable physical hazards or levels are not present.
- Processing control (e.g., use of magnets, metal detectors, sifter screens, destoners, clarifiers, air tumblers).
- Environmental control (i.e., ensuring that good manufacturing practices are followed and that no physical contamination occurs to the food through the building, facilities, work surfaces, or equipment).

Table 12
2D Health Risk Assessment Model

High	Sa	Mi	Ma	Cr
Medium	Sa	Mi	Ma	Ma
Low	Sa	Mi	Mi	Mi
Negligible	Sa	Sa	Sa	Sa
		Low	Medium	High

Significance of the hazard: Sa, satisfactory (negligible); Mi, minor; Ma, major; Cr, Critical

3.3.4. Hazard Assessment

3.3.4.1. Severity

Severity is the magnitude of a hazard or the degree of consequences that can result when a hazard exists. Disease-causing hazards can be categorized according to their severity.

- High (life-threatening). Examples include illnesses caused by *Clostridium botulinum*, *Salmonella typhi*, *Listeria monocytogenes*, *Escherichia coli* O157:H7, *Vibrio cholerae*, *Vibrio vulnificus*, paralytic shellfish poisoning, amnesic shellfish poisoning.
- Moderate (severe or chronic). Examples include illnesses caused by *Brucella* spp., *Campylobacter* spp.. *Salmonella* spp., *Shigella* spp. *Streptococcus* type A, *Yersinia entercolitica*, hepatitis A virus, mycotoxins, ciquatera toxin.
- Low (moderate or mild). Examples include illnesses caused by *Bacillus* spp., *Clostridium perfringens*, *Staphylococcus aureus*, Norwalk virus, most parasites, histamine-like substances, and most heavy metals that cause mild acute illnesses.

3.3.4.2. Risk of Hazard

Risk is a function of the probability of an adverse effect and the magnitude of that effect, consequential to a hazard(s) in food. Degrees of risk can be categorized as high (H), moderate (M), low (L), or negligible (N). By taking into account the probability of occurrence (the inverse proportion to the degree of control) and the severity of consequences, the significance of the hazard can be classified as satisfactory (Sa), minor (Mi), major (Ma), or critical (Cr) ***(24,26,27)*** (**Table 12**).

3.3.4.3. Identification of Points, Steps, and Procedures

The significance of the hazard can be differentiated as satisfactory (Sa), minor (Mi), major (Ma), or critical (Cr).

Note: Likelihood of occurrence is inversely proportional to the degree of control.

3.4. Determine Critical Control Points (Principle 2)

3.4.1. Critical Control Points

The *Codex* guidelines define a critical control point (CCP) as "a step at which control can be applied and is essential to prevent or eliminate a food safety hazard or reduce it to an acceptable level" (*see* **Notes 1** and **4**).

3.4.2. Review of Identified Hazards

Prior to determining the CCPs, a review should be conducted to verify whether any of the identified hazards are fully controlled by the application of the *Codex* General Principles of Food Hygiene, good manufacturing practices (GMPs), or good hygienic practices (GHPs) (**Table 13**).

The decision tree consists of a systematic series of four questions designed to assess objectively whether a CCP is required to control the identified hazard at a specific operation of the process.

Question 1: Do control measure(s) exist?
Question 2: Is the step specifically designed to eliminate or reduce the likely occurrence of the identified hazard to an acceptable level?
Question 3: Could contamination with the identified hazard occur in excess of acceptable levels or increase to unacceptable levels?
Question 4: Will a subsequent step eliminate the identified hazard or reduce likely occurrence to an acceptable level?

3.4.3. Identification of CCPs

CCPs should be identified numerically with a category qualifier B, P, or C for biological, physical, or chemical. For example, if the first CCP identified will control a biological hazard, it is recorded as CCP-1 (B). If the second CCP identified will control a chemical hazard, it is recorded as CCP-2 (C). If the fifth CCP will control both a biological and a chemical hazard at the same processing operation, it is recorded as CCP-5 (BC). For hazards fully controlled by application of the *Codex* General Principles of Food Hygiene, write "GMP/GHP" (**Tables 14** and **15**).

3.4.4. Parameters Attached to CCPs

Once the CCPs have been established, the next step is to report the CCPs and to document on the same form the parameters that will be monitored and controlled (**Table 16**).

Any hazards previously unaddressed are handled as shown in **Table 17**.

3.5. Establish Critical Limits for Each Critical Control Point (Principle 3)

3.5.1. Critical Limits

At each CCP, critical limits are established and specified. A critical limit represents the boundaries that are used to judge whether an operation is producing safe products. Critical limits may be set for factors such as temperature, time (minimum time exposure), physical product dimensions, water activity, moisture level, and others. These parameters, if maintained within boundaries, will confirm the safety of the product. It is essential that the person(s) responsible for establishing critical limits have a knowledge of the process and of the legal and commercial standards required for the product ***(45,46)***.

- An acidified beverage that requires a hot fill and hold as the thermal process may have acid addition as the CCP. If insufficient acid is added or if the temperature of the hot fill is insufficient, the product would be underprocessed, with potential for the growth of pathogenic spore-forming bacteria. The critical limits in this case would apply to pH and fill temperature.

Table 13
Hazard Identification: Biological Hazards

Product Name(s): Canned mushrooms
List all biological hazards related to ingredients, incoming material, processing, product flow, etc.

Identified biological hazards	Controlled at
Ingredients/Materials	
Mushrooms	CCP5B
Could contain *C. botulinum* or other pathogenic organisms, yeast, or molds	
Dry ingredients	CCP5B
Could contain bacterial spores	GMP/GHP
Could contain rodent excrement	(Sanitation)
Water	GMP/GHP
Could contain coliform or spore-forming bacteria or other microorganisms	(Premises)
Empty cans/ends	CCP4B
Could arrive with serious internal double seam or body plate defects, which could result in leakage causing post-process contamination	CCP1B/CCP4B
Could arrive with serious external double seam, body plate, or lacquer/coating defects or damage, which could result in leakage causing post-process contamination	
Process Steps[a]	
5. Refrigerated mushroom storing	GMP/GHP
Improper storage temperature and humidity could result in increase of bacterial load	(Equipment)
6. Can/end storing	CCP1B
Physical damage could result in serious double seam defects, which could result in postprocess contamination with pathogenic bacteria	GMP/GHP
Could be contaminated with rodent excrement	(Sanitation)
7. Dry ingredient storing	GMP/GHP
Could be contaminated with rodent excrement	(Sanitation)
9. Can depalletizing/inspection	
Incorrect cans, physical damage, or serious visible defects could result in leakage and postprocess contamination with pathogenic bacteria	CCP1B
11. Mushroom blanching	GMP/GHP
Improper cleaning of the blancher could result in the growth of thermophilic bacteria in mushrooms	(Sanitation) GMP/GHP
Inadequate blanching could result in insufficient removal of gases, which could cause stress on double seams and perforations and lead to postprocess contamination with pathogenic bacteria	(Equipment)
Excessive blanching could result in textural changes to the mushrooms, which could result in inadequate thermal processing	
12. Can conveying	GMP/GHP
Physical damage could result in the formation of defective double seams, which could lead to postprocess contamination with pathogenic bacteria	(Equipment)
20. Weighing	CCP2B
Overfilled cans not properly rejected for overweight could be underprocessed	

(*continued*)

Table 13 (continued)

Identified biological hazards		Controlled at
21.	Water filling	CCP5B
	Inadequate temperature could result in low initial temperature and subsequent underprocessing	
22.	Headspacing	CCP3B
	Insufficient headspace could result in excessive internal pressure during processing causing distorted seams and leakage contamination	
23.	End feeding/closing/inspecting	CCP4B
	Ends with damaged curls or other serious defects could result in leakage nd contamination with pathogenic bacteria	CCP4B
	Improperly formed double seams could result in leakage and contamination with pathogenic bacteria	
25.	Thermal processing	GMP/GHP (Records)
	Nonvalidated process or vent schedule could result in underprocessing and survival of pathogenic bacteria	GMP/GHP (Personnel)
	Improper flow patterns in processing area could result in heat-processed cans being contaminated with unclean water from unprocessed baskets of cans	CCP5B
	Improper flow design in processing area could result in retort baskets missing the retort, allowing growth of pathogenic bacteria	CCP5B
	Excessive time lapse between closing and retorting could result in excessive buildup of bacteria, some of which could survive the thermal process	CCP5B
	Lack of adherence to time, temperature, and other critical factors of the scheduled process or vent schedule could result in inadequate heat treatment, allowing the survival of pathogenic bacteria	
26.	Cooling	CCP6B
	Insufficient chlorinated cooling water could result in contamination of product during contraction of cans	CCP6B
	Excess chlorine in cooling water could result in corrosion and subsequent leakage and contamination of product	GMP/GHP (Sanitation, personnel)
	Insufficient contact time between the chlorine and water could result in contamination of product during contraction of the cans	GMP/GHP (Sanitation, personnel)
	Insufficient or excessive cooling could result in thermophilic spoilage or postprocess contamination because of leakage of corroded cans	
27.	Conveying/drying	GMP/GHP (Sanitation)
	Contaminated water from wet and unclean postprocess equipment could contaminate product	
28.	Labeling/storing	GMP/GHP (Equipment, personnel)
	Physical damage to cans could result in leakage and contamination of product	
	High temperatures could result in growth of thermophilic bacteria	GMP/GHP (Personnel)
29.	Shipping	GMP/GHP (Personnel training)
	Physical damage to cans could result in leakage and contamination of product	

DATE:______________ **APPROVED BY:** ____________________________

B, biological; C, chemical; Pk physical; GHP, good hygienic practices; GMP, good manufacturing practices.
[a]Numbers are from the flow diagram in **Table 11**.

Table 14
Hazard Identification: Chemical Hazards

Product Name(s): Canned mushrooms
List all chemical hazards related to ingredients, incoming material, processing, product flow, and so on.

Identified chemical hazards		Controlled at
Ingredients/Materials		
Mushrooms		
	Could contain pesticide residues	*See* Form 9
	Could contain heat-stable staphylococcal enterotoxin from improper handling	*See* Form 9
Water		
	Could be contaminated with dissolved heavy metals or toxic substances	GMP/GHP (Premises)
Empty cans/ends		
	Cans/ends could be contaminated with greases/oils or cleaning chemicals	GMP/GHP (Receiving, storage, and transport)
Process Steps[a]		
6.	Can/end storing Cans/ends could become contaminated with nonfood chemicals as a result of improper storage	GMP/GHP (Sanitation)
7.	Dry ingredient storing Food ingredients could become contaminated with nonfood chemicals if improperly stored	GMP/GHP (Sanitation)
11.	Mushroom blanching Cleaning-chemical residues could contaminate the mushrooms	GMP/GHP (Sanitation)
	If live steam is used, boiler water additives could carry over and contaminate the product	GMP/GHP (Sanitation)
14, 16, 19, 23.	Mushroom conveying, mushroom slicing/dicing, filling, end feeding/closing Cleaning-chemical residues or lubricants could contaminate the mushrooms	GMP/GHP (Sanitation)

DATE:________________ **APPROVED BY:**____________________________

B, biological; C, chemical; P, physical; GHP, good hygienic practices; GMP, good manufacturing practices.

[a]Numbers are from flow diagram in **Table 11**.

- Beef patties are cooked in a continuous oven. More than one critical limit is set to control the hazard of heat-resistant pathogen survival. The critical limits could be minimum internal temperature of the patty; oven temperature; time in the oven determined by the belt speed in rpm; or patty thickness. These examples illustrate that CCPs may be controlled by more than one critical limit (**Table 18**).

3.5.2. Operating Limits

If monitoring shows a trend toward lack of control at a CCP, operators can take action to prevent loss of control of the CCP before the critical limit is exceeded. The point at which operators take such action is called the *operating limit.*

The process may need to be adjusted when the operating limit is exceeded. Such actions are called *process adjustments.* A processor should use these adjustments to prevent loss of control and the need for product disposition. **Tables 19**, **20**, and **21** show examples of critical limits vs operating limits.

3.6. Establish a Monitoring System for Each Critical Control Point (Principle 4)

3.6.1. Monitoring

The *Codex Hazard Analysis and Critical Control Point (HACCP) System and Guidelines for Its Application* defines monitoring as "the act of conducting a planned sequence of observations or measurements of control parameters to assess whether a CCP is under control." Monitoring procedures performed during the operation should result in written documentation that will serve as an accurate record of the operating conditions.

3.6.2. Design of a Monitoring System

The control measures are intended to control a hazard or hazards at each CCP. The monitoring procedures will determine if the control measures are being implemented and ensure that critical limits are not exceeded.

3.6.3. What Will Be Monitored?

Monitoring may mean measuring a characteristic of the product or of the process to determine compliance with a critical limit. Examples include:

- Measurement of the time and temperature of a thermal process.
- Measurement of cold-storage temperatures.
- Measurement of pH.
- Measurement of A_w.

Monitoring may also mean observing whether a control measure at a CCP is being implemented. Examples include:

- Visual examination of sealed cans.
- Verification of vendor's certificates of analysis.

3.6.4. How Will Critical Limits and Preventive Measures Be Monitored?

Deviation from a critical limit should be detected in as short a time as possible to allow corrective action to limit the amount of adversely affected product. The monitoring procedures should provide rapid (real-time) results and should not involve lengthy analytical procedures. Microbiological testing is rarely effective for monitoring CCPs for this reason, and also because large sample sizes would be needed to find microorganisms at levels that may cause illness. Instead, physical and chemical mea-

Table 15
Hazard Identification: Physical Hazards

Product Name(s): Canned mushrooms
List all physical hazards related to ingredients, incoming material, processing, product flow and so on.

Identified physical hazards	Controlled at
Ingredients/Materials	
Mushrooms	
Could be contaminated with harmful extraneous materials,e.g., glass, metal, plastic, wood	Not applicable (not likely to get through equipment and inspection belt)
Empty cans	
Could contain metal fragments, and so on.	GMP/GHP (Receiving, storing, and transport)
Dry ingredients	
Could be contaminated with harmful extraneous materials	GMP/GHP (Receiving, storing, and transport)
Process Steps[a]	
1. Mushroom receiving Inadequate protection against harmful extraneous material could result in contamination of mushrooms	GMP/GHP (Premises)
2. Can/end receiving Inadequate protection against harmful extraneous material could result in contamination of cans and ends	GMP/GHP (Premises)
3. Dry ingredient receiving Inadequate protection against harmful extraneous material could result in contamination of ingredients	GMP/GHP (Premises)
5. Mushroom storing Inadequate protection against harmful extraneous material could result in contamination of raw mushrooms	GMP/GHP (Premises, receiving, storage, and transport)
6. Can/end storing Inadequate protection against harmful extraneous material could result in contamination	GMP/GHP (Premises, receiving, storage, and transport)
7. Dry ingredient storing Inadequate protection against harmful extraneous material could result in contamination of food ingredients	GMP/GHP (Premises, receiving, storage, and transport)
9. Can inspection/depalletizing Empty cans coming from storage could contain harmful extraneous material which could result in contamination of food product	CC1P
12. Can conveying Inappropriate design and protection against harmful extraneous material could result in contamination of food product	GMP/GHP (Equipment)
14. Mushroom conveying/inspection Inappropriate design and protection against harmful extraneous material could result in contamination of the mushrooms	GMP/GHP (Premises, equipment, and personnel)

(continued)

Table 15 (continued)

Identified physical hazards	Controlled at
16. Mushrooms slicing/dicing Product could become contaminated with metal fragments from plant machinery	GMP/GHP (Equipment)
18. Foreign object removal Inadequate monitoring of foreign object removal could allow foreign objects to contaminate the product	GMP/GHP (Equipment)
19. Filling Filled cans of mushrooms could become contaminated with metal fragments from the filling equipment	GMP/GHP (Equipment)

DATE:________________ **APPROVED BY:** ______________________________

C, chemical; P, physical; GHP, good hygienic practices; GMP, good manufacturing practices.
[a]Numbers are from flow diagram in Table 11.

surements (e.g., pH, A_w, time, temperature) are preferred, as they can be done rapidly and can often be related to the microbiological control of the process. The equipment used for monitoring CCPs will vary depending on the attribute being monitored. Examples of monitoring equipment include:

- Thermometers.
- Clocks.
- Scales.
- pH meters.
- Water activity meters.
- Chemical analytical equipment.

3.6.5. Monitoring Frequency

Monitoring can be continuous or noncontinuous. When possible, continuous monitoring is preferred; it is possible for many types of physical or chemical methods. Examples of continuous monitoring include:

- Measuring the time and temperature of a pasteurization or retorting process.
- Checking each package of frozen, mechanically chopped spinach with a metal detector.
- Monitoring the container closures on glass jars by passing them under a dud detector.

The following questions will help to determine the correct frequency:

- How much does the process normally vary?
- How close is the operating limit to the critical limit?
- How much product is the processor prepared to risk if there is deviation from the critical limit?

Table 16
CCP Determination

Process step/incoming material[a]	Category and identified hazard[b]	Question[c] 1	2	3	4	CCP no.
Mushrooms as delivered	**B**, pathogens	**Yes,** heat treatment	N/A	**Yes**	**Yes**, thermal processing ***(25)***	
	C, pesticides	**No**, control is at farms/growers				
	C, heat-stable toxins	**No**, control is at farms/growers, storage				
	P, HEM	**Yes**, visual inspection and foreign object removal	**Yes**	**No**		
Empty cans as delivered	**B**, postprocess contamination from serious internal seam defects	**Yes**, can teardown and inspection	N/A	**Yes**	**Yes**, closing and inspecting ***(23)***	
	B, postprocess contamination from serious external visible can defects	**Yes**, visual can inspection	N/A	**Yes**	**Yes**, inspecting/ depalletizing ***(9)***	
	C, cleaning chemicals (GMPs)					
	P-HEM		N/A	**Yes**	**Yes**, inspecting/ depalletizing ***(9)***	
Dry ingredients as delivered	**B**, bacterial spores	**Yes**, heat treatment	N/A	**Yes**	**Yes**, thermal processing ***(25)***	
	B, rodent excrement (GMPs)					
	P, HEM (GMPs)					
Water at intake	**B**, fecal coliform (GMPs)					
	C, heavy metals and other toxic chemicals (GMPs)					
1. Mushroom receiving	**P** - HEM (GMPs)					
2. Can/end receiving	**P** - HEM (GMPs)					
3. Dry ingredient receiving	**P** - HEM (GMPs)					
5. Mushrooms at receiving	**B**, growth of pathogens (GMPs)					
	P, HEM (GMPs)					

(*continued*)

Table 16 (continued)

Process step/incoming material[a]	Category and identified hazard[b]	Question[c] 1	2	3	4	CCP no.
6. Can/end storing	**B**, postprocess contamination because of damaged cans/ends	**Yes**, visual inspection	**No**	**Yes**	**Yes**, inspecting/ depalletizing ***(9)***	
	B, rodent excrement (GMPs) **C**, cleaning chemicals (GMPs) **P**, HEM (GMPs)					
7. Dry ingredient storing	**B**, rodent excrement (GMPs) **C**, cleaning chemicals (GMPs) **P**, HEM (GMPs)					
9. Can inspecting/ depalletizing	**B**, postprocess contamination because of incorrect cans or serious can defects	**Yes**, visual inspection	**Yes (BP)**	**Yes**		**CCP 1**
	P, HEM	**Yes**, visual inspection	**Yes**	**Yes**		
11. Mushroom blanching	**B**, growth of thermophiles, textural changes affecting thermal process (GMPs) **B**, inadequate removal of gases (GMPs) **C**, cleaning chemicals (GMPs)					
12. Can conveying	**B**, postprocess contamination because of damage (GMPs) **P**, HEM (GMPs)					
14. Mushroom conveying/ inspecting	**C**, cleaning chemicals (GMPs) **P**, HEM (GMPs)					
16. Mushroom slicing/ dicing	**C**, cleaning chemicals, lubrificants (GMPs) **P**, HEM (GMPs)					
18. Foreign object removal	**P**, metal fragments (GMPs)					
19. Filling	**C**, cleaning chemicals, lubrificants (GMPs) **P**, metal fragments (GMPs)					

(*continued*)

Table 16 (continued)

Process step/incoming material[a]	Category and identified hazard[b]	Question[c] 1	2	3	4	CCP no.
20. Weighing	**B**, product heavier than maximum fill weight in scheduled process	**Yes**, weighing	**Yes**	**Yes**	**No**	**CCP 2 (B)**
21. Water filling	**B**, inadequate temperature resulting in low initial temperature (IT) for process	**Yes**, take it just prior to thermal process	**No**	**Yes**	**Yes**, thermal processing *(25)*	
22. Head-spacing	**B**, insufficient headspace resulting in distorted, potentially leaking seams	**Yes**	**Yes**	**Yes**	**No**	**CCP 3 (B)**
23. End feeding/ closing/inspection	**B**, postprocess contamination because of damaged ends	**Yes**, visual inspection	**Yes**	**Yes**	**No**	**CCP 4 (B)**
	B, postprocess contamination because of improperly formed seams	**Yes**, visual and tear-down can inspection	**Yes**	**Yes**	**No**	
	C, cleaning chemicals, lubrificants (GMPs)					
25. Thermal processing	**B**, nonvalidated process or vent schedule could result in underprocessing and survival of pathogenic bacteria (GMPs)					
	B, improper flow patterns for process could result in crosscontamination (GMPs)					
	B, improper flow patterns for process could allow bypass of thermal process	**Yes**, use of heat-sensitive indicator	**Yes**,	**Yes**	**No**	**CCP 5 (B)**
	B, excessive delays between closing and retorting could result in excessive growth of pathogenic bacteria	**Yes**, monitor time lapse between the two operations	**Yes**	**Yes**	**No**	
	B, lack of adherence to time, temperature, and. other critical factors of scheduled process or vent schedule could result in inadequate heat treatment and growth of pathogens	**Yes**, control critical factors of scheduled process and vent schedule	**Yes**	**Yes**	**No**	

(*continued*)

Table 16 (continued)

Process step/incoming material[a]	Category and identified hazard[b]	Question[c] 1	2	3	4	CCP no.
26. Cooling	**B**, postprocess contamination during cooling/ contracting of cans because of insufficiently chlorinated cooling water	**Yes**, control chlorine level in cooling water	**Yes**	**Yes**	**No**	**CCP 6 (B)**
	B, postprocess contamination because of leakage resulting from corrosion from excessive chlorine cleaning chemicals	**Yes**, control chlorine level in cooling water	**Yes**	**Yes**	**No**	
	B, insufficient chlorine contact time could lead to contamination (GMPs)					
	B, insufficient or excessive cooling could result in thermophilic s poilage or contamination because of corrosion leakage (GMPs)					
27. Conveying/drying	**B**, unclean wet equipment could lead to contamination (GMPs)					
28. Labeling/storing	**B**, postprocess contamination because of damaged cans (GMPs)					
	B, growth of thermopiles (GMPs)					
29. Shipping	**B**, postprocess contamination because of damaged cans (GMPs)					

B, biological; C, chemical; GMP, good manufacturing practice; HEM, harmful extraneous material; NA, not available.

[a]Numbers are from flow diagram in **Table 11**.

[b]Category and identified hazard: Determine whether hazard is fully controlled by adherence to *Codex* General Principles of Food Hygiene. If **Yes**, indicate GMP, describe, and proceed to next identified hazard. If **No**, proceed to Question 1.

[c]Questions. **Question 1: Do control preventive measure(s) exist?** If **No**, this is not a CCP. Identify how the hazard can be controlled before or after the process and proceed to the next identified hazard. If **Yes**, describe and proceed to the next question. **Question 2: Is the operation specifically designed to eliminate or reduce the likely occurrence of a hazard to an acceptable level?** If **No**, proceed to Question 3. If **Yes**, this is a CCP; identify it as such in the last column. **Question 3: Could contamination with identified hazard(s) occur in excess of acceptable level(s) or could these increase to unacceptable levels?** If **No**, this is not a CCP; proceed to the next identified hazard. If **Yes**, proceed to Question 4. **Question 4: Will a subsequent operation eliminate identified hazard(s) or reduce likely occurrence to an acceptable level?** If **No**, this is a CCP; identify it as such in the last column. If **Yes**, this is not a CCP; identify the subsequent step and proceed to the next identified hazard.

Table 17
Unaddressed Hazards

Product Name(s): Canned mushrooms
List any biological, chemical, and/or physical hazards that are not controlled at the establishment

Unaddressed hazard from previous list	Identified methods of addressing the hazard[a]
C, raw mushrooms could contain pesticide residues	Upstream (farm-level) programs such as: 1. Training persons who apply pesticides 2. Purchasing registered pesticides for growers 3. Auditing growers' application of pesticides and records thereof 4. Requiring periodic pesticide residue analysis reports
C, raw mushrooms could contain heat-stable staphylococcal enterotoxin from improper grower handling	Upstream (farm-level) programs such as: 1. Training growers on handling of raw product 2. Ensuring growers' use of proper, effective refrigeration equipment 3. Ensuring prompt delivery of raw product after picking

DATE:______________ APPROVED BY:______________________________________

C, chemical.
[a] For example, cooking instructions, public education, use by date, and so on.

Table 18
Examples of Critical Limits

Hazard	CCP	Critical limit
Bacterial pathogens (nonsporulating)	Pasteurization	72°C for at least 15 s
Metal fragments	Metal detector	Metal fragments >0.5 mm
Bacterial pathogens	Drying oven	$A_w < 0.85$ for controlling growth in dried food products
Excessive nitrite	Curing room/brining	Maximum 200 ppm sodium nitrite in finished product
Bacterial pathogens	Acidification step	Maximum pH of 4.6 to control Clostridium botulinum in acidified food
Food allergens	Labeling	Label that is legible and contains a listing of correct ingredients
Histamine	Receiving	Maximum of 25 ppm histamine levels in evaluation of tuna for histamine[a]

A_w, water activity; CCP, critical control point.
[a]Regulatory action level is 50 ppm, but histamine levels may increase during processing. Therefore industry may want to set lower histamine critical limits at receiving.

Table 19
Critical Limits Versus Operating Limits

Process	Critical limit	Operating limit
Acidification	pH 4.6	pH 4.3
Drying	0.84 A_w	0.80 A_w
Hot fill	80°C	85°C
Slicing	2 cm	2.5 cm

A_w, water activity

3.6.6. Who Will Monitor?

Individuals assigned to monitor CCPs may include:

- Line personnel.
- Equipment operators.
- Supervisors.
- Maintenance personnel.
- Quality assurance personnel.

Once assigned, the individual responsible for monitoring a CCP must:

- Be adequately trained in the CCP monitoring techniques.
- Fully understand the importance of CCP monitoring.
- Have ready access (be close) to the monitoring activity.
- Accurately report each monitoring activity.
- Have the authority to take appropriate action as defined in the HACCP plan.
- Immediately report critical limit deviation.

It is important that the responsible individual report all unusual occurrences and deviations from critical limits immediately to make sure that process adjustments and corrective actions are made in a timely manner. This person should record and sign all monitoring results and occurrences associated with monitoring CCPs. Records and documents associated with monitoring CCPs should also be signed by one or more responsible reviewing officials of the company.

3.7. Establish Corrective Actions (Principle 5)

3.7.1. Establishing Corrective Actions

The *Codex Hazard Analysis and Critical Control Point (HACCP) System and Guidelines for Its Application* defines corrective action as "any action to be taken when the results of monitoring at the CCP indicate a loss of control." Loss of control is considered a deviation from a critical limit for a CCP ***(47)***.

Table 20
HACCP Plan

Product Name(s): Canned mushrooms

Process step	CCP No.	Hazard description	Critical limits	Monitoring procedures	Deviation procedures	HACCP records
9. Can inspecting/ depalletizing	1B	Postprocess contamination resulting from incorrect cans, damaged cans, and serious defects	Can manufacturer's specifications No defects			
	1P	Harmful extraneous materials (HEM), e.g., wood, glass, metal fragments	No HEM			
20. Weighing	2B	Overfilling resulting in underprocessing	Maximum fill weight as specified in the scheduled process			
22. Head spacing	3B	Insufficient headspace resulting in excessive internal pressure and distorted seams	Minimum headspace as specified in the scheduled process			
23. End feeding/ closing/inspecting	4B	Postprocess contamination resulting from damaged or defective ends or improper double seams	Can manufacturer's specifications No serious problems			
25. Thermal processing	5B	Inadequate heat treatment	Maximum time lapse between closing and retort up, minimum IT, minimum time and temperature for vent and cook as specified in the scheduled process Heat-sensitive indicator changes color			
26. Cooling	6B	Postprocess contamination of product from cooling water	Detectable residual chlorine levels to 2 ppm in the cooling water			

DATE:________________ **APPROVED BY:**________________________________

CCP, critical control point; HEM, harmful extraneous material; IT, interval time.
[a]Numbers are from flow diagram in **Table 11**.

Table 21
Examples of Control Procedures

Example of hazard	Critical control point	Control procedures
Growth of *C. botulinum*	Salting of salmon to be smoked	Required salt level: 3.0–3.5% NaCl in water phase of fish samples for testing from every batch
Contamination	Chlorination of cooling water	Continuous chlorine monitor, daily sampling of water for testing. Limit 5 ppm; tolerance 3–5 ppm
	Factory hygiene	Specification of cleaning and sanitation procedures Visual control before work start Twice weekly microbiological controls of cleaned surfaces in contact with food Limit < 100 cfu cm^{-2}; tolerance mean < 100 cfu cm^{-2}; Max. 10^3 cfu
Survival of pathogens	Cooking	Define and ensure time/temperature requirements. Continuous and automatic recording of water temperature

3.7.2. Deviation

The *Codex* guidelines for the application of the HACCP system define deviation as "failure to meet a critical limit." Procedures should be in place to identify, isolate, and evaluate products when critical limits are exceeded. Inadequate deviation procedures could result in unsafe products and the eventual recurrence of the deviation ***(48–54)*** (**Table 22**).

The producer should control deviations as follows.

3.7.2.1. Identification of Deviation

The producer should have a system in place to identify deviations when they occur.

3.7.2.2. Isolation of Affected Product

The producer should have effective procedures in place to isolate, mark clearly, and control all product produced during the deviation period, as follows:

- All affected product, i.e., that processed since the last point at which the CCP was known to be under control, should be isolated.
- Isolated product should be clearly marked, e.g., with firmly attached tags, with information including: hold number, product, amount, date held, the reason for the hold, the name of the person holding the product.
- The producer should maintain control of the product from the hold date to the date of final disposition.

Table 22
HACCP Plan

Product Name(s): Canned mushrooms

Process step	CCP No.	Hazard description	Critical limits	Monitoring procedures	Deviation procedures	HACCP records
9. Can inspecting/ depalletizing	1B	Postprocess contamination resulting from incorrect cans, damaged cans, and serious defects No defects	Can manufacturer's specifications	Continuous visual monitoring by the depalletizer operator		
	1P	HEM, e.g., wood, glass, metal fragments	No HEM	Continuous visual monitoring by the depalletizer operator		
20. Weighing	2B	Overfilling resulting in underprocessing	Maximum fill weight as specified in the scheduled process	On-line check-weigher to eject over- and underfilled cans after filling		
22. Head spacing	3B	Insufficient headspace resulting in excessive internal pressure and distorted seams	Minimum headspace as specified in the scheduled process	Headspace check done after closing on consecutive samples, at least one from each head, by seam mechanic at start-up and every hour		
23. End feeding/ closing/inspecting	4B	Postprocess contamination resulting from damaged or defective ends or improper double seams	Can manufacturer's specifications No serious problems	Continuous visual monitoring of ends by closing machine operator Visual examination of sealed cans at start-up, after severe jam-ups and after adjustments as well as every half hour, and turn-down examination every 4 h on consecutive samples, one from each head, by closing machine operator		

25. Thermal processing	5B	Inadequate heat treatment	Maximum time lapse between closing and retort up, minimum IT, minimum time, and temperature for vent and cook as specified in the scheduled process Heat-sensitive indicator changes color	QC to check on time lapse between closing and retort up (at least once per period) Retort operator to check on IT, time, and temperature for vent and cook and thermograph Busse unloader to check heat-sensitive indicator tape Busse unloader to segregate product if no indicator tape or no color change of indicator tape
26. Cooling	6B	Postprocess contamination of product from cooling water	Detectable residual chlorine levels to 2 ppm in the cooling water	Chlorine checks every hour at exit of cooling water

DATE:____________ **APPROVE BY:**____________

HEM, harmful extraneous materials; QC, quality control.

[a]Numbers are from flow diagram in **Table 11**.

3.7.2.3. Evaluation of Affected Product

Evaluation of the affected product should be adequate to detect potential hazards, i.e., it should be ensured that sampling is adequate to identify the extent of the problem, that the tests are appropriate, that the judgment is based on sound science, and that the product is not released until the evaluation has determined that no potential hazard exists.

3.7.3. Corrective Action Procedures

Corrective action should be taken following any deviation to ensure the safety of the product and to prevent recurrence of the deviation.

Corrective action procedures are necessary to determine the cause of the problem, take action to prevent recurrence, and follow up with monitoring and reassessment to ensure that the action taken is effective. If the corrective action does not address the root cause of the deviation, the deviation could recur.

The producer's corrective action program should include the following:

- Investigation to determine the cause of the deviation.
- Effective measures to prevent recurrence of the deviation.
- Verification of the effectiveness of the corrective action taken.

3.7.4. Deviation and Corrective Action Records

Records should be available to demonstrate the control of products affected by the deviation and the corrective action taken.

3.7.4.1. Deviation

- Product/code.
- Date produced/held/released.
- Amount of product held.
- Results of evaluation: amount analyzed, analysis report, number and nature of defects.
- Signature of personnel responsible for hold and evaluation.
- Disposition of held product (if appropriate).
- Signed authorization for disposition.

3.7.4.2. Corrective Action

- Corrective action taken to correct deficiency.
- Follow-up/assessment of effectiveness of corrective action
- Date.
- Signature of person responsible.

3.7.5. Deviation Procedures

The following are some examples of deviation procedures for different products.

3.7.5.1. Milk

Antibiotics in incoming raw milk are detected by a rapid screening test. The detected level exceeds the established critical limit. The milk receiver refers to the deviation procedure.

The deviation procedure states that the milk is to remain in the truck and not be unloaded. The procedure also describes the follow-up action. The processor will follow up with the milk supplier involved. All corrective actions are recorded ***(35,55–58)***.

3.7.5.2. Cooked Sausages

Cooked sausages are sliced with equipment that has not been cleaned with the specified frequency. The supervisor notices that the slicer has excessive product buildup and believes that the sausages are being subjected to excessive bacterial contamination.

The deviation procedure states that the supervisor must hold all product produced since the last recorded clean-up. The product under hold is subjected to microbiological testing and is not released until the laboratory results are received. The deviation procedure also states that the employee responsible for equipment cleaning should be questioned as to the reason for the deviation from the specified procedure and be retrained as necessary.

3.8. Establish Verification Procedures (Principle 6)

3.8.1. Verification

The *Codex* guidelines define verification as "the application of methods, procedures, tests and other evaluations, in addition to monitoring to determine compliance with the HACCP plan." Verification and auditing methods, procedures, and tests, including random sampling and analysis, can be used to determine whether the HACCP system is working correctly ***(20,47)***.

3.8.2. Description of Verification Activities

Each HACCP plan should include verification procedures for individual CCPs and for the overall plan.

3.8.2.1. HACCP Plan Validation

Validation is the act of assessing whether the HACCP plan for the particular product and process adequately identifies and controls all significant food safety hazards or reduces them to an acceptable level. HACCP plan validation should include:

- Review of the hazard analysis.
- CCP determination.
- Justification for critical limits, based for example on current good science and regulatory requirements.
- Determination of whether monitoring activities, corrective actions, record-keeping procedures, and verification activities are appropriate and adequate.

The process of validating an existing HACCP plan should also include:

- Review of HACCP audit reports.
- Review of changes to the HACCP plan and the reasons for those changes.
- Review of past validation reports.
- Review of deviation reports.

- Assessment of corrective action effectiveness.
- Review of information on consumer complaints.
- Review of linkages between the HACCP plan and GMP programs.

3.8.2.2. HACCP System Audits

Audits are systematic and independent examinations involving on-site observations, interviews, and reviews of records to determine whether the procedures and activities stated in the HACCP plan are implemented in the HACCP system.

Records to be reviewed during auditing of the HACCP plan include, for example, those demonstrating that:

- Monitoring activities have been performed at the locations specified in the HACCP plan.
- Monitoring activities have been performed at the frequencies specified in the HACCP plan.
- The affected product has been controlled and corrective actions have been taken whenever monitoring has indicated the occurrence of a deviation from critical limits.
- Equipment has been calibrated at the frequencies specified in the HACCP plan. Audits should occur frequently enough to ensure that the HACCP plan is being followed continuously. This frequency depends on a number of conditions, such as the variability of the process and product.

3.8.2.3. Calibration

Calibration involves checking instruments or equipment against a standard to ensure accuracy. Calibration should be documented, and the records should be available for review during verification.

Calibration of CCP monitoring equipment is important; if the equipment is out of calibration, then monitoring results will not be accurate and may be completely unreliable.

3.8.2.4. Targeted Sample Collection and Testing

Verification may also include targeted sampling and testing and other periodic activities. Targeted sampling and testing involves taking product samples periodically and testing them to ensure that critical limits are appropriate for product safety. When sampling and testing are used as a verification tool, the usefulness of the test often depends on how the material is sampled. The risk and level of confidence needed will determine the sample size and the method of sample collection.

3.8.3. Role of Microbiological Testing in HACCP Verification

Microbiological testing is seldom effective for monitoring CCPs and cannot be used as a means of process control because of the lengthiness of analytical procedures and the inability to provide results in real time ***(39)***.

Microbiological testing does have a role in HACCP verification, however: when critical limits are established for the elimination of pathogens or their reduction to an acceptable level, microbiological testing can be used to verify the HACCP plan's effectiveness and to ensure that the identified microbiological limits have not been exceeded.

3.8.4. Verification Frequency

Verification activities should be performed according to a pre-established schedule described in the HACCP plan or whenever there are indications that the food safety status may have changed. These indications may include:

- On-line observations that CCPs may not be operating within critical limits.
- Record reviews indicating inconsistent monitoring.
- Record reviews indicating that CCPs are repetitively operated outside critical limits.
- Consumer complaints or product rejections by customers.
- New scientific data

3.8.5. Records of Verification

Records of verification should include methods, date, individuals and/or organizations responsible, results or findings, and action(s) taken.

3.8.6. Regulatory Verification

The reasons for regulatory verification activities include, among others: government obligation in consumer protection, support to the food industry (particularly medium- and small-scale food industry), and assistance to industry in trade opportunities when certification is required.

The inspector should document the existence and implementation of the HACCP plan. Regulatory verification should also involve review and/or audit of the adherence of the processor's HACCP system to its HACCP plan. In particular, the inspector should focus on the following:

- Review of the hazard analysis.
- Review of the CCP determination.
- Verification that the critical limits are based on good science and meet regulatory requirements.
- Review of the deviation and corrective action procedures.
- Review of the verification procedures.
- Review of records to verify that the HACCP plan is being followed effectively at all times.
- Verification of the accuracy of CCP monitoring equipment.

3.9. Establish Documentation and Record Keeping (Principle 7)

3.9.1. Documentation and Record Keeping

A record shows the process history, the monitoring, the deviations, and the corrective actions (including disposition of product) that occurred at the identified CCP. It may be in any form, e.g., processing chart, written record, computerized record. The importance of records to the HACCP system cannot be overemphasized. It is imperative that the producer maintain complete, current, properly filed, and accurate records.

Four types of records should be kept as part of the HACCP program:

- Support documentation for developing the HACCP plan.
- Records generated by the HACCP system.
- Documentation of methods and procedures used.
- Records of employee training programs.

3.9.2. Support Documents

The HACCP plan support documents include information and support data used to establish the HACCP plan such as the hazard analysis and records documenting the scientific basis for establishing the CCPs and critical limits ***(31)***. Examples include:

- Data used to establish the control measures to prevent microbiological growth.
- Data used to establish the shelf life of the product (if age of the product can affect safety).
- Data used to establish the adequacy of critical limits in ensuring the safety of the product.

The HACCP plan support documents should also include a list of the HACCP team members and their responsibilities, as well as all the forms produced during the preparation of the HACCP plan, showing:

- Product description and intended use.
- Flow diagram.
- Hazard analysis.
- Identification of CCPs.
- Identification of the critical limits for each CCP, including data from experimental studies or information collected to support the critical limits.
- Documented deviation and corrective action plans
- Planned verification activities and procedures.
- Identification of the preventive measures for each hazard.

3.9.3. Records Generated by the HACCP System

HACCP system records are kept to demonstrate adherence of the HACCP system to the HACCP plan. These records are used to demonstrate control at CCPs in the food process. The records generated by the HACCP system include all activities and documentation required by the plan.

3.9.3.1. Monitoring Records for All CCPs

All HACCP monitoring records should be kept on forms that contain the following information:

- Form title.
- Time and date.
- Product identification (including product type, package size, processing line, and product code).
- Critical limits.
- Monitoring observation or measurement.
- Operator's signature or initials.
- Corrective action taken, where applicable.
- Reviewer's signature or initials.
- Date of review.

3.9.3.2. Deviation and Corrective Action Records

- Identification of the deviant lot/product.
- Amount of affected product in the deviant lot.
- Nature of the deviation.
- Information on the disposition of the lot.
- Description of the corrective action.

3.9.3.3. Verification/Validation Records

- In-house on-site inspection.
- Equipment testing and evaluation.
- Accuracy and calibration of monitoring equipment.
- Results of verification activities, including methods, date, individuals and/or organizations responsible, results or findings, and action taken.

3.9.4. Documentation of Methods and Procedures Used

The producer should maintain records of the methods and procedures used in the HACCP system.

- Description of the monitoring system for the critical limit of each CCP, including the methods and equipment used for monitoring, the frequency of monitoring, and the person performing the monitoring.
- Plans for corrective actions for critical limit violations or situations resulting in potential hazards.
- Description of record-keeping procedures, including copies of all record forms.
- Description of verification and validation procedures.

3.9.5. Records of Employee Training Programs

Records should be kept of all employee training. This is of particular importance for employees involved in monitoring critical limits for CCPs and those involved with deviation review, corrective actions, and verification. These employees must be trained to understand fully the appropriate procedures/methods and actions to be taken regarding control of CCPs (**Table 22**).

4. Notes

1. The intent of the HACCP system is to focus control at CCPs. Redesign of the operation should be considered if a hazard that must be controlled is identified but no CCPs are found.
2. The HACCP application should be reviewed and necessary changes made when any modification is made in the product, process, or any step. It is important when applying HACCP to be flexible when appropriate, given the context of the application, taking into account the nature and the size of the operation.
3. Prior to application of HACCP to any sector of the food chain, that sector should be operating according to the *Codex* General Principles of Food Hygiene, the appropriate *Codex* Codes of Practise, and appropriate food safety legislation. Management commitment is necessary for implementation of an effective HACCP system.
4. HACCP should be applied to each specific operation separately. CCPs identified in any given example in any *Codex* Code of Hygienic Practice might not be the only ones identified for a specific application or might be of a different nature.
5. During hazard identification, evaluation, and subsequent operations in designing and applying HACCP systems, consideration must be given to the impact of raw materials, ingredients, food manufacturing practices, role of manufacturing processes to control hazards, likely end-use of the product, categories of consumers of concern, and epidemiological evidence relative to food safety.

References

1. Ito, K. (1974) Microbiological critical control points in canned foods. *Food Technol.* **28,** 46–48.
2. Sperber, W. H. (1991) *The Modern HACCP System.* Chicago, Food Technology, pp. 116–118.
3. The Pillsbury Co. (1981) Food safety through the hazard analysis and critical control point system. The Pillsbury Co. apud Bryan, F. L., Hazard analysis of food Service operations. *Food Technol.* **35,** 78–87.
4. Silliker, J. H. (1986) Principles and applications of the HACCP approach for the food processing industry. In: *Proceedings of the 1986 Conference for Food Protection,* Food Protection Technology (Felix, C. W., ed.). Lewis Publications, Chelsea, MI, pp. 81–89.
5. Archer, D. L. (1990) The need for flexibility in HACCP. *Food Technol.* **May,** 174–178.
6. Bobeng, B. J. and David, B. D. (1977) HACCP models for quality control of entree production in food service systems. *J. Food Prot.* **40,** 632–638.
7. Bryan, F. L. (1984) Análises de riscos nas empresas de alimentos. *Higiene Alimentar* **3,** 92–100.
8. Bryan, F. L. (1990) Application of HACCP to ready-to-eat chilled foods. *Food Technol.* **July,** 70–77.
9. Bryan, F. L. (1981) Hazard Analysis Critical Control Point approach: epidemiologic rationale and application to foodservice operations. *J. Environ. Health.* **44,** 7–14.
10. Bryan, F. L. (1992) *Hazard Analysis Critical Control Point Evaluations.* World Health Organization, Geneva, 1992.
11. Bryan, F. L. (1985) Procedures for local health agencies to institute a hazard analysis critical control point program for food safety assurance in foodservice operations. *J. Environ. Health* **47,** 241–245.
12. Martins, J. P. S. (1983) Aplicação do Método de Análise de Risco por Pontos Críticos de Controle em Cozinhas Industriais Ministério da Agricultura, SIPA, SIF (Serviço de Inspeção Federal), Brasília.
13. Puri, S. C. (1995) Qualidade no setor alimentício. *Artigo. Rev. Controle Qualidade* **33,** 60–61.
14. Apostila do Centro Tecnológico de Madrid/Projeto ARCA Madrid Espanha (CETEMA) (1995) *Control Chart.* Yamamura Consulting Engineer Office, Japão.
15. FAO (1995) Food and Nutrition Paper No. 58. The use of hazard analysis critical control point (HACCP) principles in food control. Report of an FAO Expert Technical Meeting, Vancouver, Canada, 12–16 December, 1994, FAO, Rome.
16. ICMSF (International Comission on MicrobiologicaL Specifications for Foods). Microorganisms in foods (1988) Application of the Hazard Analysis Critical Control Point system to ensure microbiological safety and quality. Blackwell Scientific Publications, Oxford.
17. Kauffman, F. L. (1974) How FDA uses HACCP. *Food Technol.* **28,** 84.
18. Portaria 1428/Ministério da Saúde (1993) Publicado no "Diário Oficial da União/DOS".
19. Bauman, H. E. (1990) HACCP: concept, development, and application. *Food Technol.* **May,** 153–158.
20. Bauman, H. E. (1974) The HACCP concept and microbiological hazard categories. *Food Technol.* **29,** 30–74.
21. Haberstroh, C. (ed.) (1988) HACCP: making the system work. *Food Eng.* **Aug,** 70–80.
22. Ishikawa, K. (1983) *Guide to Quality Control, Industrial Engineering and Technology,* 2nd ed. Asian Productivy Organization, Tokyo.

23. Bryan, F. L. (1988) Hazard analysis critical control point: what the system is and what it is not. *J. Environ. Health* **50,** 400–401.
24. Bryan, F. L. (1974) Microbiological food hazards today based on epidemiological information. *Food Technol.* **28,** 52, 54, 58–60, 62, 64, 66, 84.
25. Bryan, F. L.(1988) Risks associated with vehicles of foodborne pathogens and toxins. *J. Food Protection* **51,** 498–508.
26. Bryan, F. L. (1988) Risks of practices, procedures and processes that lead to outbreaks of foodborne diseases. *J. Food Protection* **51,** 663–673.
27. Campos, V. F. (1992) *TQC: Controle da Qualidade Total (no estilo japonês).*4th ed. Fundação Christiano Ottoni, Belo Horizonte.
28. Campos, V. F. (1994) *TQC: Gerenciamento da Rotina do Trabalho do Dia-a-dia*, 2nd ed. Fundação Christiano Ottoni, Belo Horizonte.
29. Harrington, H. J. (1991) *Business Process Improvement: The Breakthrough Strategy for Total Quality, Productivity, and Competitiveness.* McGraw Hill, New York.
30. Hradeski, J. L. (1989) *Aperfeiçoamento da Qualidade e da Produtividade: Guia Prático para Implementação do Controle Estatístico de Processo CEP.* McGraw-Hill, São Paulo.
31. Monteiro, J. G. D. M. (1994) *Gerenciamento de Processos Empresariais: Interface direta com o Processo Produtivo.* Trabalho de Mestrado, UFSC.
32. Paladini, E. P. (1995) Gestão da Qualidade no Processo: a qualidade na Produção de Bens e Serviços. Atlas, São Paulo-SP.
33. Pinto, J. L. G. C. (1993) *Gerenciamento de Processos na Indústria de Móveis.* Trabalho de Mestrado, UFSC.
34. CCDAM (Comittee on Communicable Diseases Affecting Man) (1991) *Procedures to Implement the Hazard Analysis Critical Control Point System.* International Association of Milk, Food and Environmental Sanitarians, Ames, IA.
35. Huss, H. H. (1993) Assurance of seafood quality. FAO Fisheries Technical Paper. FAO, Rome.
36. Bryan, F. L. (1981) Hazard analysis of food service operations. *Food Technol.* **35,** 78–87.
37. de Carvalho, A. B. M. and Frosini, L. H. (1995) CEP de tendências em processos contínuos. *Artigo. Rev. Controle Qualidade* **33,** 63, 64, 66.
38. Jay, J. M. (1992) Modern Food Microbiology, 4th ed. Chapman & Hall, New York.
39. NCR. Executive summary. *An Evaluation of the Role of Microbiological Criteria for Foods and Food Ingredients.* Food Protection Commitee/Subcommitee on Microbiological Criteria/National Research Council/National Academy of Sciences, Washington DC.
40. Bauman, H. E. (1974) The HACCP concept and microbiological hazard categories. *Food Technol.* **29,** 30–74.
41. Leitão, M. F. F. (1995) Análise de perigos e pontos críticos de controle, conceitos e definições. In: *Seminário Sobre Análise de Riscos e Pontos Críticos de Controle (ARPCC) na Indústria de Pescado e Derivados.* CIAL/ITAL, Campinas, pp. 1–11.
42. Stevenson, K. E. (1990) Implementing HACCP in the food industry. *Food Technol.* **May,** 179–180.

42a. Shampton, N. (1989) Implementing a food safety programme. *Food Manufacture* **17,** 47–50.

43. Dellaretti Filho, O. and Drumond, F. B. (1994) *Itens de Controle e Avaliacao de Processos.* Fundacao Christiano Ottini, Belo Horizonte, 1994.
44. Spencer G. III, E. and Hudak-Roos, M. (1991) Development of an HACCP-based inspection system for the seafood industry. *Food Technol.* **Dec,** 53–57.

44a. Scott and Moberg (1995), 4–14.
44b. Avery, S. M., Hudson, J. A., Phillips, D. M. (1976) Use of response surface models to predict bacterial growth from time temperature histories. *Food Control* **7,** 121–128.
44c. Lund, B. M., Baird-Parker, T. C., and Gould, G. W. (2000) *The Microbial Safety and Quality of Food.* Apsen, Gaithersburg, MD.
45. Spencer G. III, E. and Hudak-Roos, M. (1990) Use of HACCP for seafood surveillance and certification. *Food Technol.* **May,** 159–168.
46. Corlett Jr., D. A. (1991) Regulatory verificationa of industrial HACCP systems. *Food Technol.* **Apr,** 144–146.
47. Peterson, A. C. Gunnerson, R. E. (1974) Microbiological Critical Control Points in frozen foods. *Food Technol.* **28,** 37–44.
48. Beauregard, M. R., Mikulak, R. J., and Olson, B. A. (1991) *A Practical Guide to Statistical Quality Improvement: Opening up the Statistical Toobox.* Van Nostrand Reinhold, New York.
49. Doty, L, A. (1990) *Statistical Process Control,* 1st ed. ASQC-Quality Press Book by Industrial Press, New York.
50. Grant, E. L. (1964) *Statistical Quality Control, International Student Edition,* 3rd ed. McGraw-Hill, New York.
51. Kume, H. (1988) *Statistical Methods for Quality Improvement.* AOTS-The Association for Overseas Technical Scholarship Japão.
52. Lourenço Filho, R. de C. B. (1982) *Controle Estatístico de Qualidade.* LTC Livros Técnicos e Científicos Editora., Rio de Janeiro. *Manual de Boas Práticas de Fabricação para Indústrias de Alimentos.* Sociedade Brasileira de Tecnologia de Alimentos (SBCTA) Campinas, SP, 1991.
53. Rihmer, J, F., Spencer, A. (1989) *Controle Estatístico de Processos (CEP).* Apostila, Joinville, SC.
54. Tompkin, R. B. (1990) The use of HACCP in the production of meat and poultry products. *J. Food Prot.* **53,** 795–803.
55. Corlett Jr., D. A. (1989) Refrigerated foods and use of hazard analysis and critical control point principles. *Food Technol.* **43,** 91–94.
56. Hajadenwurcel, J. R. (1996) Apostila do "Curso de Análise de Perigos e Pontos Críticos de Controle na Indústria de Laticínios," citado por Santos, J. A. dos. HACCP é Garantia de Qualidade. *Rev. Indústria Laticínios.* **I,** 16–20.
57. NFPA (1989) Factors to be considered in stablishing Good Manufacturing Practices for the production of refrigerated foods, National Food Processors Assn., apud Corlett, D. A.. Refrigerated foods and use of Hazard Analysis and Critical Control Point principles. *Food Technol.* **43,** 91–94.
58. USHEW (1989) Proceeding of the 1971 National Conference on Food Protection. US Department of Health, Education, and Welfare, Washington, DC, apud Corlett, D. A. (1989) Refrigerated foods and use of Hazard Analysis and Critical Control Point principles. *Food Technol.* **43,** 91–94.

23

Testing Disinfectants in the Food Factory

Phenol Coefficient Method

Anavella Gaitan Herrera

1. Introduction

Contamination of foods by the environment has direct public health and keeping quality significance. The food factory environment (with raw materials and processing) governs the numbers and types of microorganisms in finished products ***(1)***. Use of the appropriate sampling procedures permits us to discover the magnitude and type of contamination. Microbiological sampling allows objective evaluation of the disinfectants and the sanitation practices and procedures used in the food factory (***2***; *see* **Note 1**).

Disinfectants are antimicrobial pesticides that are primarily used on inanimate surfaces (such as floors, walls, and countertops) to kill infectious bacteria, fungi, and viruses. Antimicrobial pesticides are substances used to kill or suppress the growth of harmful microorganisms on inanimate objects and surfaces ***(3)***. Products intended for the control of microorganisms in or on people or animals are considered drugs, not pesticides, and are therefore regulated by the Food and Drug Administration (FDA). Antimicrobial pesticides ***(4,5)*** are divided into two broad use categories:

1. Non-public health products include those used to control the growth of algae, odor-causing bacteria, and microorganisms causing spoilage, deterioration, and fouling of materials. Examples include antimicrobials used in cooling towers, paints, and paper products ***(6)***.
2. Public health products are intended to control microorganisms infectious to people. Examples include sterilants, which are used to destroy or eliminate all forms of microbial life including fungi, viruses, and all forms of bacteria and their spores; disinfectants, which are used to destroy or irreversibly inactivate infectious fungi and bacteria, but not necessarily their spores; and sanitizers, which are used to reduce, but not necessarily eliminate microorganisms. Examples range from sterilants used to treat surgical instruments to disinfectants applied to hospital floors, walls, and bed linens and sanitizers used on carpets or in laundry additives ***(6)***.

From: *Methods in Molecular Biology, vol. 268: Public Health Microbiology: Methods and Protocols*
Edited by: J. F. T. Spencer and A. L. Ragout de Spencer © Humana Press Inc., Totowa, NJ

Products making pesticidal claims are registered by the Environmental Protection Agency (EPA). In the early 1990s, the Office of Pesticide Programs (OPP) implemented the Antimicrobial Product Testing Program to test the antimicrobial effectiveness and chemical composition of the EPA-registered products *(7,8)*.

In 1881, Robert Koch developed the first method for evaluating the activity of disinfectants *(9)*. In 1897, Kroing and Paul noted that the activity of disinfectants could be compared under controlled conditions (temperature and concentration) *(9)*. In 1903, Rideal and Walker developed the phenol coefficient method *(10)*. The novelty of this method was the standardization of components (test methods, recovery medium, microorganisms, and disinfectant stock solution). This method continues to be used today with some modifications *(3)*. Ideally, food, finished product container, and product contact surfaces of the equipment should be practically free from microorganism (*see* **Note 2**).

The purposes of microbiological monitoring can be summarized as follows *(6)*:

1. Determination of the frequency required for execution of sanitation cycles.
2. Determination of environmental contamination by foodborne pathogens.
3. Evaluation of designs of food processing facilities and equipment in terms of sanitation.
4. Discovery of environmental sources of spoilage organisms in response to shelf life problems.
5. Determination of the efficiency of sanitation procedures and cycles.
6. Determination of the frequency required for special maintenance procedures (changing of air filters to reduce airborne mold contamination.

The Phenol Coefficient Method given here is applicable to testing disinfectants miscible with water or for standard resistance of test bacteria. The 95% confidence limits are ± 12% *(3)*.

2. Materials

2.1. Culture Media

1. Nutrient broth (*see* **Note 3**): Boil 5 g beef extract, 5 g NaCl and 10 g peptone in 1 L H_2O for 20 min and adjust to pH 6.8. Filter through paper. Place 10-mL portions in 20 × 150-mm test tubes. Autoclave for 20 min at 121°C.
2. Synthetic broth:
   a. Solution A: dissolve 0.05 g L-cystine, 0.37 g DL-methionine, 0.4 g L-arginine·HCl, 0.3 g DL-histidine·HCl, 0.85g L-lysine·HCL, 0.21 g L-tyrosine, 0.5 g DL-threonine, 1.0 g DL-valine, 0.8 g L-leucine, 0.44 g DL-isoleucine, 0.06 g glycine, 0.61 g DL-serine, 0.43 g DL-alanine, 1.3 g L-glutamic acid·HCl, 0.45 g L-aspartic acid, 0.26 g DL-phenylalanine, 0.05 g DL-tryptophan, and 0.05 g L-proline in 500 mL H_2O containing 18 mL 1 N NaOH.
   b. Solution B: dissolve 3.0 g NaCl, 0.2 g KCl, 0.1 g $MgSO_4 \cdot 7\,H_2O$, 1.5 g KH_2PO_4, 4.0 g Na_2HPO_4, 0.01 g thiamine·HCl, and 0.01 g niacinamide in 500 mL H_2O. Mix solution A and solution B, dispense in 10-mL portions in test tubes (16 × 150 mm), and autoclave (20 min at 121°C). Before using for daily transfers of test cultures, aseptically add 0.1 mL sterile 10% glucose solution per test tube. Grow cultures with tube slanted 8° from horizontal.

3. Nutrient agar: Dissolve Bacto Agar to 1.5% in nutrient broth (or synthetic broth) and adjust to pH 7.2 or 7.4, add 10-mL portions per tube (13 × 100 mm), autoclave (20 min at 121°C), and slant.
4. Letheen broth (*see* **Note 4**): Dissolve 0.7 g lecithin and 5.0 g polysorbate 80 or Tween-80 in 400 mL of hot H_2O and boil until clear. Add 600 mL solution of 5.0 g beef extract, 10.0 g peptone, and 5.0 g NaCl in H_2O and boil for 10 min. Adjust pH to 7.0 ± 0.2 (with 1 *N* NaOH or 1 *N* HCl). Filter through coarse paper, transfer 10-mL portions to test tubes (16 × 150 mm), and autoclave (20 min at 121°C).
5. Thioglycolate medium USP XX (*see* **Note 4**): Mix 0.5 g L-cystine, 0.75 g agar-agar, 2.5 g NaCl, 5.5 g glucose·H_2O, 5.0 g H_2O-soluble yeast extract, and 15.0 g pancreatic digest of casein with 1 L H_2O. Heat in H_2O bath to disolve, add 0.5 g sodium thioglycolate (or 0.3 g thioglycolate acid), and adjust to pH 7.1 ± 0.2 (with 1 *N* NaOH). Filter if necessary when hot (without boiling) through moistened filter paper. Add 1.0 mL of 0.1% sodium resarzurin solution (freshly prepared). Transfer 10 mL medium to 16 × 150-mm tubes and autoclave (20 min at 121°C). Cool to 25°C and store at 20–30°C (protected from light).

2.2. Reagents and Apparatus

1. Test organism (*see* **Note 2**):
   a. *Salmonella typhi*, ATCC 6539: Maintain stock culture on nutrient agar slants and transfer monthly. Incubate new stock transfer for 2 d at 2–5°C. Inoculate tubes of nutrient broth, from stock culture and make at least four consecutive daily transfers (≤30) in nutrient broth, incubating at 37°C (before using culture for testing). Culture the organism for 22–26 h in nutrient broth at 37°C in test tubes. Shake and let settle for 15 min before using.
   b. *Staphylococcus aureus*, ATCC 6538: Use 22–26-h cultures of *S. aureus*, having at least the resistance indicated in **Table 1** at 20°C. Maintain stock culture on nutrient agar slants. Incubate both 18–24 h at 37°C. (Store slants at 2–5°C). Inoculate Trypticase soy agar (TSA) plates, and Mannitol Slants Agar (MSA). Incubate overnight at 37°C. Colonies of *S. aureus* are round, shiny, and yellow. Gram staining shows Gram positive cocci. Store the cultures for 30 d (2–5°C).
   c. *Pseudomonas aeruginosa* ATCC 15442: Use 22–26-h cultures of *P. aeruginosa*, having at least the resistance indicated in **Table 2**, at 20°C. Inoculate Cystina trypticase agar (CTA). Incubate tube 18–24 h at 37°C and store at 2–5°C. Inoculate TSA and Pseudosel Petri plates; incubate overnight at 37°C. Colonies of *P. aeruginosa* are flat, greenish-yellow, and opaque. Gram staining shows Gram-negative rods. Store cultures for 30 d (2–5°C).
2. Phenol stock solution (*see* **Note 5**): Solution should be 5% w/v.
   a. Weigh 50 g phenol (USP), which congeals at ≥40°C, in beaker.
   b. Dissolve in H_2O, rinse solution in 1-L volumetric flask, and dilute to volume.
   c. Standardize with 0.1 *N* KBr-$KbrO_3$ solution (*see* **Subheading 2.2.3.**).
   d. Transfer 25 mL stock solution to 500 mL volumetric flask and dilute to volume with H_2O.
   e. Transfer 15-mL aliquot of diluted solution to 500 mL I_2 flask and add 30 mL standard KBr-$KBr0_3$ solution.
   f. Add 5 mL HCl and immediately insert stopper. Shake frequently (for 30 min) and let stand for 15 min.

Table 1
Phenol Coefficient Method of Testing Disinfectants Against *Staphylococcus aureus*[a]

Phenol dilution	5 min	10 min	15 min
1–60	+ or 0	+ or 0	0
1–70	+	+	+

[a]O, organism not present; +, organism present

From 5% stock phenol solution (1:20) dilute further to make 1:60 and 1:70 dilutions. 5 min, 10 min, and 15 min correspond to a set of transfers; for example, after 30 s, transfer loop from second subculture tube and continue process for each successive dilution; 5 min after making first transfer, repeat for 10 and 15 min intervals.

Table 2
Phenol Coefficient Method of Testing Disinfectants Against *Presudomonas aeruginosa*[a]

Phenol dilution	5 min	10 min	15 min
1–80	+ or 0	+ or 0	0
1–90	+	+	+

[a]O, organism not present; +, organism present.

From 5% stock phenol solution (1:20) dilute further to make 1:80 and 1:90 dilutions. 5 min, 10 min, and 15 min correspond to a set of transfers; for example, after 30 s, transfer loop from second subculture tube and continue process for each successive dilution; 5 min after makeing first transfer, repeat for 10 and 15 min intervals.

g. Remove stopper just enough to quickly add 5 mL 20% KI solution, taking care that no Br_2 vapors escape, and immediately stopper the flask.
h. Shake thoroughly, remove stopper, and rinse it and neck of flask with a little H_2O so that washings flow into flask.
i. Titrate with 0.1 *N* $Na_2S_2O_3$, using starch indicator. (Mix approx 2 g finely powdered potato starch with cold H_2O to a thin paste, add approx 200 mL boiling H_2O, stirring constantly, and immediately discontinue heating.) Add approx 1 mL Hg, shake, and let stand over the Hg (1 mL 0.1 *N* KBr-$KbrO_3$ = 0.001569 g phenol).

$$\text{\% phenol in stock solution} = (30 \text{ mL } 0.1\ N\ Na_2S_2O_3 \text{ solution from tritation}) \times 0.001569 \times 1333 \times 100/1000$$

where: 30 mL 0.1 N KBr-$KbrO_3$ solution added.
0.001569 = g phenol equivalent to 1 mL 0.1 N KBr-$KbrO_3$ solution.
1333 = dilution factor.
1000 = original volume phenol stock solution.

j. If necessary, adjust solution to 5.00 ± 0.05% phenol by adding H_2O or phenol.
k. Keep in well-stoppered amber bottles in cool place, protected from light.

3. Potassium bromide-bromate solution 0.1 *N*: Standardization: Transfer 30 mL to I_2 flask, and add 25 mL H_2O, 5 mL 20% KI solution and 5 mL HCl. Shake thoroughly and titrate with 0.1 N $Na_2S_2O_3$, using starch indicator.
4. Transfer loop: Make 4-mm id single loop at the end of a 50–75-mm or 4-mm loop fused on a 75-mm shaft. Then bend the loop at 30° angle with stem.
5. Water bath: Maintain a constant temperature of maintaining 20 ± 0.2°C.
6. Racks.
7. Glassware: Sterilize all glassware for 2 h in hot air at 180°C. (Place pipets in closed metal containers before sterilizing.)

3. Methods

3.1. Technique

1. Make a 1% stock solution of the disinfectant to be tested. (You can use other concentrations; the convenient dilution depends on the anticipated dilution.) Keep in a glass-stoppered cylinder.
2. Make final dilutions, from 1% stock dilution, into test tubes and remove all in excess of >5 mL. The range of dilutions should cover killing limits of disinfectant in 5–15 min and should at the same time be close enough for accuracy.
3. From 5% stock phenol solution (1:20), dilute further to make 1:90 and 1:100 dilutions, and place in medication tubes.
4. Place these tubes (containing 5 mL each of final dilutions of disinfectants and phenol) and tubes containing test culture in an H_2O bath at 20°C and leave 5 min.
5. Add 0.5 mL of the test culture to each of the dilutions at time intervals corresponding to intervals at which transfers are to be made. Thus, by time 10 tubes have been seeded at 30-s intervals, 4.5 min have elapsed, and a 30-s interval intervenes before transfer to the subculture begins.
6. Add culture from graduated pipet large enough to see all tubes in any one set. In inoculating test tubes, hold them in a slanted position after removal from bath, insert the pipet to just above the surface of the disinfectant, and run in the culture without letting the tip touch the disinfectant.
7. After adding culture, agitate gently but thoroughly to ensure even distribution of bacteria, and replace in the bath 5 min after seeding. For the first test tube, transfer one loopful of the mixture of culture and diluted disinfectant from the test tube to the corresponding subculture tube. To facilitate transfer of uniform drops of antibacterial mixture, hold the tube at a 60° angle, and withdraw the loop so that the plane of the loop is parallel with the surface of the liquid.
8. After 30 s, transfer one loopful from the second test tube to the second subculture tube, and continue the process for each successive dilution; 5 min after making the first transfer, begin the second set of transfers for a 10-min period, and finally repeat for a 15-min period.
9. Gently agitate the test tubes before taking each interval loop subsample. Before each transfer, heat the loop to redness in a flame, and flame the mouth of every tube. Sterilize the loop immediately after each transfer (before replugging tubes) to allow time for cooling. Use care in transferring and seeding to prevent the pipet or needle from touching the sides or mouth of the test tube, and see that no cotton threads adhere to the inner sides or mouth of the tubes.
10. Incubate the subculture for 48 h at 37°C and read the results.

Table 3
Example of Methods for Calculating Phenol Coefficient Number[a]

Disinfectant (X) Dilution	5 min[b]	10 min[b]	15 min[b]
1:300	0	0	0
1:325	+	0	0
1:350	+	0	0
1:375	+	+	0
1:400	+	+	+
Phenol			
1:90	+	0	0
1:100	+	+	+

[a]Phenol coefficient number for *S. typhi* ($PCN_{S.\ typhi}$) = 350/90 = 3.89.
350, Highest dilution of disinfectant that kills test organism in 10 min but not in 5.
90, Highest dilution of phenol that kills test organism in 10 min but not in 5.
3.89, Dilution of disinfectant capable of killing S. typhi in 10 min but not in 5.
[b]0, organism not present; +, organism present.

11. Thoroughly agitate individual subculture tubes before incubation. Macroscopic examination is usually sufficient. Occasionally a 3-d incubation period, agar streak, microscopic examination, or agglutination with specific antiserum may be necessary to determine feeble growth or suspected contamination.

3.2. Calculation

Express results in terms of phenol coefficient number, or highest dilution that kills the test organism in 10 min but not in 15 min, whichever most accurately reflects the germicidal value of the disinfectant. The phenol coefficient is the number obtained by dividing the numerical value of the greatest dilution (denominator of fraction expressing dilution) of disinfectant capable of killing the microorganism in 10 min but not in 5 min by the greatest dilution of phenol showing the same results. Examples of how to calculate the phenol coefficient number are given in (**Tables 3 and 4**).

If none of the dilutions of disinfectant shows growth in 5 min and killing in 10 min, estimate a hypothetical dilution only when any three consecutive dilutions show the following results: first, no growth in 5 min; second, growth in 5 and 10 min but not in 15 min; and third, growth in 5, 10, and 15 min (**Table 5**).

4. Notes

1. All methods used to evaluate antimicrobial efficacy are influenced by factors that can affect the activity of a disinfectant, such as ***(5)***:
   a. Efficacy under test conditions.
   b. Activity in the target food process.
   c. Factors inherent in the microorganism (phenotypes and genotypes from the type test satrain) ***(11)***.
   d. The antimicrobial agent.
   e. The test medium.

Table 4
Satisfactory Readings for Phenol Control[a]

Phenol dilution	5 min	10 min	15 min
1:90	+ or 0	+ or 0	0
1:100	+	+	+ or 0

[a]0, organism not present; +, organism present.

Table 5
Conditions Needed to Establish Hypothetical Dilution[a]

Disinfectant (X) Dilution	5 min[b]	10 min[b]	15 min[b]
1:300	0	0	0
1:350	+	+	0
1:400	+	+	+
Phenol			
1:90	0	0	0
1:100	+	+	+

[a]Establish hypothetical dilution: PCN = 325/95 = 3.42. 325, Highest dilution of disinfectant that killing test organism in 10 min but not in 5. 95, Highest dilution of phenol that killing test organism in 10 min but not in 5. 3.42, Dilution of disinfectant capable of killing microorganism in 10 min but not in 5.

[b]0, organism not present; +, organism present

f. The test procedure itself.
g. The size and physical state of the inoculum ***(12)***.
h. The cell physiology and concentrations of microorganism(s) tested. One should identify the factors capable of influencing results.

2. The microorganism(s) selected for use in the disinfectant evaluation should reflect the flora of the target food process.
3. Nutrient broth: The colorimetric method is adjusted to give dark green with bromothymol blue. Use this broth for daily transfers of test cultures ***(3)***.
4. For subculture media: Fluid thioglycolate medium USP XX will usually give the lowest result with oxidizing products and products formulated with toxic compounds containing certain heavy metals like Hg. Letheen broth will usually give the lowest results with products containing cationic surface active materials. Letheen broth can be obtained from special laboratories ***(3)***.
5. It is a commonly accepted criterion that disinfectants be at a dilution equivalent in germicidal efficiency to phenol against *S. typhi* by calculating 20X the *S. typhi* coefficient to determine the number of parts of H_2O in which 1 part disinfectant may be mixed; this must be regarded as presumptive ***(3)***.

References

1. American Health Public Association (1978) *Standard Methods for the Examination of Dairy Products*, 14th ed. (Marth, E. H., ed.). Washington, DC, American Health Public Association.
2. Barry, A. L. (1976) *The Antimicrobic Susceptibility Test: Principles and Practices. Lea & Febiger, Philadelphia.*
3. *Beloian, A., ed. (1995) Disinfectants. In:* Official Methods of Analysis, *15th ed., vol. 1 (Helrich, K., ed.). Arlington, VA, Association of Official Analytical Chemists.*
4. *NCCLS (1991)* Antimicrobial Susceptibility Testing, *3rd ed. Villanova, PA, National Committee for Clinical Laboratory Standards.*
5. Parish, M. E. and Davidson, P. M. (1993) Methods for evaluation. In: *Antimicrobials in Foods*, 2nd ed. (Davidson, M. and Branes, A. L., eds.). Dekker, New York, pp 589–615.
6. Davidson, P. M. and Parish, M. E. (1989) Methods for testing the efficacy of food antimicrobials. *Food Technol.* **43,** 148.
7. International Commission on Microbiological Specifications for Foods (1974) *Microorganisms in Foods 2. Sampling for Microbiological Analysis: Principles and Specific Applications.* Toronto, University of Toronto Press.
8. International Commission on Microbiological Specifications for Foods (1980) *Microbial Ecology of Foods. Factors Affecting Life and Death of Microorganisms*, vol. 1. Toronto, University of Toronto Press.
9. Block, S. S. (1983) Historical review. In: *Disinfection, Sterilization and Preservation*, 3rd ed. (Block, S. S., ed.). Lea & Febiger, Philadelphia.
10. Cremieux, A. and Fleurette, J. (1983) Methods of testing disinfectants. In: *Disinfection, Sterilization and Preservation*, 3rd ed. (Block, S. S., ed.). Lea & Febiger, Philadelphia, p. 918.
11. Tenover, F. C. (1986) Studies of antimicrobial resistance genes using DNA probes. *Antimicrob. Agents Chemother.* **29,** 721–725.
12. Piddock, L. J. (1990) Techniques used for determination of antimicrobial resistance and sensitivity in bacteria. *J. Appl. Bacteriol.* **68,** 307.

24

Nontraditional Method of Evaluating Disinfectants

With Isolated Microorganisms From the Food Factory

Anavella Gaitan Herrera

1. Introduction

Cleaning and disinfection in the food industry are critical in the production process, and the efficacy of the disinfectants used is frequently debated *(1)*. Several factors are involved in the effectiveness of a disinfectant agent. It is important to consider the number and type of microorganism present as well as the physical and chemical characteristics of the water; these factors vary from industry to industry and they determine efficacious disinfection *(2)*.

In the laboratory it is possible to evaluate disinfectants to be used in a particular factory, even though these are different from those reported by international organizations *(3)*. Some useful practices are:

1. To use cultures of microorganisms isolated in one's own lab instead of reference cultures.
2. To use as a diluter the water that is used daily in the factory under question.
3. To compare different disinfectant products under identical conditions of time and temperature.

This methodology is useful for microorganisms that are ubiquitous in the food industry *(4)*.

2. Materials

1. Trypticase Soy broth in tubes.
2. Cultures to be tested.
3. Tubes with 2.0 mL of water from the factory, to be used as diluters.
4. Pipets of 2.0 and of 10 mL.
5. Petri plates with trypticase soy agar.
6. Disinfectants to be evaluated.
7. Microorganisms to be tested.

From: *Methods in Molecular Biology, vol. 268: Public Health Microbiology: Methods and Protocols*
Edited by: J. F. T. Spencer and A. L. Ragout de Spencer © Humana Press Inc., Totowa, NJ

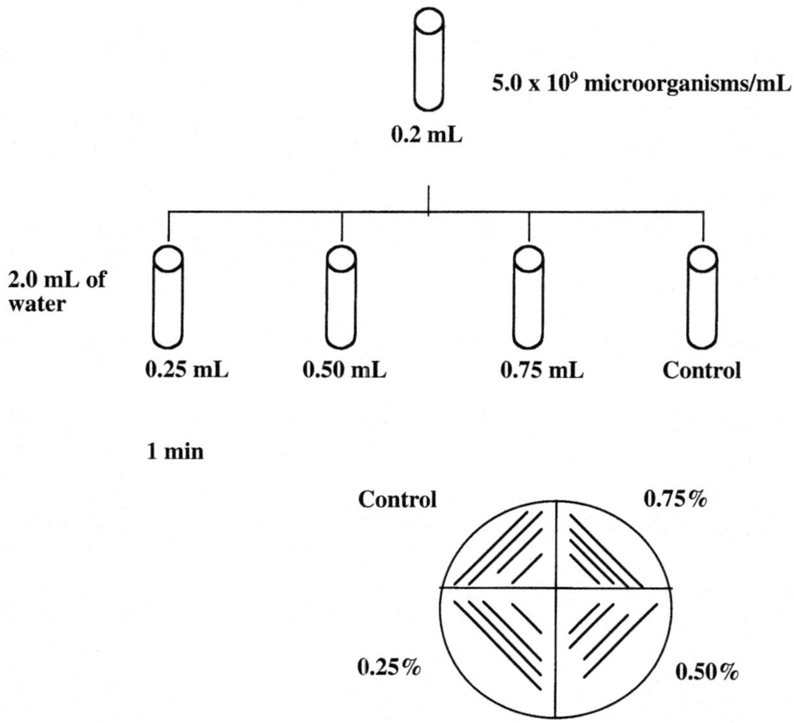

Fig. 1. Scheme for the evaluation of disinfectant.

3. Methods

1. Prepare a series of three dilutions of the disinfectant to test for concentrations above and below those recommended by the maker, using water from the plant free of chlorine (*see* **Note 1**).
2. Place 2.0 mL of each concentration in a test tube
3. Prepare as many series as there are microorganisms to be evaluated.
4. Prepare a suspension from each equivalent microorganism of 5.0×10^9 microorganisms/mL according to MacFarland's scale (*see* **Note 2**).
5. Inoculate 0.2 mL of the suspension from the microorganism into each of the dilutions.
6. Prepare four Petri dishes for each time interval (1, 2, 5, and 10 min) so that the disinfectant (at concentrations of 0.25, 0.50, and 0.75%) will be in contact with the microorganism. use plates of trypticase soy agar (if the microorganism is a bacterium) or Saboureaud agar (if the microorganism under evaluation is a mold). Using a gaged handle inoculate the agar by making four grooves in the surface, one for each concentration, at 1, 2, 5, and 10 min after initial inoculations of each concentration (*see* **Fig. 1**).
7. Repeat the same procedure at 2, 5, and 10 min.
8. Incubate the plates sown with bacteria at 35°C for 24–48 h and those sown with molds at 22–24°C for 3–5 d.

9. Carry out the reading comparatively with the control, considering the number of grooves that show growth and the density, on a numeric scale from 0 to 4.
10. Interpretation: The product to choose will be the one that inhibits growth at the smallest concentration in the shortest time of contact with the evaluated agent.

4. Notes

1. It should be free of chlorine since the residual action of chlorine can interfere with evaluation of the disinfectant being tested.
2. Prepare $BaSO_2$ suspensions corresponding to a bacterial concentration of 5.0×10^9 bacterial count/mL. Mix 4.0 mL of 2% $BaCl_2$ solution (v/v) + 96.0 mL of 1% H_2SO_4 solution (v/v).

References

1. Code of Federal Regulations (1983) Title 21, Food and Drugs. Par 129, Processing and Bottling of Bottled Drinking Water. Washington, DC, US Govt. Printing Office,
2. Weber, G. R. and Black, L. A. (1948) Inhibitors for neutralizing the germicidal action of quaternary ammonium compounds. *Soap Sanitary Chem.* **24,** 137–142.
3. American Public Health Association (1978) *Standard Methods for the Examination of Dairy Products*, 14th ed. (Marth, E. H., ed.). Washington DC, American Public Health Association.
4. American Public Health Association (1984) *Compendium of Methods for the Microbiological Examination of Foods*, 2nd ed. (Speck, M. L., ed.). Compiled by the APHA Technical Commitee on Microbiological Methods for Foods. Washington DC, American Public Health Association.

25

Microbiological Analysis of Cosmetics

Anavella Gaitan Herrera

1. Introduction

Cosmetics are products of chemical or natural origin dedicated specifically for use in skin and mucosa *(1)*. The constant development of the cosmetic industry has generated the necessity to carry out microbiological analysis on the raw materials used in the industrial production of cosmetics as well as the final products, with the purpose of obtaining products of good microbiological quality *(2)*.

Cosmetic products are recognized to be substrates for the survival and development of a large variety of microorganisms, since they possess some of the nutrients that facilitate growth such as: lipids, polysaccharides, alcohol, proteins, amino acids, glucosides, esteroids, peptides, and vitamins. Also, the conditions of readiness (oxygenation, pH, temperature, osmotic degree, superficial activity, perfume, and essential oils) present in the cosmetic products favor microbial multiplication *(3)*.

Routine analyses to determine the microbiological quality of a cosmetic product include the following:

Count of mesophilic aerobic microorganisms.
Most probable number (MPN) of total coliforms.
Count of molds and yeasts.
Absence/presence of *Staphylococcus aureus* probe.
Absence/presence of *Pseudomonas aeruginosa* probe.

2. Materials

1. Flask with 90 mL of TAT broth (tripticase azolectin Tween) or Letheen broth (*see* **Note 1**).
2. Tubes with 9 mL TAT broth or Letheen broth.
3. Tripticase soy agar + lecithin and Tween-80 fused and cooled 45°C.
4. Nine tubes of Brilliant Green Bile broth, with fermentation tube (Durham).
5. Oxytetracycline-glucose-yeast (OGY) agar or potato glucose agar.
6. Petri plates with Vogel and Johnson agar.
7. Petri plates with Cetrimide (AlphaBisciences) or Presudosel agar.

From: *Methods in Molecular Biology, vol. 268: Public Health Microbiology: Methods and Protocols*
Edited by: J. F. T. Spencer and A. L. Ragout de Spencer © Humana Press Inc., Totowa, NJ

8. Pipets of 1 mL.
9. Tubes, 16 × 150 mm.
10. Sterile Petri plates.
11. Water bath at 50°C.
12. Incubator.
13. Glass spatula (Drigalsky).

3. Methods

1. Weigh 10 g of the product under aseptic conditions and add the sample to 90 mL of TAT or Leethen broth to obtain a dilution of 10^{-1} (*see* **Note 1**).
2. Make dilutions of 12^{-2} and 10^{-3} in tubes with 9 mL of the same diluter.

Starting from the dilutions, perform the following counts.

1. Count of mesophilic aerobic microorganism (standard plate-count method).
   a. Incubate at 35°C for 48 h. Carry out the reading.
   b. Calculate the colony-forming units (CFU)/mL.
   c. To duplicate sets of Petri plates, pipet 1-mL aliquots from the 10^{-1} dilution and add tripticase soy agar with lecithin and polysorbate 80.
   d. Agitate gently and appropriately.
   e. Add additional layer of agar.
   f. Incubate at 35°C for 24–48 h.
   g. Calculate and report the result per gram or milliliter of analyzed cosmetic product.
2. Test for total coliforms MPN.
   a. Inoculate three replicate tubes of brilliant green bile (BGB) broth per dilution with 1 mL of the previously prepared 1:10, 1:100, and 1:1000 dilutions.
   b. Incubate tubes for 24 and 48±2 h at 35±0.5°C.
   c. Observe all tubes for gas production either in the inverted vial (Durham) or by effervescence produced. Read tubes for gas production after 24 h. Reincubate negative tubes for an additional 24 h.
   d. Report the results as the MPN of total coliforms per gram or milliliter of sample.
3. Count of molds and yeasts.
   a. Pipet 1-mL aliquots from 10^{-1} to 10^{-5} dilutions. Promptly put into Petri plates 10–15 mL of oxytetracycline gentamicin yeast extract glucose (OGY) agar, melted and tempered to 45°C.
   b. Mix and leave to solidify, adding an additional layer of agar (4–5 mL).
   c. Incubate at 22–24°C for 3–4 d.
   d. Count all colonies on plates containing 30–300 colonies; compute the number of yeasts and molds per gram or milliliter of product. Report as CFU per gram or milliliter of sample.
4. Test of absence/presence of *Staphylococcus aureus* (coagulase positive).
   a. Starting from the TAT broth, that was incubated for 4–6 h, inoculate 0.25 mL of a 10^{-1} dilution on each one of two dishes of Vogel and Johnson agar.
   b. Spread the inoculate using a Drigalsky spatula.
   c. Incubate at 35°C for 24 h.
   d. If there is growth of suspicious colonies, confirm by coagulase test or thermostable endonuclease.

5. Test of absence/presence of *Pseudomonas aeruginosa.*
   a. Repeat the previous procedure, sowing in Cetrimide.
   b. Observe the growth and typical characteristics and confirm with an oxidase test (positive).
   c. Report the results as a test of abscence or presence.
   d. Starting from the TAT broth at a dilution of 10:1 that was incubated for 4–6 h, as a revivification procedure, inoculate 0.25 mL on the surface of each of four Petri dishes with Cetrimide or Pseudosel agar.
   e. Carefully spread the inoculate with a Drigalsky spatula.
   f. Incubate for 24 h at 37°C.
   g. Confirm characteristics compatible with *P. aeruginosa* with an oxidase test.

4. Notes

1. Letheen broth: Dissolve 0.7 g lecithin and 5.0 g polysorbate 80 or Tween-80 in 400 mL of hot H_2O and boil until clear. Add 600 mL solution of 5.0 g beef extact, 10.0 g peptone, and 5.0 g NaCl in H_2O, and boil for 10 min. Adjust pH to 7.0 ± 0.2 (with 1 *N* NaOH or 1 *N* HCl). Filter through coarse paper, transfer 10-mL portions to test tubes (16 × 150 mm), and autoclave (20 min at 121°C).

References

1. American Public Health Association (1984) *Compendium of Methods for the Microbiological Examination of Foods*, 2nd ed. (Speck, M. L., ed.). Compiled by the APHA Technical Commitee on Microbiological Methods for Foods. Washington, DC, American Public Health Association.
2. American Public Health Association (1978) *Standard Methods for the Examination of Dairy Products*, 14th ed. (Marth, E. H., ed.). Washington, DC, American Public Health Association.
3. Code of Federal Regulations (1983) Title 21, Food and Drugs. Par 129, Processing and Bottling of Bottled Drinking Water. Washington, DC, US Govt. Printing Office.

26

Helicobacter pylori and Food Products

A Public Health Problem

Anavella Gaitan Herrera

1. Introduction

Helicobacter pylori is a major human pathogen causing gastritis and chronic superficial infection (CSG). It colonizes the stomach of more than 50% of humans and causes disease *(1)*. This microorganism is associated with the gastric antral epithelium in patients with active chronic gastritis, peptic (gastric) or duodenal ulcers, and gastric adenocarcinoma *(2–4)* *H. pylori* is present in feces, sewage, and water *(5)* but is killed by routine chlorination. Therefore, in developing countries *(6)*, consumption of sewage-contaminated drinking water and vegetables may pose a risk *(7)*; properly cooking foods and chlorinating water reduces the risk of transmitting *H. pylori* to humans (*see* **Note 1**). In South America the consumption of raw vegetables fertilized with human feces has been found to be a risk factor for infection, and consumption of water from a municipal supply has been suggested as a risk factor for children *(8)*.

Epidemiological studies have found that *H. pylori* organisms colonize the stomach and duodenum of humans and many animal species and family clusters *(9)*; it is believed to be orally transmitted person to person *(10)*. This transmission is the major, if not exclusive, source of infection.

H. pylori has been detected in the mouth from dental plaque. Recent observations in persons infected with *H. pylori* caused to vomit or have diarrhea showed that an actively unwell person with these symptoms could spread *H. pylori* in the immediate vicinity by aerosol, splashing of vomitus, infected vomitus, and infected diarrhea. In summary, *H. pylori* is usually spread by the fecal-oral route but possibly also by the oral-oral route and the spread of contaminated secretions. Thus, in developing countries, individuals catch *H. pylori* at a very young age from other persons (children) in their environment. In developed countries, *H. pylori* is more difficult to acquire and is usually transmitted from one family member to another, possibly by the fecal-oral

From: *Methods in Molecular Biology, vol. 268: Public Health Microbiology: Methods and Protocols*
Edited by: J. F. T. Spencer and A. L. Ragout de Spencer © Humana Press Inc., Totowa, NJ

route, or by the oral-oral route, e.g., kissing, vomitus. On occasion, transmission occurs from person to person via contaminated endoscopes. Other gastric *Helicobacter*-like organisms have now been observed in a variety of animals ***(11)***, including rodents, primates, swine, and ferrets, but, with the exception of primates and possibly cats, these isolates are clearly different from human isolates ***(12)***. Foodborne transmission would not be unusual.

In the developed world, up to approx 50–90% of the adult population will be infected by the microorganism at the age of 50 yr, *H. pylori* affects about 20% of persons younger than 40 yr old. Children acquire the infection soon after being weaned. A 10% acquisition rate per year for children between 2 and 8 yr of age, for example, is found in West Africa; 80% of the children may be infected with the organism by age 5 yr. The overriding association with *H. pylori* is lower socioeconomic status during childhood.

The genus *Helicobacter* has grown steadily in the past few years; some of the new species have been reassigned from *Campylobacter*. The genus now includes *H. pylori*, *H. heilmanni*, *H. muridarum*, *H. nemestrinae*, *H. mustelae*, *H. felis*, *H. acinonyx*, *H. pamatensis*, *H. bilis*, *H. canis*, *H. hepaticus*, *H. pullorum*, *H. cinaedi*, and *H. fennelliae*. However, the only human pathogenic species described so far are *H. pylori* and *H. heilmanni*.

Helicobacter cells have blunted/rounded ends in gastric biopsy specimens. *H. pylori* organisms are microaerophilic, nonsporulating (characteristics that facilitate penetration and colonization of mucosal environments), Gram-negative spiral bacteria or curved rods, 3.5 μm long and 0.5–1 μm wide, with a spiral periodicity in fresh cultures on agar medium and spherical (coccoid) forms in older or prolonged cultures (*see* **Notes 2** and **3**). *H. pylori* further differs from *Campylobacter* species in having multiple polar sheathed flagellae; *H. pylori* are highly motile by means of lophotrichous flagella (tufts of flagella at poles of cell). *H. fennellae* and *H. cinaedi* have a single polar flagellum (electron microscopy reveals between four and six flagella positioned on one pole), a unique composition of cell wall with unusual fatty acids, and a smooth surface ***(11)***.

Taxonomy is based on 16S rRNA sequencing, DNA hybridization, genus-specific probes, cell wall protein and lipid characterization, serological and biochemical analyses. *Campylobacter* and *Helicobacter* (formerly *Campylobacter*) are the most clinically important members of the rRNA superfamily Recently defined family members of Campylobacteriaceae include the genera *Campylobacter*, *Arcobacter*, *Helicobacter*, *Wolinella*, and *Flexispira*, now considered a phylogenetically distinct family, and as yet unnamed *Helicobacter* (17 species by rRNA sequencing) human pathogens. Only three species are currently considered to be human pathogens: *H. pylori*, *H. cinaedi*, and *H. fennelliae*, formerly called *Campylobacter*-like organisms (CLOs).

2. Materials

1. Oxidase test.
2. Chocolate or blood agar plates. Blood agar containing an antifungal agent.
3. Medium containing urea and phenol red indicator.

4. Bismuth salts.
5. Sensidisc of erythromycin, rifampin, tetracycline, metronidazole, nalidixic acid, and sulfonamides.
6. Samples of endoscopic antral gastric biopsy material, oral secretions, dental plaque, and human feces. Gastric tissue or samples from biopsies of esophageal or duodenal tissue containing gastric metaplasia.
7. Transport media for culture: Nutrient, thioglycolate, *Brucella*, brain-heart infusion, supplemented tryptic soy broths, and semisolid agars (e.g., Wang's blood-supplemented medium) have all been used for transport culture. Insensitive complex basal media (agar or broth) supplemented with whole blood, heme, serum, charcoal, cornstarch, or egg yolk emulsion *(5)*.
8. Microaerobic conditions (C-rich atmosphere for broth culture 15–17% oxygen and 6–10% carbon dioxide).
9. Columbia agar with 5% horse blood and triphenyltetrazolium.

3. Methods

3.1 Culture Characteristics

1. *H. pylori* has been cultured from both the oral cavity and feces, but the typical site is the gastric mucosa. It can be isolated from the stomach in individuals whose fecal and oral cultures are negative, but it is always found in the stomach when the other two sites are positive.
2. Samples can be simply smeared onto blood agar containing an antifungal agent (which is required to avoid fungal overgrowth) and incubated in a humid atmosphere.
3. The culture conditions are similar to those used for *Campylobacter*, i.e., a reduced oxygen atmosphere of about 5% with 10% carbon dioxide and the balance either nitrogen or hydrogen (*see* **Note 4**). Hydrogen may offer better recovery of injured organisms.
4. Cultures should be incubated at 37°C for a minimum of 5 d. *H. pylori* produces urease and does not grow when incubated below 30°C.
5. Growth is best on chocolate or blood agar plates after incubation for 2–5 d; for liquid media, either a blood or a hemin source appears to be essential. On Columbia agar, small colonies appear after 5 d of incubation. The colonies have a yellow sheen, which can aid recognition during primary isolation using nonselective media such as chocolate agar, or antibiotic-containing selective media, such as those of Skirrow or Goodwin. Spiral organisms that are oxidase-, catalase-, and urease-positive can be identified as *H. pylori*. Culture allows determination of antimicrobial susceptibilities.
6. As transport media for culture, nutrient, thioglycolate, *Brucella*, brain-heart infusion, supplemented tryptic soy broths, and semisolid agars (e.g., Wang's blood-supplemented medium) have all been used. Insensitive complex basal media (agar or broth) supplemented with whole blood, heme, serum, charcoal, cornstarch, or egg yolk emulsion may also be used.
7. Grow optimally under microaerobic conditions (C-rich atmosphere for broth culture).

3.2. Biochemical Characteristics

H. pylori is oxidase positive but nonsaccharolytic and does not reduce nitrate. Two enzymes of *H. pylori*, urease ***(13)*** and catalase, are unusual among medically important bacteria because of their high activity and can be used in the identification of *H. pylori*.

Abundant quatities of urease are produced only by gastric strains. Abundant quantities of mucinase and catalase are produced.

3.3. Detection Methods

1. Detection of urease activity in *H. pylori* can be used as a bedside test by applying a gastric biopsy sample to a simple medium containing urea and phenol red indicator.
2. In the presence of *H. pylori* urease the ammonia released by splitting urea raises the pH of the medium, and this is detected by a color change of the phenol red.
3. The speed with which this reaction occurs in *H. pylori* is highly unusual, and results are obtained in minutes at room temperature.
4. In vitro susceptibility to erythromycin, rifampin, tetracycline, and metronidazole is present.
5. Resistance to nalidixic acid and sulfonamides is present.

3.4. Laboratory Identification

1. *Helicobacter* organisms can be recovered from or detected in endoscopic antral gastric biopsy material; multiple biopsies should be taken.
2. Characteristically shaped organisms can be seen at high magnification with silver-stained hematoxylin and eosin-stained or methylene blue-stained biopsy specimens of the gastric antrum (antral biopsies) in the mucosa adherent to the surface epithelium or pit epithelial cells deep within the crypts.
3. Organisms are always found adjacent to gastric mucosal cells.

4. Notes

1. Susceptibility to removal or inactivation by conventional water treatment processes. There is no direct evidence for the effectiveness of treatment processes for removal of *H. pylori*, although they are likely to be removed to a similar extent as other bacteria. On the basis of their similarity to *Campylobacter* organisms, they should be susceptible to inactivation by disinfectants such as chlorine and ozone.
2. A coccal form of *H. pylori* has been described, although speculation continues as to the function, if any, and indeed the actual viability of these forms. They may mimic the so-called nonculturable but viable forms.
3. *H. pylori* changes from helical to coccoid when exposed to adverse conditions (oxygen or upon prolonged culture). The coccoid forms remain viable, at least for a short time.
4. *H. pylori* can grow, although relatively poorly, in atmospheric oxygen.

References

1. Marshall, B. J., Barrett, L., Prakash, C., McCallum, R. W., and Guerrant, R. L. (1990) Urea protects *Helicobacter* (*Campylobacter*) *pylori* from the bactericidal effect of acid. *Gastroenterology* **99,** 697–702.
2. Blaser, M. J. (1997) The versatility of *Helicobacter pylori* in the adaptation to the human stomach [Review]. *J. Physiol. Pharmacol.* **48,** 307–314.
3. Marshall, B. J. and Warren, J. R. (1984) Unidentified curved bacilli in the stomach of patients with gastritis and peptic ulceration. *Lancet* **1,** 1311–1315.
4. Parsonnet, J., Shmuely, H., and Haggerty, T. (1999) Fecal and oral shedding of Helicobacter pylori from healthy infected adults. *JAMA* **282,** 2240–2245.

5. Westblom, T. U., Madan, E., and Midkiff, B. R. (1991) Egg yolk emulsion agar, a new medium for the cultivation of Helicobacter pylori. *J. Clin. Microbiol.* **29,** 819–821.
6. Shimoyama, T., Tominaga, Y., Sakagami, T., and Fukuda, Y. (1999) Epidemiological study for infection with *H. pylori* in Japan compared with that in USA, Europe and Asian Pacific area [Review]. *Nippon Rinsho* (Jpn J. Clin. Med.) **57,** 11–16.
7. Hulten, K., Han, S. W., Enroth, H., et al. (1996) *Helicobacter pylori* in the drinking water in Peru. *Gastroenterology* **110,** 1031–1035.
8. Webb, P. M., Knight, T., Greaves, S., et al. (1994) Relation between infection with Helicobacter pylori and living conditions in childhood: evidence for person to person transmission in early life. *BMJ* **308,** 750–753.
9. Samuels, A. L., Windsor, H. M., Ho, G. Y., Goodwin, L. D., and Marshall, B. J. (2000) Culture of Helicobacter pylori from a gastric string may be an alternative to endoscopic biopsy. *J. Clin. Microbiol.* **38,** 2438–2439.
10. Li, C., Ha, T., Ferguson, D. A., Jr., et al. (1996) A newly developed PCR assay of *H. pylori* in gastric biopsy, saliva, and feces. Evidence of high prevalence of *H. pylori* in saliva supports oral transmission. *Dig. Dis. Sci.* **41,** 2142–2149.
11. Lee, L., Phillips, M. W., O'Rourke, J. L., et al. (1992) *Helicobacter muridarum* sp. nov., a microaerophilic helical bacterium with a novel ultrastructure isolated from the intestinal mucosa of rodents. *Int. J. Syst. Bacteriol.* **42,** 27–36.
12. Dore, M. P., Sepulveda, A. R., Osato, M. S., Realdi, G., Graham, D. Y. (1999) *Helicobacter pylori* in sheep milk. *Lancet* **354,** 132.
13. Luck, J. M. and Seth, T. N. (1924) Gastric urease. *Biochem. J.* **18,** 1227–31.

27

Microbiological Test for Sanitation of Equipment in the Food Factory

Anavella Gaitan Herrera

1. Introduction

Microbiological sampling of utensils, tableware, and kitchenware, in addition to equipment, permits objective evaluation of sanitation practices and procedures used for these items from food service operations *(1)*.

RINSE SOLUTION METHOD

2. Materials

1. Sterile Petri dishes.
2. Sterile pipets.
3. Stock phosphate buffer solution.
4. 70% Ethanol.
5. Plate count agar (PCA), Violet Red Bile agar (VRBA), MF Endo broth.
6. Buffered rinse solution: add 1.25 mL of stock phosphate buffer solution, 5 mL of 10% aqueous sodium thiosulfate ($Na_2S_2O_3 \cdot 5H_2O$), 4 g azolectin (*see* **Note 1**), and 10 g Tween-20 (or Tween-80) to distilled water, and make up to 1 L. Dispense in screw-capped vials made up to contain, after sterilization, 20 mL, 100 mL, or other volumes needed. Weigh the powder and dissolve rapidly by heating over boiling water (*see* **Note 2**).
7. Sodium thiosulfate solution 10%: dissolve 100 g of $Na_2S_2O_3 \cdot 5H_2O$ in distilled water, and make up to 1 L, filter, and store in the refrigerator or dark place (or in an amber bottle).

3. Methods

1. Remove container cartons or containers from the conveyor line.
2. Add 20 mL of sterile buffered rinse solution to each container and aseptically reap the container (*see* **Note 3**).
3. Shake each container vigorously 10 times through a 20-cm arc.
4. Turn the container 90° and repeat the horizontal shaking treatment. Turn the containers 90° twice more and repeat the horizontal shaking.

From: *Methods in Molecular Biology, vol. 268: Public Health Microbiology: Methods and Protocols*
Edited by: J. F. T. Spencer and A. L. Ragout de Spencer © Humana Press Inc., Totowa, NJ

5. Swirl the container vigorously 20 times in a small circle with the long axis in the vertical position; then invert and repeat.
6. Rinse solutions for container samples are analyzed by distributing a total of 10 mL of the rinse among three sterile Petri dishes. In addition, seed two other Petri dishes with 1-mL portions of the rinse.
7. Pour the plates with 15–18 mL of the desired medium (PCA, VRBA) using appropriate incubation conditions for each (*see* **Note 4**).
8. After incubation, count the colonies. If 20 mL of rinse solution was used, then the total number of colonies on the three plates that received 10 mL of rinse is multiplied by 2, or the total number of colonies on the two plates that received 1 mL each, multiplied by 10, yields the number of colonies per container. When the plates that received 1 mL each show more than 20 colonies, report the counts as more than 5000 per container.

3.1. Interpretation

For interpretations of results, take into consideration the numbers and types of microorganism *(1,2)*.

1. Types of microorganism. The type is important in terms of potential to cause spoilage. The number should be very low; however, under certain conditions (such as aseptic packaging), the microbiological condition of the container is a critical control point.
2. Bacterial standards have been published for multiuse and single containers used for:
   a. Packaging pasteurized milk or milk products. Four containers from any given day are sampled, and three of four samples must meet this standard.
   b. Bottled water. State that at least once each 3 mo, bacteriological swab or rinse samples are to be taken from at least four containers and closures selected just prior to filling and sealing.

The standards require that such containers have a residual bacteria count of ≤ 1 colony/mL of capacity or not >1 colony/cm^2 of product contact surface. No coliform organisms may be present.

3. Processing assemblies (tanks, pipelines, fillers, and so on). Rinse water for large volume rising of equipment may be sterilized or may be sanitized by chlorinating to a residual concentration of 25 mg/L, holding for 10 min, and then neutralizing by adding an excess of sterile 10% sodium thiosulfate solution. Tap water may be used after sterilizing by membrane filtration followed by the addition of sterile 10% w/v sodium thiosulfate to inactivate residual disinfectant (*see* **Note 5** and **6**).

A sufficient volume of treated rinse water is added to the system at the upstream end of the assembly and then pumped or allowed to flow by gravity through the assembly. A control sample (1 L) of the treated water rinse is taken before using the water for rising. Samples of water rinse are collected from the discharge end of the assembly from the first, middle, and final portions of the rinse water. Average the number of colonies obtained from rinse samples taken at the beginning, middle, and end of drainage and subtract the number of colonies obtained from the control samples. Calculate the ratio of sample volume to rinse volume and multiply by the corrected yield to obtain an indication of numbers of organisms present in the entire system *(3)*.

4. Notes

1. Azolectin is hygroscopic and should be stored in a desiccator.
2. Nutrient broth may also be applied as a rinse since it effectively neutralizes residual active chlorine or quaternary ammonium compound. It may be used instead of buffered rinse solution.
3. For containers larger than 1 gallon, use 100 mL of rinse solution.
4. The presence of molds and yeast, proteolytic bacteria, and other microorganisms may be determined by the use of appropriate differential media and incubation temperatures and times.
5. When 100 mL or greater portions of rinse solutions are used, membrane filtration procedures for analysis should be followed (if low levels of contamination are expected)
6. Membrane filtration may be used for analysis of the 20-mL rinse samples.

References

1. American Public Health Association (1978) *Standard Methods for the Examination of Dairy Products*, 14th ed. (Marth, E. H., ed.). Washington DC, American Public Health Association.
2. American Public Health Association (1984) *Compendium of Methods for the Microbiological Examination of Foods*, 2nd ed. (Speck, M. L., ed.). Compiled by the APHA Technical Commitee on Microbiological Methods for Foods. Washington DC, American Public Health Association.
3. Code of Federal Regulations (1983) Title 21, Food and Drugs. Par 129, Processing and Bottling of Bottled Drinking Water. Washington, DC, US Govt. Printing Office.

Surface Contact Methods

1. Introduction

Swab and replicate organism direct agar contact (RODAC) is the customary and appropriate method for microbiological examination of surfaces.

Swab procedures should be used for areas on equipment such as cracks, corners, or services and for sampling utensils, tableware, kiltchenware, and so on. RODAC should only be utilized on flat, impervious surfaces that are easy to clean and disinfect.

The levels of microorganisms should not exceed some number of colonies per sampling site; the type of microorganism may be more important than the number in terms of potential spoilage of the finished product. For example, the presence of very low numbers (1 or <10 CFU) of *Pediococcus* spp. or *Xantomonas* spp. in a brewery may be highly significant with respect to potential spoilage of the finished product.

The United States Public Health Service recommends standards for utensils and equipment of not more than 100 colonies per utensil or surface area of equipment swabbed.

Swab Contact Method 1

This method is used for sampling by the cellulose ***(1)*** sponge swab technique. Large areas, such as cleaned and sanitized floors and other areas (where high residual microbial levels are expected), may be sampled with a cellulose sponge, a method useful for detecting *Salmonella* and other pathogens on food plant equipment and in the environment (*see* **Note 1**). This technique can be used to evaluate the efficacy of cleaning and sanitizing procedures for environment when pathogens microorganisms such as *Salmonella* exist ***(2–4)***. A 4–5 log cycle reduction in the residual microbial level should be obtained on surfaces after cleaning and sanitizing. The results should be negative after the correct application of sanitation procedures (*see* **Note 2**).

2. Materials

1. Cellulose sponges cut into 5 × 5-cm pieces placed in individual bags and autoclaved.
2. Sterile buffered solution for the rinsing agent (*see* **Note 3**).
3. Sterile plastic bags suitable to contain the sponge sample.
4. Sterile crucible tongs, and sterile gloves to hold the sponge aseptically during sampling.

3. Methods

1. Hold the sponge aseptically with sterile gloves and swab the surface to be sampled by vigorously rubbing the sponge over the specific area.
2. Area of several square meters may be effectively sampled with the swabbed method (*see* **Note 4**).
3. Aseptically place the sponge into a sterile plastic bag.
4. Transport under refrigeration.
5. Moisten the sponge with 10 mL of the chosen rinsing agent.
6. Quantitative analysis: take 50–100 mL of diluent, and add it to the bag containing the sponge. Mix vigorously for 1 min, plate, and incubate under specific conditions and appropriate selective differential media.
7. Calculation. Calculate the number of microorganisms per unit of surface (on the basis of the size of aliquot plated, the amount of diluent used, and the area swabbed).
   Example: 50 colonies are obtained from a 1-mL aliquot derived from the sponge in 100 mL of the rising agent.
   Swabbed area: 1 m^2
   The result is 5000 CFU/m^2.

4. Notes

1. If the surface to be sampled is flat, the rinse solution may be applied directly to the surface and then taken up into the sponge by the rubbing action.
2. This method would be used for development of a program to eradicate pathogenic microorganisms from the environment of the food factory.
3. Nutrient broth or 0.1% peptone water should be used as the rinsing agent.
4. When the surface to be sampled contains fatty materials, use as the rinsing agent 0.5% tergitol anionic or 0.5–1.0% Tween-80.

References

1. Silliker, J. H. and Gabis, D. S. (1975) A cellulose sponge sampling technic for surfaces. *J. Milk Food Technol.* **38,** 504.
2. American Public Health Association (1978) *Standard Methods for the Examination of Dairy Products*, 14th ed. (Marth, E. H., ed.). Washington DC, American Public Health Association.

3. American Public Health Association (1984) *Compendium of Methods for the Microbiological Examination of Foods*, 2nd ed. (Speck, M. L., ed.). Compiled by the APHA Technical Commitee on Microbiological Methods for Foods. Washington DC, American Public Health Association.
4. Code of Federal Regulations (1983) Title 21, Food and Drugs. Par 129, Processing and Bottling of Bottled Drinking Water. Washington, DC, US Govt. Printing Office.

Swab Contact Method 2

2. Materials

1. Screw-capped (7–10-cm-long) tubes to contain 5 mL of buffered rinse solution *(1)*, used for swabs *(2)* made from calcium alginate fibers. All organisms dislodged from swabbed surfaces are thus liberated (*see* **Note 1**).
2. Sterile nonabsorbent cotton swabs with the head firmly twisted to 0.5 cm in diameter by 2 cm long on an applicator stick 15 cm long (*see* **Note 2**).

3. Methods

3.1. Sampling Procedure (3,4,5–8)

3.1.1. Equipment Surfaces

1. Open the sterile swab (*see* **Note 3**), grasp the end of a stick, not touching any portion that might be inserted into the vial, and remove the swab aseptically (*see* **Note 4**).
2. Open a buffered rinse solution, and moisten the swab head (pressing out the excess solution against the interior wall of the vial with a rotating motion).
3. Make a 30° angle contact with the surface. Rub the swab head slowly and thoroughly over 50 cm^2 of surface three times, reversing direction between successive strokes. Move the swab on a path 2 cm wide by 25 cm long or other dimensions to cover an equivalent area.
4. Return the swab head to the solution. Rinse briefly in the solution and press out the excess (*see* **Note 5**).

3.1.2. Unmeasured Surface Areas

Pump impellers, gaskets, rings, valve seats, filler nozzles, and so on, have been swabbed, the results are reported on the basis of the entire sampling site instead of a measured area.

3.1.3. Utensils To Be Examined

1. At least glasses, cups, and spoons *(3,5)* should be examined. Four of each should be selected at random from where clean utensils are stored. If a direct check of dishwashing efficacy is desired, utensils should be selected from those recently washed.
2. Use one swab for each four or more similar utensils.
3. The significant surfaces of utensils consist of the upper 1.5 cm of the inner and outer rims of cups and glasses and the entire inner and outer surfaces of the bowls and spoons.

3.1.4. Forks and Surfaces of Dishes

The area to be swabbed should include the entire inner and outer surfaces of the tines of forks and the inner surfaces of plates and blows.

After the fifth area has been swabbed, position the swab head in the vial, and break or cut it with sterile scissors or another device *(2)*, leaving the swab head in the vial. Replace the screw cap, put the vial in a waterproof container packed in cracked ice or other suitable refrigerant, deliver to the laboratory, and analyze within 24 h.

3.2. Plating Swab Rinse Solutions

1. Remove the vial from refrigerated storage.
2. Shake, making 50 complete cycles of 15 cm in 10 s.
3. Plate 1.0- and 0.1-mL portions of rinse solution (with appropriate media, depending on the organisms of interest).
4. Incubate plates, count colonies, and calculate the number of colonies recovered from 50 cm^2 (equivalent to 1 mL of rinse).
5. For utensils, report the count as the residual bacterial count per utensil examined.

3.3. Calculation

1. Swab three utensils.
2. Sample 1 mL of the diluent (of the 5 mL).
3. Results are of 150 colonies.
4. Count per utensil is 175.
5. For satisfactory results in adequately cleaned and sanitized food service equipment on which five areas of 50 cm^2 have been swabbed, residual bacterial count should not exceed 500.

4. Notes

1. The swabs made of calcium alginate are soluble in aqueous solutions containing 1% of sodium hexametaphosphate, sodium glycerophosphate, sodium citrate, or any mixture of these.
2. Rayon and dacron swabs may also be used.
3. The swabs may be sterilized in the laboratory.
4. Swab four more 50-cm^2 areas of the surface being sampled, rinsing the swab in the solution after each swabbing (removing the excess).
5. Care should be taken to prevent contamination by handling during sampling.

References

1. Buchbinder, L. T. C., Buck, R. Jr., Phelps, P. M., Stone, R. V., and Tiedman, W. D. (1947) Investigation of the swab rinse technic for examination eating and drinking utensils. *Am. J. Public Health* **37,** 373–378.
2. Buck, T. C. Jr. and Kaplan, E. (1944) A sterile cutting device for swab vial outfits utilizing wood applicators. *J. Milk Technol.* **7,** 141–142.
3. American Public Health Association (1978) *Standard Methods for the Examination of Dairy Products*, 14th ed. (Marth, E. H. ed.). Washington DC, American Public Health Association.
4. American Public Health Association (1984) *Compendium of Methods for the Microbiological Examination of Foods*, 2nd ed. (Speck, M. L. ed.). Compiled by the APHA Technical Commitee on Microbiological Methods for Foods. Washington DC, American Public Health Association.

5. Code of Federal Regulations (1983) Title 21, Food and Drugs. Par 129, Processing and Bottling of Bottled Drinking Water. Washington, DC, US Govt. Printing Office.
6. Tiedman, W. D., chairman (1984) Technic for the bacteriological examination of food utensils. Committee report. Am. J. Public Health Yearbook 1947–48. Part 2, pp. 68–70.
7. US Department of Health, Education, and Welfare. Public Health Service (1967) *Procedure for the Bacteriological Examination of Food Utensils and/or Food Equipment Surfaces*. Public Health Service Publication No. 1631, Technical Information Bulletin No. 1. Washington, DC, Food and Drug Administration,
8. Walter, W. G. and Potter, J. (1963) Bacteriological field studies on eating utensils and fat surfaces. *J. Environ. Health* **26,** 187–197.

Agar Contact Method
Replicate Organism Direct Agar Contact (RODAC)

1. Introduction

The RODAC plate method is a simple and valuable technique for estimating the sanitary quality or surfaces in food factories and ideally should be used on previously cleaned and sanitized surfaces *(1,2)*. This method is indicated when quantitative date are sought from flat, impervious surfaces (It should not be used for irregular, pervious, or creviced surfaces) *(3)*. Colony counts can be made only on plates with 200 or fewer colonies *(4)*.

2. Materials

1. Plastic RODAC plates with 15.5–16.5 mL of the specific media. Plate count agar is used for total counts, and differential medium may be used if qualitative values are sought (*see* **Note 1**).
2. When carried out on surfaces previously subjected to chemical microbicide treatment, added neutralizers are needed for the media *(5,6)*: 0.5% polysorbate, which neutralizes some substituted phenolic disinfectants, and Tween-80 plus 0.07% soy lecithin, which neutralizes quaternary ammonium compounds.

3. Methods

1. Remove the plastic cover and carefully press the agar surface to the surface being sampled. Make certain that the entire agar meniscus contacts the surface, using a rolling uniform pressure on the back of the plate to effect contact (*see* **Note 1**).
2. Replace the cover and incubate in an inverted position for 24–48 h. A 24-h incubation period when heavy contamination is present and 48 h when contamination is light should be allowed (*see* **Note 2**).
3. Colonies should be recorded as the number of colonies per RODAC plate or number of colonies per cm^2.

4. Notes

1. The meniscus of the agar should rise above the rim of the plate to give a slightly convex surface. This is very important so the agar makes proper contact with the surface to be sampled.
2. Following preparation of the plates, incubate at 32°C for 18–24 h as a sterility check.

They should be used within the succeeding 12 h unless they are wrapped and refrigerated.

References

1. American Public Health Association (1978) *Standard Methods for the Examination of Dairy Products*, 14th ed. (Marth, E. H., ed.). Washington DC, American Public Health Association.
2. American Public Health Association (1984) *Compendium of Methods for the Microbiological Examination of Foods*, 2nd ed. (Speck, M. L., ed.). Compiled by the APHA Technical Commitee on Microbiological Methods for Foods. Washington DC, American Public Health Association.
3. Angeloti, R. J., Wilson, L., Litsky, W., and Walter, W. G. (1964) Comparative evaluation of the cotton swab and RODAC methods for the recovery of *Bacillus subtilis* spore contamination from stainless steel surfaces. *Health Lab. Sci.* **1,** 289–296.
4. Code of Federal Regulations (1983) Title 21, Food and Drugs. Par 129, Processing and Bottling of Bottled Drinking Water. Washington, DC, US Govt. Printing Office.
5. Engley., F. B. and Dey., B. P. (1970) A universal neutralizing medium for antimicrobial chemicals. Chemical Specialities Manufacturer´s Association, Proceedings of the 56th Meeting, New York.
6. Weber, G. R. and Black, L. A. 1948. Inhibitors for neutralizing the germicidal action of quaternary ammonium compounds. *Soap Sanitary Chem.* **24,** 137–142.

28

Spectrophotometric Determination of Histamine in Fisheries Using an Enzyme Immunoassay Method

Tânia L. P. Pessatti, José D. Fontana, and Marcos L. Pessatti

1. Introduction

The biogenic amines are low-weight organic bases that exhibit variable biological activity. Approximately 30 vasoactive and psycoactive amines have been found in foodstuffs *(1)*. Histamine, tyramine, phenylethylamine, tryptamine, cadaverine, putrescine, spermine, and spermidine are examples of biogenic amines detectable in meat and its derivatives *(2)*. These amines occur naturally in animals, plants, and microorganisms owing to their natural metabolic processes, and are usually formed by decarboxylation of amino acids.

Histamine, 5-imidazol-ethylamine (MW 111), is produced by oxidative decarboxylation of histidine, and decarboxylase-positive microorganisms are able to produce it as a byproduct of their action on host tissues. This process, associated with enterobacteria such as *Escherichia*, *Salmonella*, *Clostridium*, and *Bacillus*, can be fast if the meat storage conditions are inadequate *(3)*.

Additionally, because of its nonvolatile behavior, histamine may bestow toxicity on the product even before it is considered decayed or organoleptically unacceptable. Secondary amines such as piperidine and pyrrolidine can be produced in the catabolism of histamine, and they, in turn, are precursors of *nitrosamines*, substances with known carcinogenic activity *(4)*. Some cases of histamine toxicity have been reported *(5–7)*.

Therefore, the histamine content of fishery products has been proposed as an index for microbial deterioration, caused by low-quality fish or inadequate processing. The evaluation of histamine levels has been demonstrated to be an important indicator in the quality control of both fresh and processed fish *(8,9)*, particularly fish of the Scombridae family, which exhibit elevated levels of histidine in their tissues.

From: *Methods in Molecular Biology, vol. 268: Public Health Microbiology: Methods and Protocols*
Edited by: J. F. T. Spencer and A. L. Ragout de Spencer © Humana Press Inc., Totowa, NJ

For these reasons, legal regulations were implemented to monitor the histamine content of fish, with the maximum limit being set from 50 to 200 mg histamine/kg fish (50–200 ppm) of sample, varying by country ***(10)***. The values reported for fish in an advanced stage of decomposition varied from 100 to > 4000 mg/kg ***(11)***.

Some analytical methods for the determination of histamine have been developed employing two main techniques: spectrofluorometry ***(9,12)*** and/or chromatography (liquid ***[13]***, liquid-gas ***[1]***, or in thin layer ***[14]***).

Fluorometric determination, based on histamine derivation with *o*-phthalaldehyde (OPA) ***(11,15)***, has no specificity since histidine can also react with the fluorogenic reagent, which thus interferes with the reaction. Even though these methods have been modified and evolved, they are still time-consuming and expensive.

In chromatographic determination, the reagents also do not exhibit specific reactivity and therefore require a laborious extraction step to eliminate possible interfering substances, specially histidine itself, peptides with histidine, and other amines ***(16,17)***. This can make this methodology less practical, particularly when a fast, trustworthy result is needed and when combining for a large number of samples is required.

More recently, a method of histamine determination by immunoassay has been developed. It consists of competitive binding between a monoclonal antibody coated on the wells in a microplate and modified histamine (acylated) and a conjugate (histamine bound to alkaline phosphatase). Upon alkaline phosphatase catalysis, the *p*-nitrophenyl phosphate substrate (*p*-NPP) releases free phenol, which develops a yellow coloration of an intensity inversely related to the histamine concentration in the sample. This immunoassay method has been shown to be more efficient, less expensive, faster, and sensitive (1 mg histamine/kg fishery = 1 ppm).

This chapter deals with the use of an immunoassay method (the Histamarine® kit) to evaluate the levels of histamine in skipjack tuna samples (*Katsuwonus pelamis*), in order to monitor the effect of storage temperature on histamine levels, color determination being carried out by spectrophotometry.

2. Materials

1. The object of this study was a marine fish species of economic interest and widespread commercialization, the skipjack tuna (*Katsuwonus pelamis*). The fish were caught by a local fishery in the Atlantic Ocean in the zone of Itajaí Port, Santa Catarina State, Southern Brazil and maintained on crushed ice for 5 d.
2. Once transported to the laboratory, tuna samples were immediately cut longitudinally to get only the dorsal muscle. Each piece was transversely sectioned in three slices of approx 2 cm each and then separately stored in sterile plastic bags at –20°C or 0°C or 8°C, respectively. The procedures for manipulating the samples were aseptic to avoid cross-contamination.
3. All reagents used for sample acylation and immunoassay were from the Histamarine® kit (Immunotech, France) (*see* **Note 1**).
4. Other materials required for the immunoassays but not provided by the kit:
   a. Precision balance (e.g., AG204 from Mettler Toledo, Switzerland).
   b. Homogenizer (*see* **Note 2**).

c. Spectrophotometer (e.g., CE1020 from CECIL, Cambridge) or microplate reader (λ_{max} = 405–414 nm).
d. Plastic tubes (*see* **Note 3**).
e. Variable volume micropipets (20, 200, and 1000 µL).
f. Gloves (*see* **Note 4**).
g. 25-mL and 1-L Graduated cylinders.

3. Methods

3.1. Sample Preparation and Histamine Extraction

1. Take a 10-g aliquot of each slice from fish dorsal muscle for analysis at time zero, and store the remainder at –20°C, 0°C, and 8°C, respectively.
2. For histamine extraction, put each sample into a sterile bag with 80 mL of iced distilled water (the amount of muscle sample may vary between 1 and 10 g, but the relationship sample(g): distilled water [mL] must be 1:8) and homogenize with a BagMixer® 400 (Interscience, France) for 10 min. Filter the homogenate with a paper filter (2×), which results in a translucent liquid (filtrate) that is then used in the histamine analysis. Alternatively, the double-filtration step may be replaced by centrifugation (10,000*g* for 5 min).

3.2. Sample Acylation

Acylate the filtrate or supernatant using the reagent supplied in the commercial kit, as follows: Add 180 µL of the acylation buffer to a sterile (or sufficiently clean) plastic tube, followed by 20 µL of the extracted sample and 50 µL of the acylation reagent. This reaction is instantaneous, and the product is quite stable (for 48 h at 4°C).

3.3. Enzyme Immunoassay

1. Wash procedure.
   a. Put 50 µL of either the acylated sample or the standard in the microplate wells coated with the high-affinity specific monoclonal antibody (antihistamine), followed by 200 µL of the alkaline phosphatase-histamine conjugate.
   b. After 30 min, wash the wells with buffered water wash solution. For this washing step, turn the microplates upside down and shake vigorously over the sink; use a micropipet (or a squeeze bottle) to fill the wells carefully with wash solution. Repeat this procedure three times (*see* **Note 5**).
   c. Finally, tap the wells firmly on clean absorbent paper, still in the upside-down position.
   d. The wash procedure may also be performed with an automated microplate washer machine.
2. After the washing step, add 200 µL of the *p*-NPP substrate and incubate the solution for 30 min at room temperature (18–25°C), covering the microplate to avoid evaporation. Stop the reaction with 50 µL of 1 *N* sodium hydroxide.
3. To determine the absorbance of the solutions in the wells, siphon the respective volumes individually using a micropipet and then read them in a spectrophotometer at 405 nm using a microcuvette of silica (quartz) with a maximum capacity of 200 µL. Ideally, the solution absorbances must be determined in a microplate reader at 405–414 nm. Then compare each absorbance reading with the calibration curve using a semi-log scale (**Fig. 1**) to obtain the histamine concentration directly from the graph and/or by the equation $y = 2.3609\,x^{-0.3101}$ ($R^2 = 0.9949$), derived from the data. Reading the microplates can be delayed up to a maximum of 2 h.

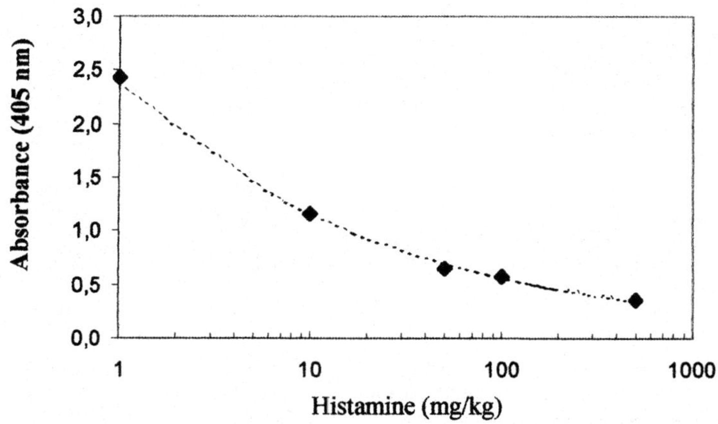

Fig. 1. Calibration curve of histamine at 405 nm. The exponential equation calculated was $y = 2.3609\, x^{-0.3101}$, with $R^2 = 0.9949$.

4. Analysis of results.
   a. **Figure 1** shows that the sensitivity limit of the calibration curve is 1 ppm (1 mg histamine/kg fish), even though the detection limit of the method is supposed to be 0.5 ppm.
   b. As the histamine levels must be below 50–200 ppm (mg/kg), depending on local legislation, the calibration curve demonstrates that this immunotechnique is an efficient methodological tool to control fishery quality in terms of onset of putrefaction.
   c. To provide examples, histamine analyses were done on fish samples (*Katsuwonus pelamis*) stored at temperatures of –20°C, 0°C, and 8°C (**Fig. 2**).
   d. The results show that temperature is a determining factor in the quality of fishery products, especially those that exhibit a flora positive for histidine decarboxylase, such as the Scombridae (tuna).
   e. Furthermore, the results show that at freezing temperatures (–20°C and 0°C) histamine levels were almost constant during the 50-d storage. In the fish kept at 8°C, the histamine levels increased quickly. From the 10th d onwards the histamine level exceeded the legal limit and reached nearly 1000 ppm, meaning that fish stored at 8°C could not be considered safe for human consumption.

4. Notes

1. The Histamarine kit contains the following: microplate, standards (1, 10, 50, 100, and 500 mg/kg [ppm], dimethyl sulfoxide [DMSO]), acylation reagent (to be dissolved in DMSO), acylation buffer (stored at 2–8°C), conjugate (dissolved in distilled water), wash solution (diluted in distilled water), substrate buffer, substrate tablets of *p*-NPP (to be diluted in substrate buffer; the tablets must be handled with forceps, since contact with skin may affect the product adversely), and stop solution (1 *N* NaOH). All solvents, reagents, resulting solutions, and microplates are stored at 2–8°C or frozen at –20°C until further use, at which time they are carefully thawed to room temperature. All reagents in the unopened kit are stable at 2–8°C until the expiration date, and it is recommended to allow the components of the kit to reach room temperature before use.

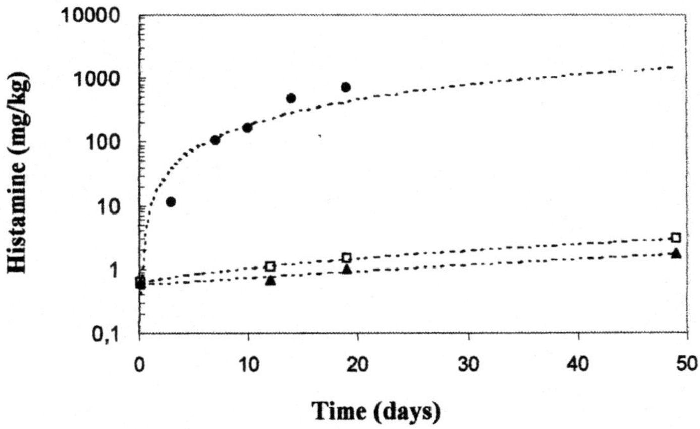

Fig. 2. Determination of histamine levels in samples of skipjack tuna (*Katsuwonus pelamis*) stored at –20°C (▲), 0°C (□), and 8°C (●).

2. Homogenization is carried out in a solid sample blender, e.g., BagMixer 400, but a common blender may be substituted.
3. Care should be taken to avoid glass tubes since histamine adsorbs to glass.
4. Direct contact of samples with the naked hands should be avoided since histamine may be present in human secretions such as sweat.
5. Careful washing is important since incomplete removal of unbound conjugate will lead to high background levels.

Acknowledgments

The authors thank Central de Produtos Enzimáticos e Imunológicos Ltda. (CPEI) for support with the Histamarine kit, CCTMAR-UNIVALI-SC for funding, and Prof. A. Barreto for reading the manuscript.

References

1. Zee, J. A., Simard, R. E., and L'Heureux, L. (1983) Evaluation of analytical methods for determination of biogenic amines in fresh and processed meat. *J. Food Protection* **46,** 1044–1049.
2. Halász, A., Ágnes, B., Simon-Sarkadi, L., and Holzapfel, W. (1994) Biogenic amines and their production by microorganisms in food. *Trends Food Sci. Technol.* **5,** 42–49.
3. Wei, C. I., Chen, C.-M., Koburger, J. A., Otwell, W. S., and Marshall, M. R. (1990) Bacterial growth and histamine production on vacuum packeged tuna. *J. Food Sci.* **55,** 59–63.
4. Silla Santos, M. H. (1996) Biogenic amines: their importance in foods. *Int. J. Food. Microbiol.* **29,** 213–231.
5. Arnold, S. H. and Brown W. (1978) Histamine toxicity from fish products. *Adv. Food Res.* **24,** 113–154.
6. Muller, G. J., Lamprecht, J. H., Barnes, J. M., De Villiers, R. V., Honeth, B. R., and Hoffman, B. A. (1992) Scombroid poisoning. Case series of 10 incidents involving 22 patients. *S. Afr. Med. J.* **81,** 427–430.

7. Becker, K., Southwick, K., Reardon, J., Berg, R., and MacCormack, J. N. (2001) Histamine poisoning associated with eating tuna burgers. *JAMA* **285,** 2977–2978.
8. Mietz, J. L. and Karmas, E. (1978) Polyamine and histamine content of rockfish, salmon, lobster, and shrimp as an indicator of decomposition. *J. AOAC* **61,** 139–145.
9. Vidal-Carou, M. C., Veciana-Nogués, M. T., and Mariné-Font, A. (1990) Spectrofluorometric determination of histamine in fish and meat products. *J. AOAC* **73,** 565–567.
10. Glória, M. B. A. and Soares, V. F. M. (1993) Comparison of fluorometric methods for the determination of histamine in fish. *Arq. Biol. Tecnol.* **36,** 229–235.
11. Staruszkiewicz, Jr., W. F. (1977) Fluorometric determination of histamine in tuna: collaborative study. *J. Assoc. Off. Anal. Chem.* **60,** 1131–1136.
12. Middlebrooks, B. L., Toom, P. M., Douglas, W. L., Harrison, R. E., and Mcdowell, S. (1988) Effects of storage time and temperature on the microflora and amine development in Spanish Mackerel (*Scomberomorus maculatus*). *J. Food Sci.* **53,** 1024–1029.
13. Rönnberg, A. L., Hansson, C., and Hakanson, R. (1984) High-performance liquid chromatographic determination of histamine in biological samples after derivation with *o*-phthalaldehyde. *Anal. Biochem.* **139,** 338–344.
14. Thadhani, V. M, Jansz, E. R., and Peiris, H. (2002) Effect of exogenous histidine and *Garcinia cambogia* on histamine formation in skipjack (*Katsuwonus pelamis*) homogenates. *Int. J. Food Sci. Nutr.* **53,** 29–34.
15. Shore, P. A., Burkhalter, A., and Cohn, Jr., V. H. (1959) A method for the fluorometric assay of histamine in tissues. *J. Pharmacol. Exp. Ther.* **127,** 182–186.
16. Taylor, S. L., Lieber, E. R., and Leatherwood, M. (1978) A simplified method for histamine analysis of foods. *J. Food Sci.* **43,** 247–250.
17. Soares, V. F. M., Vale, S. R., Junqueira, R. G., and Glória, B. A. (1998) Teores de histamina e qualidade físico-química e sensorial de peixe congelado. *Cienc. Tecnol. Aliment.* **18,** 1–15.

29

Flavonoids From Argentine *Tagetes* (Asteraceae) With Antimicrobial Activity

María L. Tereschuk, Mario D. Baigorí, Lucia I. C. de Figueroa, and Lidia R. Abdala

1. Introduction

The flavonoids, constituting one of the most numerous and widespread groups of natural plant constituents, are important to humans not only because they contribute to plant colors but also because many members are physiologically active. These low-molecular-weight substances, found in all vascular plants, are phenylbenzopyrones. Over 4000 structures have been identified in plant sources, and they are categorized into several groups (**Fig. 1**). Primarily recognized as pigments responsible for the autumnal burst of hues and the many shades of yellow, orange, and red in flowers and food, the flavonoids are found in fruits, vegetables, nuts, seeds, stems, flowers, and leaves as well as tea and wine and are important constituents of the human diet *(1)*. They are prominent components of citrus fruits and other food sources. Flavonols (quercetin, myricetin, and kaempferol) and flavones (apigenin and luteolin) are the most common phenolics in plant-based foods. Quercetin is also a predominant component of onions, apples, and berries. Such flavanones as naringin are typically present in citrus fruit, and flavanols, particularly catechin, are present as catechin gallate in such beverages as green or black tea and wine. Some major sources of flavonoids are outlined in **Table 1** *(2)*. The daily intake of flavonoids in humans has been estimated to be approx 25 mg/d, a quantity that could provide pharmacologically significant concentrations in body fluids and tissues, assuming good absorption from the gastrointestinal tract *(3)*.

Biological activity of flavonoids was first suggested by Szent-Gÿorgyi 1938 *(4)*, who reported that citrus peel flavonoids were effective in preventing the capillary bleeding and fragility associated with scurvy. The broad spectrum of biological activity within the group and the multiplicity of actions displayed by a certain individual members make the flavonoids one of the most promising classes of biologically active compounds *(5)*.

From: *Methods in Molecular Biology, vol. 268: Public Health Microbiology: Methods and Protocols*
Edited by: J. F. T. Spencer and A. L. Ragout de Spencer © Humana Press Inc., Totowa, NJ

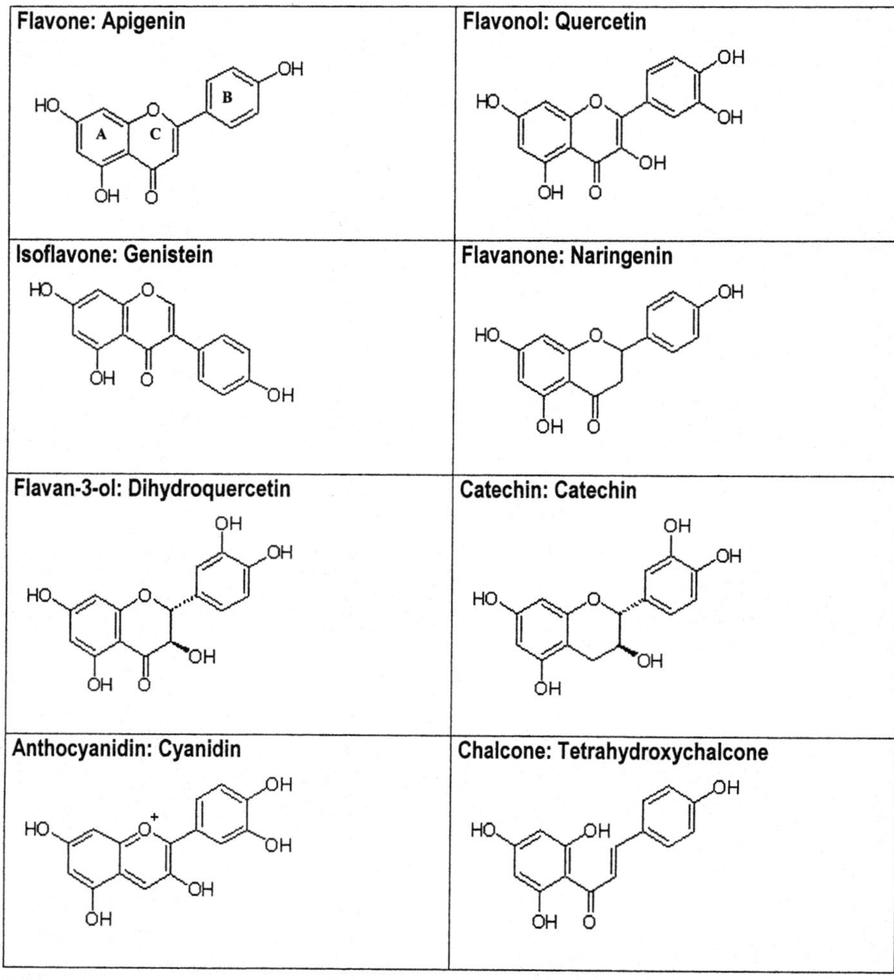

Fig. 1. Basic structure of some flavonoids.

Flavonoids possess important effects in plant biochemistry and physiology, acting as antioxidants, enzyme inhibitors, precursors of toxic substances, pigments, and light screens. In addition, these compounds are involved in photosensitization and energy transfer, actions of plant growth hormones and growth regulators, control of respiration and photosynthesis, morphogenesis and sex determination, and defence against infection ***(6)***. Several reports ***(7–9)*** indicate that plant flavonoids activate bacterial nodulation genes involved in control of nitrogen fixation, suggesting important relationships between particular flavonoids and activation and expression of genes.

Table 1
Dietary Sources of Flavonoids and Phenolic Acids

Flavonoid	Source
Catechins	Tea, red wine
Flavanones	Citrus fruits
Flavonols (e.g., quercetin)	Onions, olives, tea, wine, apples
Anthocyanidins	Cherries, strawberries, grapes, colored fruits
Caffeic acid	Grapes, wine, olives, coffee, apples, tomatoes, plums, cherries

Flavonoids have been recognized to possess antiallergic, anti-inflammatory, antiviral, antiproliferative, and anticarcinogenic activities; they also affect some aspects of mammalian metabolism ***(6)***. Epidemiological studies have indicated that high flavonoid consumption is associated with reduced risk of chronic diseases such as cardiovascular diseases ***(1,10)***.

The resurgence of interest in traditional medicine during the past two decades, together with efforts in expanded pharmacognosy has rekindled an interest in flavonoids and the need to understand their interaction with mammalian cells and tissues ***(6)***.

Plants have been used by humans since the beginning of human culture for various purposes, connected to survival, including medicine. Such folk or ethnomedical uses represent leads that may shortcut the discovery of modern therapeutic drugs, either directly from the plants or from their synthetic analogs. In fact, 74% of the 121 biologically active plant-derived compounds currently in use worldwide have been discovered through follow-up research to verify the authenticity of folk or ethnomedical uses of plants ***(11)***.

The World Health Organization (WHO) estimates that 80% of people living in developing countries use traditional medicine almost exclusively. Medicinal plants form the principle component of traditional medicine. This means that some 3,300 million people use medicinal plants on a regular basis; they should therefore be studied in terms of safety and efficacy ***(12)***.

Medicinal components from plants also play an important role in conventional western medicine. In 1984, Farnsworth et al. ***(13)*** identified 119 secondary plant metabolites that are used globally as drugs. It has been estimated that 14–28% of higher plant species are used medicinally, but only 15% of all angiosperms have been investigated chemically, and 74% of pharmacologically active plant-derived components were discovered after following up on ethnomedicinal use of the plant ***(14)***. Because the numbers of resistant strains of microbial pathogens have been growing since methycillin-resistant *Staphylococcus aureus* appeared, it is critically important to develop new antimicrobial compounds for these and other microorganisms ***(15)***.

In Argentina, native people have been using plants as remedies for infectious diseases for centuries. Many plant preparations have been used in the treatment of diarrhea and respiratory diseases, and these are still being used by rural populations ***(16)***.

Species of the family Asteraceae are some of the most useful, and some South American members were reported previously to show pharmacological activities ***(17–19)***. Leaf infusions from members of *Tagetes* (Asteraceae) have been used in folk medicine to treat stomach and intestinal diseases ***(20)***, and several *Tagetes* species have been found to possess biological activity ***(16,21–23)***.

The genus *Tagetes* (family: Asteraceae; tribe: Tageteae; subtribe: Tagetinae, *sensu* Strother ***[24]***) comprises 56 strongly aromatic species, some of them known as marigolds ***(25)***. In Argentina, the genus *Tagetes* is composed of 12 species, some of which are used in popular medicine; their use has been supported by several studies on their biological activity ***(16,23)***.

Nine species of *Tagetes* are widely distributed in the northwest Argentina. Three members of this group are used frequently in popular medicine in Tucumán: *T. minuta*, *T. pusilla*, and *T. terniflora*. *T. minuta* is a native plant known as *suico* or *chinchilla* that has been used as an antimicrobial, an antihelminthic, a diuretic, and an antispasmodic, and against intestinal diseases ***(20,26)***. *T. pusilla*, known as *anís del cerro*, is used as a condiment and its infusion as an antisyphilitic, carminative, diuretic, and tonic ***(20)***. *T. terniflora* is also a native plant known as *suico-suico* or *quichia*. It is used by people in Argentina and Ecuador as a condiment in soups ***(27)***. Other studies on the biological activity of the genus *Tagetes* showed active thiophenes with larvicidal activity ***(28)*** and 3-OCH_3 flavone derivatives as inhibitor agents of picornavirus and vesicular stomatitis virus ***(29)***.

Experiments that functioned on the antimicrobial activity of methanolic extracts and fractions (ethyl acetate and aqueous) from *T. minuta*, *T. pusilla*, and *T. terniflora* have been carried out against Gram-positive and Gram-negative bacteria, with interesting results. The same fractions were inactive against *Lactobacillus*, *Zymomonas*, and *Saccharomyces* species. An absence of antimicrobial activity against nonpathogenic human bacteria could be beneficial for intestinal disease treatments, in which the intestinal flora must be preserved.

6-OH and 6-CH_3O flavonoid derivatives (quercetagetin and patuletin) from leaves and flowers of *T. minuta*, *T. pusilla*, and *T. terniflora* has been identified ***(30)***. *T. minuta* and *T. pusilla* contain the following major flavonoid compounds: quercetagetin, quercetagetin-7-*O*-arabinosyl-galactoside, quercetagetin-3-*O*-arabinosyl-galactoside, quercetagetin-7-*O*-glucoside, patuletin, patuletin-7-*O*-glucoside, and isorhamnetin. However *T. terniflora* has a different flavonoid pattern: quercetagetin-7-*O*-arabinosyl-galactoside, quercetin, quercetin-3-*O*-arabinoside, quercetin-3-*O*-galactoside, isorhamnetin-3,7-diglucoside and isorhamnetin.

The major component of the extract isolated from leaves of *T. minuta*, the flavonol quercetagetin-7-*O*-arabinosyl-galactoside, showed significant antimicrobial activity on the pathogenic microorganism tested ***(16)***. The recently isolated flavonol quercetagetin-7-*O*-glucoside from *T. pusilla* and *T. minuta* exhibited better antimicrobial activity than the compound cited above ***(31)***. Although *T. pusilla* had an identical flavonoid compound profile compared with *T. minuta*, antimicrobial activity against almost all microorganisms was lower than that of *T. minuta* ***(23)***.

Recent studies with flavonoid glycosylated derivatives show that monohydroxylated flavonoids in the B-ring are absorbed without cleavage of the β-glycosidic bond because of the resistance of this bond to the stomach HCl and the action of pancreatic enzymes action, whereas diglycoside derivatives are absorbed in the colon after deglycosylation by microorganisms. The authors also noted the low absorption efficiency of the related aglycone (quercetin) ***(32)***. These studies and the antimicrobial activity found in quercetagetin 7-*O*-glucoside allow us to consider this flavonoid a promising antibacterial agent with good bioavailability in humans ***(31)***.

The object of this chapter is to help readers in the beginning of studies of antimicrobial activity of flavonoids isolated from three native *Tagetes* species from Argentina. The separation procedure is described but identification is not included. The methodology to determine antimicrobial activity against Gram-positive, Gram-negative bacteria as well as yeasts is also described.

2. Materials

2.1. Plant Material

Leaves (50 g) of *Tagetes minuta* (HBK), *Tagetes terniflora* (HBK), and *Tagetes pusilla* (HBK) were collected from different areas of Tucumán, northwest Argentina. Sample specimens were deposited in the Teodoro Meyer Herbarium (LIL).

2.2. Extraction of Flavonoids (see *Note 1*)

1. 80% Methanol.
2. 50% Methanol.

2.3. Separation of Flavonoids

1. TBA: reagent-grade tertiary butanol/glacial acetic acid/water (3:1:1). TBA is used for both paper chromatography (PC) and thin-layer chromatography (TLC; cellulose).
2. 15% and 30% Acetic acid, for PC and TLC (cellulose).
3. Acetic acid/chloride acid/water (30:3:10), for PC and TLC (cellulose).
4. Methanol.
5. 80% Methanol.
6. 70% Methanol.
7. Ethyl acetate/glacial acetic acid/formic acid/water (100:11:11:26), for TLC (silica).
8. Visualization by spraying with natural product reagent (1% methanolic solution of diphenylboric acid ethylamino ester) followed by 5% ethanolic polyethylene glycol 4000 ***(33)***.
9. Chloroform.
10. Ethyl acetate.
11. Trifluoracetic acid (TFA; solution A) for high-performance liquid chromatography (HPLC) analysis.
12. Acetonitrile (solution B) for HPLC analysis.
13. Separatation funnels (125 and 250 mL) with glass stopcock.
14. Rotary evaporator with temperature-controlled water bath, attached to a water pump (Büchi Rotavapor R 110).
15. Whatman 3MM paper (46 × 57-cm sheets).
16. Chromatographic cabinet (Chromatocab).

17. Chromatography column, glass with solvent reservoir, and glass stopcock (1 × 50 cm).
18. Sephadex LH-20 (Pharmacia-Biotech 17-0090-10).
19. Fraction collector (Gilson FC203B).
20. TLC aluminum sheets (20 × 20 cm), silica gel 60 (without fluorescent indicator) with concentrating zone (Merck 5582).
21. TLC aluminium sheets (20 × 20 cm) cellulose F, thickness 0.1 mm (Merck 5574).
22. Glass tank with glass cover (Desaga).
23. HPLC equipment.
24. Analytical C_{18} column (Phenomenex).
25. Ultraviolet viewing lamp (366 nm).
26. Recording UV-VIS spectrophotometer (Beckman DU 640).
27. Quartz cuvets (0.5 mL).
28. Ungraduated pipet.

2.4. Microorganisms

The microorganisms to be used are Gram-positive and Gram-negative bacteria as well as yeasts (*see* **Note 2**).

2.5. Antimicrobial Assays

2.5.1. Culture Media and Devices for Microorganisms Growth

Media to be used must correspond to the specific microorganism.

1. Commercial Müeller-Hinton broth (Biokar), 23 g/L, is used for almost all Gram-positive and Gram-negative bacteria as a suitable medium for antimicrobial screening.
2. de Man-Rogosa-Sharpe (MRS) broth: 10g/L polypeptone, 10 g/L meat extract, 5 g/L yeast extract, 20 g/L glucose, 5 g/L sodium acetate, 2 g/L ammonium citrate, 2 g/L KH_2PO_4, 0.25 g/L $MgSO_4 \times 7H_2O$, 0.058 g/L $MnSO_4 \times 4H_2O$, and 1.08 mL/L Tween-80 at a final pH 6.4. Sterilize at 121°C for 20 min. This medium is used for *Lactobacillus* sp.
3. Zym broth containing: 50 g/L glucose, 10 g/L yeast extract, 1 g/L KH_2PO_4, 1 g/L $MgSO_4 \times 7H_2O$, 1 g/L $(NH_4)_2SO_4$, final pH 5.3. Sterilize at 121°C for 20 min. This medium is used for *Zymomonas mobilis*.
4. YEPD: 10 g/L peptone, 20 g/L glucose, and 10 g/L yeast extract, final pH 4.5. Sterilize at 121°C for 20 min. This medium is used for *Saccharomyces cerevisiae*.
5. Incubator shaker (rotary) with temperature and speed control.
6. Recording UV-VIS spectrophotometer (Beckman DU 640).

2.5.2. Antimicrobial Assays in Solid Medium

1. 2 and 4% Commercial Müeller-Hinton agar (Biokar) for *Proteus vulgaris*.
2. 0.6% Commercial Müeller-Hinton agar (Biokar).
3. 2 and 0.6% MRS agar.
4. 2 and 0.6% Zym agar.
5. 2 and 0.6% YEPD agar.
6. Chloramphenicol as standard antibiotic.
7. Magnetic stirrer.
8. Teflon-coated magnetic stirring bars.
9. Petri dishes: glass (100 × 15 mm) or plastic (90 × 15 mm).
10. Pipets.
11. Sterile tips.

2.5.3. Antimicrobial Assays in Liquid Medium

1. Commercial Müeller-Hinton broth (Biokar).
2. Magnetic stirrer.
3. Teflon-coated magnetic stirring bars.
4. Vortex mixer.
5. Pipets.
6. Sterile tips.

3. Methods

3.1. Extraction of Flavonoids From Plants

The plant material should be macerated prior to extraction. Dried aerial parts of the plant (50 g) are successively extracted at room temperature for 3 d with cold 80% and 50% aqueous methanol (400 mL). After the plant material is separated by filtration, the solvent is removed under vacuum, yielding a sticky green residue of 10.8 g (*see* **Note 3**).

3.2. Separation of Flavonoids

3.2.1. Separation of Flavonoids by Paper Chromatography ***(34–38)***

1. Dissolve about 1 g of the dried residue in 80% aqueous methanol (1 mL) and spot it on the lower right-hand corner of a sheet of Whatman 3 MM chromatographic paper (semipreparative). Use a flow of warm air for solvent evaporation between repeated applications of the solutions to the paper. The final spot, which appears deep purple when viewed under a 366-nm UV lamp, is about 2 cm in diameter and 8 cm from each edge of the paper.
2. Develop the chromatogram containing flavonoids in a descendent mode in the longest dimension into a chromatocab using TBA as the first solvent (*see* **Note 4**).
3. Remove the chromatogram from the cabinet, dry it, trim off the folded portion, and then refold it for descending chromatography in the second dimension with AcOH 15% solvent (*see* **Note 5**).
4. The dry two-dimensional developed chromatogram should be revealed under UV light alone and in the presence of ammonia fumes. It can be also be revealed with NA reagent in order to detect the ortho dihydroxylated system in the B-ring (*see* **Note 6**).

3.2.1.1. Fractionation of Total Extract in Different Polarity Solvents ***(37)***

1. Partition the total flavonoid extract obtained (50–100 mL) between methanol/water (8:2) and chloroform (50–100 mL) in a separation funnel. Separate the chloroform phase (lower phase) and repeat the extraction three times with 25 and 12.5 mL, respectively. Concentrate the chloroform phases and store for antimicrobial assays on the rotary evaporator. This fraction contains chlorophyll, lipids, terpenoids, and less polar aglycones.
2. Add to the rest of the methanolic extract an equal volume of ethyl acetate in a separation funnel and proceed as in **step 1**. In this case the upper phase should be separated. Glycosides in this fraction, are mono- and diglycosylated in the B-ring.
3. The residue extract exhausted with solvents represents the aqueous fraction, which contains diglycosides of flavonoids, tryglycosides, sulphated flavonoids, alkaloids, amino acids, phenolic acids, and cumarines.

3.2.1.2. Preparative 2D Paper Chromatography Isolation, and Purification of Flavonoids

Cut out the area containing the flavonoid alone on a two-dimensional chromatogram, about 20–50/sheet, to obtain sufficient material for identification of the flavonoid (*see* **Note 7**).

3.2.1.3. Partial Purification of Flavonoid: 1D Paper Chromatography

1. After a two-dimensional run (TBA and AcOH 15 or 30%), make a selection of the most effective solvent system to separate the mixture. The selected one will be the solvent to be used in one-dimensional paper chromatographic purification of the flavonoid.
2. Separate the flavonoid mixture from the two-dimensional run (about 20 sheets) by cutting out the related area, dissolve it in MeOH 80%, and apply it as a band (*see* **Note 8**).
3. A descending run in the selected solvent gives UV-detectable bands. Cut out the major one and elute it with spectroscopic methanol for a few minutes for spectral analysis according to the method developed by Mabry et al. *(34)*.

3.2.2. Purification of Flavonoids by Column Chromatography (CC)

The classical chemical procedure for obtaining organic constituents from dried plant tissue, when working on the preparative scale, involves continuously extracted powdered material in a Soxhlet apparatus with a range of solvents, starting with *n*-hexane and chloroform and then using methanol, ethanol or ethyl acetate. The extracts obtained are concentrated in a vacuum; they may deposit crystals. Mixtures of substances present in those crystals must be redissolved in a suitable solvent and separated from the constituents by chromatography *(37)*. We now describe a protocol for purification by CC.

Sephadex LH-20 is used for the final clean-up of flavonoid aglycones and glycosides previously separated by paper, cellulose, silica, or polyamide (*see* **Note 9**) *(36)*. MeOH is generally an adequate solvent, although some water may be needed initially to dissolve the flavonoid. Presoaking of the gel in MeOH is necessary *(37)*. To avoid a long and tedious run, an automated fraction collector could be used

Concentrate each cut-out spot from the paper on a rotary evaporator, dissolve the residue in a minimum of MeOH, vortex the mixture and purify the individual components as follows:

1. Apply 1 mL of the mixture to a 50 × 1-cm LH20 column equilibrated with MeOH and elute with decreased quantities of MeOH in water (9:1, 8:2, and 7:3) flowing down three column volumes for each system.
2. Elute the flavonoids from the column slowly (0.1 mL/min) to allow good resolution. If the band is not visible or detectable under UV light (366 nm), then analyze each fraction by TLC (silica G60 or cellulose) to determine which fractions should be combined. Run an analytical TLC sheet (4 × 10 cm) in a glass tank with 50 mL of suitable solvent (described in **Subheading 2.3., items 3** and **7**), previously equilibrated for 20 min. Remove the plates from the glass tank, mark the solvent front with a pencil, and allow them to dry in the fume hood for approx 15 min. Spray the plates with NA reagent or visualize the spots under UV light (366 nm) alone and under ammonia vapors.
3. Concentrate each eluate and test it for antimicrobial activity.
4. Each active fraction must be rerun in an LH20 column to isolate the active compound. Repeat **steps 2** and **3**.

3.2.3. Separation of Flavonoids by HPLC

When separation by the methods described above is not convenient for the type of flavonoid, then HPLC separation should be used. The technique described here is based on a previous procedure described by Corzier et al. ***(39)*** and the chromatographic conditions given in ref. ***40*** with modifications ***(31)***. Classical HPLC procedures for flavonoids are carried out after glycoside hydrolysis ***(39,40)***. However, the following methodology is for mono- and diglycosylated flavonol without previous hydrolysis.

1. After LH-20 column separation, concentrate the eluate in a rotary evaporator, resuspend it in the initial phase (15% B/85% A; 0.1 mg/L) and filter through a Millipore membrane before injection.
2. Inject 100 μL of solution into the HPLC apparatus (Gilson) and run in a linear gradient 0–25 min of 15–35% B. The column is a C18 (Phenomenex). The solvents used are given in **Subheading 2.3.**, **items 11** and **12**.
3. Monitor the flavonoids at 365 nm under a UV detector.
4. Concentrate the fractions containing pure flavonoids and prepare them for antimicrobial assays.

3.3. Antimicrobial Assays

3.3.1. Microbial Culture Conditions

The different strains must be cultured twice for activation in Erlenmeyer flasks (125 mL) containing 25 mL of Müeller-Hinton broth at 37°C for 12 h in a shaker, except for *Lactobacillus rhamnosus*, *L. plantarum*, *Zymomonas mobilis*, and *Saccharomyces cerevisiae* (*see* **Note 10**). Stop the cultures at 0.8–1 OD_{600} (equivalent to 10^8 CFU), dilute until 0.1 OD_{600}, and use for antimicrobial testing (*see* **Note 11**).

3.3.2. Detection of Antimicrobial Activity in Solid Medium

The agar well diffusion method, previously described ***(41,42)***, with some modifications ***(16)***, is used to detect antimicrobial activity of plant extracts.

1. Inoculate 3 mL of 0.6% M-H agar media with 30 μL of an overnight culture of each microorganism prepared as described in **Subheading 3.3.1.**
2. Overlay onto plates containing 2% Müeller-Hinton agar media, except for *Proteus vulgaris*, for which 4% agar must be used.
3. Make wells of 3 mm in the agarized medium after inoculation with the microorganism to be tested. More than five wells per plate are not recommended.
4. Fill the wells with 20 μL of total extract of flavonoids, fractions, or pure compound at a range of concentrations from 5 to 1.000 μg/mL. The solvent for resuspending the extract should be MeOH 80% instead of pure MeOH because the latter produces irregular inhibition zones.
5. Carry out a negative control with each solvent.
6. Maintain the culture Petri dishes at 37°C, and do not invert them until the solvent is evaporated. After that, invert the plates to avoid water vapors condensation.
7. Measure the inhibition zones from one side to the other of the circle three times and discount the well diameter (**Figs. 2** and **3**).
8. Use chloramphenicol, as a standard antibiotic, at a range of 0–250 μg/mL in order to make a calibration curve. The results are the averages of three individual experiments.

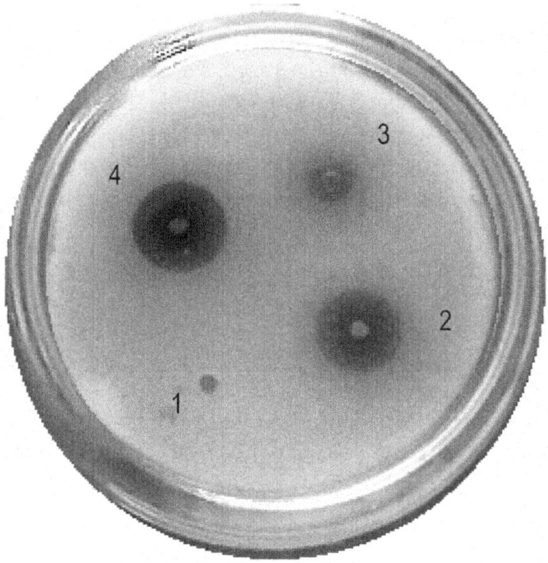

Fig. 2. Inhibition zones of total extract from *Tagetes terniflora* (2) *T. pusilla* (3), and *T. minuta* (4) against *E. coli*. Solvent as negative control (1).

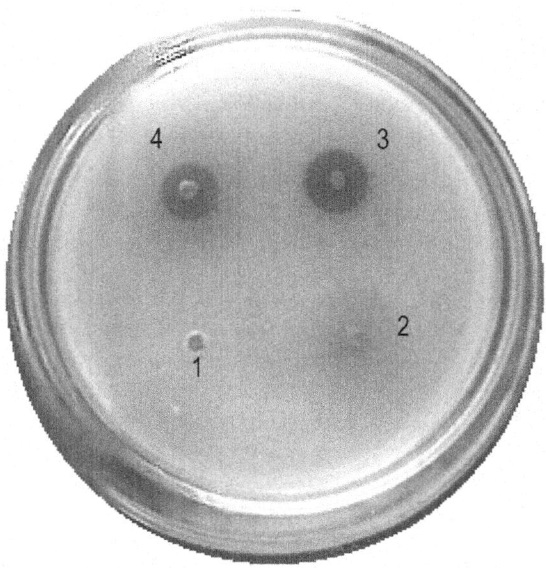

Fig. 3. Inhibition zones of fractions from *T. minuta* against *E. coli*. Solvent as negative control (1); chloroform fraction (2); ethyl acetate fraction (3); and aqueous fraction (4).

Table 2
Microorganisms Suitable for Antimicrobial Screening

Microorganism	Species
Gram-negative rods	*Escherichia coli*
	Salmonella enteriditis
	Shigella sp.
	Pseudomonas aeruginosa
	Proteus vulgaris
	Zymomonas mobilis
Gram-positive rods	*Bacillus subtilis*
	Lactobacillus rhamnosus
	Lactobacillus plantarum
Gram-positive cocci	*Staphylococcus aureus*
	Staphylococcus epidermidis
	Micrococcus luteus
Yeast	*Saccharomyces cerevisiae*

3.3.3. Minimum Inhibitory Concentration (MIC) in Liquid Medium

MIC values against bacterial strains are performed using the Ericcson and Sherris broth dilution method ***(43)***. Inocula suspensions are prepared from 6-h broth cultures and adjusted to 0.5 McFarland turbidity equivalents. Substances and extracts must be sterilized by Millipore filtration (0.45 μm) and add to the MH broth medium. MIC determination are performed using serial dilutions of each extract and pure substance at a range of 0–1.000 μg/mL.

4. Notes

1. **Caution:** chloroform is irritating to the skin, eyes, mucous membranes, and respiratory tract. Chloroform is a carcinogen and may damage the liver and kidneys. Methanol is poisonous and can cause blindness if ingested in sufficient quantities. Adequate ventilation is necessary to limit exposure to vapors. Glacial acetic acid is volatile. Concentrated acids should be handled with great care. Wear gloves and safety glasses and work in a chemical fume hood.
2. Microorganisms suitable for antimicrobial screening are shown in **Table 2**.
3. The possible enzyme action occurring during this early period of isolation, leading in particular to hydrolysis of glycosides, may be avoided by plunging the plant material into boiling solvent or by rapid drying prior to extraction. The success of the extraction is directly related to the extent that chlorophyll is removed from the solvent and when the tissue debris, on repeated extraction, is completely free of green colors.
4. Depending on the room temperature, it will take the solvent approx 24 h at 30°C to approach 2 cm from the end of sheet.
5. The run is developed in about 3–4 h. It is remarkable that with 15% or 30% AcOH the run typically forms a rounded front effect.

6. Spots detected should be circled with a pencil, and R_f values are determined as follows:

$$R_f = \frac{\text{Distance between origin and the centre of concentration of the flavonoid spot}}{\text{Distance between origin and solvent front}}$$

7. Is important to minimize contamination of a flavonoid to be used for UV spectral studies. For purification of the flavonoid used in the spectroscopic investigation, it should be extracted from the chromatographic paper with spectroscopic methanol for a few minutes only.
8. The band should be located at approx 8 cm from the top of the chromatographic paper.
9. Flavonoids isolated by PC may still contain some soluble polysaccharide impurities; it is best to purify them further by Sephadex LH-20 chromatography.
10. *Lactobacillus* strains are cultured aerobically in MRS broth at 30°C for 12 h at 250 rpm. *Zymomonas mobilis* should be grown in Zym broth at 30°C for 12 h without agitation. *Saccharomyces cerevisiae* should be cultured at 30°C for 12 h at 200 rpm.
11. Prior to carrying out the antimicrobial screening, let the microbial dilution sit for 10 min at 37°C without agitation. This culture could be maintained at 4°C for a week for further antimicrobial analyses.

References

1. Middleton, E. and Kandwasami, C. (1994) The Impact of plant flavonoids on mammalian biology: implications for immunity, inflammation and cancer. In: *The Flavonoids: Advances in Research Since 1986* (Harborne, J. B. ed.). Chapman and Hall, London, pp. 619–621.
2. Croft, K. (1998) The chemistry and biological effects of flavonoids and phenolic acids. *Ann. NY Acad. Sci.* **20,** 435–442.
3. Hertog, M. G. L., Fesens, E. J. M, Hollman, P. C. H., Katan, M. B., and Kromhout, D. (1993) Dietary antioxidants: flavonoids and the risk of coronary heart disease: the Zutphen Elderly Study. *Lancet* **342,** 1007–1011.
4. Szent-Gÿorgyi, A. (1938) *Physiol. Chem.* **225,** 126.
5. Pathak, D, Pathak, K., and Singla, A. K. (1991) Flavonoids as medicinal agents. Recent advances. Review. *Fitoterapia* **LXII,** 371.
6. Harborne, J. B. (1994) *The Flavonoids. Advances in research since 1986.* Chapman and Hall, London.
7. Djordjevic, M. A., Redmond, J. W., Batley, M., and Rolfe, B. G. (1987) Clovers secrete specific phenolic compounds which either stimulate or repress *nod* gene expression in *Rhyzobium trifolii. EMBO J.* **6,** 1173.
8. Firmin, J. L., Wilson, K. E., Rossen, L., and Johnston, A. W. B. (1986) Flavonoid activation of nodulation genes in *Rhyzobium* reversed by other compounds present in plants. *Nature* **324,** 90.
9. Welton, A. F., Hurley, J., and Will, P. (1988) Flavonoids and arachidonic acid metabolism. *Prog Clin Biol Res* **280,** 301–312.
10. Hertog, M. G. L, Kromhout, D., and Aravanis, C (1995) Flavonoid intake in long-term risk of coronary heart disease and cancer in seven countries study. *Arch. Intern. Med.* **155,** 381–386.
11. Soejarto, D. D. and Farnsworth, N. R. (1989). Tropical rain forest: potential source of new drugs? *Perspect Biol. Med.* **32,** 244–256.
12. Farnsworth, N. R. (1994) Ethnopharmacology and drug development. In: *Ethnobotany and the Search for New Drugs* (Prance, G. T., ed.). Wiley, Chichester (Ciba Foundation Symposium 185), pp. 42–59.

13. Farnsworth, N. R. (1984) The role of medicine plants in drug development. In: *Natural Products and Drug Development* (Krogsgaard-Larsen, P., Christensen, S. B., and Kofod, H., eds.). Balliere, Tindall and Cox, London, pp. 8–98.
14. Farnsworth, N. R and Soejarto, D. D. (1991) Global importance of medicinal plants. In: *Conservation of Medicinal Plants* (Akerele, O., Heywood, V., and Synge, H., eds.). Cambridge University Press, Cambridge, pp. 25–51.
15. Eloff, J. N. (1998) Which extractant should be used for the screening and isolation of antimicrobial components from plants? *J. Ethnopharmacol.* **60,** 1–8.
16. Tereschuk, M. L., Quarenghi de Riera, M, Castro, G. R. and Abdala, L. R. (1997) Antimicrobial activity of flavonoids from leaves of *Tagetes minuta. J. Ethnopharmacol.* **56,** 227–232.
17. González, A., Vázquez, A., Ferreira, F., and Moyna, P. (1993) Biological screening of Uruguayan medicinal plants. *J. Ethnopharmacol.* **39,** 217–220.
18. Quarenghi, M. V., Tereschuk, M. L., Baigorí, M. D., and Abdala, L. R. (2000) Antimicrobial activity of flowers from *Anthemis cotula. Fitoterapia* **71,** 710–712.
19. Penna, C. A., Radice, M., Gutkind, G. O., et al. (1994) Antibacterial and antifungal activities of some Argentinean plants. *Fitoterapia* **65,** 172–174.
20. Zardini, E. M. (1984). Etnobotánica de las compuestas argentinas con especial referencia a su uso farmacológico. *Acta Farm. Bonaerense* **3,** 169–194.
21. Cáceres, A., Figueroa, L., Taracena, A., and Samayoa, B. (1993) Plants used in Guatemala for treatment of respiratory diseases. 2: Evaluation of activity of 16 plants against gram-positive bacteria. *J. Ethnopharmacol.* **39,** 77–82.
22. Cáceres, A., Torres, M., Ortiz, S., Cano, F., and Jauregui, E. (1993) Plants used in Guatemala for the treatment of gastrointestinal disorders. IV. Vibriocidal activity of five American plants used to treat infections. *J. Ethnopharmacol.* **39,** 73–75.
23. Tereschuk , M. L., Gallegos Serruto, S., Quarenghi, M. V., and Abdala, L. R. (1999) Comparative assays of the antimicrobial activity between two species of *Tagetes* (Astaeraceae). *Acta Horticulturae* **501,** 223–225.
24. Strother, J. L. (1977) Tageteae—systematic review. In: *The Biology and Chemistry of the Compositae* (Heywood, V., Harborne, J. B., and Turner, B. L., eds.). Academic, London, pp. 769–796.
25. Soule, J. A. (1996) Novel annual and perennial *Tagetes*. In: *Progress in New Crops* (Janick, J., ed.). ASHS Press, Arlington, VA, pp. 546–551.
26. Amat, A. G. (1983) Investigación farmacológica de los taxones mayoritarios de compuestas bonaerenses. *Acta Farm. Bonaerense* **2,** 23–36.
27. Neher, R. T. (1966) Monograph of the genus *Tagetes* (Compositae). PhD Thesis, Indiana University, Bloomington, IN.
28. Macedo, M. E., Consoli, R. A., Grandi, T. S., et al. (1997) Screening of Asteraceae (Compositae) plant extracts for larvicidal activity against Aedes fluviatilis (Diptera:Culicidae). *Mem. Inst. Oswaldo Cruz* **92,** 565–570.
29. Abad, M. J., Bermejo, P., Sánchez Palomino, S., Chiriboga, X., and Carrasco, L. (1999) Antiviral activity of some South American medicinal plants. *Phytother. Res.* **13,** 142–146.
30. Abdala, L. R. and Seeligmann, P. (1983) Distribución de los flavonoids de hojas de tres especies de *Tagetes* (Compositae) y su significado quimiosistemático. *Lilloa* **36,** 5–14.
31. Tereschuk, M. L., Vázquez, A., Pezaroglo, H., Baigorí, M., and Abdala, L. R. (2002) Determinación estructural de compuestos con actividad antibacteriana en especies de *Tagetes* del noroeste argentino. I Congreso Latinoamericano de Fitoquímica-IV Meeting de la Sociedad Latinoamericana de Fitoquímica, Buenos Aires, Argentina.

32. Hollman, P. Bijsman, M., Gameren, Y, Cnossen, E., de Vries, J., and Katan, M. (1999) The sugar moiety is a major determinant of the absorption of dietary flavonoid glycosides in man. *Free Rad. Res.* **31,** 569–573.
33. Wagner, H. and Bladt, S. (1996) *Plant Drug Analysis: A Thin Layer Chromatography Atlas*, 2nd ed. Springer Verlag, Berlin, Heidelberg, New York, pp. 169–224.
34. Mabry T. J., Markhan K. R., and Thomas M. B. (1970) *The systematic Identification of Flavonoids*. Springer-Verlag, New York, p.11.
35. Markham K. R. (1982) *Techniques of Flavonoids Identification*. Academic, New York.
36. Imperato, F. (1998) Two new flavonol glycosides from the fern *Pteridum aquilinum*. IInd International Electronic Conference on Synthetic Organic Chemistry (ECSOC-2), no. 074.
37. Harborne, J. B. (1998) *Phytochemical Methods*, 3rd ed. Chapman and Hall, London, pp. 63–82 and pp. 235–255.
38. Abdala, L. R., Del Pero de Martínez, M. A., and Seeligmann,P. (1991) Myricetin in *Tagetes* (Asteraceae). Chemosystematic significance. *Phytochemistry* **30,** 4037–4038.
39. Crozier, A., Jensen, E., Lean, M. E. L., and McDonald, M. S. (1997) Quantitative analysis of flavonoids by reversed-phase high-performance liquid chromatography. *J. Chromatogr. A.* **761,** 315–321.
40. Merken, H. and Beecher, G. (2000) Measurement of food flavonoids by high-performance liquid chromatography: a Review. *J. Agric. Food Chem.* **48,** 577–599.
41. Bennet, J. V., Brodie, J. L., Benner, J. L., and Kirby, W. M. M. (1966) Simplified accurate method for antibiotic assay of clinical specimens. *Appl. Microbiol.* **14,** 170–177.
42. Vanden Berghe D.A and Vlietink, A.J. (1991) *Methods in Plant Biochemistry*, vol. 6. Academic Press, London, p. 47.
43. Ericcson, H. M. and Sherris, J. C. (1971) Antibiotic sensitivity testing: report of an international collaborative study. *Acta Pathol. Microbiol. Scand.* **217(suppl).**

30

Purification of Bacteriocins Produced by Lactic Acid Bacteria

Lucila Saavedra, Patricia Castellano, and Fernando Sesma

1. Introduction

Bacteriocins are antibacterial substances of a proteinaceous nature that are produced by different bacterial species. Lactic acid bacteria (LAB) produce biologically active peptides or protein complexes that display a bactericidal mode of action almost exclusively toward Gram-positive bacteria and particularly toward closely related species. Generally they are active against food spoilage and foodborne pathogenic microorganisms including *Bacillus cereus*, *Clostridium perfringens*, *Staphylococcus aureus*, and *Listeria monocytogenes* ***(1,2)***.

There is an increased tendency to use natural occurring metabolites to prevent the growth of undesirable flora in foodstuffs. These metabolites could replace the use of chemical additives such as sorbic acid, sulfur dioxide, nitrite, nitrate, and others. For instance, bacteriocins produced by LAB may be promising for use as biopreservaties ***(3,4)***.

Bacteriocins of lactic acid bacteria are typically cationic, hydrophobic peptides and differ widely in many characteristics including molecular weight, presence of particular groups of amino acids, pI, net positive charge, and post-translational modifications of certain amino acids. This heterogeneity within the LAB bacteriocins may explain the different procedures for isolation and purification developed so far ***(8,9)***. The methods most frequently used for isolation, concentration, and purification involve salt precipitation of bacteriocins from culture supernatants, followed by various combinations of gel filtration, ion-exchange chromatography, and reverse-phase high-performance liquid chromatography (RP-HPLC).

In this chapter, a protocol is described that combines several methods used in our laboratory for the purification of two cationic bacteriocins, Lactocin 705AL and Enterocin CRL10, produced by *Lactobacillus casei* CRL705 and *Enterococcus mundtii* CRL10, respectively.

From: *Methods in Molecular Biology, vol. 268: Public Health Microbiology: Methods and Protocols*
Edited by: J. F. T. Spencer and A. L. Ragout de Spencer © Humana Press Inc., Totowa, NJ

2. Materials

2.1. Growth Media

1. Growth medium for *Enterococcus* sp.: LAPTg broth: 15g/L Peptone, 10g/L Tryptone, 10g/L yeast extract, 10g/L glucose, and 1 mL/L Tween-80, final pH 6.5. Sterilize at 121°C for 20 min.
2. Growth medium for *Listeria* sp.: BHI (brain-heart infusion, BioKar Diagnostics), final pH 6.5. Sterilize at 121°C for 20 min (*see* **Note 1**).
3. Growth medium for *Lactobacillus* sp.: de Man-Rogosa-Sharpe (MRS) broth (BioKar Diagnostics), final pH 6.5. Sterilize at 121°C for 20 min.
4. Medium for bacteriocin activity assays: BHI base agar (1.5% agar) and BHI top agar (0.7% agar).

2.2. Bacteriocin Purification

2.2.1. Obtaining the Crude Extract

1. Ammonium sulfate (Sigma).
2. Refrigerated centrifuge.
3. Distilled water.

2.2.2. Solid-Phase Extraction Protocols

1. Octadecyl (C_{18}) Impaq RG1080 columns, pore diameter 100 Å, particle size 80 μm (Sigma), or C_8 columns from Supelco (Supelclean LC-8) with 3-mL tubes (or similar; *see* **Note 2**).
2. Acetonitrile and isopropanol (HPLC grade; Merck).
3. Turvo-Vap Evaporator (Zymark, Hopkinton, MA).
4. Distilled water.

2.2.3. Ion Exchange Chromatography

1. SP-Sepharose Fast Flow strong cation exchanger (Amersham Pharmacia Biotech).
2. 0.1 *M* Phosphate buffer, pH 7.0.
3. 5 *M* NaCl.
4. Refrigerated centrifuge.
5. Distilled water.

2.2.4. Reverse-Phase Chromatography

1. Mobile phase.
   a. Enterocin CRL10 mobile phase.
      Solution A: 0.1% trifluoroacetic acid (TFA) in distilled water.
      Solution B: 99% acetonitrile/0.1% TFA.
   b. Lactocin 705AL mobile phase.
      Solution A: 5% isopropanol/0.1% TFA.
      Solution B: 99.9% isopropanol/0.1% TFA.
2. Column for Enterocin CRL 10: μBondapack C_{18}, pore diameter 125 Å, particle size 10 μm; 3.9 × 300 mm.
   Column for Lactocin 705AL:Vydac 208 TP C8, particle size 10μm, 4.6 mm × 250 mm.
3. Turbo-Vap evaporator.
4. Distilled water.

2.3. Antimicrobial Activity Assays

1. BHI soft agar (0.7% agar).
2. BHI agar (1.5% agar).
3. Sterile hollow punch, 0.5-cm diameter.
4. Indicator or sensitive strain.
5. Distilled water.

3. Methods

3.1. Isolation and Purification of Enterocin CRL10

1. Grow the bacteriocin producer strain in 2 L of LAPTg broth (16-h static culture).
2. Remove the cells by centrifugation (12,000*g* for 15 min) and bring the supernatant to a final concentration of 40% (w/v) saturation of ammonium sulfate. Stir for 12 h at 4°C (*see* **Note 3**).
3. Centrifuge the mixture at 17,000*g* for 20 min at 4°C. Harvest the pellet and the floating material and dissolve them in 100 mL of distilled water. Test the activity of the crude extract using the bacteriocin activity assay (*see* **Subheading 3.6.**).

3.2. Reverse-Phase Solid-Phase Extraction Columns

1. Equilibrate the column with distilled water.
2. Apply the sample (crude extract) to the column. (*see* **Note 4**).
3. Wash steps (*see* **Note 5**):

   4 vol of 5% acetonitrile.
   4 vol of 10% acetonitrile.
   4 vol of 20% acetonitrile.
4. Elution step:

   4 vol of 40% acetonitrile.
5. Test the activity of the fractions using the bacteriocin activity assay.
6. Concentrate the active fraction in a rotary evaporator at 40°C for 2 h.
7. Additional wash steps to clean the column:

   4 vol of 70% isopropanol.
   4 vol of 99% isopropanol.

3.3. Ion Exchange Chromatography

1. Equilibrate the column with 10 m*M* phosphate buffer, pH 7.0.
2. Apply the sample very slowly.
3. Wash steps (*see* **Notes 5** and **6**):

   4 vol of 10 m*M* of phosphate buffer, pH 7.0, 50 m*M* NaCl.
   4 vol 10 m*M* phosphate buffer pH 7.0, 100 m*M* NaCl.
   4 vol 10 m*M* phosphate buffer pH 7.0, 150 m*M* NaCl.
4. Elution step:

   4 vol 10 m*M* phosphate buffer, pH 7.0, 200 m*M* NaCl.
   4 vol 10 m*M* phosphate buffer, pH 7.0, 500 m*M* NaCl.
5. Collect the fractions containing the bacteriocin activity, desalt them with an octadecyl (C_{18}) Impaq column, eluting with 2 vol of 40% acetonitrile. Concentrate the sample in the TurboVap or rotary evaporator.

3.4. Reverse-Phase Chromatography

1. Perform RP-HPLC on the active extract (*see* **Note 7**).
2. Equilibrate the column using solution A at a flow rate of 1 mL/min.
3. Elute the antimicrobial peptide (injection volume 2000 μL) with the following gradient:

Minutes	Flow (mL/min)	% Solution B
0	0	0
0.5	0.4	0
1	0.4	0
15	0.4	0
16	0.8	0
20	0.8	0
30	0.8	30
38	0.8	30
48	0.8	100
50	0.8	100
57	0.8	0
74	0.8	0
75	0	0

4. Monitor the elution spectrophotometrically at 220 nm.
5. Pool the active fractions and concentrate them by evaporation using nitrogen as the carrier gas (TurboVap evaporator).

3.5. Isolation and Purification of Lactocin 705AL

The procedure is similar to that used for purification of Enterocin CRL 10, with the following modifications:

1. The culture medium to grow *Lactobacillus* is MRS.
2. The final concentration of ammonium sulfate is 44%. Dissolve the resultant precipitate in 100 mL of isopropanol 5%.

3.5.1. Solid-Phase Extraction Columns

1. Equilibrate the column with 2 mL of 5% isopropanol.
2. Apply the sample (crude extract) to the column (*see* **Note 4**).
3. Wash step (*see* **Note 5**) 2 mL 10% isopropanol.
4. Elution step: 2 mL 30% isopropanol.
5. Concentrate the active fraction in a TurboVap evaporator at 50°C for 30 min.
6. Additional wash steps to clean and reuse the column.

   4 vol of 70% isopropanol.
   4 vol of 99% isopropanol.

3.5.2. RP-HPLC

1. Subject the active extract to RP-HPLC (*see* **Note 7**).
2. Equilibrate the column using solution A at a flow rate of 1 mL/min.
3. Elute the peptide with a 30 min linear gradient of 5–100% of isopropanol in aqueous 0.1% TFA (v/v)
4. Monitor the elution spectrophotometrically at 220 nm.

3.6. Bacteriocin Activity Assays

The bacteriocin activity of both strains should be determined in all the steps of the purification process using the following assay:

1. Prepare serial twofold dilutions of cell-free supernatant and of all the fractions obtained in the purification steps, in sterile dH_2O or sterile culture media (1:2, 1:4, 1:8, 1:*n*).
2. Prepare Petri dishes with 10 mL of BHI base agar. Let solidify and dry.
3. Mix gently 10 μL of an overnight culture of the sensitive strain (approx 1.10^6 CFU/mL) with 7 mL of molten BHI soft agar (45°C) and pour onto the base BHI layer. Slide the plate in circles on the bench top immediately to spread the top agar over the plate. Let solidify and dry.
4. Cut wells 0.5 cm in diameter with sterile hollow punchs in the agar. In each well, seed 30 μL of undiluted bacteriocin and each dilution of it.
5. Incubate the plates (with the agar side up) overnight at 30°C.
6. Bacteriocin activity is detected by the presence of growth inhibition zones of the indicator strain around the wells.
7. The titer of the bacteriocin is expressed in arbitrary units (AU/mL), as the reciprocal of the highest dilution showing a definite zone of inhibition.

4. Notes

1. TSA (trypticase soy broth) and TSA agar (trypticase soy agar) are also used for maintaining and culturing *Listeria* sp.
2. This column is recommended for the separation of hydrophobic peptides and proteins, including membrane proteins. Use of an Octyl-Sepharose CL-4B column has also been reported ***(5,6)***.
3. A two-step ammonium sulfate (0–25% and 25–70%) precipitation was reported for mundticin purification ***(4)***, and 70–75% was also used for Enterocin A ***(5)***.
4. The sample could also be applied using a vacuum pump. Apply the sample until the column is saturated. This step could be repeated using a small amount of sample each time.
5. In this step it is very important to know the volume of the column, e.g., for a 5-mL column use 20 mL of the wash solution.
6. The final concentration of NaCl for the elution step may vary between 0.2 and 1 *M* depending on the bacteriocin that is being purified ***(7)***.
7. ISCO HPLC system (2360 Gradient programmer, 2350 HPLC Pump, and V^4 Absorbance detector; ISCO, Lincoln, NE).

References

1. Jack, R., Tagg, J., and Ray, B. (1995) Bacteriocins of Gram-positive bacteria. *Microbiol. Rev.* **59,** 171–200.
2. Nes, I. and Holo, H. (2000) Class II antimicrobial peptides from lactic acid bacteria. *Biopolymers* **55,** 50–61.
3. De Vuyst, L. and Vandamme, J. (1994) Lactic acid bacteria and bacteriocins: their practical importance. In: *Bacteriocins of Lactic Acid Bacteria. Microbiology, Genetics and Applications* (De Vuyst, L. and Vandamme, J., eds.). Blackie , London, pp. 1–13.
4. De Vuyst L. and Vandamme, J. (1994) Antimicrobial potential of lactic acid bacteria. In: *Bacteriocins of Lactic Acid Bacteria. Microbiology, Genetics and Applications* (De Vuyst, L. and Vandamme, J., eds.). Blackie, London, pp. 91–143.

5. Bennik, M. H. J., Vanloo, B., Brasseur, R., Gorris, L., and Smid, E. (1998) A novel bacteriocin with YGNGV motif from vegetable-associated *Enterococcus mundtii*: full characterization and interaction with target organisms. *Biochem. Biophys. Acta* **1373,** 47–58.
6. Aymerich T.; Holo H.; Havarstein L.S.; Hugas M.; Garriga M; Nes I. (1996) Biochemical and genetic characterization of enterocin A from *Enterococcus faecium*, a new antilisterial bacteriocin in the pediocin family of bacteriocins. *Appl. Environ. Microbiol.* **62,** 1676–1682.
7. Cintas, L. M., Casaus, P., Havarstein, L. S., Hernandez, P., and Nes, I. (1997) Biochemical and genetic characterization of enterocin P, a novel sec-dependent bacteriocin from *Enterococcus faecium* P13 with a broad antimicrobial spectrum. *Appl. Environ. Microbiol.* **63,** 4321–4330.
8. Uteng, M., Hauge, H. H., Brondz, I., Nissen-Meyer, J., and Fimland, G. (2002) Rapid two-step procedure for large-scale purification of pediocin-like bacteriocins and other cationic antimicrobial peptides from complex culture medium. *Appl. Environ. Microbiol.* **68,** 952–956.
9. Guyonnet, D., Fremaux, C., Cenatiempo, Y., and Berjeaud, J. M. (2000) Method for rapid purification of class IIa bacteriocins and comparison of their activities. *Appl. Environ. Microbiol.* **66,** 1744–1748.

31

Production of Antimicrobial Substances by Lactic Acid Bacteria I

Determination of Hydrogen Peroxide

María Silvina Juárez Tomás, María Claudia Otero, Virginia Ocaña, and María Elena Nader-Macías

1. Introduction

Restoration of the balance of different ecological niches has been proposed as a way to control the income of pathogenic microorganisms ***(1)***. The genus *Lactobacillus* has been used in different human and animal tracts as probiotic microorganisms with this objective in mind ***(2)***. The characteristics of the strains proposed as probiotics have been published ***(3)*** or patented under the process of elaboration of different types of products ***(4–6)***. One of the mechanisms suggested to control the vaginal ecosystem is the production of antagonistic substances (lactic acid, bacteriocins, or H_2O_2) ***(7–9)***. The H_2O_2-producing microorganisms present in the vagina of healthy women have been suggested as some of the bacteria responsible for maintenance of ecological balance, mainly in pregnant women ***(10,11)***. The absence of these microorganisms is related to a higher risk of: bacterial vaginosis ***(12)***, recurrent urinary tract infections by *Escherichia coli* ***(13)***, and acquisition of human immunodeficiency virus type 1 (HIV-1) ***(14)***. Bauer ***(15)*** has proposed that H_2O_2-producing lactobacilli also might exert control over vaginal cancer through specific interactions of reactive oxygen species, such as superoxide anion, hydroxyl radicals, and hypochlorous acid. The conversion of H_2O_2 into more toxic compounds during the oxidative process is potentiated by peroxidase and halures. This enzyme and some halures, such as chloride and bromide, are present in vaginal washes in sufficient amounts ***(16)*** to allow an optimal environment for successful inhibition of pathogens.

In vitro tests provide an approach for determining the ability of lactobacilli to produce H_2O_2. The H_2O_2 amounts produced in such systems are probably not a direct reflection of what happens in the vaginal tract of women or animals, which is not yet

From: *Methods in Molecular Biology, vol. 268: Public Health Microbiology: Methods and Protocols*
Edited by: J. F. T. Spencer and A. L. Ragout de Spencer © Humana Press Inc., Totowa, NJ

know. However, there is a registered patent ***(4)*** with an H_2O_2-generating *L. crispatus* strain, also supporting the use of H_2O_2-producing lactobacilli to restore the vaginal ecosystem

The detection of H_2O_2 can be performed with different methods, including a number of plate and spectrophotometric tests ***(17)***. As the results reported in the references are quite different, we optimized two methods to perform the screening and study of the kinetics of production of H_2O_2 by *Lactobacillus* strains isolated from human and bovine vagina. The qualitative method on plate ***(18)*** employs horseradish peroxidase incorporated in agar medium, which oxidizes the chromogenic substrate (tetramethylbenzidine) to a purple-blue pigment in those colonies that produce H_2O_2. The quantitative spectrophotometric method uses horseradish peroxidase and involves oxidation of *o*-dianisidine in the presence of the H_2O_2 produced during the growth of lactobacilli in broth media ***(19)***. In previous papers, the isolation and identification of human and bovine vaginal lactobacilli were reported ***(20,21)***, and the microorganisms with potential probiotic use were selected by applying the methodology described in the present chapter ***(22–24)***. The techniques had the following objectives:

1. Screening for the production of antimicrobial substances among human and bovine vaginal *Lactobacillus* strains, by using a plate diffusion technique ***(8)***.
2. Detection of H_2O_2-producing lactobacilli by a qualitative plate method ***(18)*** and a modified quantitative spectrophotometric method using *o*-dianisidine, horseradish peroxidase ***(19)***.
3. Study of the kinetics of production of hydrogen peroxide by selected *Lactobacillus* strains.

2. Materials

2.1. Screening of Antimicrobial Substances Generated by Lactic Acid Bacteria

2.1.1. Microorganisms

1. *Lactobacillus* strains isolated from vaginal swabs from humans and cows. The methods applied for the isolation and identification of bovine and human vaginal lactobacilli were previously reported ***(20,21)***.
2. Human vaginal pathogens: *Escherichia coli*, *Klebsiella* sp., group B *Streptococcus* sp., *Enterococcus faecalis*, *Staphylococcus aureus*, and *Candida* sp. (from the Culture Collection of the Instituto de Microbiología "Luis C. Verna" of the Universidad Nacional de Tucumán, Argentina). *Streptococcus agalactieae* ATCC 1020 (from the American Type Culture Collection) was also employed.
3. Bovine pathogens (isolated from clinical material of metritis): *Escherichia coli* 99/14 and *Corynebacterium pyogenes* 96/393 (from the Culture Collection of Estación Experimental Agropecuaria Balcarce, INTA).

2.1.2. Storage of Microorganisms

1. LAPTg broth: 1.5% peptone, 1% tryptone, 1% glucose, 1% yeast extract, 0.1% Tween-80 (pH 6.5). Autoclave at 121°C for 15 min and store at refrigeration temperature. The chemicals for LAPTg preparation were obtained from Britania Laboratories (Argentina).
2. Brain-heart infusion (BHI) agar: 20% brain infusion, 20% heart infusion, 1% peptone, 0.2% glucose, 2.5% Na_2HPO_4, 1.5% agar (pH 7.4). Autoclave at 121°C for 15 min and

store at refrigeration temperature. Dehydrated BHI medium was obtained from Britania Laboratories (Argentina).
3. Milk-yeast extract: 10% skim milk, 1% glucose, 0.5% yeast extract, pH 6.0. Autoclave for 20 min at 115°C (3/4 atm) and store at refrigeration temperature.
4. BHI broth with glycerol: the same as **step 2** of this subheading without agar, added with 25% glycerol.

2.1.3. Plate Diffusion Technique

1. 2 *N* NaOH: Sterile distilled water, NaOH (anhydrous, Anedra, Argentina). Store at room temperature.
2. Catalase (EC 1.11.1.6) used at 1000 U/mL, (from bovine liver, Sigma, St. Louis, MO). Store at <0°C. **Caution**: Avoid contact and inhalation.
3. 0.22 µm Membrane filters, 25-mm dia. (Millipore, Bedford, MA). Autoclave at 121°C for 15 min.
4. Peptone water (0.1% meat peptone). Autoclave at 121°C for 15 min and store at refrigeration temperature.
5. LAPTg agar (*see* **Subheading 2.1.2.**) with 1% agar.
6. BHI agar (*see* **Subheading 2.1.2.**) with 1% agar.
7. Plastic straws, cut in 7–8-cm pieces and sterilized in glass flasks.

2.2. Screening of Hydrogen Peroxide-Generating Lactic Acid Bacteria

2.2.1. Qualitative Method on Plate

1. de Man-Rogosa-Sharpe (MRS) agar: Dehydrated MRS was obtained from Biokar Diagnostics (Beauvais, France). Composition (w/v): 1% peptone, 1% meat extract, 0.5% yeast extract, 2% glucose, 0.1% Tween-80, 0.2% dipotassium phosphate, 0.5% sodium acetate, 0.2% ammonium citrate, 0.02% magnesium sulfate, 0.005% manganese sulfate, and 1.5% agar (pH 6.5). Autoclave at 121°C for 15 min and store at refrigeration temperature.
2. 17.34 m*M* TMB: (3,3',5,5'-tetramethyl-benzidine (Sigma). TMB is air- and light-sensitive; store at 2–8°C. Caution: Irritating to eyes, respiratory system and skin; possible risk of irreversible effects, possible mutagen.
3. Methanol pro-analysis (Merck, Darmstadt, Germany).
4. 0.5 mg/mL (100 U/mL) Peroxidase (EC 1.11.1.7), type II: From horseradish, 200 U/mg solid (Sigma), stored at <0°C.
5. Sterilized distilled water.

2.2.2. Quantitative Method: Spectrophotometric Method Modified by Using o-Dianisidine Horseradish Peroxidase

1. 3.2 m*M* *o*-Dianisidine: 3,3'-dimethoxybenzidine (Fast Blue B, dihydrochloride, from Sigma) light-sensitive, stored at 2–8°C.
2. 0.4% Triton X-100: *t*-octyl-phenoxy-polyethoxy-ethanol, nonionic surfactant (Sigma), stored at room temperature.
3. 120 µg/mL (24 U/mL) Peroxidase (*see* **Subheading 2.2.1., item 4**).
4. 10 m*M* Phosphate buffer pH 7.4: NaH_2PO_4 and Na_2HPO_4 (Anedra, Argentina) stored at 4°C.
5. 30% H_2O_2, pro-analysis solution (Cicarelli, Argentina), stored at 4°C.

2.3. Kinetics of Hydrogen Peroxide Production by Lactobacilli

1. Saline solution (0.85% NaCl) autoclave at 121°C for 15 min and store at refrigeration temperature.
2. 250-mL Erlenmeyer flasks containing 100 mL of LAPTg (*see* **Subheading 2.1.2.**) or MRS broth (*see* **Subheading 2.2.1.**, but without agar), both at initial pH of 5.0, 6.5, and 8.0.
3. Water baths (Vicking S.R.L. Masson model and Dubnoff model Shaker, Industria Argentina).
4. Digital spectrophotometer (model 250; Gilford).
5. Glass cuvets (10-mm light path).
6. Blanks for the spectrophotometric determinations (sterile growth media).
7. Digital pHmeter (Digimeter IV, Luftman, Argentina).
8. Reference standard buffers (pH 7.00 and 4.00, from LQT, Argentina).
9. Plastic bottles to receive the samples.
10. 0.1% Peptone water (*see* **Subheading 2.1.3.**, **item 4**, dilution medium).
11. LAPTg agar (*see* **Subheading 2.1.2.**, **item 1** with 1.5% agar) and MRS agar (*see* **Subheading 2.2.1.**, **item 1**)
12. Incubator at a temperature of 37°C (Forma Scientific, model 3185).
13. Reagent solution for the spectrophotometric method modified by using *o*-dianisidine horseradish peroxidase (*see* **Subheading 2.2.2.**).

3. Methods

3.1. Screening of Antimicrobial Substances Generating Lactic Acid Bacteria

3.1.1. Storage of Microorganisms

1. Screwtop-glass tubes containing 5 mL LAPTg broth were inoculated with a colony of each of the lactobacilli isolated and incubated at 37°C for 24 h. A 150-μL aliquot of these cultures was subcultured twice at 37°C for 12 h in 5 mL of the same medium.
2. The pathogenic microorganisms were subcultured three times in LAPTg broth for 12–14 h at 37°C, except for *Corynebacterium pyogenes*, which was subcultured two times in BHI agar, for 24 h at 37°C in a 5% CO_2 atmosphere (*see* **Note 1**).
3. The third culture (obtained in **step 1**) was centrifuged (10 min, 2000*g*), and the spent supernatant was discarded.
4. The pellet was resuspended in 2 mL of milk-yeast extract for lactobacilli and pathogen microorganisms, or in BHI with 25% glycerol for *Corynebacterium pyogenes*. This bacterial suspension was distributed in 1-mL cryovials and stored at –20°C.

3.1.2. Plate Diffusion Technique

1. Subcultivation of the lactobacilli and vaginal pathogens was performed as described in **Subheading 3.1.1.** (**steps 1** and **2**), but from the microorganisms stored at –20°C in milk-yeast extract or BHI/glycerol. The third lactobacilli culture was centrifuged (10 min, 2000*g*), and the spent supernatants were separated for detection of inhibitory substances.
2. The supernatants (obtained as described in **step 1**) were filter-sterilized (0.22 μm) and fractioned in three aliquots. The first aliquot was not treated, the second was neutralized with 2 *N* NaOH, and the third was neutralized and treated with catalase (1000 U/mL) for 1 h at 25°C. The treated supernatants were stored at 4°C while the plates with the pathogens microorganisms were prepared.
3. Overnight cultures of pathogens, grown as described in **Subheading 3.1.1.**, were diluted in peptone water to obtain a concentration of 10^7–10^8 CFU/mL (*see* **Note 2**).

Table 1
Results of the Plate Diffusion Techinque

Treatment of supernatant	Inhibition halo	Antagonistic substance
Neutralized	Absent	Organic acids
	Present	H_2O_2 or bacteriocin
Neutralized and treated with catalase	Absent	H_2O_2
	Present	Bacteriocin

4. Samples of 0.5 mL of each pathogen dilution were mixed with 15 mL of 45°C melted LAPTg agar or BHI agar for *Corynebacterium pyogenes* and poured on Petri dishes. Holes of 4 mm in the pathogen plates were punched with sterilized straws, and 25 µL of lactobacilli supernatant aliquots (obtained as described in **step 1** of this subheading) were added to the holes.
5. The plates were incubated for 5 h at room temperature, and then for 24 h at 37°C under aerobic conditions, except for *Corynebacterium pyogenes*, which was incubated under 5% CO_2 atmosphere. The diameters of the inhibition halos were measured after the incubation. If an inhibition halo is present in a nontreated supernatant, the antagonistic substance can be organic acid, H_2O_2, or bacteriocin.
6. Interpretation of the results is shown in **Table 1**.

3.2. Screening of Hydrogen Peroxide-Generating Lactic Acid Bacteria

3.2.1. Qualitative Method on Plate

1. Solution A was prepared with 12.5 mg of TMB (*see* **Note 3**) and 3 mL of methanol (*see* **Note 4**; 17.34 m*M* TMB), with shaking to complete dissolution.
2. Solution B was prepared with 0.5 mg of peroxidase and 1 mL sterilized distilled water (0.5 mg/mL or 100 U/mL peroxidase; *see* **Note 5**).
3. Preparation of TMB-agar plates: MRS medium was melted at 45°C. Then 10 mL of MRS agar were mixed with a 0.6-mL aliquot of solution A and a 0.2-mL aliquot of solution B. This mixture was added to plates and allowed to solidify (*see* **Note 6**). The final concentrations were: 1 m*M* TMB and 10 µg/mL (2 U/mL) peroxidase.
4. Assay: The lactobacilli, grown as described in **Subheading 3.1.2., step 1**, were spread over the surface of TMB agar plates. The plates were then incubated microaerophically at 37°C, for 48 h. After 48 h, plates were exposed to air. Colonies able to produce hydrogen peroxide turn blue or brown (*see* **Notes 7–9**).

3.2.2. Quantitative Method: Spectrophotometric Method Modified by Using o-Dianisidine Horseradish Peroxidase

3.2.2.1. Preparation of Reagent Solution

1. Solution C was prepared with 1 mg of *o*-dianisidine (*see* **Note 10**) per mL of 0.4% Triton X 100 (3.2 m*M* *o*-dianisidine).
2. Solution D was prepared with 120 µg of peroxidase per mL of 10 m*M* phosphate buffer (120 µg/mL or 24 U/mL peroxidase; *see* **Note 11**).

3. The reagent solution (20 mL) was prepared by using: a 1-mL aliquot of solution C, a 0.3 mL aliquot of solution D, 0.2 mL of 0.4% Triton X-100, and 18.5 mL of 10 m*M* phosphate buffer (pH 7.4). The final concentrations were: 0.16 m*M* *o*-dianisidine and 1.8 μg/mL (0.36 U/mL) peroxidase (*see* **Notes 12** and **13**).

3.2.2.2. Standard Curve for the Quantification of H_2O_2

1. An H_2O_2 solution (8.8 m*M*) was prepared with 5 μL of 30% H_2O_2 solution filled up to 5 mL with distilled water (*see* **Note 14**). The solutions with different H_2O_2 concentrations (from 0.00176 to 2.64 m*M*) were prepared from 8.8 m*M* H_2O_2 solution (*see* **Note 15**).
2. An 0.8-mL aliquot of reagent solution was mixed with a 0.2-mL aliquot of each H_2O_2 concentration, or with 0.2 mL of distilled water (reference). These mixtures were incubated at 25°C for 15 min.
3. The absorbance was measured at 430 nm, subtracting the optical density (OD) of the reference.
4. The OD values were plotted against the H_2O_2 concentrations (in m*M*). The slope of the linear part of the standard curve is the transformation factor of OD values to millimolar H_2O_2 concentrations.

3.2.2.3. Detection of the Levels of H_2O_2 Produced During the Growth of Microorganisms

1. The microorganisms were grown twice in LAPTg broth without agitation at 37°C and a third time (15 h, 37°C) with agitation (at 50 opm).
2. The supernatant fluids of the third cultures were separated by centrifugation (2000*g*, 10 min) (*see* **Note 16**).
3. A 0.2-mL aliquot of the supernatant fluids of lactobacilli was mixed with a 0.8-mL aliquot of reagent solution, and the assay was performed as described above (*see* **Subheading 3.2.2.1.**) (*see* **Note 17**).

3.3. Kinetics of Hydrogen Peroxide Production by Lactobacilli

3.3.1. Inoculum Buildup for the Growth Experiments

1. Lactobacilli subcultivation: *L. crispatus* CRL 1266, *L. paracasei* CRL 1251 and 1289 (from human vagina), and *L. acidophilus* 1012 and 2014 (from bovine vagina). These *Lactobacillus* strains were selected for hydrogen peroxide production and inhibition of pathogenic microorganisms, according to the screening performed. The lactobacilli were grown as described in **Subheading 3.1.2., step 1**.
2. The third culture was centrifuged (10 min, 2000*g*), and the spent supernatant was discarded. The pellet was washed with 3 mL saline solution and then resuspended in the same solution to reach a final OD of 1.4 at 540 nm. This bacterial cell suspension was used as the inoculum for the growth experiments.

3.3.2. Kinetics of Hydrogen Peroxide Production

1. The lactobacilli were inoculated (2% v/v) into 100 mL of each medium (250-mL Erlenmeyer flasks) and then incubated in a water bath with and without agitation, at constant temperatures (30, 37, and 44°C). Samples were withdrawn at different times from lactobacilli culture (every 2 or 3 h) to perform the analytical procedures (*see* **Note 18**).
2. Optical density measurements (OD at 540 nm): The initial turbidity of the growth media was corrected to absorbance value 0 (zero) with the corresponding blanks of culture media placed in glass cuvets.

3. The samples of lactobacilli cultures were placed in small plastic bottles to determine the pH with a digital pHmeter, previously calibrated with the reference standard buffers.
4. Determination of colony forming units (CFUs): The samples were serially diluted (10-fold) in peptone water, from 1:10 to 1:1,000,000. Aliquots of each dilution were placed on the plates, and 12–15 mL of 45°C melted culture medium (LAPTg or MRS agar, depending on whether the previous lactobacilli growth was in LAPTg or MRS broth) were added. The plates were incubated at 37°C for 48 h (*see* **Note 19**).
5. Quantification of hydrogen peroxide: Samples taken at different times of lactobacilli culture were centrifuged at 2000*g* for 10 min, and the supernatant fluids were used to perform the assay (spectrophotometric method modified by using *o*-dianisidine horseradish peroxidase), as described in **Subheading 3.2.2.3.** (*see* **Note 20**).

4. Notes

1. To enhance the growth of *Corynebacterium*, BHI medium can be enriched with: 1:10 v/v hemine solution (hemine 50 mg, 1 mL 1 *N* NaOH, in 100 mL of solution) and 0.02:100 v/v vitamin K.
2. Before performing the plate diffusion technique, is necessary to set up a calibration curve of optical density vs CFU/mL, to select the best inoculum of pathogen microorganisms to achieve 10^7–10^8 CFU/mL in the plates. In the case of *Corynebacterium pyogenes*, the bacterial suspension ($OD_{540} = 0.24$) was made from the microorganisms grown on the surface of the agar plates and then diluted 1:10.
3. **Caution**: TMB is irritating to the eyes, respiratory system, and skin, there is a possible risk of irreversible possibly mutagenic, effects.
4. **Caution**: Methanol is flammable and toxic. Avoid contact with the skin, and store far from flames or spark sources.
5. Solutions A and B were prepared immediately before the TMB agar plate preparation.
6. The TMB agar plates can be stored at 2–8°C for 1 wk maximum.
7. When performing these assay, an appropriate dilution of lactobacilli must be spread over the surface of TMB agar plates to obtain isolated small numbers of colonies. Is possible to use a surface dissemination method with a loop to obtain single colonies at the end of the dissemination.
8. The exposure of the plates to air is a very important step to allow rapid maximum color development of the colonies over the TMB agar plates.
9. The colonies that turn blue after exposure to air and TMB agar are those microorganisms that produce higher amounts of H_2O_2 when they grow in liquid media.
10. **Caution**: *o*-dianisidine is toxic and may cause cancer or heritable gene damage. It is irritating to the eyes, respiratory system, and skin; wear suitable protective clothing, gloves, and eye/face protection, and do not breathe the dust.
11. Solutions C and D were prepared immediately before the reagent solution preparation.
12. The final concentration of peroxidase in the reagent solution, modified from the original technique ***(19)***, was adjusted to 1.8 μg/mL.
13. The reagent solution should be stored in a dark flask at 4°C for 2 mo maximum.
14. **Caution**: 30% H_2O_2 is corrosive and flammable; avoid contact with the skin, and wear suitable protective clothing, gloves, and eye/face protection.
15. The solutions containing different concentrations of H_2O_2 were prepared on the day on which the standard curve was performed, because storage of H_2O_2 can result in complete loss after a short time.

16. The third lactobacilli culture, for the screening performed with the spectrophotometric method modified by using *o*-dianisidine horseradish peroxidase, must always be grown with shaking, because there are no detectable levels of H_2O_2 in nonagitated cultures.
17. To determine the H_2O_2 produced during the growth of microorganisms, the reference solution and the solutions with different H_2O_2 concentrations used to perform the standard curve were prepared with the corresponding sterilized growth medium (LAPTg or MRS broth). If the OD of the assay mixture is out of the linear range of the standard curve, the samples should be diluted with the growth medium to perform the assay again, and the H_2O_2 concentration values obtained should be multiplied by the dilution factor.
18. The presence of oxygen in the environment of lactobacilli growth can influence their physiology. In fact, some lactic acid bacteria possess oxidases that reduce oxygen to hydrogen peroxide, oxidizing substrates such pyruvate or NADH *(25)*. The H_2O_2-producing lactobacilli were grown in agitated and nonagitated cultures in order to determine the factors that affect the production of this metabolite and the most favorable conditions to obtain the highest biomass. The vigorous shaking provided permanent aeration to cultures, whereas no shaking was used for nonaerated cultures. The results obtained showed that O_2 must be supplied through agitation to detect the H_2O_2 in spent culture supernatant fluids, whereas there was no detectable production (within the sensitivity range of the assay) of the metabolite in nonagitated cultures.
19. In aerated conditions, it is recommended to spread an aliquot of undiluted lactobacilli culture on the surface of agar plates because the growth of microorganisms can be inhibited by the H_2O_2 produced. The toxicity of H_2O_2 is caused by an effect exerted by the molecule itself or by superoxide anion and hydroxyl radicals, which cause oxidative stress in the cells, decreasing the number of viable cells *(25)*.
20. The samples must be withdrawn from the flasks every 2–3 h to minimize the time in which the flasks are outside the incubator. The supernatants separated by centrifugation can be stored at 4°C for 2–3 h up to the H_2O_2 determination in order to avoid the loss of this metabolite.

Acknowledgments

This paper was supported with grants from CONICET (Consejo Nacional de Investigaciones Científicas y Técnicas de Argentina), PIP 359, and Beca Ramón Carrillo-Arturo Oñativia para Investigadores formados (Ministerio de Salud Pública de Argentina).

References

1. Reid, G. (1999) The scientific basis for probiotic strains of *Lactobacillus*. *App. Env. Microbiol.* **65,** 3763–3766.
2. Redondo-López,V., Cook, R. L., and Sobel, J. D. (1990) Emerging role of lactobacilli in the control and maintenance of the vaginal bacterial microflora. *Rev. Infect. Dis.* **12,** 856–872.
3. Reid, G. and Bruce, A. W. (2001) Selection of *Lactobacillus* strains for urogenital probiotic applications. *J. Infect. Dis.* **183(suppl 1),** S77–S80.
4. Chrisope, G. L. (2000) Vaginal *Lactobacillus* medicant. US patent & trademark office, United States Patent, GyneLogix, Inc., Boulder, CO, no. 093,394-6-21.
5. Larsen, B. (1999) Vaginal pharmaceutical compositions. US patent & trademark office, United States Patent, Marshall University Research Corp. Huntington, WV, no. 5,958,461-7-34

6. Reid, G. and Bruce, A. W. (1997) *Lactobacillus* and skim milk compositions and methods for preventing microbial urogenital infections. US patent & trademark office, United States Patent, Research Corporation Technologies, Inc., Tucson, AZ, no. 5,643,830-1-1.
7. Eschenbach, D., Davick, P., Williams, S., Klebanoff, S., Young-Smith, C., and Holmes, K. K. (1989) Prevalence of hydrogen peroxide-producing *Lactobacillus* species in normal women with bacterial vaginosis. *J. Clin. Microbiol.* **27,** 251–256.
8. Jack, R. W., Tagg, J. R., and Ray, B. (1995) Bacteriocins of Gram-positive bacteria. *Microbiol. Rev.* **59,** 171–200.
9. Aroutcheva, A., Gariti, D., Simon, M., et al. (2001) Defense factors of vaginal lactobacilli. *Am. J. Obstet. Gynecol.* **185,** 375–379.
10. McGroarty, J. (1993) Probiotic use of lactobacilli in the human female urogenital tract. *FEMS Immunol. Med. Microbiol.* **6,** 251–264.
11. Goffeng, A. R., Holst, E., Milsom, I., Lindstedt, G., Lundberg, P. A., and Andersch, B. (1997) Fetal fibronectin and microorganisms in vaginal fluid of women with complicated pregnancies. *Acta Obstet. Gynecol. Scand.* **76,** 521–527.
12. Hawes, S. E., Hillier, S. L., Benedetti, J., et al. (1996) Hydrogen peroxide-producing lactobacilli and acquisition of vaginal infections. *J. Infec. Dis.* **174,** 1058–1063.
13. Gupta, K., Stapleton, A. E., Hooton, T. M., Roberts, P. L., Fennell, C. L., and Stamm, W. E. (1998) Inverse association of H_2O_2-producing lactobacilli and vaginal *Escherichia coli* colonization in women with recurrent urinary tract infections. *J. Infect. Dis.* **178,** 446–450.
14. Martin, H. L. Jr., Richardson, B. A., Nyange, P. M., et al. (1999) Vaginal lactobacilli, microbial flora, and risk of human immunodeficiency virus type 1 and sexually transmitted disease acquisition. *J. Infect. Dis.* **180,** 1863–1868.
15. Bauer, G. (2001) Lactobacilli-mediated control of vaginal cancer through specific reactive oxygen species interaction. *Med. Hypotheses.* **57,** 252–257.
16. Klebanoff, S. J., Hillier, S. L., Eschenbach, D. A., Waltersdorph, A. M. (1991) Control of the microbial flora of the vagina by H_2O_2-generating lactobacilli. *J. Infect. Dis.* **164,** 94–100.
17. Juven, B. J. and Pierson, M. D. (1996) Antibacterial effects of hydrogen peroxide and methods for its detection and quantification. *J. Food Proteins* **11,** 1233–1241.
18. Berthier, F. (1993) On the screening of the hydrogen peroxide-generating lactic acid bacteria. *Lett. Appl. Microbiol.* **16,** 150–153.
19. Nuñez de Kairuz, M. S., Olazábal, M. E., Oliver, G., de Ruiz Holgado, A. P., and Farías, R. N. (1988) Fatty acid-dependent hydrogen peroxide production in *Lactobacillus*. *Biochem. Biophys. Res. Commun.* **152,** 113–121.
20. Ocaña, V. S., Bru, E., de Ruiz Holgado, A. P., and Nader-Macías, M. E. (1999) Surface characteristics of lactobacilli isolated from human vagina. *J. Gen. Appl. Microbiol.* **45,** 203–212.
21. Otero, C., Silva de Ruiz, C., Ibáñez, R., Wilde, O., Ruiz Holgado, A. and Nader-Macías, M. E. (1999) Lactobacilli and enterococci isolated from the bovine vaginal during the estrous cycle. *Anaerobe* **5,** 305–307.
22. Ocaña, V.S., P.de Ruiz Holgado, A., and Nader-Macías, M. E. (1999) Selection of vaginal H_2O_2-generating *Lactobacillus* for probiotic use. *Curr. Microbiol.* **38,** 279–284.
23. Ocaña, V.S., P.de Ruiz Holgado, A., and Nader-Macías, M. E. (1999) Growth inhibition of *Staphylococcus aureus* by H_2O_2-producing *Lactobacillus paracasei* subsp. *paracasei* isolated from the human vagina. *FEMS Immunol. Med. Microbiol.* **23,** 87–92.

24. Juárez Tomás, M. S., Bru, E., and Nader-Macías, M. E. (2002) Comparison of the growth and H_2O_2-production by vaginal probiotic lactobacilli under different culture conditions. *Am. J. Obstet. Gynecol.* **188,** 35–44.
25. Marty-Teysset, C., De La Torre, F., Garel, J. R. (2000) Increased production of hydrogen peroxide by *Lactobacillus delbrueckii* subsp. *bulgaricus* upon aeration: Involvement of an NADH oxidase in oxidative stress. *Appl. Env. Microbiol.* **66,** 262–267.

32

Production of Antimicrobial Substances by Lactic Acid Bacteria II

Screening Bacteriocin-Producing Strains With Probiotic Purposes and Characterization of a Lactobacillus *Bacteriocin*

Virginia S. Ocaña and María Elena Nader-Macías

1. Introduction

Bacteriocins have been defined as proteinaceous, bactericidal substances synthesized by bacteria, which usually have a narrow spectrum of activity, only inhibiting strains of the same or closely related species *(1)*. The term *bacteriocin-like substance* is applied to antagonistic substances that are not completely defined or do not fit the typical criteria of bacteriocins. They have been reported to inhibit a wide range of both Gram-positive and Gram-negative bacteria as well as fungi *(2)*.

Lactobacillus species are the dominant microorganisms isolated from the vagina of healthy premenopausal women *(3)*. In this environment, they exert a protective effect against pathogenic microorganisms by different mechanisms such as production of antimicrobial agents, which include organic acids, hydrogen peroxide, and probably bacteriocins *(2,4–6)*. The production of bacteriocins by vaginal lactobacilli has been demonstrated in vitro; however, it is not yet well established whether they are produced in vivo as another antagonistic mechanism exerted by the normal microflora.

Bacteriocin-producing bacteria as well as bacteriocins *per se* are of growing interest as biological controls in the manufacture of beverages and fermented products, mainly in the area of dairy products *(7)*. These bacteria have also been proposed as probiotic candidates for human or animal use.

The objectives of the present chapter are to describe the methods employed for:

1. Detection of production of bacteriocins among vaginal *Lactobacillus* strains.
2. Characterization of the bacteriocin or bacteriocin-like substances.
3. Study of the kinetics of production and mode of action of bacteriocins.
4. Determination of the inhibition of pathogenic microorganisms by bacteriocin-producing strains in mixed cultures.

From: *Methods in Molecular Biology, vol. 268: Public Health Microbiology: Methods and Protocols*
Edited by: J. F. T. Spencer and A. L. Ragout de Spencer © Humana Press Inc., Totowa, NJ

2. Materials

2.1. Microorganisms

1. *Lactobacillus* strains. The microorganisms employed for this study were 134 *Lactobacillus* strains isolated from human vaginal swabs. The methods applied for the isolation and identification of lactobacilli have been previously reported *(8)*.
2. Pathogenic microorganisms. These were kindly provided by the Culture Collection of the Instituto de Microbiología "Luis C. Verna" (Universidad Nacional de Tucumán, Argentina). They had been isolated from vaginal swabs and identified by standard techniques. *Streptococcus agalactiae* ATCC 27956, *Enterococcus faecium* ATCC 19434, *Enterococcus faecalis* ATCC 19433 (all from American Type Culture Collection), *Enterococcus faecium* CRL 318 and *E. faecalis* CRL 341 (both from Cerela Culture Collection) were also used.

2.2. Culture, Dilution, and Conservation Media

1. LAPTg broth: 1.5% peptone, 1% tryptone, 1% glucose, 1% yeast extract, 0.1% Tween-80, pH 6.5. Sterilize at 121°C for 15 min (autoclaved), and store at refrigerated temperature.
2. LAPTg agar: 1.5% peptone, 1% tryptone, 1% glucose, 1% yeast extract, 0.1% Tween-80, 1.5% agar-agar, pH 6.5. Sterilize at 121°C for 15 min and store at refrigerated temperature.
3. LEL: 10% low fat milk, 0.5% yeast extract, 1% glucose. Sterilize at 121°C for 15 min and store at refrigerated temperature.
4. LBS agar: 1% peptone from casein, 0.5% yeast extract, 2% glucose, 0.6% potassium dihydrogen phosphate, 0.2% ammonium citrate, 0.1% Tween-80, 1.5% sodium acetate, 0.0575% magnesium sulfate, 0.0034% iron (II) sulfate, 0.012% manganese sulfate, 1.5% agar-agar, pH 5.5. Sterilize at 121°C for 15 min and store at refrigerated temperature.
5. Blood agar: Brain-heart infusion agar supplemented with 2% human blood. Add the blood to the 45°C melted agar and pour into Petri dishes. Store the plates at 4°C.
6. Chocolate agar: Brain-heart infusion supplemented with 2% human blood. Add the blood to the melted agar, heat at 70°C, and pour into Petri dishes. Store the plates at refrigerated temperature.
7. *Streptococcus faecalis* (SF) agar: Streptococcus Selective Agar (Merck, Germany), prepared according to the manufacturer's indications, final pH 7.4. Sterilize by autoclaving and store at refrigerated temperature.
8. Peptone water: 1% meat peptone in distilled water. Store at room temperature.

2.3. Proteolytic Enzymes and Other Reagents Employed for the Characterization of Bacteriocins

All of these were obtained from Sigma (St Louis, MO).

1. 10 mg/mL Proteinase K, trypsin, type II, and type XV protease in phosphate buffer, pH 7. Filter-sterilize the solutions (0.22-µm pore size membranes; Millipore) and store at –20°C.
2. 10mg/mL Chymotrypsin in Tris-HCl buffer, pH 8.0, with 0.01 *M* $CaCl_2$. Filter-sterilize the solution (0.22-µm pore size membranes; Millipore) and store at –20°C.
3. 10mg/mL Pepsin solution in citrate buffer, pH 3.0. Filter-sterilize the solution (0.22-µm pore size membranes; Millipore) and store at –20°C.
4. 1 *M* α-Amylase in phosphate buffer, pH 6.9. Filter-sterilize the solution (0.22-µm pore size membranes; Millipore) and store at –20°C.
5. 0.1 *M* Sodium *m*-periodate in acetate buffer, pH 4.5. Filter-sterilize the solution (0.22-µm pore size membranes; Millipore) and store at –20°C.

6. 0.1 g/mL Lipase solution in PBS, pH 7.7, with 0.01% $CaCl_2$. Filter-sterilize the solution (0.22-µm pore size membranes; Millipore) and store at –20°C.
7. Chloroform (Merck, Germany).
8. Ethanol 96° (Merck).
9. 1 *N* NaOH (Merck).
10. 10% HCl (Merck).

2.4. Accessories

1. Plastic straws, cut in 7–8-cm pieces and sterilized in glass flasks.
2. Sterile cotton swabs, used to take vaginal samples and to spread *Neisseria* and *Gardnerella* on agar plates.
3. 0.22-µm Membrane filters (Millipore).

3. Methods

3.1. Microorganisms

1. *Lactobacillus* storage: Isolated lactobacilli are grown in 5 mL of LAPTg for 12 h at 37°C, subcultured twice in the same media under identical conditions, and then separated by centrifugation at 2000*g* for 15 min. The pellets are resuspended in 5 mL of LEL and stored at –20 °C.
2. *Lactobacillus* growth: Frozen lactobacilli are subcultured tree times in LAPTg broth and incubated for 12–14 h at 37°C prior to performing the studies. The inocula for the cultures were 2%. Overnight cultures grown in LAPTg broth are centrifuged at 2000*g* for 30 min, and the supernatants were separated for detection of bacteriocins.
3. Pathogen growth and storage: *Enterococcus*, *Staphylococcus*, *Listeria*, *Candida* and *Streptococcus* species are grown in LAPTg broth and stored in milk-yeast extract at –20°C. They are subcultured three times in LAPTg for 12–14 h at 37°C prior to perform the inhibition studies. The inocula for the cultures are of 2%. *Neisseria gonorrhoeae* strains are grown in chocolate agar and *Gardnerella vaginalis* in blood agar, both incubated microaerophilically at 37°C. They are not stored but immediately employed for the inhibition assays.

3.2. Detection of Antimicrobial Activity

To perform screening for bacteriocin production, the following steps were followed:

1. The *Lactobacillus* supernatants (obtained as described in **Subheading 3.1., step 2**) were neutralized with 2 *N* NaOH, treated with catalase (1000 U/mL) for 1 h at 25°C, and filter-sterilized (Millipore, 0.22-µm pore size). The treated supernatants were stored at 4°C while the plates with the indicator strains were prepared.
2. Preparation of plates with indicator strains: The pathogens (grown as described in **Subheading 3.1., step 3**) were diluted in peptone H_2O (1:10 or 1:100 dilutions to obtain concentrations of 10^7–10^8 CFU/mL, mixed with 10 mL of 45°C melted LAPTg agar, and poured onto Petri dishes. Colonies of microorganisms isolated in blood or chocolate agar are transferred to the same media by using cotton swabs to test the antimicrobial activity of lactobacilli supernatant in pure cultures.
3. Effect of *Lactobacillus* supernatants on pathogen growth: 25 µL of lactobacilli supernatant aliquots (obtained as described in **Subheading 3.2., step 1**) were assayed on pathogen plates into which 4-mm holes had been punched with sterilized straws and that were filled with the supernatant. The plates are incubated for 5 h at room temperature (*see* **Note 1**) and then for 24 h at 37°C. The diameters of the inhibition halos are measured after the incubation.

3.3. Characterization of the Bacteriocin

Among 134 vaginal lactobacilli, *Lactobacillus salivarius* subsp. *salivarius* CRL 1328 (CERELA Culture Collection) was selected for further study because it was able to inhibit the growth of *Enterococcus*, *Listeria*, *Neisseria*, and *Gardnerella* strains in the plaque assay for a substance different from acid or hydrogen peroxide.

3.3.1. Effect of Temperature

1. Heat resistance of the antimicrobial substance is studied by heating aliquots of the filter-sterilized supernatant to 60°C, 80°C, and 100°C for 10 min and also after autoclave treatment (121°C at 15 min).
2. The heated supernatant aliquots were cooled down and tested for activity as described before (*see* **Subheading 3.2., step 3**).

3.3.2. Effect of pH

1. The pH of supernatant aliquots was adjusted to 2.0–8.0 with 2 *N* NaOH or 10% HCl prior to the filtration process.
2. Different pH supernatants were filter-sterilized and tested for activity as described before (*see* **Subheading 3.2., step 3**).

3.3.3. Activity of the Antimicrobial Substance After Treatment With Proteolytic Enzymes, Sodium m-Periodate, and Organic Solvents

3.3.3.1. Preparation of Reagent Solutions

1. Proteinase K, trypsin, type II, and type XV protease solutions.
   a. 10 mg of each enzyme was dissolved in 1 mL of phosphate buffer, pH 7, filter-sterilized (0.22-μm pore size membranes; Millipore) and stored at –20°C.
   b. To treat the spent lactobacilli supernatants, the pH was adjusted to 7, and 10 μL of the enzyme solution were mixed with 90 μL of supernatants, to obtain a final enzyme concentration of 1 mg/mL.
2. Chymotrypsin solution.
   a. 10 mg of enzyme were dissolved in 1 mL of Tris-HCl buffer, pH 8.0, with 0.1 *M* $CaCl_2$. The $CaCl_2$ must be dissolved in Tris-HCl buffer and then the enzyme can be added.
   b. The enzyme solution was filter-sterilized (0.22-μm pore size membranes; Millipore) and stored at –20°C.
   c. To treat the lactobacilli supernatants, the pH was adjusted to 8, and 10 μL of the enzyme solution were mixed with 90 μL of supernatants, to obtain a final enzyme concentration of 1 mg/mL (with a final $CaCl_2$ concentration of 0.01 *M*).
3. Pepsin solution.
   a. 10 mg of the enzyme were dissolved in 1 mL of citrate buffer, pH 3.0, filter-sterilized (0.22-μm pore size membranes; Millipore), and stored at –20°C.
   b. To treat the lactobacilli supernatants, the pH 3.0 was adjusted with 10% HCl, and 10 μL of the enzyme solution were mixed with 90 μL of supernatants, to obtain a final enzyme concentration of 1 mg/mL.

4. α-Amylase solution.
   a. The enzyme was resuspended in phosphate buffer, pH 6.9, to obtain a solution of a 1 *M* concentration.
   b. The enzyme solution was filter-sterilized (0.22-μm pore size membranes; Millipore) and stored at –20°C.
   c. To treat the lactobacilli supernatants, the pH was adjusted to 6.9 with 1 *N* NaOH, and 10 μL of the enzyme solution were mixed with 90 μL of supernatants, to obtain a final enzyme concentration of 0.1 *M*.
5. Sodium *m*-periodate solution.
   a. Sodium *m*-periodate was dissolved in acetate buffer, pH 4.5, to obtain a 0.1 *M* solution.
   b. The enzyme solution was filter-sterilized (0.22-μm pore size membranes; Millipore) and stores at –20°C.
   c. To treat the spent lactobacilli supernatant, the pH was adjusted to 4.5 with 10% HCl, and 10 μL of the enzyme solution were mixed with 90 μL of supernatants, to obtain a final enzyme concentration of 10 m*M*.
6. Lipase solution.
   a. The enzyme was dissolved in PBS pH 7.7 with 0.01% $CaCl_2$ to obtain a final concentration of 0.1 g/mL.
   b. The solution was filter-sterilized (0.22-μm pore size membranes; Millipore) and stored at –20°C.
   c. To treat the lactobacilli supernatants, the pH was adjusted to 7.7 with 1 *N* NaOH, and 10 μL of the enzyme solution were mixed with 90 μL of the supernatant, to obtain a final enzyme concentration of 10 mg/mL (containing 0.001% $CaCk_2$).

3.3.3.2. Treatment of Lactobacilli Supernatant With Enzymes and Sodium *m*-Periodate and Study of the Remaining Antimicrobial Activity

1. The enzymes and sodium *m*-periodate were diluted 1:10 in the spent lactobacilli supernatant (*see* **Note 2**) with a pH adjusted according to the maximal enzyme activity (as described in **Subheading 3.3.3.1.**) (*see* **Note 3**).
2. The mixtures are incubated at room temperature for 1 h and the remaining activity is tested as described in **Subheading 3.2.**
3. The effect of the enzyme solution diluted 1:10 in LAPTg (at the same final concentration as in spent supernatants) is also tested on the indicator plaques as controls (*see* **Note 4**).

3.3.3.3. Treatment of Lactobacilli Supernatant With Organic Solvents and Study of the Remaining Antimicrobial Activity

1. The supernatant is also treated with 10 and 25% chloroform and 25% ethanol. For chloroform treatment the steps are as follows
2. For chloroform treatment the steps were as follows:
   a. The organic solvent was added to the spent supernatant containing the bacteriocin (10 μL of chloroform and 90 μL of spent supernatants, as well as 25 μL of the organic solvent with 75 μL of the supernatant).
   b. The mix was vortex agitated for 1 min, and the phases were separated.
   c. The organic phase was evaporated by heating at 70°C in a water bath (*see* **Note 5**).
   d. The remaining dry residue was resuspended in LAPTg broth.

e. Both the residue pellet (resuspended in the same original volume of LAPTg broth) and the aqueous supernatant under extraction are assayed for bacteriocin activity as described in **Subheading 3.2.**

3. The effect of ethanol was study in a similar way.

a. 10 μL of 96° ethanol was added to 90 μL of spent supernatant.

b. After a 1-min vortex agitation, the ethanol is eliminated from the mixture by evaporation at 70°C.

c. The remaining pellet was dissolved in the original LAPTg volume and tested for activity as described in **Subheading 3.2.**

3.4. Kinetics of Bacteriocin Production

1. LAPTg broth was inoculated at 2% with the bacteriocin-producing lactobacilli (grown in LAPTg for 12 h, which is the early stationary phase).
2. The culture was incubated at 37°C and samples are taken every 3 h to determine the titer of the bacteriocin and the number of CFU/mL.
3. For titration, supernatants are separated from 2 mL of culture by centrifugation at 2000*g* for 30 min, filter-sterilized, and serially diluted in LAPTg broth. Samples of 25 μL of each dilution were poured into the holes of LAPTg agar plates containing the chosen indicator strain (in this example, *E. faecalis*; *see* **Note 6**), and inhibition is measured as described in **Subheading 3.2., step 2**. The titer was expressed in arbitrary units (AU), defined as the inverse of the dilution (in mL) that produced inhibition.
4. The number of CFU/mL was determined by serial dilution of samples taken from the culture at different times using peptone H_2O for the dilutions and LAPTg agar plates to determine the number of colonies grown.

3.5. Mixed Cultures of L. salivarius and E. faecalis

1. The effect of bacteriocin-producing *L. salivarius* on *E. faecalis* growth can be studied in mixed cultures of both microorganisms performed in LAPTg broth at 37°C.
2. The initial inocula is 10^6 CFU/mL for lactobacilli and 10^5 or 10^7 CFU/mL for enterococci.
3. The number of CFU is determined by the serial dilution method using peptone H_2O for the dilutions and LBS or SF agar plates to count viable microorganisms.
4. The plates are incubated for 48 h at 37°C.

3.6. Mode of Action of the Bacteriocin

1. Supernatant fluid of a 12-h *L. salivarius* culture is obtained by centrifugation at 2000*g* for 30 min and filter-sterilized as described in **Subheading 3.1., step 1**.
2. Aliquots of 5 mL are lyophilized, resuspended in the same volume of LAPTg broth, inoculated with *E. faecalis* (10^3, 10^5, and 10^7 CFU/mL), and incubated at 37°C (*see* **Note 7**).
3. Samples are taken at different times, and the CFU/mL of *Enterococcus* were determined-by the serial dilution method using H_2O peptone for the dilutions and SF agar plates to count the viable microorganisms. *Enterococcus* growth in LAPTg is used for controls.

4. Notes

1. Incubation at room temperature is an important step, which allows diffusion of the supernatant into the media prior to the incubation step, which in turn allows growth of the indicator microorganisms to begin.

2. The *Lactobacillus* supernatant was always obtained as described in **Subheading 3.2., step 1**.
3. The spent supernatant pH can be adjusted over a wide range after testing the pH resistance of the bacteriocin.
4. It was observed, for all cases, that the enzyme solutions (tested in controls) do not produce any kind of inhibition. In any case, control of their effect on the growth of indicator microorganisms in recommended.
5. **Caution**: Do not try to evaporate the solvents directly in the flame.
6. The indicator strain should grow in the same broth as the lactobacilli and should be chosen from sensitive species that are easily grown.
7. The purpose of lyophilizing and resuspending the spent lactobacilli supernatant in LAPTg broth instead of using the spent supernatant directly, is to ensure that inhibition is not caused by a deficiency of any metabolite. It is possible to work using this strategy because the bacteriocin is resistant to lyophilization.

Acknowledgments

This paper was supported by grants from CONICET (Consejo Nacional de Investigaciones Científicas y Técnicas), PID-BID 395, and the Ministerio de Salud Pública de la República Argentina (Beca Carrillo Oñativia para investigadores formados).

References

1. Jack, R. W., Tagg, J. R., and Ray, B. (1995) Bacteriocins of Gram-positive bacteria. *Microbiol. Rev.* **59,** 171–200.
2. McGoarty, J. (1993) Probiotic use of lactobacilli in the human female urogenital tract. *FEMS Immunol. Med. Microbiol.* **6,** 251–264.
3. Larsen, B. (1993) Vaginal flora in health and disease. *Clin. Obstet. Gynecol.* **36,** 107–121.
4. Klebanoff, S. J., Hillier, S. L., Eschenbach, D. A., and Waltersdorph, A. M. (1991) Control of the microbial flora of the vagina by H_2O_2-generating lactobacilli. *J. Infect. Dis.* **164,** 94–100.
5. McGoarty, J. A., Tomeczek, L., Pond, D., Reid, G., and Bruce, A. (1992) Hydrogen peroxide production by *Lactobacillus* species: correlation with susceptibility to the spermicidal compound nonoxynol-9. *J. Infect. Dis.* **165,** 1142–1144.
6. Khedkar, C. D., Dave, J. M., and Sannabhadti, S. S. (1990) Antibacterial activity of human strains of *Lactobacillus acidophilus* grown in milk against selected pathogenic and spoilage type bacteria. *Cultured Dairy Products J.* **25,** 29–31.
7. Okereke, A. and Montville, T. (1991) Bacteriocin-mediated inhibition of *Clostridium botulinum* spores by lactic acid bacteria at refrigeration and abuse temperatures. *Appl. Environ. Microbiol.* **57,** 3423–3428.
8. Ocaña, V. S., Bru, E., de Ruiz Holgado, A. P., and Nader-Macías, M. E. (1999) Surface characteristics of lactobacilli isolated from human vagina. *J. Gen. Appl. Microbiol.* **45,** 203–212.

33

Statistical Models to Optimize Production of Probiotic Characteristics

María Silvina Juárez Tomás, Elena Bru, and María Elena Nader-Macías

1. Introduction

The use of probiotic microorganisms has been widely promoted in the last 20 years *(1)*. Probiotics have applied to the gastrointestinal tract as capsules or as fermented milks *(2)*. The characteristics of the strains proposed as probiotics have been published *(3)* or patented after elaboration of the product.

The first step in designing a probiotic product is to isolate and characterize strains with beneficial properties. Our group has reported the isolation and characterization of 134 strains of lactobacilli from the vagina of women from our city *(4)*. Surface properties and the production of antagonistic substances has also been published. From these studies, certain strains were selected because they shared probiotic characteristics *(5–7)*.

The second step in the design of a pharmaceutical product is to determine the optimal conditions for obtaining the highest amount of viable microorganisms, as well as the best conditions to produce antagonistic substances *(8)*. To perform these types of experiments, laboratory assays can be carried out to determine the optimum conditions in which the microorganisms can be cultured. Growth parameters and optimal growth conditions can be studied by the classical methodology *(9)* or by the application of alternative methodologies *(10–12)*. Alternative methods have been widely used in the food industry, including experimental designs to test multiple conditions at the same time, complemented by the use of statistical models to evaluate and compare all the conditions tested *(13–15)*.

The object of this chapter is to describe the experimental designs used to study the technological characteristics of vaginal probiotic lactobacilli as well as the application of statistical models for evaluation of the different conditions tested for the production of biomass and antagonistic substances in these bacteria. With this objective in mind, in the present chapter, the following areas are explored:

1. Growth experiments performed to study the effect of temperature, pH, agitation, and culture media on the growth and production of antagonistic substances by vaginal lactobacilli.

From: *Methods in Molecular Biology, vol. 268: Public Health Microbiology: Methods and Protocols*
Edited by: J. F. T. Spencer and A. L. Ragout de Spencer © Humana Press Inc., Totowa, NJ

2. Fractional and complete factorial experimental designs used to perform the growth experiments.
3. Statistical models applied to evaluate the growth parameters and their modifications under the conditions assayed.

2. Materials

2.1. Growth Experiments

2.1.1. Inoculum Buildup for the Growth Experiments

1. Probiotic microorganisms: *Lactobacillus salivarius* subsp. *salivarius* CRL 1328, *L. crispatus* CRL 1266, and *L. paracasei* subsp. *paracasei* CRL 1289.
2. LAPTg broth: 15 g/L peptone, 10 g/L tryptone, 10 g/L glucose, 10 g/L yeast extract, and 1 mL/L Tween-80, pH 6.5. Autoclave at 121°C for 15 min. Store at refrigeration temperature. The chemicals for LAPTg preparation were obtained from Britania (Argentina).
3. MRS broth: Dehydrated MRS was obtained from Biokar Diagnostics (Beauvais, France). Composition: 10 g/L peptone, 10 g/L meat extract, 5 g/L yeast extract, 20 g/L glucose, 1.08 g/L Tween-80, 2 g/L dipotassium phosphate, 5 g/L sodium acetate, 2 g/L ammonium citrate, 0.2 g/L magnesium sulfate, and 0.05 g/L manganese sulfate, pH 6.5. Autoclave at 121°C for 15 min. Store at refrigeration temperature.
4. Saline solution (0.85% NaCl). Autoclave at 121°C for 15 min. Store at refrigeration temperature.
5. Spectrophotometer (Spectronic 20, Bausch & Lomb, Miltou Roy). Determinations of optical density (OD) in screw-top glass tubes.

2.1.2. pH Adjustment of Culture Media

1. Solutions for pH adjustment: 1 *N* HCl and 1 *N* NaOH. Sterile distilled water, HCl ACS (reagent grade, purity 37%; Anedra, Argentina), NaOH (anhydrous, Anedra). The reagent for the solutions and the 1 *N* solutions are stored at ambient temperature.
2. Digital pH meter (Digimeter IV, Luftman, Argentina).
3. Standard reference buffers:
   a. pH 7.00 (± 0.01 at 25°C): 50 mL 0.1 *M* KH_2PO_4 + 29.1 mL 0.1 *M* NaOH, diluted to 100 mL with distilled water; color-coded yellow (LQT, Argentina). Store at room temperature.
   b. pH 4.00 (± 0.01 at 25°C): 50 mL 0.1 *M* KH phthalate + 0.1 mL 0.1 *M* HCl, diluted to 100 mL with distilled water; color-coded red (LQT, Argentina). Store at room temperature.
4. Accessories: Distilled water, blotting paper, and plastic bottles to receive the samples.
5. Culture media: LAPTg and MRS broths (described in **Subheading 2.1.1.**), both at initial pH of 5.0, 6.5, and 8.0.

2.1.3. Incubation Systems

2.1.3.1. Cultures Without Agitation

1. 250-mL Erlenmeyer flasks containing 100 mL of each culture media.
2. Sterile screw-top glass tubes.
3. Water baths (Vicking S.R.L., Masson model, Industria Argentina) at incubation temperatures of 30, 37, or 44°C.

2.1.3.2. Cultures With Agitation

1. 250-mL Erlenmeyer flasks containing 100 mL of each culture media.
2. Water baths (Vicking S.R.L. Shaker, Dubnoff model, Industria Argentina, speed of agitation: 50 opm), incubation temperatures of 30, 37, or 44°C.

2.1.4. Obtaining Samples

1. Automatic pipets.
2. Laminar flow chamber (Clean Room Products).
3. Screw-top glass tubes and Eppendorf tubes.

2.1.5. Optical Density Measurements

1. Digital spectrophotometer (model 250; Gilford Instrument Lab).
2. Glass cuvets with a 10-mm light path.
3. Blanks for the spectrophotometric determinations: Screw-top glass tubes with sterile LAPTg or MRS broths, both to pH 5.0, 6.5, or 8.0 (*see* **item 1** of **Subheading 2.1.2.**).

2.1.6. Determination of Colony-Forming Units (CFUs)

1. Dilution medium: Peptone water (0.1% meat peptone). Autoclave at 121°C for 15 min. Store at refrigeration temperature.
2. Automatic pipets (for 100 and 1000 μL), glass tubes containing peptone water (stored at refrigeration temperature), vortex (Vicking, Argentina).
3. Culture media: LAPTg agar or MRS agar (as described in **Subheading 2.1.1.**, but with 15 g/L agar).
4. Incubation system: Incubator, temperature of 37°C (Forma Scientific, model 3185).

2.1.7. Registration of the Experimental Data

Sheets can be obtained using Microsoft Excel for Windows.

2.1.8. Determination of Antagonistic Substances (H_2O_2 and Bacteriocin)

The experimental details of these determinations are described in Chapter 31 (hydrogen peroxide) and Chapter 32 (bacteriocin).

2.2. Experimental Designs

1. Complete factorial 2 × 3 × 3 design. To perform the growth and bacteriocin-production experiments for *L. salivarius* CRL 1328, a design of three factors with 2, 3, and 3 levels, respectively, is used. This design has in 18 different treatments total (2 × 3 × 3).
2. Fractionated factorial design. The screening of growth curves and H_2O_2 production by *L. crispatus* CRL 1266 and *L. paracasei* CRL 1289 is performed by using a factorial fractional 2^{5-1} design, with eight central points. A total of 24 different combinations are tested. The fractionated factorial design was obtained by using the MINITAB software.

2.3. Estimation of Growth Parameters

2.3.1. Description of the Bacterial Growth Curves

A modified four-parameter-Gompertz model was used ***(10)***.

2.3.2. Method for Estimation of the Growth Curves

1. Ordinary least squares: Statistical software STATISTICA 5.0. (STATISTICA for Windows [Computer program manual], Stat Soft Inc., 1995). For the algorithm of minimization of the loss function, the Rosenbrock/quasi-Newton procedure was applied.
2. Bootstrap method: SPSS 10 for Windows and S-PLUS 200. To plot the growth curves and also to estimate their parameters, the bootstrap method is used.

2.3.3. Analysis of the Variance (ANOVA)

The statistical software MINITAB was used.

3. Methods

3.1. Growth Experiments

3.1.1. Inoculum Buildup for the Growth Experiments

1. Probiotic microorganisms: Lactobacilli were isolated from vaginal swabs of women from Tucumán, Argentina. The microorganisms were identified by the methods described in *Bergey's Manual of Determinative Bacteriology* ***(16)***, standard techniques used in the laboratory, and API 50 CHL (Biomerieux-France) ***(4)***. Some of the selected strains have good surface properties and probiotic characteristic, as follows: *L. salivarius* subsp. *salivarius* CRL (Centro de Referencia para Lactobacilos Culture Collection) 1328, producer of a bacteriocin ***(5)*** and *L. crispatus* CRL 1266 and *L. paracasei* subsp. *paracasei* CRL 1289, producers of H_2O_2 ***(6,7)***.
2. Before the growth experiments, 5 mL of either MRS or LAPTg broths (tubes) are inoculated with 30 µL of lactobacilli stored at –20°C in milk-yeast extract and incubated at 37°C for 24 h. A 150-µL aliquot of these precultures is cultivated twice at 37°C for 12 h in 5-mL of the same media (*see* **Note 1**).
3. The third culture is centrifuged (10 min, 2000*g*), and the spent supernatant is discarded. The pellet is washed with 3 mL saline solution and then resuspended in the same solution to reach a final OD of 1.4 at 540 nm. This bacterial cell suspension is used as the inoculum for the growth experiments (*see* **Note 2**).

3.1.2. pH Adjustment of Culture Media

1. 1 *N* HCl and 1 *N* NaOH solutions were aseptically prepared with sterile distilled water (*see* **Note 3**).
2. Before the pH determinations of culture media, the pH meter must be calibrated with the reference standard buffers. The electrode is washed with distilled water and dried with blotting paper; 2 mL of each buffer is placed in small plastic bottles, and the pH meter is calibrated first at pH 7.00 and then at pH 4.00. When the pH meter is not used, the electrode should be submerged in 3 *M* KCl solution (*see* **Note 4**).
3. Before inoculation, the initial pH of the LAPTg and MRS broths is aseptically adjusted to 5.0 with 1 *N* HCl, or to 6.5 and 8.0 with 1 *N* NaOH, determining the pH with the pH meter as described above in **step 2** (*see* **Note 5**).

3.1.3. Incubation Systems

3.1.3.1. Cultures Without Agitation

1. 100 mL of each medium (in 250-mL Erlenmeyer flasks) is inoculated (2% v/v) with precultures of each *Lactobacillus* strain, prepared as described above in **Subheading 3.1.1.** (*see* **Note 6**).

2. The flasks are agitated and later distributed (in 5 mL samples) into glass tubes for growth determinations (*see* **Note 7**).
3. The tubes are incubated in a water bath in aerobic conditions, without agitation, at a constant temperature of 30, 37, or 44°C (*see* **Note 8**).
4. This incubation system is used for all the growth conditions corresponding to the complete factorial experimental design used to study growth and bacteriocin production of *L. salivarius* CRL 1328, and for the nonaerated conditions tested in the fractional factorial experimental design for growth and H_2O_2-production of *L. paracasei* CRL 1289 and *L. crispatus* CRL 1266.

3.1.3.2. Cultures With Agitation

1. The lactobacilli are inoculated (2% v/v) into 100 mL of each medium (250-mL Erlenmeyer flasks).
2. The cultures are incubated in a water bath with agitation (*see* **Note 9**).
3. This incubation system is used for evaluation of the effect of aerated conditions tested in the fractional factorial experimental design used for both growth and H_2O_2-production of *L. paracasei* CRL 1289 and *L. crispatus* CRL 1266.

3.1.4. Obtaining Samples

1. For determination of antagonistic substances, pH, and CFU/mL, the samples are withdrawn every 3 h; the OD measurements are performed every one hour.
2. For nonagitated cultures, at each time point a tube is removed from the incubator; for agitated cultures, the samples are aseptically withdrawn from the flasks (4 mL), placed in sterile tubes, and labeled (*see* **Note 10**). In both cases, different aliquot are taken, as follows:
   a. First, 500 μL are placed in sterile tubes, and the supernatants are separated by centrifugation (5 min at 6000*g* and 20°C) and stored at –20°C (up to 1 mo) or at 4°C (2–3 h) for the bacteriocin and H_2O_2 determinations, respectively.
   b. Then 100 μL are diluted with peptone water in glass tubes and stored at 4°C for the viable count. They cannot be kept under these conditions for more than 1 h (*see* **Subheading 3.1.7.**).
   c. The remaining volume is used for the OD and pH determinations (*see* **Subheadings 3.1.5.** and **3.1.6.**, respectively).

*3.1.5. Optical Density Measurements (see **Note 11**)*

1. The selected wavelength of the spectrophotometer is 540 nm (*see* **Note 12**).
2. The initial turbidity of the growth media must be corrected to absorbance value 0 (zero) with the corresponding blanks of culture media placed in glass cuvets (*see* **Note 12**).

3.1.6. pH Determination of Culture Samples

The pH of the culture media during the lactobacilli growth is determined as described in **Subheading 3.1.2.**

3.1.7. Determination of Colony-Forming Units

1. The samples are serially diluted (10-fold) in peptone water, from 1:10 to 1:1,000,000, using glass tubes with peptone water and automatic pipets; each dilution tube was vigorously mixed with a vortex.

2. A 0.5-mL aliquot of each sample dilution is placed on a plate, and 12–15 mL of melted culture medium (LAPTg or MRS agar, depending on whether the previous lactobacilli growth was in LAPTg or MRS broth) are added and mixed carefully while the plate is rotated for an even distribution of the bacteria. The agar media should be melted and maintained at a temperature of 45–50°C before use.
3. The plates are incubated at 37°C for 48 h. Then the number of CFUs was determined by selecting data for the plates that contained between 30 and 300 CFU. The results are expressed as CFU/mL.

3.1.8. Registration of the Experimental Data

Registration of the experimental data and some transformations of these original data were performed using Excel sheets.

3.2. Experimental Designs (see Note 14)

3.2.1. Complete 2 × 3 × 3 Factorial Design

1. To test the growth and bacteriocin production of *L. salivarius* CRL 1328, three factors are evaluated: culture media (LAPTg and MRS broth), temperature (30, 37, and 44°C), and initial pH (5.0, 6.5, and 8.0).
2. Ten curves were performed experimentally per day, two corresponding to the central points and eight to the remaining combinations (**Table 1**).
3. The experiments are repeated twice.
4. The complete design can be finished in 4 d (blocks). A total of 40 curves with 18 different combinations are made. The day number (block) in which each one of the culture conditions is tested is indicated in each cell of **Table 1**.

3.2.2. Fractional Factorial Design

1. To test the growth and H_2O_2 production by vaginal lactobacilli, five factors were evaluated: *Lactobacillus* strains (*L. crispatus* CRL 1266 and *L. paracasei* CRL 1289), culture media (LAPTg and MRS broth), temperature (30, 37, and 44°C), initial pH (5.0, 6.5, and 8.0), and agitation (nonagitated and agitated cultures).
2. This design was obtained by using the menu DOE (Design of Experiments) of the MINITAB software.
3. The experiments corresponding to the central points are repeated four times, and the vertex points (conditions of the 2^{5-1} design) are carried out once.
4. Six curves are experimentally performed per day, corresponding to two vertex points and four central points of the design.
5. The complete design can be finished in 8 d (blocks). A total of 48 curves with 24 different combinations are made, 16 of them corresponding to the factorial design of two levels and 8 to the central points (**Table 2**). The day number (blocks) in which each one of the culture condition is tested is indicated in each cell of **Table 2**.

3.3. Estimation of Growth Parameters

3.3.1. Descriptin of the Bacterial Growth Curves

1. The bacterial growth curves were described using the modified four-parameter Gompertz model ***(10)***.

Table 1
Complete Factorial Experimental Design Used for Screening of Growth Conditions and Bacteriocin Production by Vaginal *L. salivarius* CRL 1328[a]

	pH					
	5.0		6.5		8.0	
Temperature (°C)	LAPTg	MRS	LAPTg	MRS	LAPTg	MRS
30	1, 3	1, 3	2, 4	2, 4	1, 3	1, 3
37	2, 4	2, 4	**1, 2, 3, 4**	**1, 2, 3, 4**	2, 4	2, 4
44	1, 3	1, 3	2, 4	2, 4	1, 3	1, 3

[a]Bold numbers correspond to the central points of the experimental design. 1–4, days or block in which each condition was tested.

Table 2
Fractional Factorial Experimental Design Used for Screening of Growth Conditions and H_2O_2 Production by Vaginal Lactobacilli[a]

		L. paracasei CRL 1289				*L. crispatus* CRL 1266			
		LAPTg		MRS		LAPTg		MRS	
Temp (°C)	pH	Without	With	Without	With	Without	With	Without	With
30	5	1			2		7	8	
	8		3	4		5			5
37	**6.5**	**1, 4, 5, 8**	**3, 2, 6, 8**	**2, 4, 6, 7**	**2, 3, 5, 7**	**2, 4, 5, 8**	**1, 3, 6, 7**	**1, 4, 6, 8**	**1, 3, 6, 8**
44	5		5	6		3			4
	8	7			7		1	2	

[a]Bold numbers correspond to the central points of the experimental design. The others represent the vertex of the experimental design. Without and With, without and with agitation. 1–8, days or block in which each condition was tested.

2. At time t, the reparametrized Gompertz model is expressed by the following function:

$$D_t = D_0 + A\exp\{-\exp\,[(\mu \times e/A)\,(\lambda - t) + 1]\} \qquad (1)$$

where D_t is the OD at time t, t is the time of growth in hours, D_0 is the OD at $t = 0$, A is the difference between the initial and final ODs, μ is the maximum specific growth rate (h), λ is the duration time of lag phase in hours, and e is the base of the neperian logarithm (2.718281828).

3.3.2. Method of Growth Curve Estimation

1. Ordinary least squares: STATISTICA 5.0 software; the module used is Nonlinear Regression (*see* **Note 15**). For the algorithm of minimization of the loss function, the Rosenbrock/quasi-Newton algorithm is used (*see* **Note 16**).
2. Bootstrap method (software S-PLUS): *see* basis explanation in **Note 17**. To calculate the standard errors of the parameters estimations, the bootstrapping technique is applied using repeated samples obtained from the original data, by sampling with replacement ***(17,18)***. The number of bootstrap samples was chosen as 1000 or 100 for each set of data of OD and CFU/mL obtained from the growth curve, respectively. SPSS 10 for Windows is used to apply all the calculations, and S-PLUS 200 is used to formulate the corresponding graphics (*see* **Note 17**).

3.3.3. Analysis of Variance (ANOVA)

To determine the statistically significant differences between the growth curves (log CFU/mL) or between the kinetics of antagonistic substances production under different culture conditions, the method of the analysis of variance (ANOVA) with repeated measurements was performed by using the statistical software MINITAB (*see* **Note 18**).

4. Notes

1. Before each growth experiment, one cryovial from the freezer must be removed. The unfreezing procedure should be fast enough to decrease the damage produced by the intracellular ice. The microorganisms must always be observed under the optical microscope and also checked for purity. To subcultivate the microorganisms in the broth in which the growth curves are performed, the time and temperature of each subculture must be carefully maintained. To obtain the third preculture, at least three tubes are inoculated.
2. The spectrophotometer should be stabilized 15 min prior to use. Before centrifugation, the absorbance of the third preculture is determined (blank: sterile growth media) in order to gain an approximate notion of the cellular density. This step will allow rapid adjustment of the OD of the bacterial cell in the saline solution (blank: sterile saline solution). The OD_{540} of the bacterial cell suspension (in saline solution) used as the inoculum is adjusted at 1.4 (equivalent to 10^8–10^9 CFU/mL) in order to begin with the same cellular concentration for the growth experiments.
3. HCL causes burns that are irritating to respiratory system. **Caution:** Work under a bell, with a propipet.
4. For the pH meter calibration, only the necessary amount of standard buffers is poured in the plastic bottles, and this volume is not returned to the original recipient. Between each pH determination, the pH meter should be on "stand-by," and the electrode should be washed and dried to avoid errors in the sample pH caused by an unclean electrode.
5. Check the initial pH of the LAPTg and MRS carefully before inoculation (or the previous day of the growth experiments).
6. The initial lactobacilli inoculum is 2% v/v (equivalent to 10^6–10^7 CFU/mL). This inoculum was previously selected from a tested range (from 0.1 to 10%), because the lag phase was much longer for those experiments performed with inocula lower than 2%. Although for inoculum values higher than 2% there was a decrease in the lag time, the growth curves were not acceptable from the practical point of view, because the final OD did not increase significatively, and also because the times needed to reach those values were not significantly shorter ***(19)***.
7. Inoculation of the *Lactobacillus* suspension in flasks containing the culture media is performed in a random order. The cultures are distributed into tubes from the flasks. The tubes are numbered from 1 to 18 and are later randomly chosen to perform the growth and bacteriocin determinations. For every point of the curve, one tube is removed from the water bath, to avoid the agitation and temperature variations of the flask that occurred when the samples are withdrawn from the same recipient.
8. The temperature of each incubator is controlled with a thermometer within 0.1°C of the set point. For the growth conditions corresponding to the fractional and complete experimental designs, three water baths are needed (at 30, 37, and 44°C).
9. In this incubation system, the temperature variation of lactobacilli culture is minimal because the samples are withdrawn every 3 h. The amount of shakers used depends on the number of blocks of the fractional design, with three being maximal.

10. For nonagitated cultures, every hour a tube is removed from the incubator and the OD measurements are performed in order to cover the entire growth curve time, but focusing mainly on the range in which the absorbance response changes significantly. For the agitated cultures, samples are withdrawn from the flasks every 3 h to minimize the time in which the flasks are outside the incubator.
11. The OD determination is a rapid, nondestructive, and inexpensive technique to measure bacterial growth, and it is useful in many areas of microbiology in which different growth conditions must be tested. The growth parameters can be accurately estimated from absorbance measurements ***(11,12)***.
12. The Gilford spectrophotometer was selected to obtain the absorbance data for lactobacilli growth, because of the following characteristics: absorbance range from 0.000 to 3.000 unit of OD (UOD), accuracy of the measurement of ± 0.005 UOD, and reproducibility ± 0.002 UOD. The OD measurements performed with this spectrophotometer are more accurate than those performed with the Spectronic 20.
13. The correction of initial turbidity of the growth media to absorbance value 0 (zero) should be performed before the OD determinations of lactobacilli cultures corresponding to each growth medium or pH, using different glass cuvets. The initial absorbance values of lactobacilli cultures are approx 0.090 ± 0.010 (above the detection threshold of the spectrophotometer used). The nonlinearity of the absorbance measurements was not corrected by diluting the cultures, because of the nonsignificant effect of this factor on the growth rate values previously quantified. This results are in agreement with studies performed by other authors ***(11,12)***.
14. An experimental design is an organized and efficient way to obtain data with different treatments. These designs also allow us to establish statistically whether the variations in answers of interest obtained through the application of the different treatments were produced by the action of these treatments, or whether they could be attributed to the randomization of the experimentation. In a factorial design, there are variables, called factors, with levels (at least two) controlled by the researcher to evaluate the variation of the observed characteristic when the value of one or more factors is modified. Other variables, with worthless effects and able to produce nonidentifiable perturbations on the observations, were not controlled by the fixation of levels, but by the randomization of the treatments to the experimental units. A complete factorial design is one used to explore all the possible combinations or treatments of the factors under study and their levels. For example, if a design has four factors with two levels in each one, the number of treatments of the complete design is $2 \times 2 \times 2 \times 2 = 2^4 = 16$. A fractional factorial design, on the other hand, allows the utilization of only one fraction of the total of treatments, $^1/_2$, $^1/_4$, and so on. In the above example, instead of the 16 treatments used, only 8 or 4 can be used, selected in such a way that the results reflect an "equilibrated" design. Tables are published in books ***(20)*** or software (such as MINITAB) that allow one to obtain all the details of the designs for which the number of factors are consigned, for all the effects that can be studied, at two or three levels, and also at the central points.
15. The method of ordinary least squares uses as a loss function the sum of squares of differences between the observed and estimated values. In this function, the constants are the experimental determinations, and the variables are the parameters whose values we are attempting to determine. The least squares method proposes as estimators those parameter values that minimize the loss function.
16. Algorithm of minimization of the loss function: Rosenbrock/quasi-Newton algorithm. To find the minimum of a function, it is first necessary to calculate the first derived function with respect to each parameter and to set the results equal to zero. As the addition of

squares sometimes results in an expression with a derived function that cannot be calculated explicitly, the computational algorithms are used to look for the minimal (maximal) of a function modifying their parameters according to pre-established norms. This technique rotates the space of the parameters and aligns one of the co-ordinated axes with the group of points that are around a minimal (maximal). The other axes are situated perpendicular to this one. If the loss function is unimodal and has groups of points around the minimal, the method will have very low error to this point. However, if the function has many peaks, the algorithm can take the results very far from the searched value.

17. The bootstrap method (software S-PLUS) is one of the estimation methods known as resampling, because it uses the sample values to generate a highest number of artificial samples. In the regression situation, the curve is first estimated from the original experimental data. Then the residual values (observed value less estimated value) are assigned randomly to the estimated values of the dependent variable. In this way, a sample different from the observations is obtained, which allows one to estimate another curve, which, if the residues present a regular behavior, should not be substantially different from the one obtained with experimental values. The new curve generates other residual values, which are again randomly assigned to the new estimated values. This process generates a third sample, and the process is again repeated, for example, 1000 times, in such a way that the empirical distributions of the parameter estimators are obtained. The mean value obtained will be used as the estimated value of the parameter, and its error will be calculated as the standard deviation of the values divided by the square root of the total number of generated values. By applying this method, it is possible to calculate the confidence intervals and to perform the test of hypothesis of the parameters by applying the empirical distribution, without making a restrictive hypothesis of normality.
18. The ANOVA method allows one to know whether the growth curves or the production of antagonistic substances for the different combinations of strains, growth media, agitation, pH, and temperatures present similar shapes throughout. If the answer is positive, an analysis is performed to detect if there are statistically significant differences between the data obtained from the different treatments. If the plots are statistically different throughout the time, data from a more reduced group are later compared.

Acknowledgments

This work was supported by grants from CONICET (Consejo Nacional de Investigaciones Científicas y Técnicas), PIP 359, and from the Ministrio de Salud Pública de la República Argentina (beca Carrillo-Oñativia).

References

1. Redondo-López, V., Cook, R. L., and Sobel, J. D. (1990) Emerging role of lactobacilli in the control and maintenance of the vaginal bacterial microflora. *Rev. Infect. Dis.* **12,** 856–872.
2. Reid, G. (1999) The scientific basis for probiotic strains of *Lactobacillus*. *Appl. Env. Microbiol.* **65,** 3763–3766.
3. Reid, G. and Bruce, A. W. (2001) Selection of *Lactobacillus* strains for urogenital probiotic applications. *J. Infect. Dis.* **183(suppl 1),** S77–S80.
4. Ocaña, V. S., Bru, E., de Ruiz Holgado, A., and Nader-Macías, M. E. (1999) Surface characteristics of lactobacilli isolated from human vagina. *J. Gen. Appl. Microbiol.* **45,** 203–212.

5. Ocaña, V. S., de Ruiz Holgado, A., and Nader-Macías, M. E. (1999) Selection of vaginal H_2O_2-generating *Lactobacillus* for probiotic use. *Curr. Microbiol.* **38,** 279–284.
6. Ocaña, V. S., de Ruiz Holgado, A., and Nader-Macías, M. E. (1999) Growth inhibition of *Staphylococcus aureus* by H_2O_2-producing *Lactobacillus paracasei* subsp. *paracasei* isolated from the human vagina. *FEMS Immunol. and Med. Microbiol.* **23,** 87–92.
7. Ocaña, V. S., de Ruiz Holgado, A., and Nader-Macías, M. E. (1999) Characterization of a bacteriocin-like substance produced by a vaginal *Lactobacillus salivarius* strain. *Appl. Environ. Microbiol.* **65,** 5631–5635.
8. Mäyra-Mäkinen, A. and Bigret, M. (1998) Industrial use and production of lactic acid bacteria. In: *Lactic Acid Bacteria, Microbiology and Functional Aspects* (Salminen, S. and von Wright, A., eds.). Marcel Dekker, New York, pp. 73–102.
9. Beal, C., Deschamps, N., Juillard, V., De Roissart, H., Richard, J., and Saraux, B. (1994) Cinétiques de croissance et d´acidification des bacteries lactiques. In: *Bacteries Lactiques I. Aspects Fondomentaux et Technologiques* (De Roissart, H. and Luquet, F. M., eds.). Lorica, Chemin de Saint Georges, France, pp. 367–401.
10. Zwietering, M. H., Jongenburger, I., Rombouts, F. M., and van't Riet, K. (1990) Modeling of the bacterial growth curve. *Appl. Environ. Microbiol.* **56,** 1875–1881.
11. Dalgaard, P., Ross, T., Kamperman, L., Neumeyer, K., and McMeekin, T. A. (1994) Estimation of bacterial growth rates from turbidimetric and viable count data. *Int. J. Food Microbiol.* **23,** 391–404.
12. Dalgaard, P. and Koutsoumanis, K. (2001) Comparison of maximum specific growth rates and lag times estimated from absorbance and viable count data by different mathematical models. *J. Microbiol. Methods* **43,** 183–196.
13. McClure, P. J., Blackburn, C. D., Cole, M. B., et al. (1994) Modelling the growth, survival and death of microorganisms in foods: the UK Food Micromodel approach. *Int. J. Food Microbiol.* **23,** 265–275.
14. Nerbrink, E., Borch, E., Blom, H., and Nesbakken, T. (1999) A model based on absorbance data on the growth rate of *Listeria monocytogenes* and including the effects of pH, NaCl, Na-lactate and Na-acetate. *Int. J. Food Microbiol.* **47,** 99–109.
15. Gorret, N., Maubois, J. L., Engasser, J. M., and Ghoul, M. (2001) Study of the effects of temperature, pH and yeast extract on growth and exopolysaccharides production by *Propionibacterium acidi-propionici* on milk microfiltrate using a response surface methodology. *J. Appl. Microbiol.* **90,** 788–796.
16. Kandler, O. and Weiss, N. (1986) The genus Lactobacillus. In: *Bergey's Manual of Systematic Bacteriology*, Vol 2 (Sneath, P., ed.). Williams and Wilkins, Baltimore, MD, pp. 1209–1234.
17. Efron, B. (1982) The bootstrap. In: *The Jackknife, the Bootstrap and other Resampling Plans*. SIAM (Society for Industrial and Applied Mathematics), Philadelphia, pp. 27–36.
18. Huet, S., Bouvier, A., Gruet, M. A., and Jolivet, E. (1996) Accuracy of estimators, confidence intervals and tests. In: *Statistical Tools for Nonlinear Regression*. Springer-Verlag, New York, pp. 29–59.
19. Juárez Tomás, M. S., Bru de Labanda, E., de Ruiz Holgado, A. P., and Nader-Macías, M. E. (2002) Estimation of vaginal probiotic lactobacilli growth parameters with the application of the Gompertz model. *Can. J. Microbiol.* **48,** 82–92.
20. Lorenzen, T. J. and Anderson, V. L. (1993) *Design of Experiments: A Non-Name Approach*. Marcel Dekker, New York.

34

Meat-Model System Development for Antibacterial Activity Determination

Graciela Vignolo and Patricia Castellano

1. Introduction

Bacteriocins are antibacterial substances produced by many different bacterial species. Although the bacteriocins form a heterogenous group with respect to production of bacteria, antibacterial spectrum, mode of action, and chemical properties, they are by definition proteinaceous compounds that are bactericidal toward organisms taxonomically close to the producer *(1)*. The ability of lactic acid bacteria (LAB) to compete and finally dominate in mixed fermentations has been attributed to the production of several antimicrobial metabolites such as organic acids, carbon dioxide, hydrogen peroxide, diacetyl, and bacteriocins *(2,3)*. The antimicrobial activities of the LAB have long been known, but their bacteriocins have received limited attention until recently *(4)*. Numerous strains of lactic acid bacteria associated with food systems are capable of producing bacteriocins, or antibacterial proteins with activity against foodborne pathogens and contaminants. Recently, considerable enphasis has been placed on the physicochemical, biochemical, and genetic characterization of these proteins.

Many methods for the detection of bacteriocin production as well as the determination of the potency of bacteriocin preparations have been described. All the usual techniques are based on the fact that bacteriocins can diffuse in solid or semisolid culture media, which are subsequently inoculated with a suitable indicator strain. This method has long been shown to have a good perfomance in the characterization of new bacteriocins.

Although results obtained from broth systems show that bacteriocins inhibit target organisms, applied studies must be performed to confirm their effectiveness in food. As many lactic acid bacteria associated with meat products were described to be important natural bacteriocin producers, it has been necessary to assay their inhibitory efficacy in meat or meat products. Because of the complexity of these kinds of foods a simplified meat-model system was developed to determine the inhibitory activity of a bacteriocin.

From: *Methods in Molecular Biology, vol. 268: Public Health Microbiology: Methods and Protocols*
Edited by: J. F. T. Spencer and A. L. Ragout de Spencer © Humana Press Inc., Totowa, NJ

2. Materials

2.1. Meat Slurry

1. Stomacher® 400 Circulator (Seward, London, UK) and stomacher bags.
2. Fresh lean beef/pork muscle aseptically minced after 24 h of exsanguinization and stored at –30°C.
3. Distilled water (dH_2O).

2.2. Preparation of the Bacteriocin or the Bacteriocinogenic Strain

1. LAB bacteriocin producer strain.
2. de Man-Rogosa-Sharpe (MRS) broth *(5)*: 10 g/L peptone, 10 g/L meat extract, 5 g/L yeast extract, 20 g/L glucose, 5 g/L sodium acetate, 1.08 g/L Tween-80, 2 g/L ammonium citrate, 2 g/L K_2HPO_4, $MgSO_4 \cdot 7H_2O$ (0.2g/L), 0.05 g/L $MnSO_4 \cdot 4H_2O$, pH 6.5. Sterilize in an autoclave at 121°C for 15 min.
3. Active lactic acid bacteria culture grown in MRS broth for about 16 h at 30°C.
4. 2 *N* NaOH.
5. Centrifuge.
6. TurboVap LV evaporator (Zymark).

2.3. Bacteriocin Activity Determination

1. Distilled and sterile water.
2. MRS soft agar (0.7% agar).
3. MRS agar (1.5% agar).
4. Sterile hollow punch, 0.5 cm in diameter.
5. Indicator or sensitive strain.

2.4. Inoculation of the Meat Slurry

1. Meat slurry previously prepared.
2. An active culture of the sensitive strain (required volume to reach the desirable colony-forming units [CFU]/mL in the slurry) (*see* **Note 1**).
3. Concentrated bacteriocin (required volume to reach the desirable activity in the slurry) or bacteriocinogenic strain (required volume to reach the desirable CFU/mL in the slurry) (*see* **Note 2**).
4. Stomacher 400 blender to homogenize.
5. Incubator.

3. Methods

3.1. Meat Slurry Preparation

1. Mince aseptically fresh lean beef/pork muscle, obtained from a local abattoir. Store at –30°C (*see* **Note 3**).
2. Mix 10-g portions of minced meat with dH_2O in a stomacher bag. The final volume will be adjusted by separate addition of the cell suspension of the bacteriocin producer strain or the concentrated bacteriocin and the cell suspension of the indicator strain.
3. Stomach the bags for 1 min to homogenize.

3.2. Obtaining the Bacteriocin From the Producer Strain

1. Centrifuge (14,000*g* × 5 min) the overnight culture of the LAB bacteriocin producer strain.
2. Remove the supernatant and neutralize to pH 7.0 with 5 *N* NaOH.
3. Heat-sterilize (100°C for 15 min).

4. Concentrate the bacteriocin in the supernatant using an evaporator (TurboVap).
5. Store the concentrated bacteriocin at 5°C for no longer than 30 d.

3.3. Use of the LAB Bacteriocinogenic Strain

1. Inoculate the LAB bacteriocin producer in the appropriate medium and grow at the usual temperature for 16–18 h. Calculate the medium volume necessary to have a desired activity in the meat slurry.

3.4. Bacteriocin Activity Determination

1. Centrifuge an overnight culture of the bacteriocin producer strain (14,000*g* × 5 min).
2. Remove the supernatant and neutralize to pH 7.0 with 5 *N* NaOH.
3. Sterilize by filtration through a membrane of 0.45 µm (*see* **Note 4**).
4. Prepare serial twofold dilutions of the bacteriocin in sterile dH_2O (1:2, 1:4, 1:8, 1:*n*).
5. Prepare a Petri dish with 10 mL of MRS base agar. Let solidify and dry.
6. Mix gently 70 µL of an overnight culture of the sensitive strain with 7 mL of molten MRS soft agar (45°C) and pour onto the base MRS layer. Slide the plate in circles on the bench top immediately to spread the top agar over the plate. Let solidify and dry.
7. Cut wells 0.5 cm in diameter with sterile hollow punches in the agar.
8. In each well, seed 30 µL of undiluted bacteriocin and each dilution of it.
9. Incubate the plates (with the agar side up) overnight at 30°C.
10. Bacteriocin activity is detected by the presence of growth inhibition zones (halos) of the indicator strain around the wells.
11. Calculate bacteriocin activity expressed in arbitrary units (AU) per milliliter, as follows: AU/mL = 1/0.03 mL × maximal dilution, which inhibits growth of the indicator strain.

3.5. Antibacterial Activity Determination in the Meat Slurry

1. Prepare the meat slurry in stomacher bags.
2. Add the sensitive strain.
3. Add the concentrated bacteriocin or the bactriocinogenic strain.
4. Adjust the final volume to 20 mL with dH_2O.
5. Mix the stomacher bags thoroughly.
6. Incubate the bags at the desired temperature.
7. Take samples at time intervals.
8. Determine CFU/mL of the sensitive strain.
9. Determine the AU/mL of the bacteriocin in the slurry (*see* **Note 5**).

4. Notes

1. The cell concentration of the sensitive strain has to be previously optimized. The cell lawn in the agar layer must show a marked inhibition zone (halo) around the seeded well.
2. The bacteriocin concentration must be calculated taking into acount the final volume of the slurry and the required AU/mL for the study.
3. Before mincing the beef/pork muscle, remove fat and connective tissue in aseptic conditions and cut the lean muscle into small cubes (3 cm).
4. Filtration through a membrane of 0.45 µm will not produce detectable loss of activity. If the bacteriocin is sufficiently thermostable to withstand temperatures that kill the producer cells, it can be sterilized by autoclaving.
5. An aliquot of the slurry must be centrifuged to eliminate meat particles; the supernatant will be used to prepare the serial twofold dilutions.

References

1. Tagg, J. R., Dajani, A. S., and Wannamaker, L. W. (1976) Bacteriocins of gram-positive bacteria. *Bacteriol. Rev.* **40,** 722–756.
2. Piard, J. C. and Desmazeaud, M. (1991) Inhibiting factors produced by lactic acid bacteria, 1. Oxygen metabolites and catabolism end-products. *Lait* **71,** 525–541.
3. Piard, J. C. and Desmazeaud, M. (1992) Inhibiting factors produced by lactic acid bacteria. 2. Bacteriocins and other antibacterial substances. *Lait* **72,** 113–142.
4. Klaenhammer, T. R. (1988) Bacteriocins of lactic acid bacteria. *Biochimie* **70,** 337–349.
5. de Man, J., Rogosa, M., and Sharpe, M. (1960) A medium for the cultivation of lactobacilli. *J. Appl. Bacteriol.* **23,** 130–135.

VI

In Vivo Studies in Mice

35

Colonization Capability of Lactobacilli and Pathogens in the Respiratory Tract of Mice

Microbiological, Cytological, Structural, and Ultrastructural Studies

Rosa Cangemi de Gutierrez, Viviana M. Santos, and María Elena Nader-Macías

1. Introduction

Respiratory tract infections are among the bacterial infections that affect humans with higher frequency. Those produced by *Streptococcus pneumoniae* are reported to have the highest incidence in the world, affecting both children and old people. As a 2001 report from the World Health Organization *(1)* expressed it, the basic fight of children under 5 yr old is to survive. Five different conditions (acute respiratory infections, diarrhea, measles, palludism, and undernutrition) directly produce more than 50% of the deaths in this age group. Respiratory tract infections in the developing countries in the Americas are among the first three causes of death in children under 1 yr and between the first and second cause in children between 1 and 4 yr old. Pneumonia is responsible for 85 and 90% of deaths in children under 5 yr old (approx 150,000 annually), 95% of them occurring in the developing countries in the Americas.

There is an increased worldwide tendency to use preventive measures and to consume products that help to maintain the health status of the individual. Thus the use of probiotics has increased systematically during the last decade, and the scientific literature trying to demonstrate the positive effect of such preparations has also increased. The term *probiotic* ***(2)*** has been applied to products that (1) contain live microorganisms, freeze-dried or included in fermented products or (2) improve the health status of humans and animals, exerting effects in the mouth or gastrointestinal tract (included in foods or capsules), in the respiratory tract (as aerosols), or in the urogenital tract (by local application)

From: *Methods in Molecular Biology, vol. 268: Public Health Microbiology: Methods and Protocols*
Edited by: J. F. T. Spencer and A. L. Ragout de Spencer © Humana Press Inc., Totowa, NJ

Having in mind the high incidence of respiratory tract infections, and looking for preventive measures as well as the possible applications of probiotics, the aim of this chapter was to use mice as experimental models to determine whether members of the genus *Lactobacillus* were able to colonize and give protection from infections after inoculation by the intranasal route. To this end, the following procedures were carried out:

1. Screening of the predominant bacterial species in respiratory organs *(3)*.
2. Study of the kinetics of colonization of the different groups of microorganisms from 15 d up to adult (2 mo) *(3)*.
3. Screening of the probiotic characteristics of all the isolated strains.
4. Determination of the colonization ability of the selected *Lactobacillus* strains after intranasal inoculation, as well as the optimal dose and amount of microorganisms *(4)*.
5. Experiments to evaluate whether intranasal administration of *Lactobacillus fermentum* produces some type of modification at the structural or ultrastructural level in the different organs of the respiratory tract *(5)*.
6. Determination of whether *L. fermentum* could prevent an infection with *Streptococcus pneumoniae* in the respiratory tract when given by the intranasal route *(6)*.

The first procedures were developed by applying in vitro techniques described in previously published papers *(3,4)*, and also in some other chapters of this book. In this chapter, a detailed description of the methodology applied to animals is given.

2. Materials

2.1. Microorganisms

The microorganisms used in this study were *L. fermentum* and *S. pneumoniae*. Lactobacilli were isolated from the respiratory tract of mice, as described previously *(3,4)*. *S. pneumoniae* was isolated from a human source and characterized as described in a previous paper *(6)*.

2.2. Culture Media

1. Storage: Milk-yeast extract: 10% skim milk, 1% glucose, 0.5% yeast extract, pH 6.5. Sterilize by autoclaving and store at refrigeration temperature. L. fermentum was stored at –70°C in milk-yeast extract.
2. Supplemented Brain-Heart Infusion (BHI): BHI: 5 mg/mL hemin, 0.005% cystein chlorhydrate, 0.5% yeast extract, 100 mg/mL kanamycin, 1 mg/mL vitamin K, final pH 7.4 ± 0.2. Sterilize by autoclaving prior to the addition of antibiotics and vitamins and store at refrigeration temperature. At the time of use, the sterilized basal media is melted and the antibiotic and vitamins added before pouring onto the plates. The vitamins and antibiotics were sterilized by filtration and stored at freezing conditions. S. pneumoniae was stored in glycerol-BHI broth at –70°C.

2.3. Preparation of Microorganisms for Inoculation Into Mice

1. LAPTg broth *(7)*: 15 g/L meat peptone, 10 g tryptone, 10 g yeast extract, 10 g glucose, 1 mL Tween-80, final pH 6.5. Sterilize by autoclaving at 121°C for 15 min). Store at refrigeration temperature. All the media ingredients were obtained from Britania Laboratories (Buenos Aires, Argentina).

2. LAPTg + rifampicin: 15 g meat peptone, 10 g tryptone, 10 g yeast extract, 10 g glucose, 1 mL Tween-80, 1.5% agar, final pH 6.5, prepared as described in **step 1**. The antibiotic rifampicin was prepared by dissolving the powder in ethyl alcohol and sterilizing by filtration. It was stored for 1 wk at freezing temperature. At the time of use, the antibiotic was defrozen and added aseptically to the melted media for a concentration of 1 mg/mL.
3. Blood-agar: Columbia agar (Britania Laboratories) with 5% human blood. The agar base was sterilized by autoclaving and stored at refrigeration temperature. At the time of use, sterile blood was added for a concentration of 5%.
4. Columbia agar with 5% heated blood: The agar base was sterilized by autoclaving. At the time of use it was melted, the human blood was added for a concentration of 5%, and it was later heated for 15 min at 70°C. This media was prepared prior to use.
5. de Man-Rogosa-Sharpe (MRS) agar *(8)*: 10 g/L polypeptone, 10 g/L meat extract, 5 g/L yeast extract, 20 g/L glucose, 0.8 g/L Tween-80, 2 g/L dipotassium phosphate, 5 g/L sodium acetate, 2 g/L ammonium citrate, 0.2g/L magnesium sulfate, 0.05 g/L manganese sulfate, 15 g/L bacteriological agar, pH 6.5.

2.4. Animal Maintenance Conditions

1. Mice.
   a. Adult male 2 $^{1}/_{2}$-mo-old Balb/c mice weighing 25–30 g were obtained from the inbred colony maintained in our laboratory at the Microbiology Institute of the National University of Tucumán.
   b. They were housed in plastic cages, fed *ad libitum* with a conventional balanced diet for rodents, given water *ad libitum*, and maintained under constant environmental conditions.
   c. The experimental protocol was previously approved by the Ethical Committee of Animal Care at CERELA.
   d. Each experimental assay was carried out with groups of 25–30 mice.
   e. A group of control animals of the same size was similarly inoculated with a solution without microorganisms (placebo).
   f. Mice were randomly assigned to the different experimental groups.
2. Keeping standard conditions in the mice colony.
   a. The ideal room temperature is around 18°C with small fluctuations.
   b. The cages were cleaned daily, and the wood-chip bed was changed.
3. Other materials.
   a. Spatulas.
   b. Pipets.
   c. Petri dishes.
   d. Glass tubes.
   e. Calibrated loops.
   f. Tuberculin syringes.
   g. Sterile plastic tips.
   h. Gloves.
4. Mice sacrifice.
   a. Dissection boards.
   b. Scissors.
   c. Material for dissection.

2.5. Histological Technique

1. Baker phormol-calcium reagent: 10% anhydrous sodium chloride (Cicarelli), 100 mL distilled water, pH 6.5 (Biopack), 10 mL formaldehyde 40% (stabilized in methanol; Biopur). Store at 15–25°C and prepare a new set of reactive every 30 d ***(9)***.
2. Activated hematoxylin for histological purposes (Biopur-Argentina), ready for use; does not require the addition of acid. Store at 15–25°C.
3. Alcoholic yellow eosin (Biopur-Argentina), ready to use. Store at 15–25°C ***(9)***.
4. Ziehl fucsin diluted with distilled water (1:5) (Biopack); prepared at the time of use. Store at 15–25 °C.
5. Indigo picrocarmin (Biopack).
6. Watery acuose solution of acetic acid (0.2%) (Biopur) ***(9)***.
7. Graduated ethyl alcohols (50°, 70°, 80°, 96°, and 100°; Cicarelli). Store at 15–25°C; prepared from pure ethanol (B6) with demineralized water added, pH 6.5 (Biopack).
8. Xylol ACS. Store at 15–25°C (Cicarelli).
9. Other materials:

    a. Coplin flasks.
    b. Glass bottles for organ fixation.
    c. Glass cuvets for staining slides.
    d. Stove for impregnation (54–58°C).
    e. Glass slides.
    f. Glass cover slides.
    g. Canadian natural balsam (Biopack).
    h. Paraffin microtome, Minot type.
    i. Thermostatized iron for the paraffin slides.
    j. Light microscope.

2.6. Cytological Technique

1. Pure methyl alcohol. Store at 15–25°C (Cicarelli).
2. Giemsa stain: Azur II eosin, methylene blue (Merck, ready to use). Store up to the date recommended on the label ***(10)***.
3. Graduated ethyl alcohols (50°, 70°, 80°, 96°, and 100°; Cicarelli). Store at 15–25°C. They were prepared from pure ethanol with demineralized water, pH 6.5 (Biopack).
4. Xylol. Store at 15–25°C (Cicarelli).
5. Other materials:

    a. Glass bottles.
    b. Cover slides.
    c. Glass slides.
    d. Glass cuvets for staining.
    e. Natural Canadian balsam (Biopack).
    f. Ethyl alcohols (Cicarelli).
    g. Light microscope.

2.7. Electron Microscopy

1. 3.5% Glutaraldehyde solution buffered with 0.1 *M* phosphate buffer, pH 7.4 (Biopack). Store at 5–10°C in a cold room.
2. 1% Osmium tetroxide.

3. Aqueous solution of 2% uranyl acetate.
4. Ethyl alcohols of increasing strength (Cicarelli).
5. Spurr resin.
6. Uranyl acetate (Pelco).
7. Lead citrate (Pelco).
8. Zeiss EM 109 transmission electron microscope at 50 kV.

2.8. Microbiological Techniques

1. Glass tubes for samples.
2. Peptone water.
3. Sterile plastic syringes.
4. Culture media for lavage.
5. Homogeneizor (Omni-Mixer 17106, Sorvall).
6. Glass tubes for diluting samples.
7. MRS and MRS agar with 100 μg/mL rifampin (Sigma). Incubator.
8. For Gram stain: Standard reactants or kit.
9. For microbiological identification in enriched, selective, and differential culture media: MRS, LBS, BHI, blood agar, and chocolate agar.
10. Biochemical tests to identify the microorganisms:
    a. Standard biochemical tests described in the *Bergey's Manual of Bacteriological Procedures* ***(11)*** and API Strepto.
    b. API Staph and API 50 CHL (Biomerieux, France).
    c. Supplementary test to identify Lactobacilli (*Bergey's Manual* ***[11]***)
    d. Additional tests for streptococci: production of urease, growth in media with sodium azide and 6.5% NaCl, esculin bilis reaction.
11. For the optoquine test: 5 μg optoquin (Difco; Bacto-differentiation disks, Optoquin BBL, or Taxo P disks), blood agar plates, surface spreaders.
12. For oxgall solubility: 10% Sodium deoxycholate (BBL or Difco).
13. For the Quellung test: Glass slides, specific antisera, methylene blue.

2.9. Antibody Determinations

1. Suspension of *L. fermentum* as antigen: *L. fermentum* was killed by heating at 100°C for 1 h.
2. Suspension of *S. pneumoniae*: By tyndalization (heating at 60°C for 1 h for three consecutive days).
3. For the agglutination reaction: Glass tubes, saline solution, pipets.

3. Methods

3.1. Microorganisms

1. Isolation of Lactobacilli: *L. fermentum* was isolated from the respiratory tract (pharynx) of adult Balb/c mice. Lactobacilli were identified by the methods described in *Bergey's Manual of Determinative Bacteriology* ***(11)***, standard techniques used in the laboratory, and API 50 CHL (Biomerieux-France).
2. Rifampicin-resistant *L. fermentum*.
    a. *L. fermentum* mutants resistant to antibiotics were obtained after determining their sensitivity to different antibiotics (by the monodisc diffusion test and minimal inhibitory concentration [MIC]).

b. Mutants were grown on MRS-R. *L. fermentum* RR (resistant to rifampyn) showed the same phenotypical properties as the parental strain (tested by the API System).
c. To obtain the suspensions for inoculation, strains were subcultured in LAPTg broth *(7)* at 37°C three times, alternatively in media with and without antibiotic.
d. They were centrifuged, washed with saline solution (SS), and resuspended to the desired concentration.

3. *S. pneumoniae.*
   a. *S. pneumoniae* was isolated from human pneumonia-suffering subjects and identified by standard techniques.
   b. The virulence of this microorganisms was tested in mice.
   c. Pathogenicity in mice was increased by inoculating *S. pneumoniae* intraperitoneally.
   d. The following day, mice were bled by heart puncture, and capsulated *S. pneumoniae* was isolated from blood agar plates.
   e. The pathogen was stored in glycerol-BHI broth at –70°C.
   f. Prior to challenge, *S. pneumoniae* was plated onto blood agar plates, and a suspension prepared from the microorganisms was grown on the surface.
   g. Mice inoculated with *S. pneumoniae* without passing through animals as described in **Subheading 3.2., item 1a**, were not susceptible to the pathogen and its subsequent colonization and infection.
   h. The *S. pneumoniae* strain belongs to the 6A serotype. Serotyping was performed in the Servicio de Bacteriología, Instituto Nacional de Enfermedades Infeccionsas-ANLIS (Dr. Carlos Malbrán, Buenos Aires, Argentina).
   i. The technique applied was the Quellung techique using antisera produced by the Statens Serum Institute (Copenhagen, Denmark).
4. The number of lactobacilli was determined by the serial dilution method; the aliquots were plated in LAPTG agar plates and incubated at 37°C for 48 h.
5. The number of *S. pneumoniae* was determined by the serial dilution method; the aliquots were plated in blood agar plates, and incubated microaerophilically at 37°C for 48 h.

3.2. Inoculation Procedure

1. Intranasal inoculation of mice with *L. fermentum.*
   a. Mice were intranasally inoculated with 50 μL of *L. fermentum* in SS (1×10^9 CFU/mL) every 12 h.
   b. The exposure time was approx 1 min per animal, with one intranasal inoculation of 10 μL each time.
   c. Inoculation was performed by using an automatic micropipetor equipped with a nose tip, and the suspension was sniffed in by the animals with normal tidal breathing.
   d. To facilitate downward migration of the inoculum, mice were held in a vertical position for 2 min.
   e. Different groups of mice (three to four animals each) were inoculated with one, two, or four doses of 1×10^9 CFU/mL (50 μL each), administered every 12 h.
   f. Other groups of mice were reinoculated 10 d after the first inoculation with four doses (50 μL) of the same *L. fermentum* suspension in SS (1×10^9 CFU/mL).
2. Intranasal inoculation with *S. pneumoniae.*
   a. Inoculation with *S. pneumoniae* was performed with 50 μL of a suspension containing 1×10^9 CFU.
   b. The suspension of the pathogenic microorganisms was obtained as described in **Subheading 3.3**.

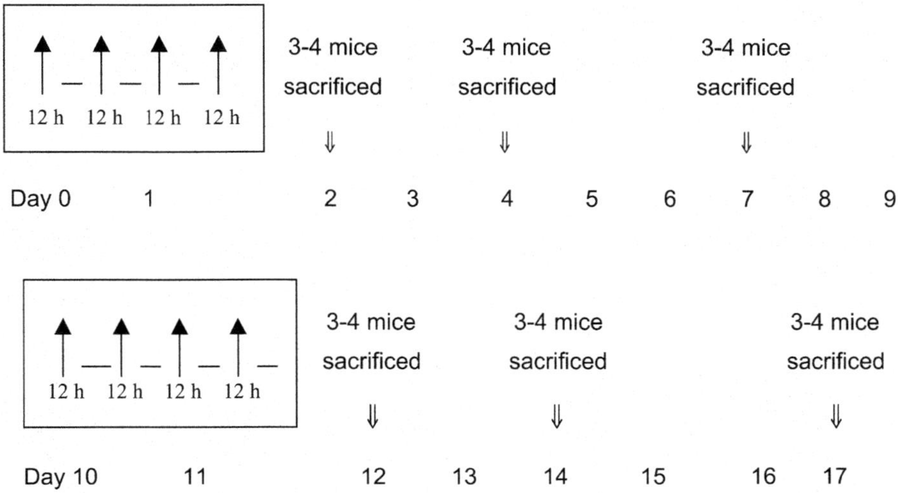

Fig. 1. Scheme of the inoculation and killing steps. Black arrows indicate the intranasal inoculation of lactobacilli, performed every 12 h. Open arrows indicate the days of mice sacrifice. The upper scheme indicates the regular protocol, while the lower figure indicates the reinoculatio procedure.

3. Intranasal inoculation with *L. fermentum* and *S. pneumoniae.*
   a. Mice were intranasally inoculated with *L. fermentum*, receiving four doses of 10^7 CFU every 12 h, as described in **step 1**.
   b. The next day, *S. pneumoniae* (10^9 CFU/mouse) was also intranasally inoculated. This was the experimental group for studing the preventive effect of treatment with *L. fermentum* prior to *S. pneumoniae* challenge.
   c. Control animals were inoculated with sterile 0.1% peptone water intranasally.
4. The inoculation and sacrifice protocol is shown in **Fig. 1**.

3.3. Experimental Procedure to Obtain Samples From Mice

1. Mice were sacrificed by cervical dislocation on d 2, 4, and 7 after inoculation depending on the experimental group under study.
2. Nasally instilled pharynx, trachea, bronchia, and lungs were extracted aseptically and transferred to 0.1% peptone water.
3. The organs were homogenized with a Teflon pestle in a tissue homogenizor.
4. Aliquots of the samples, or their dilutions, were added to selective media to determine the numbers of viable microorganisms.
5. Plates containing between 30 and 300 bacteria were counted.
6. Identification of isolated microorganisms was checked either by Gram stain or by biochemical tests of single colonies.
7. The organs were also added to 10% formaldehyde to be processed for histological purposes.

3.4. Weight Determinations

Mice were weighed every day throughout the experiment to determine whether *S. pneumoniae* challenge produces some type of clinical condition that makes the animals lose weight. Weight was determined before sacrifice by weighing the animals in an OHAUS balance, Precision Standard Model T.S 400 D.

3.5. Histological Studies

1. From each experimental group, two mice were randomly selected for histological examination.
2. Trachea, bronchia, and lungs were separated for further studies after cervical dislocation.
3. The areas studied were:
   a. The high trachea located in the neck base near the thyroid gland.
   b. The bronchia, where they divide.
   c. The lungs (left and right lobule): terminal bronchiole and alveolar wall with their capillary and connective tissue to which the alveolar macrophages are associated.
4. The samples were fixed with 10% paraformaldehyde in glass flasks with 10–20 vol more than the sample for 24 h at room temperature.
5. The samples ere passaged in alcohol 70° (1 h each), and then embedded in paraffin for 24 h according to the method standardized at our laboratory.
6. The samples were cut into 5-µm-thick sections.
7. On the histological slides, different techniques were used to allow analysis of the modifications produced in the lamina propria of the trachea by the increase in the number of lymphocytes.
8. Slides were stained with H&E and by the Ramón y Cajal technique and then processed for light microscopy (40× objective).
9. Soon after the organs were obtained, they were fixed in Baker's fixative (sample size 2–3 cm and volume of fixation liquid approx twice the sample size).
10. Dehydration was performed with graduated alcohols: 50°, 70°, 80°, 96°, and 100°, keeping the samples in each one for 2–3 min.
11. The organs were embedded in paraffin at 56–58°C.
12. Slides (5–6 µm wide) were obtained with a sliding microtome.
13. The staining technique used (either H&E or Ramón y Cajal technique *[9]*) depends on the type of cell of interest.

3.5.1. H&E Technique

1. The slides underwent the following steps:
   a. Xylene: two passages of 3–5 min.
   b. Ethyl alcohol 100° to take out the excess xylene: 1 min.
   c. Ethyl alcohols 96°, 70°, and 50°: 1 min each.
   d. Distilled water: 1 min.
2. Staining was performed with Erlich hematoxylin for 2–5 min.
3. The slides were washed with distilled water and transferred for the differentiation step (top water with concentrated HCl).
4. Slides were allowed to change color until they were violet and stored for 10 min.
5. Alcoholic eosin was applied for 3 min.
6. Slides were washed in distilled water.

7. For dehydration and mounting, the slides were passed very quickly through:
   a. 70° Ethyl alcohol.
   b. 96° Ethyl alcohol: 2 min.
   c. 100° Ethyl alcohol: two different passages at 1 min each.
   d. Xylene: two passages of 1–2 min each.
   e. The slides can be mounted in Canadian balsam.
9. The results were as follows:
   a. Hematoxylin stains the nuclear components blue or purple, the intensity depending on the concentration of nucleic acids (DNA and RNA). The cytoplasm can show an aggregation of substances stained blue with hematoxylin; this basophilic reaction also depends on the presence of nucleic acids in the cytoplasm.
   b. Eosin stains the cytoplasm red or pink, the degree depending mainly on the amount of structural proteins.

3.5.2. Ramón y Cajal Stain

This method can be used after any previous method of fixation. It is useful for the differentiation of cellular categories.

1. Soon after the organs were obtained, they were fixed in Baker's fixative (sample size of 2–3 cm, fixation liquid volume approx twice the sample size).
2. Dehydration was performed with graduated alcohols: 50°, 70°, 80°, 96°, and 100°, keeping the samples in each for 2–3 min.
3. The organs were embedded in paraffin 56–58°C.
4. Slides (5–6 μm wide) were obtained with a sliding microtome.
5. Slides were stained with Ziehl's fuscin for 10 min.
6. Slides were washed with running water.
7. Slides were washed with acetic water to eliminate fucsin excess.
8. Slides were stained with indigo picrocarmine in a bottle for 10 min.
9. Slides were washed with acetic water for 10 min and dehydrated with absolute alcohol to eliminate the red (excess).
10. Dehydration was performed in alcohols of increasing gradation for 2–3 min.
11. Slides were washed with pure xylene for 3–4 min.
12. Slides were mounted in Canadian balsam for longer storage (*see* **Note 2**).
13. To quantify the lymphocyte population, an area of the epithelium corresponding to 50 epithelial cells was taken as a reference. This area correlates with the number of lymphocytes present in the lamina propria area (*see* **Note 3**).

3.6. Cytological Studies

1. The left lobule was used to perform lung cytological slides. Before light micrography, the organ was immersed in 9% saline solution. Macerated cellular material was spread on the glass slides, which had been previously immersed in 96° ethyl alcohol and air-dried.
2. The glass slides with the lung impress were fixed with methyl alcohol for 3 min.
3. Later they were placed into the Giemsa stain (diluted 1:10 with tap water), pH 5.0 for 10 min.
4. Then the slides were washed several times to prevent contamination by the stains.
5. The slides were dried in air.
6. Slides were observed with a light microscope (Leitz, Germany).

7. Numbers of cells on the lung slides were counted based on the method used for bone marrow counts, according to Grignaschi et al. *(10)*. A total of 500 cells was counted in different areas of the slide and according to the proportion of cells, the percentage was later determined.

3.7. Electron Microscopy Techniques

1. Higher trachea and lung samples (bronchia, bronchioli, and alveoli) were obtained after sacrificing mice by cervical dislocation.
2. Samples were fixed in 3.5% glutaraldehyde solution buffered with 0.1 *M* phosphate buffer (pH 7.4) for 3 h and postfixed in 1% osmium tetroxide with the same buffer overnight.
3. Samples were treated with an aqueous solution of 2% uranyl acetate for 40 min.
4. After fixation, tissues were gradually dehydrated in a series of alcohols of increasing strength, passed through acetone, and embedded in Spurr resin.
5. Ultrathin sections were stained with uranyl acetate and lead citrate.
6. Observation was with a Zeiss EM 109 transmission electron microscope at 50 kV (*see* **Note 4**).

3.8. Microbiological Techniques

1. Determination of the number of Microorganisms in the respiratory tract organs.
   a. Nasal and pharynx instillations were obtained by washing each open cavity with 1 mL of peptone water and later scraping them gently with mini-sterile cotton tips.
   b. Trachea, bronchia, and lung were aseptically removed and homogenized in 1 mL peptone water with a Teflon pestle.
   c. The number of microorganisms was determined by the serial dilution method.
   d. The organs were diluted in peptone water (1% peptone) and inoculated on MRS-rifampin (15 μg/mL) agar plates (Biokar Diagnostics, Beauvais, France).
   e. Plates were incubated at 37°C for 48–72 h in a microaerobic environment.
   f. Lactobacilli resistant to rifampin were counted after that period.
   g. The number of microorganisms was determined, taking into account those plates showing between 30 and 300 CFU/plate.
2. Identification of microorganisms was performed by Gram staining and biochemical tests. The streptococci were also identified by the test described next, in **step 3**.
3. Direct examination: Gram stain. Direct observation of the samples was performed after Gram staining the slides. The technique and reactants are those standards set up in the laboratory for routine Gram stain.
4. Microbiological identification in enriched, selective, and differential culture media. The microorganisms present in the respiratory tract of mice were isolated by standard techniques used in the laboratory. The culture media used were BHI, blood agar, and chocolate agar.
5. Biochemical tests to identify the microorganisms.
   a. Identification was performed by standard morphological and biochemical tests used for clinical diagnosis. The criteria applied for identification were (1) observation of B-hemolytic colonies grown in blood agar with typical morphology and (2) variable predominance.
   b. Identification of the following microorganisms was performed: α-hemolytic *Streptococcus*, *Staphylococcus*, *Micrococcus*, *Corynebacterium*, Enterobacteriaceae,

Peptocococcus, *Veillonella*, *Clostridium*, *Bacteroides*, *Citrobacter*, *Escherichia*, *Klebsiella*, and *Aeromonas*.

c. The phenotypic properties studied were (1) standard biochemical tests as described in the *Manual of Bacteriological Procedures* and (2) API Strepto, API Staph, and API 50 CHL (Biomereux, France).

6. Lactobacilli identification.
   a. Additional tests performed for the identification of lactobacilli were the following: catalase test, Gibson's test, fermentation of calcium gluconate, growth at different temperatures (15° and 45°C), and growth in skim milk.
   b. Carbohydrate fermentation was evaluated from lactose, mannitol, raffinose, sorbitol, and inulina.
   c. Hydrolysis was used for arginine, esculine, and hippurate.
7. Streptococci identification: Streptococci were evaluated by production of urease, growth in media with sodium azide and 6.5% NaCl, and the esculin bilis reaction.
8. *S. pneumoniae* identification: The optoquin and oxgall test was performed as follows:
   a. The colonies were spread into the surface of blood agar, where the optoquin discs were added aseptically. The *S. pneumoniae* colonies are inhibited by the optoquin.
   b. Colonies morphologically suspected to be *S. pneumoniae* must also be tested for oxgall solubility.
   c. Oxgall discs were also used.
9. Oxgall solubility: This test is based on the characteristics of *S. pneumoniae*, because the colonies are lysed when incubated in 10% sodium deoxycholate (BBL or Difco).
10. Quellung test.
    a. Serological identification of *S. pneumoniae* is based on the reaction of specific antisera against capsular polysaccharides using the Quellung reaction. It can be applied directly to the clinical samples (sputum, pleural exudates, cerebrospinal fluid, or other exudates) for fast identification of pneumococci. It is also valuable as a confirmatory procedure for the identification of *S. pneumoniae* from cultures.
    b. One small drop of the clinical material or the young culture in broth is added to the glass slide.
    c. Three drops of the antiserum are added, which are mixed with the bacterial suspension.
    d. Then, one drop of saturated solution of methylene blue is added, and the slide is covered with a glass cover slip.
    e. After 15 min, the slides are observed under an immersion microscope.
    f. *S. pneumoniae* stains dark blue, surrounded by a very strong transparent halo.

3.9. Serum Antibodies

1. Mice were bled from the retro-orbital venous plexus at different days post *S. pneumoniae* challenge.
2. Serum was kept frozen until determination of antiserum level.
3. Antibody titration was performed using an agglutination reaction with *L. fermentum* and *S. pneumoniae* suspensions (1×10^9 microorganisms/mL each).
4. The agglutination reaction was performed with serial dilutions of the sera and microorganism suspensions.
5. Tubes were incubated for 1 h at 37°C and then kept at 4°C for 12 h.
6. Antibody titers were defined as the inverse of the highest dilution presenting a positive agglutination reaction.

3.10. Statistical Analysis

Experiments were performed at least three times, and the results obtained were used to calculate the mean and SD. Results are shown in the tables and graphs. Student's *t*-test was used to determine the statistical significance of the differences between data.

4. Notes

1. Screening of the surface characteristics of the microorganisms allowed selection of some lactobacilli strains from the respiratory tract of mice.
2. Inoculation of *S. pneumoniae* by the intranasal route in mice was hard to assess. No respiratory infections were produced in mice. Only after animalization of the microorganisms (which means ip inoculation of young mice and bleeding after 24 h by a heart puncture) was respiratory tract infection produced in the experimental mice model.
3. Inoculation of the microorganisms by the intranasal route must be performed very carefully to allow the small drop to be sniffed into the nasal cavity. The mice must be kept in a vertical position for some minutes to allow the microorganisms to enter the nasal area.
4. When performing the histological technique, care must be taken before immersing the trachea and lung into the fixation step. All blood must be extracted from the tissues by washing twice with buffered solution.
5. The macerated lung is soaked in sterile saline solution. The impress must be performed very soon after extraction of the organ, always using a wet fixation to avoid cell retraction.
6. For trachea samples for electronic microscopy purposes, it is recommended to perform a glutaraldehyde (fixate) passage with a tuberculin syringe in the central area of the trachea, and later to put the trachea into the fixative.

Acknowledgments

This work was supported by grants from CONICET (Consejo Nacional de Investigaciones Científicas y Técnicas).

References

1. World Health Organization (2001) *Global Consultation on Child and Adolescent Health and Development. Acute Respiratory Infections*. WHO, Geneva.
2. Reid, G. (1999) The scientific basis for probiotic strains of *Lactobacillus*. *Appl. Environ. Microbiol.* **65,** 3763–3766.
3. de Gutiérrez, R. C., de Nader, O. M., de Ruiz Holgado, A. P., and Nader-Macias, M.-E. (1999) Microbial flora variations in the respiratory tract of mice. *Mem. Inst. Oswaldo (Rio de Janeiro)*, **94,** 701–707.
4. de Gutiérrez, R. C., Santos de Araoz, V., and Nader- Macías, M.-E. (2000) Effect of intranasal administration of *Lactobacillus fermentum* on the respiratory tract of mice. *Biol. Pharrm. Bull.* **23,** 973–978.
5. Santos, V., Cangemi, R., Winik, B., and Nader, M.-E. (2001) Ultrastructural studies on the respiratory tract of mice inoculated intranasally with *L. fermentum. Biocell* **25,** 121–129.
6. de Gutiérrez, R. C., Santos, V., and Nader-Macías, M.-E. (2001) Protective effect of intranasally inoculated *Lactobacillus fermentum* against *Streptococcus pneumoniae* challenge on the mouse respiratory tract. *FEMS Inmunol. Med. Microbiol.* **31,** 187–195.
7. Raibaud, P., Galpin, J. V., Ducluzeau, R., Mocqout, G., and Oliver, G. (1963) The *Lactoba-*

cillus genus in the digestive tract of the rat. II. Characteristics of heterofermentative strains isolated from holo and gnotoxenic rats. *Ann. Microb. (Inst. Pasteur)* **124,A,** 223–225.
8. De Man, J. C., Rogosa, M., and Sharpe, M. E. (1960) A medium for the cultivation of lactobacilli. *J. Appl. Bacteriol.* **23,** 130–135.
9. Martoja, R. and Martoja-Pierson, M. (1970). *Initiations aux Techniques de l' Histologie Animale.* Masson, Paris.
10. Grinaschi, V., Diaz, N., Alonso, M., Lardo, M., and Lucero, G. (1983) Citomorfología y citoquímica hemáticas. Britania, Buenos Aires.
11. Kandler, O. and Weiss, N. (1986). The genus Lactobacillus. In: *Bergey's Manual of Systematic Bacteriology*, Vol 2 (Sneath, P., ed.). Williams and Wilkins, Baltimore, MD, pp. 1209–1234.

36

Effects of Estrogen Administration on the Colonization Capability of Lactobacilli and *Escherichia coli* in the Urinary Tracts of Mice

Clara Silva, Rosario Rey, and María Elena Nader-Macías

1. Introduction

The use of probiotic microorganisms has been widely promoted in the last 20 yr *(1)*. They have been used in the gastrointestinal tract as capsules or as fermented milks *(2)*. The characteristics of the strains proposed as probiotics have been published *(3)* or patented under an elaboration process *(4–6)*. The first step in designing a probiotic product is to isolate and characterize strains with some beneficial properties. The second step is to determine the optimal conditions to obtain the highest amount of viable microorganisms, together with the study of the best conditions to produce antagonistic substances *(7)*.

Urinary tract infections (UTIs) constitute a common cause of illness in pre- and postmenopausal women. It was estimated that 40–50% of adult women suffer a cystitis during their life. Ninety percent of acquired ambulatory UTIs and 30% of nosocomial infections are produced by *Escherichia coli (8)*. The healthy human urinary tract is free of microorganisms, except for the anterior urethra, which is colonized by indigenous microbiota. The vaginal environment is a dynamic and complex ecological system with a highly heterogeneous microflora; thus favorable conditions exist for the colonization process, which is also affected by factors external to the tissues. The distal urethra and periurethral areas are separated ecological niches, both covered by the vaginal secretions that contain approx 10^9 microorganisms/mL *(9)*. In these secretions, members of the genus *Lactobacillus* are predominant. Bacterial colonization does not increase because of the urinary flux, which clears the bacterial cells from the outer surfaces, as well as other factors such as pH, osmolarity, and urea concentration *(10,11)*.

From: *Methods in Molecular Biology, vol. 268: Public Health Microbiology: Methods and Protocols*
Edited by: J. F. T. Spencer and A. L. Ragout de Spencer © Humana Press Inc., Totowa, NJ

The work of Redondo-Lopez et al. ***(12)*** and Hillier et al. ***(13)*** supports the fact that the ecological stability of the microflora, its integrity in equilibrium with the different host tissues, and the effects exerted by various mechanisms and factors all protect against incoming pathogenic microorganisms. In the same way, loss or disruption of the normal genital microbiota, particularly the lactobacilli species, implies that an increase in urogenital infections will follow. The composition of the vaginal microbiota is influenced by estrogen levels, as reported by some researchers who have shown that this hormone stimulates glycogen deposition in the epithelial tissue ***(14)***. Not much is known about the microbial implications of the steroid-hormonal effect on the urogenital tract. Different characteristics can be influenced by hormones involved in the maintenance of the microecosystem ***(15)***. The reproductive hormones produce a number of physiological changes in the host, including increased phagocytic activity, which can influence the response to an infectious process ***(16)***. However, modifications in carbohydrate and protein metabolism by these hormones can result in changes of the indigenous microflora of the host, owing to alteration of the colonized tissues ***(17)***.

One of the biggest challenges for UTI therapy is recurrence in adult women: even though one episode can be rapidly eliminated, it is soon followed by another, sometimes caused by a different microorganism ***(18)***. Different methodologies have been adopted for the treatment of recurrences: local hygiene with disinfectants, long- or short-term prophylaxis with oral antibiotics using low doses, oral or topical therapy with estrogens, daily consumption of cranberry juice, and others ***(19)***. Ingestion or application of products containing viable lactobacilli is currently accepted to maintain or modify the normal microbiota. In some cases partially effective results were achieved, even though the optimal strategy was not applied ***(20)***.

In previous papers, an experimental model was set up to study the effect of lactobacilli in the urogenital tract of mice. Autochthonous lactobacilli were isolated from the vagina of normal BALB/C mice. The strains were identified by their morphology and by biochemical and physiological testing and later classified according to *Bergey's Manual*. In vitro adhesion was tested by their ability to adhere to uroepithelial cells, according to Reid et al. ***(12)***. One strain, *L. fermentum* CRL 1058, was selected by the adherence test and also by other surface characteristics and because it was largely prevalent in our isolates. To study the colonization capability of this *L. fermentum* strain in the urinary tract, adult female BALB/c mice were used as the experimental model. To prevent the microorganisms from being cleared by the urine flux, they were prepared in agarose beads and inoculated intraurethrally. The optimal dose and the physical forms were studied to determine the colonization capability of this probiotic strain. Effective colonization was obtained by the intraurethral inoculation of agarose beads in three doses containing 10^7 colony-forming units (CFU)/12 h ***(22,23)***. In other studies, in vivo experiments were performed to assess the effect of antibiotics and hormones on colonization of both lactobacilli and uropathogenic *E. coli* ***(24,25)***.

The aim of the present work was to study the effect of estradiol on the kinetics of colonization of *L. fermentum* CRL 1058 and a uropathogenic *E. coli* strain in the urinary tract and the histological modifications produced by them. The following areas were investigated:

1. Determination of microorganisms, the methods to be applied for their use, and preparation for inoculation.
2. Determination of the animals to be used.
3. The amount of hormones in sera.
4. The number of microorganisms in the tissue homogenates.
5. Cytological, histological, and ultrastructural studies performed to determine the effect of the hormones on the colonization capability of *L. fermentum* and *E. coli*.

2. Materials

2.1. Microorganisms

1. *L. fermentum* CRL 1058: isolated from the vagina of 2-mo-old Balb/c mice, by using the procedure described previously ***(20)***.
2. Uropathogenic *E. coli*: isolated from an adult female patient with a diagnostis of pyelonephritis (from the Collection of Culture Strains, Bacteriology Department, Instituto de Microbiología; courtesy of Dr. L.C. Verna, Facultad de Bioquímica, Química y Farmacia, Universidad Nacional de Tucumán.) The strain had the following characteristics related to pathogenicity: mannose-resistant hemagglutination (MRHA), hemolysin producer, pyelonephritogenic capability (assayed by intraurethral inoculation in 2-mo-old female BALB/c mice and later isolated in different organs of the urinary tract; *see* **Note 1**).

2.2. Preparation of Beads for Inoculation

1. Phosphate-buffered saline (PBS), pH 7.0.
2. Vaseline.
3. Peptone water: 0.1% meat peptone in distilled water.
4. LAPTg broth: 15 g/L peptone, 10 g/L tryptone, 10 g/L glucose, 10 g/L yeast extract, 1 mL/L Tween-80, pH 6.5. Autoclave at 121°C for 15 min. Store at refrigeration temperature. The chemicals for LAPTg preparation were obtained from Britania Laboratories (Argentina).
5. Agarose (1% in PBS; *see* **Note 2**).

2.3. Culture Media

1. Milk-yeast extract: 10% skim milk, 1% glucose, 0.5% yeast extract, pH 6.0. Autoclave for 20 min at 115°C ($^3/_4$ atm).
2. LAPTg broth.
3. LAPTg agar: The same as LAPTg broth with 15 g/L agar.
4. MRS broth: Dehydrated MRS was obtained from Biokar Diagnostics (Beauvais, France). Composition: 10 g/L peptone, 10 g/L meat extract, 5 g/L yeast extract, 20 g/L glucose, 1.08 g/L Tween-80, 2 g/L dipotassium phosphate, 5 g/L sodium acetate, 2 g/L ammonium citrate, 0.2 g/L magnesium sulfate, 0.05 g/L manganese sulfate, pH 6.5. Autoclave at 121°C for 15 min. Store at refrigeration temperature.
5. MRS agar: The same as MRS broth with 15 g/L agar.
6. McConkey's agar (Merck): 17.0 g/L casein peptone, 3.0 g/L meat peptone, 5.0 g/L sodium chloride, 10.0 g/L lactose, 1.5 g/L oxgall salt, 0.03 g/L neutral red, 0.001 violet crystal, 13.5 g/L agar, pH 7.1 ± 0.2.
7. Cystine lactose-deficient electrolyte Britania: 4.0 g/L peptone, 3.0 g/L meat extract, 0.128 g/L L-cystine, 4.0 g/L tryptamine, 0.02 bromothymol blue, 10.0 lactose, 15.0 agar, pH 7.3 ± 0.2.

8. 0.1% Peptone water.
9. Brain-heart infusion (BHI; Britania): 200 g/L goal brain infusion, 250 g/L bovine heart, 10 g/L peptone, 5g/L sodium cloride, 2.0g/L glucose, 2.5 g/L disodium phosphate. pH 7.4 ± 0.2.
10. Other materials.
    a. Sterile tubes.
    b. Pipets.
    c. Petri dishes.
    d. Incubator for 37°C (Forma Scientific, model 3185.

2.4. Animals

1. Mice: 2-mo-old female BALB/c mice from the inbred colony of the Instituto de Microbiologia at the Universidad Nacional de Tucumán (weighing approx 30 g). Animals were housed in plastic cages and fed ad libitum; environmental conditions were kept constant. Each experiment was carried out with groups of 20–25 mice. The Centro de Referencia para Lactobacilos (CERELA) Committee of Ethics approved the protocol used for animal studies (*see* **Note 3**).
2. Mice Sacrifice.
    a. Gloves.
    b. Bonnet string.
    c. Dashboard.
    d. Pasteur pipets for blood extraction.
    e. Pins.
    f. Scissors.
    g. Nippers.
    h. Petri dishes.
    i. Sterile tubes containing 2 mL 0.1% peptone water for tissue homogenization.
    j. Homogenizer.
    k. Incinerator.

2.5. Hormones

1. Estradiol valerate (Progynon depot, Schering): 10 mg in 1-mL ampules. Hormones were inoculated at different doses, to determine which dose produced some type of effect. Evaluation of the hormonal effect was determined by serum estrogen level and by vaginal cytology, which permits observation of cell changes during the estral cycle. The protocol used is schematized in **Fig. 1**.

2.6. Cytological Studies

1. Saline solution (0.15 *M* NaCl) for vaginal washes. Sterilize by autoclaving and store at refrigeration temperature.
2. 76 × 26-mm Glass slides microscope.
3. 96° Ethanol (Biopack, Argentine). Store at room temperature.
4. Glass bottles for slides.
5. Graduated ethyl alcohols: 96°, 80°, 70°, 50°, 100°.
6. Harris hematoxylin (Biopack, Argentina). Store at room temperature.
7. Aqueous eosine (E.A.36; Biopur, Argentina). Store at room temperature.

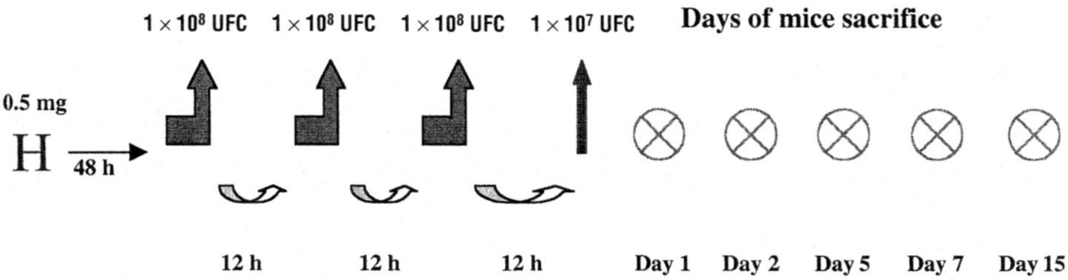

Fig. 1. Hormone inoculation scheme. H, hormone inoculation. The three bent arrows are the lactobacilli inoculation, and the single straight arrow is the *E. coli* inoculation. Days of the mice sacrifice were counted from the day of *E. coli* challenge.

8. Orange G (Biopur). Store at room temperature.
9. Xylene (Biopack). Store at room temperature.
10. Natural Canada balsam (Biopack). Store at room temperature.
11. 24 × 48-mm Glass slides.
12. Cover slides.

2.7. Histological Studies

1. 10% Formaldehyde (Biopack). Store at room temperature.
2. Increasing graduated ethyl alcohols (Biopack): 50°, 70°, 80°, 96°, 100°.
3. Xylene (Biopack). Store at room temperature.
4. 56°C Incubator.
5. Histoplast Fusion Point 56°–58°C (Biopack).
6. Sliding microtome (Zeiss).
7. Microtome blades.
8. 76 × 26-mm Glass microscope slides (Zeiss).
9. Decreasing graduated ethyl alcohols (Biopack): 100°, 96°, 80°, 70°, 50°.
10. Hematoxylin stain (Biopack). Store at room temperature.
11. Eosin yellow (Biopack).
12. Distilled water.
13. Natural Canadian balsam (Biopur).
14. 22 × 48-mm Glass cover slides.
15. Light microscope.

2.8. Electron Microscopy

1. Glutaraldehyde: 3.16% in 0.1 *M* phosphate buffer, pH 7.4 (Pelco).
2. Washing buffer: 0.1 *M* phosphate buffer, pH 7.4.
3. Osmium tetroxide.
4. Epoxy-type resin: Epon 812 (Spurr, Pelco).
5. Ultramicrotome.
6. Uranyl acetate.
7. Pb citrate.
8. Electron microscope (Zeiss 109).

3. Methods

3.1. Microorganisms

1. *L. fermentum* CRL 1058 stored in milk-yeast extract.
   a. Screw-top glass tubes containing 5 mL LAPTg broth were inoculated with the isolated lactobacillus and incubated at 37°C for 24 h.
   b. A 50-μL aliquot of these cultures was subcultured twice at 37°C for 12 h in 5 mL of the same medium.
   c. The lactobacilli culture was centrifuged for 10 min at 2000*g*), and the spent supernatant was discarded.
   d. The pellet was washed twice with 0.85% saline solution and suspended in 5 mL milk-yeast extract.
   e. This bacterial suspension was distributed in 1-mL screw-top cryovials and stored at –20°C.
2. Uropathogenic *E. coli*: The microorganisms were identified by the methods described in *Bergey's Manual of Determinative Bacteriology*, standard techniques used in the laboratory, and API 50 CHL (Biomerieux-France).
   a. Screw-top glass tubes containing 5 mL LAPTg broth were inoculated with the isolated *E. coli* and incubated at 37°C for 24 h.
   b. The *E. coli* culture was centrifuged for 10 min at 2000*g*, and the spent supernatant was discarded.
   c. The pellet was washed twice with saline solution (0.85% NaCl) and suspended in BHI + 25% glycerol.
   d. This bacterial suspension was distributed in 0.3 mL screw-top cryovials and stored at –70°C.
   e. They were subcultured in LAPTg broth for later use.

3.2. Bacterial Incorporation Into Beads

The agarose beads were prepared according to the method described by Cash et al. *(2)* with certain modifications, as follows:

1. *L. fermentum* was grown in LAPTg broth for 12 h at 37°C.
2. Organisms were harvested by centrifugation, washed twice with PBS, pH 7.0, and resuspended in PBS.
3. A suspension of 10^8 colony-forming units (CFU)/mL was mixed with the same volume of 1% agarose in PBS and maintained at 37–40°C and three volumes of vaseline added at the same temperature.
4. The mixture was gentle vortexed for 3 min and after resting at room temperature for 2 min, cooled in an ice bath and maintained at 0°C for 7–10 min.
5. The beads were washed by centrifugation at 1000*g* with 0.1% peptone water to remove excess vaseline.
6. The supernatant was taken out with a Pasteur pipet.
7. The beads were observed under the microscope to check adequate size and shape.
8. Viability was determined by culturing the beads in MRS and LAPTg agar medium.
9. The beads were distributed in 3-mL volume tubes in and stored at –20°C.

3.3. *E. coli* Suspension

1. The uropathogenic strain was grown in either BHI or LAPTg broth.
2. Microorganisms were harvested by centrifugation, washed twice with PBS, pH 7.0, and suspended in 0.1% peptone water.

3.4. Estrogenic Stimulation

1. Estrogenic stimulation was performed by im inoculation of one dose (0.5 mg estradiol valerate for human use (Progynon depot). The dose was calculated according to the weight of the mouse (approx 30 g).
2. The hormone was inoculated for 48 h before inoculation with *L. fermentum* and *E. coli*.

3.5. Quantification of Serum Estriol

1. Mice were bled from the retro-orbital venous plexus before sacrifice at 24 h to 15 d after stimulation
2. Serum was separated by centrifugation and frozen at –20°C.
3. Quantitative determination of estriol was performed by using the microparticle enzyme inmunoassay (MEIA; Imx System, Abbott).

3.6. Microbiological and Histological Studies in Mice

3.6.1. Vaginal Samples

Vaginal washing was performed with 50 μL of saline solution by using a syringe with a canule and aspirating the liquid. Samples were processed for the following studies:

1. Direct Gram staining.
2. Vaginal cytology.
3. Culture in selective media.

3.6.2. Bacterial Inoculation

The following procedure was used for microorganism inoculation:

1. Mice were anesthetized with sodium pentobarbital before intraurethral inoculation of *L. fermentum* in agarose beads (three doses of 1.5×10^8 CFU or 2.0×10^8 CFU 12 h between each dose, and 12 h later one dose of *E. coli* of 1.0×10^7 CFU or 1.0×10^8 CFU). A plastic catheter coupled to a syringe was used for this purpose.
2. After each inoculation, animals were returned to their cages.
3. During the experimental days indicated, the following samples were taken or studies performed:
   a. Urine, for quantification of microorganisms.
   b. Organ homogenates, to determine the degree of colonization, represented by the number of microorganisms in each organ.
   c. Histological studies, to determine whether certain types of collateral effects or modifications could be seen, at the structural level.
   d. Ultrastructural studies by transmission electron microscopy (TEM) to determine whether certain types of modifications had occurred at this level.

3.7. Urine Culture

Before sacrifice, urine was collected aseptically with a loop to (1) observe the presence of red blood cells, polymorphonuclear granulocytes, epithelial cells, and mucus and (2) quantify the numbers of microorganisms in CLED (*see* **Note 4**).

3.8. Bacterial Counts in Tissue Homogenates

1. Animals were sacrificed by cervical dislocation.
2. Their urinary tract organs (urethra, bladder, ureters, and kidneys) were removed aseptically, placed in 2 mL of 0.1% peptone water, and homogeneized with a Teflon pestle.
3. The urethra and ureters had been previously cut longitudinally with a scissors to allow release of microorganisms.
4. Sometimes homogenization of the bladder is difficult because of the fibrous nature of the tissue, and the bladder must be homogenized longer.
5. The samples were serially diluted (10-fold) in peptone water, from 1:10 to 1:1,000,000, using glass tubes with peptone water and automatic pipets; each dilution tube mixed vigorously in a vortex.
6. A 0.5-mL aliquot of each sample dilution was plated in duplicate in the following culture media: *Lactobacillus* selective agar (LBS), MRS, and McConkey's agar (differential for Gram-negative bacilli); 12–15 mL of melted agar (*see* **Note 5**).
7. LBS agar, MRS, and McConkey agar were added and mixed carefully while the plate was rotated, for even distribution of the bacteria. The agar media should be melted and maintained at a temperature of 45–50°C before use.
8. The plates were placed in the incubator at 37°C, for 48 h.
9. The number of CFUs was determined, selecting the data for plates that contained 30–300 isolated colonies (*see* **Note 6**).
10. Viable cells counts were made for each organ.
11. Results were expressed as means ± SD of the data obtained from three or four animals.

3.9. Cytological Studies

A syringe loaded with 0.9% physiological solution was introduced into the mouse vagina and aspiration was performed.

3.9.1. Fixation

The aspiration was spread over a glass slide and immediately fixed in 96° alcohol (*see* **Note 7**). The smears were left to dry in the open air for Papanicolaou smear.

3.9.2. Papanicolaou Smear Technique (25)

1. The smear should be hydrated by successive passages in alcohol with decreasing concentrations, as follows:
    a. Alcohol 96° for 1 min.
    b. Alcohol 80° for 1 min.
    c. Alcohol 70° for 1 min (*see* **Note 8**).
    d. Alcohol 50° for 1 min.
2. Distilled water for 1 min (*see* **Note 9**).
3. Harris's hematoxylin stain for 5 min.
4. Differentiation with acid water (water with the addition of 1 drop of HCl).
5. Alcohol 96° for 1 min.

6. Orange G for 1 min.
7. Alcohol 96° for 1 min.
8. Alcohol 96° for 1 min.

3.9.3. Dehydration (see **Note 10**)

Successive passages of alcohols of increasing concentration were made (*see* **Note 8**), as follows:

1. Ethyl alcohol 96°, two passages of 1 min.
2. Ethyl alcohol 100°, two passages of 1 min.
3. Xylene, two passages of 1 min.

3.9.4. Mounting and Observation (see **Note 9**)

1. The slides were mounted with Canadian balsam, and covered with a glass cover slide for microscopic observation.
2. The cytological slide was observed under light microscopy (ocular 10X and objective 40X), with different colors. The cells can be of three types: eosinophil surface cells, cyanophilic surface cells, cyanophilic intermediate cells, cyanophilic parabasal cells, leukocytes, or mucus. At least 200 cells are counted to determine the percentage of each of the cellular types observed.
3. *An estrogenic state* is characterized by the presence of eosinophils surface cells with or without a pycnotic nucleus. The background of the slide does not show inflammatory elements.
4. *A progesteronic state* is characterized by the presence of cyanophilic intermediate cells with vesicles nucleus. The background of the smear is dirty because of larger amounts of leukocytes and mucus.

3.10. Histological Technique (26)

1. Animals were sacrificed by cervical dislocation.
2. Samples of the different organs from the tract were aseptically extracted (kidneys, urethers, bladders, and urethra), washed with saline solution, and immediately introduced into fixative.
3. Fixation was with 10% formaldehyde for at least 24 h (*see* **Note 11**).
4. The organ must be dehydrated before paraffin embedding, by transfer through increasingly graduated alcohols (*see* **Notes 12** and **13**), as follows:
   a. Ethyl alcohol 50° for 1 h.
   b. Ethyl alcohol 70° for 1 h.
   c. Ethyl alcohol 80° for 1 h.
   d. Ethyl alcohol 96°, two passages of 1 h each.
   e. Ethyl alcohol 100°, two passages of 1 h each.
   f. Xylene, two passages of 1 h each.
   g. Xylene-paraffin (1:1) in an incubator at 56° for 1 h.
   h. Paraffin I in an incubator at 56° for 8 h.
   i. Paraffin II in an incubator at 56° for 1 h.
   j. Paper boxes were previously prepared to receive the organs in paraffin. The paraffin solidifies when brought to room temperature, and the block is ready. After a wood piece is added, the blocks are ready to be cut in the microtome.

5. The slides must be as narrow as possible; the ideal width is between 2 and 4 mμm. The tissue slides are put into warm water, to be ironed. Later, they are added to glass slides and incubated at 37°C to finish the extension process (*see* **Note 14**).
6. The paraffin must be removed; it does not permit staining of the tissue, because it is insoluble in alcohol or aqueous solutions. The slides were immersed in xylene (two successive baths of 1 min each).
7. As the stain used is suspended in aqueous solution, the slides must be progressively hydrated in successive baths, as follows:
   a. Ethyl alcohol 100°, 1 min.
   b. Ethyl alcohol 100°, 1 min.
   c. Ethyl alcohol 96°, 1 min.
   d. Ethyl alcohol 70°, 1 min.
   e. Ethyl alcohol 50°, 1 min.
   f. Distilled water, 1 min.
8. Hematoxyline: 1 min to stain the nucleus. The slides were stained red.
9. Tap water: 20 min. The alkaline salts modify the media and produce a change in color from red to blue (*see* **Note 15**).
10. Alcoholic eosin: 0.5% eosin solution in ethyl alcohol 50° to stain the cytoplasm.
11. The slides were transferred successively through different ethyl alcohol gradations (*see* **Note 10**).
    a. Ethyl alcohol 96°, two passages of 1 min each.
    b. Ethyl alcohol 100°, two passages of 1 min each.
    c. Xylene, two passages of 1 min each.
12. The slides were covered with 1 drop of Canadian balsam, covered with a glass coverslip, and dried in the air for 24 h (*see* **Note 10**).
13. Observation by light microscopy: The nucleus, particularly the chromatin, is stained blue at different intensities and the cytoplasm is stained red.

3.11. Electron Microcopy Technique

Immediately after it was obtained, the material was cut into small pieces of less than 1 mm and fixed in 6% glutaraldehyde in cacodylate buffer. The material was later fixed in osmium teroxide. Epoxy resines were used as inclusion media. Ultra-narrow slides 400–600 Å wide were obtained by using the ultramicrotome. The slides were mounted on Cupper supports covered with Parlodion and stained with Pb citrate. Tissues were observed under TEM, and the photos were taken with 35-mm Kodak film.

4. Notes

1. For adequate use of microorganisms, the same experimental protocol must always be used. The original culture must be prepared and fractionated in several tubes, which must be kept in freezing conditions. For each experiment, the culture must be obtained from this freezing culture. No more than three subcultures were performed, to avoid changes in adhesion properties.
2. To eliminate excess of Vaseline, we have also tried to wash the beads with a detergent such as sodium deoxycholate (0.125, 0.25, and 0.50% in PBS). We have also studied storage and stability using different temperatures and washing liquids. Different washing liquids such as PBS, peptone water, or a mixture of both, were assayed; better results were obtained with peptone water in terms of more stable beads.

3. The mice used for the experiment must always come from the same inbred, closed colony and be of the same age and weight, to avoid genetic interference. All the control groups must be assayed at the same time. When a certain condition is changed, it must be changed for all the groups.
4. Urine: as the urine volume obtained from each mouse is very low, it is not always possible to perform all the assays.
5. *Lactobacillus* selection agar is a selective medium for lactobacilli; inhibits the accompanying microbial flora, allowing the growth of colonies with different morphology.
6. Morphological variation in the colonies obtained after growth in plates, isolated from tissue homogenates, was sometimes observed. The colonies were identified by morphological and biochemical tests (not genetic tests). We did not study whether these variations would correspond to rough or smooth variants of the pathogenic microorganisms or whether they were caused by stress on the microorganisms because of host defenses.
7. The slides must be immersed in the fixation liquid for at least 10 min. The time of fixation must not exceed 24 h in any case.
8. The time in which the slides must be kept in each alcohol must be about the indicated time in each step.
9. The slides must be mounted in a very quiet environment to avoid interference of any type before adding the cover slips.
10. The dehydration step is very important, because the transparency of the slide depends on it, as well as the possibility of false-positive reactions when the slides are observed cytologically.
11. The amount of fixation liquid must be enough to go into the anatomical piece. The tissues must be fixed as soon as possible so that the slides do not become dry.
12. The dehydration step is very important, because it determines that the tissue and the structures are conserved without modifications, the same as at the time of fixation.
13. Graduation of the different alcohols was determined by measuring the amount of 96° alcohol and adding distilled water until 96° was reached. For example, 80 mL of 96° alcohol plus 16 mL of distilled water results in alcohol 80°.
14. The ironing step is very important, because modifications or folding of the slides can interfere in staining and observation.
15. Differentiation implies the smear changing from red to blue. The differentiation step is very important. The tissues must be in this liquid for at least 20 min to allow the cells to take the appropriate color.

Acknowledgments

This work was supported by grants from CONICET (Consejo Nacional de Investigaciones Científicas y Técnicas de Argentina), PIP 359, and from the Ministerio de Salud Pública de la República Argentina (Beca Carrillo Oñativia para Investigadores Formados).

References

1. Redondo-López, V., Cook, R. L., and Sobel, J. D. (1990) Emerging role of lactobacilli in the control and maintenance of the vaginal bacterial microflora. *Rev. Infect. Dis.* **12,** 856–872.
2. Reid, G. (1999) The scientific basis for probiotic strains of *Lactobacillus*. *Appl. Environ. Microbiol.* **65,** 3763–3766.
3. Reid, G. and Bruce, A. W. (2001) Selection of Lactobacillus strains for urogenital probiotic applications. *J. Infect. Dis.* **183(suppl 1),** S77–S80.

4. Chrisope, G. L. (Boulder, CO) (2000) Vaginal *Lactobacillus* medicant. US patent and trademark office, US Patent, GyneLogix, Inc., 093,394-6-21.
5. Larsen, Bryan (Huntington, WV) (1999) Vaginal pharmaceutical compositions. US patent and trademark office, US Patent, Marshall University Research Corporation, Huntington, WV, 5,958,461-7-34.
6. Reid, G. (London, Canada), Bruce, A. W. (Toronto, Canada) (1997) *Lactobacillus* and skim milk compositions and methods for preventing microbial urogenital infections. US patent and trademark office, US Patent, Research Corporation Technologies, Inc., Tucson, AZ, 5,643,830-1-1.
7. Mäyra-Mäkinen, A. and Bigret, M. (1998) Industrial use and production of lactic acid bacteria. In: *Lactic Acid Bacteria, Microbiology and Functional Aspects* (Salminen, S. and von Wright, A., eds.). Marcel Dekker, New York, pp. 73–102.
8. Kunin, C. M. (1994) Urinary tract infections in females. *Clin. Infect. Dis.* **18,** 1–12.
9. Bollgren, I., Vaclavinkova, V., Hurvell, B., and Bergquivst, G. (1978) Periurethral aerobic microflora of pregnant and non-pregnant women. *BMJ* **1,** 1314–1317.
10. Eschenbach, D., Davick, P., Williams, S., Klebanoff, S., Young-Smith, C., and Holmes, K. K. (1989) Prevalance of hydrogen-peroxide-producing *Lactobacillus* species in normal women and women with bacterial vaginosis. *J. Clin. Microbiol.* **27,** 251–256.
11. Reid, G., McGroarty, J. A., Tomeczek, L., and Bruce, A. (1996) Identification and plasmid profiles of *Lactobacillus* species from the vagina of 100 healthy women. *FEMS Immunol. Med. Microbiol.* **15,** 23–26.
12. Redondo-Lopez, V., Cook, R. L., Sobel, J. D. (1990) Emerging role of lactobacilli in the control and maintenance of vaginal bacterial microflora. *Rev. Infect. Dis.* **12,** 856–872.
13. Hillier, S., Krohn, M. Rabe, L., Klebanoff, S., and Eschebanch, D. (1993) The normal vaginal flora, H_2O_2 producing lactobacilli, and bacterial vaginosis in pregnant women. *Clin. Infect. Dis.* **16(suppl. 4),** S273–S281.
14. Stahl, C. and Hill, G. B. (1986) Microflora of the female genital tract. In: *Infectious Diseases in the Females Patient* (Galask, R. P. and Larsen, B., eds.). Springer-Verlag, New York, pp. 16–42.
15. Okada, H. and Hillery, A. M. (1978) Chap. 11. *The Human Vagina* (Hafez E. S. E. and Evans, T. N., eds.). Elsevier/North-Holland, Amsterdam, pp. 303.
16. Larsen, B., Markovetz A. J., and Galask R. P. (1976) Quantitative alterations in the genital microflora of female rats in relation to the estrous cycle. *J. Infect. Dis.* **134,** 486–489.
17. Larsen, B., Markovetz, A. J., and Galask R. P. (1977) Role of estrogen in controlling the genital microflora of female rats. *Appl. Environ. Microbiol.* **34,** 534–540.
18. McGroarty, J., Tomeczek, L., Pond, D., Reid, G., and Bruce, A. (1992) Hydrogen peroxide production by *Lactobacillus* species: correlation with susceptibility to the spermicidal compound nonoxynol-9. *J. Infect. Dis.* **165,** 1142–1144.
19. Reid, G., Bruce, W., Cook, L., and Llano, M. (1990) Effect on urogenital flora of antibiotic therapy for urinary tract infection. *Scand. Infect. Dis.* **22,** 43–47.
20. Nader de Macías, López de Bocanera, M. E., M. E., Silva de Ruiz, C., de Ruiz Holgado, A. A. P. (1992) Isolation of lactobacilli from the urogenital tract of mice–elaboration of beads for their inoculation. *Microbiol. Alim. Nutr.* **10,** 43–47.
21. Silva de Ruiz, C., de Macías, M. E. N., de Bocanera, M. E. L., de Ruiz Holgado, A. P. (1993) *Lactobacillus fermentum* administered in suspensions and in agarose beads to mice: a comparative study. *Microbiol. Alim. Nutr.* **11,** 391–397.

22. Nader de Macías, M. E., de Ruiz, C. S., de Bocanera, M. E. L., de Ruiz Holgado, A. A. P. (1996) Preventive and therapeutic effect of lactobacilli on urinary tract infections in mice. *Anaerobe* **2,** 85–93
23. de Ruiz, C. S., de Bocanera, M. E., de Ruiz Holgado, A. A. P. (1996) Effect of lactobacilli and antibiotic on *E. coli* urinary infections in mice. *Biol. Pharm. Bull.* **19,** 88–93.
24. Silva de Ruiz C., del Rosario Rey, M., and Nader-Macías, M. E. (2001) Experimental administration of estradiol on the colonization of *Lactobacillus fermentum* and *Escherichia coli* in the urogenital tract of mice. *Biol. Pharm. Bull.* **24,** 127–134.
25. Takahashi, M. (1982) Papanicolaou cytological technique. In: *Cytology of Cancer*, Takahashi Atlas. INGAKU-SHOIN Ltd, Tokyo, Japan; Medica Panamericana, Buenos Aires, Argentina, p. 78.
26. Martoja, R. and Martoja-Pierson, M. (1970) *Animal Histological Technique.* Toray-Masson, Barcellona, chapters I–V, pp. 2–90.
27. Martoja, R. and Martoja-Pierson, M. (1970) *Animal Histological Technique.* Toray-Masson, Barcellona, chapter VI, pp. 71–116.

37

Effect of Lactobacilli Administration in the Vaginal Tract of Mice

Evaluation of Side Effects and Local Immune Response by Local Administration of Selected Strains

Elisa Vintiñi, Virginia Ocaña and María Elena Nader-Macías

1. Introduction

Lactobacilli are the predominant microorganisms in the vaginal tract of human and some homeothermic animals *(1)*. They can maintain the ecological equilibrium of the tract by protecting against pathogenic microorganisms *(2)*. In the last few years, there has been an increased tendency to use probiotic microorganisms to restore the ecological equilibrium and to protect against infections *(3)*. This principle has been widely applied to the gastrointestinal tract *(4)*. More recently, some other studies have reported the application of probiotics in different tracts, for example, the urogenital *(5)* or respiratory tract *(6)*.

One of the objectives of our group is to design probiotic products for the urogenital tract. With this purpose, lactobacilli were isolated from the human vagina *(7)*, and later some of them were selected for their probiotic characteristics (production of antagonistic substances or adhesion capability) *(8–11)*.

The application of probiotic products in the vaginal tract has been approached empirically; some pharmaceuticals containing these microorganisms *(12)* are available in the United States or Europe or are protected under the patent process or intellectual property rights *(13–15)*. There are not enough studies in humans or animals to determine whether their administration can produce some type of collateral or adverse effect.

Using Balb/c mice as the experimental model, the object of the present work was to study (1) whether intravaginal administration of human lactobacilli can produce colonization of the tract; (2) whether such administration produces some type of adverse or

From: *Methods in Molecular Biology, vol. 268: Public Health Microbiology: Methods and Protocols*
Edited by: J. F. T. Spencer and A. L. Ragout de Spencer © Humana Press Inc., Totowa, NJ

collateral effect; and (3) whether probiotics are able to stimulate the local immune system. Keeping in mind that hormones can affect the colonization or persistence ability of microorganisms, and with the purpose of having all animals at the same point in the sexual cycle, animals were cycled with estradiol 48 h before inoculation with lactobacilli. They were then inoculated im with hormones 48 h before beginning microorganism inoculations. Later they were intravaginally inoculated with the appropriate dose of each *Lactobacillus* strains. The animals were sacrificed on different days after inoculation to perform the following studies:

1. Microbiological assays: To determine the number of lactobacilli in the tract (in vaginal washes or in organ homogenates), by plating the samples in selective media containing antibiotic (to differentiate the resident flora from those administered experimentally).
2. Histological studies: To evaluate whether the local administration of lactobacilli produced some type of modification at the structural level in the tract.
3. Immunofluoresce assays: To study whether local administration of lactobacilli produced stimulation of cells involved in the immune response at the local level.

2. Materials

2.1. Microorganisms

1. The different *Lactobacillus* strains are given in **Table 1**. They were isolated from human vaginal samples as described before ***(6)*** and used for the experiments as described in **Subheading 2.3.** Their taxonomic identification and relevant characteristic are also shown in **Table 1**.
2. Storage (in milk-yeast extract): 10% skim milk, 0.5% yeast extract, 1% glucose. Sterilize at 121°C for 15 min and store at refrigerated temperature.
3. Subcultivation (in LAPTg broth): 1.5% meat peptone, 1% tryptone, 1% yeast extract, 1% glucose, 0.1% Tween-80, final pH 6.5. Sterilize by autoclaving and store at refrigeration temperature.

2.2. Animals

1. Characteristics and conditions.
   a. Adult female Balb/c mice were fed ad libitum and kept in metallic cages, with wooden beds.
   b. The room temperature was between 19 and 23°C.
   c. The light conditions were between 5 and 60 lux in the cages and between 250 and 300 lux in the room.
   d. Inbred 7-wk-old female Balb/c mice weighing between 25 and 30 g were used. They came from a closed colony and were randomly paired.
   e. All the animals were fed a conventional diet, including the control group.
2. Hormone inoculation: Animals were cycled to the same point of the sexual cycle by using estradiol valerate (0.05 mL Progynon Depot, Schering, Germany).

2.3. Experiments in Mice

1. Peptone-water agar: 1% peptone, 1.5% agar. Sterilize at 121°C for 15 min (by autoclaving) and store at refrigerated temperature.
2. Diluent media (peptone water): 0.1% meat peptone (Britania, Argentina).

Table 1
Strains and Characteristics of Lactobacilli

Microorganism	CRL no.	Probiotic characteristic
L. crispatus	1266	H_2O_2 producer
L. paracasei ss. *paracasei*	1251	H_2O_2 producer
L. paracasei ssp *paracasei*	1289	H_2O_2 producer
L. acidophilus	1294	Self- and co-aggregating
L. salivarius ss. *salivarius*	1328	Bacteriocin producer
L. helveticus	1259	Lactic acid producer

3. Culture media: de Man-Ragosa-Sharpe (MRS) agar: 1% peptone, 1% meat extract, 0.5% yeast extract, 0.5% sodium acetate, 0.2% ammonium citrate, 0.2% potassium phosphate, 0.02% Mg sulfate, 0.05% Mn sulfate, 2% glucose, 0.1% Tween-80, 1.5% agar, final pH 6.5. Sterilize by autoclaving at 121°C for 15 min and store at refrigeration temperature.
4. MRS agar with antibiotic: chloramphenicol (Sigma, St. Louis, MO) stored at refrigerated temperature.
5. Agarized peptone water: 1% peptone, 1.5% agar. Sterilize by autoclaving and stored at refrigerated temperature.
6. Distilled water.
7. Other materials.
   a. Glass tubes.
   b. Nitrocellulose filter membranes (type GS, Sigma), pore size 0.22 μm, diameter 25 mm.
   c. Automatic pipets.
   d. Povidone iodine (10% topical solution of iodated povidone).
   e. Petri dishes.
   f. Tissue homogenizer with Teflon pestle (MSE, Sussex, UK).
   g. Glass tubes.
   h. Glass slides.
   i. Water bath (Vicking S.R.L. Masson, Argentina).
   j. Sterile tips for pipets.
   k. Dissection board.
   l. Pins.
   m. Scissors.
   n. Alcohol.
   o. Dissection pliers.
   p. Glass bottles.

2.4. Microbiological Assays

1. Vaginal washes.
   a. Sterile saline solution (0.85% sodium chloride), sterilized by autoclaving and stored at refrigeration temperature.
   b. Automatic micropipets.
   c. Glass tubes with sterile saline solution (sterilized by autoclaving at 121°C for 15 min and stored at refrigeration temperature).

   d. Glass slides.
   e. Petri dishes.
   f. MRS agar.
   g. Crystal violet solution.
   h. Iodine solution.
   i. Safranin solution (Gram staining kit. Biopack, Argentina).
2. Homogenization.
   a. Glass tubes with sterile saline solution (sterilized by autoclaving at 121°C for 15 min and stored at refrigeration temperature).
   b. Petri dishes.
   c. MRS agar.
   d. Glass tubes containing peptone water (sterilized by autoclaving at 121°C for 15 min and stored at refrigeration temperature).
   e. Tissue homogenizer with Teflon pestle (MSE, Sussex, UK).
   f. Glass slides.
3. MRS agar (described before in **Subheading 2.3., item 3.**).
4. MRS agar with chloramphenicol (described before in **Subheading 2.3., item 4**).
5. Incubator jars for microaerophilic conditions.
6. Gram stain evaluation.
   a. Glass slides.
   b. Crystal violet solution.
   c. Iodine solution.
   d. Safranin solution (Gram staining kit, Biopack) prepared according to the manufacturer's instructions.
7. Optical microscope.

2.5. Histological Technique

1. Fixation: Ethyl alcohol 96° maintained at refrigeration temperature (Biopack).
2. Dehydration: Ethyl alcohol 100°, always at 10°C.
3. Clearing step: Cold pure xylol (Biopack). Store at room temperature
4. Paraffin baths: Paraffin (Cicarelli, Argentina). Melting point 56–58°C, melted at 60°C for 48 h before use and filtered for 24 h before tissues addition.
5. Other materials: Wood pieces for mounting, paper boxes, spatulas, boxes.
6. The tissues were cut into 2-μm-wide slides, by using disposable microtome blades (Leica)
7. Hematoxylin-eosin stain.
   a. Mayer's activated hematoxyline: Stabilized solution for histological use (Biopur).
   b. Eosine: 0.5% aqueous solution (w/v) (Biopur).
   c. Ethyl alcohol at concentrations of 50°, 70°, 80°, 96°, and 100° (Biopack) prepared by dilution of absolute alcohol in distilled water. The different ethanol dilutions are stored in screw-top flasks at room temperature.
   d. Xylol: Pure, pro-analysis stored at 4°C (Cicarelli).
8. Optical microscope.

2.6. Indirect Immunofluorescence

1. Phosphate-buffered saline (PBS): 6% disodium phosphate, 1% potassium phosphate, 68% sodium chloride, pH 7.2. Store at 4°C.

2. Anti-IgA monospecific antiserum.
   a. α-Chain obtained from goat, labeled with fluorescein isothiocyanate (FITC; Sigma).
   b. A bottle containing 1 mL antiserum was fractioned in Eppendorf tubes (20 μL in each one) and stored in the freezer.
   c. At the time of use, an adequate dilution was prepared (previously titrated in our experimental conditions, by diluting in saline solution).
   d. Once diluted, the antiserum was stored at 4°C for no more than 48 h.
3. Anti-IgM monospecific antiserum.
   a. α-Chain obtained from goat, labeled with FITC (Sigma).
   b. Prepare as just described in **step 2**. Preparation and dilution of the fluorescent labeled antiserum must be in the dark.
4. Ethyl alcohol at different gradations: 50°, 70°, 80°, 96°, and 100° (Biopack). Store at 4°C for not more than 1 wk.
5. Xylol: pure p.a. (Cicarelli) stored at 4°C.
6. Mounting liquid: 9 parts neutral glycerine and 1 part buffered saline solution, pH adjusted to 9.0 with 0.5 *M* Na_2CO_3. Store at room temperature.
7. Incubation: In humid chamber.
8. Other materials.
   a. Glass flasks to perform fixation of the material.
   b. Incubators.
   c. Funnels.
   d. Filter papers.
   e. Glass slides.
   f. Cover-slips.
   g. Canada balsam.
   h. Optical microscope.
   i. Thermostatized water bath.

3. Methods

3.1. Microorganisms

1. The lactic acid bacteria were placed in sterile tubes containing 1 mL of milk-yeast extract and stored at –20°C for 6 mo.
2. The microorganisms were transferred from the milk-yeast extract tubes to tubes containing 5 mL of LAPTg broth. Three subcultures were performed, incubating each one for 12 h at 37°C (*see* **Note 1**).

3.2. Animals

1. Hormone inoculation: The animals were cycled to the same stage of the sexual cycle; by im inoculation (in the internal part of the leg) with 0.5 mg of estradiol valerate (0.05 mL Progynon Depot). All animals from all experimental groups were inoculated in this fashion 48 h before beginning the experiments (inoculation of lactobacilli by the intravaginal route) (*see* **Notes 2** and **3**).

3.3. Experiments in Mice

1. The third subculture of lactobacilli in LAPTg broth was centrifuged for 10 min at 2000*g* and washed with 1 mL of saline solution. These suspensions were centrifuged again and resuspended in 50 μL of agarized peptone water melted previously and cooled at 45°C (*see* **Note 4**).

2. The final concentration of microorganisms was determined by the plate dilution method using serial dilutions of peptone water. Aliquots were plated in MRS agar and incubated at the adequate temperature. The suspension of microorganisms was between 10^7 and 10^8 CFU/mL.
3. To differentiate the inoculated lactobacilli from those of the normal microbial flora, their resistance to different antibiotics was determined. The antibiotic sensitivity of the different lactobacilli strains was assayed by the disc diffusion agar method and also the minimal inhibitory concentration (MIC) method *(16)*. For each strain of lactobacilli, one antibiotic was selected, according to the sensitivity/resistance pattern. The method employed to determine these patterns of sensitivity is the one recommended by the National Committee for Clinical and Laboratory Standards (NCCSL) *(16)*. After having the previous results available, the following assays were performed at the laboratory level (for space reasons, only one example will be described). Different concentrations were used in order to determine the antibiotic quantities that allow the growth of the lactic bacteria inoculated in the culture media. The antibiotic was prepared at different concentrations (2.5, 5, 10, 50, and 100 μg/mL) and added to the culture media. Control experiments were performed to determine whether lactobacilli grow in the same numbers in media with and without antibiotic. The antibiotic was prepared by dissolving different amounts of the adequate solvent and was later sterilized by filtration. They were divided into small tubes and stored in the freezer.
4. Lactobacili inoculation: The lactobacilli strains were inoculated intravaginally to different experimental groups of mice. Inoculations were performed twice every day (8 h in between) for 2 d (according to the scheme shown in **Table 2**). The microorganisms were inoculated in agarized peptone water (prepared as described in **Subheading 2.3.**). The agarized bacterial suspension was introduced into the mouse vaginas by loading automatic pipetor tips with the desired amount of agarized suspension.
5. Sacrifice: Mice were sacrificed after 24 h and 2, 5, 7, and 11 d post lactobacilli inoculation and used for the different studies. They were sacrificed by cervical dislocation and placed on the dissection board.
6. Samples.
   a. The mice were dissected to obtain the vaginas by using specific dissection material.
   b. Once killed, they were immobilized on the dissection board and their extremities were fixed with metallic pins.
   c. The skin was first decontaminated with povidone iodine and later cut with scissors in aseptic conditions.
   d. The different tissue layers were cut or moved to separate the vagina, which was extracted (The extraction was made by cutting just above the endocervix and the distal extremity).
   e. The organ was transferred to a glass bottle containing alcohol 96°C (for histological and immunological techniques) or cut longitudinally with sterile scissors and transferred to sterile tubes containing 1 mL of peptone water (for microbiological assays).
7. Microbiological assays; samples from washes and homogenates. Determination of the number of lactobacilli in the mice vagina (in both vaginal washes and homogenates) was made by plating aliquots of the samples (or their dilutions) in selective media containing antibiotic (to differentiate the resident flora from those administered experimentally), as explained in **Subheading 2.3.**
8. Histological assays were performed to determine whether the local administration of lactobacilli produces some type of modification at the histological structural level in the tract, as described in **Subheading 3.5.**

Table 2
Inoculation Schedule Used for the Colonization Studies in BALb/c Mice

	Day							
	–2	–1	0	1	2	5	7	10
H	+							
Col			2 doses	2 doses				
S					+	+	+	+

H, inoculation of 0.5 mg of estradiol valeratel; C, control mice inoculated with peptone water without lactobacilli; L, inoculation of lactobacilli in peptone water; S, days of mice killing and sample processing.

9. Immunofluorescence assays were carried out to evaluate whether local administration of lactobacilli produces stimulation of cells involved in the immune response at the local level, as detailed in **Subheading 3.7.**

3.4. Microbiological Assays

1. To determine the number of microorganisms in the vagina of mice, samples were obtained by using automatic pipets with tips loaded with 100 μL of sterile saline solution. Aliquots (50 μL) of the washing samples were added to the surface of MRS agar with antibiotics and later incubated at 37°C for 48–72 h to determine the number of viable microorganisms. The samples were also spread on glass slides for different staining techniques, including direct Gram stain.
2. The organs were opened longitudinally in aseptic conditions with a sterile scissors, and transferred to the homogenizor tube containing 1 mL of peptone water. The Teflon pestle connected to the homogenizor was used, applying a speed of 1400 rpm. Different aliquots (100 μL and 200 μL) of the organ homogenates were added to Petri dishes, where melted MRS agar plus antibiotic was added. The method of the successive dilutions was also performed by diluting 0.5 mL of the organ homogenate in successive dilutions of peptone water, plating later aliquots of the respective dilutions in agar plates. The plates were incubated in microaerophilic conditions at 37° C for 48–72 h. Then, the number of microorganisms was determined by counting plates showing between 30 and 300 colonies. Some colonies were picked and Gram-stained to confirm the lactobacilli characteristics.
3. Vaginal washings were Gram-stained, and the characteristics of the microflora were observed, looking for the characteristic shape of the lactobacilli, which is sometimes very different from the indigenous Gram-positive bacilli. This observation gave us an idea of the direct microbiological results before the results of the plates were available.

3.5. Histological Technique

1. After the organs were extracted, they were transferred to 96° ethyl alcohol at 4°C for 15 h to perform fixation of the material and to maintain the cellular structure without modifications (*see* **Note 5**).
2. Dehydration was performed by passing the tissues successively three times through alcohol 100° at 4°C for 2 h, to dehydrate the samples.
3. As the next step was the addition to paraffin, a substance nonsoluble in alcohol, the method required the alternative passage of the tissues in an intermediate reactive sub-

stance able to solubilize both alcohol and paraffin. With this objective, the tissues were passed through three different glass bottles containing xylol, the two first cold (4°C) and the last one at room temperature.

4. Three paraffin passages were performed, each one for 2 h at 60°C, to obtain material with uniform consistency.
5. Microtome slides were performed in a 2-µm-wide Minot-type microtome. The slides/cuts were stretched in a thermostatized water bath maintained at 37°C. The tissue slides were put into glass slides to be later stained and evaluated by microscopy.

3.6. Staining: H&E Stain

1. The paraffin was taken out of the histological slides by two successive passages in xylol for 2 min each.
2. The material was hydrated by submerging the tissues in flasks containing different decreasingly gradated alcohols (two transfers from absolute alcohol and one in 96°, 80°, 70°, and 50° ethyl alcohols) for 2 min each at room temperature.
3. The staining procedure applied was a double staining that give color to both the nucleus and the cell cytoplasm. The slides were submerged for 10 min in Mayer's hematoxylin and then washed three times with water (distilled watter: 2 min; tap water: 15 min; distilled water: 1 min).The passage through tap water was an important step because it helped to develop the color of the stain. Then the slides were submerged in eosin (0.5% aqueous solution w/v) for 5 min and immediately put in a distilled water bath.
4. The object of dehydration was to make the material ready to be included in Canada balsam, by passing successively through ethyl alcohols of increasing concentrations (80°, 96°, 100°, and 100°) for 2 min each. Then two xylol passages were performed, for 2 min each.
5. Canada balsam (50 µL) was added to the material on the surface of the glass slides, covering the tissues impregnated in xylol.
6. The number of cells and their characteristics, from epithelial layer and lamina propria were evaluated by using an optical microscope at 100 and 40X.

3.7. Immunofluorescence

This technique was applied to identify and localize the antigens by using fluorescence-labeled antibodies such as FITC with ultraviolet wavelengths. The specificity of this technique allows recognition of all cells with surface inmunoglobulins expressed on their surface ***(17)*** (*sees* ***Note 6***).

1. The paraffin was removed from the histological slides by three successive passages in xylol for 5 min each one at 4°C. Then three alcohol baths were performed (96°, 70°, and 40°) for 2 min each at 4°C. The material was then transferred to PBS, pH 7.2 (two transfers for 10 min each).
2. FITC-conjugated antibody (0.1 mL) diluted with PBS was added to each slide, and the mixture was incubated at 37°C for 30 min in a humid chamber. Then the slides were washed for 10 min with PBS at 4°C. One drop of mounting liquid was then added to the slides, which were covered with a cover slip. The number of cells that were IgA+ or IgM+ in the lamina propia was determined.

4. Notes

1. Before carrying out the inoculation of the microorganisms, the purity of the lactobacilli under study must be checked carefully, as for any other microbiological assay. Also, in terms of the process used to obtain the microorganisms they must be derived from stock kept in freezing conditions, with the same number of subcultures, the same culture media, and all other requirements kept the same.
2. The strain of mice used under all the experimental conditions must be the same. They must also come from an inbred colony to ensure the same genetics. If some of the experimental conditions must be modified, all the controls must be set up again.
3. When intramuscular inoculation is performed, it is convenient to use the internal part of the mouse paw, because correct inoculation of the hormone can be easily observed.
4. Intravaginal inoculation of lactobacilli must be performed as fast as possible, because the agarized peptone water cools down quickly.
5. The paraffin must be filtered every time the technique is performed. The material must be kept in the alcohol bath room for the time indicated. Do not keep the material longer than indicated, because it can harden, making it very difficult to cut later in the appropriate way. Paraffin must be added carefully when the blocks are prepared to avoid the formation of bubbles, which can hinder cutting.
6. Before the fluorescent labeled antisera is diluted, the room must be prepared for darkness; dilution and all subsequent steps must be performed in the dark to avoid inactivation of the fluorescence.

Acknowledgements

This work was supported by grants from the Ministerio of Salud Pública de la República Argentina (Beca Ramón Carrillo) and CONICET (PIP 359).

References

1. Larsen, B. (1993) Vaginal flora in health and disease. *Clin. Obstet. Gynecol.* **36,**107–121.
2. Redondo López, V., Cook, R. L., and Sobel, J. D. (1999) Emerging role of Lactobacilli in the control and maintenance of the vaginal bacterial microflora. *Rev. Infect. Dis.* **12,** 856–872.
3. Reid, G. (2001) Probiotic agents to protect the urogenital tract against infection. *Am. J. Clin. Nutr.* **73(suppl),** 437S–443S.
4. Reid, G. and Bruce, A. (2001) Selection of Lactobacillus Strains for Urogenital Probiotic Applications. *J. Infect. Dis.* **183(suppl 1),** S77–S80.
5. Gorbach S. L. (2000) Probiotic and gastrointestinal health. *Am. J. Gastroenterol.* **95 1(suppl),** S2–S4.
6. Cangemi de Gutierrez, R., Santos, V., Nader-Macias, M. E. (2001) Protective effect of intranasally inoculated *Lactobacillus fermentum* against *Streptococcus pneumoniae* challenge on the mouse respiratory tract. *FEMS Immunol. Med. Microbiol.* **1341,** 1–2.
7. Ocaña, V. S., Bru, E., Ruiz Holgado, A. P., and Nader-Macías, M. E. (1999) Surface characteristics of Lactobacilli isolated from human vagina. *J. Gen. Appl. Microbiol.* **45,** 203–212.
8. Ocaña, V. S., Bru, E., Ruiz Holgado, A. P., and Nader-Macías, M. E. (1999) Growth inhibition of *Staphylococcus aureus* by H_2O_2-producing *Lactobacillus paracasei* subsp *paracasei* isolated from the human vagina. *FEMS Immunol. Med. Microbiol.* **23,** 87–92.
9. Ocaña, V. S., Ruiz Holgado, A. P., and Nader-Macías, M. E. (1999) Selection of vaginal H_2O_2-generating Lactobacilli species for probiotic use. *Curr. Microbiol.* **38,** 279–284.

10. Ocaña, V. S., Ruiz Holgado, A. P., and Nader-Macías, M. E. (1999) Characterization of a bacteriocin-like substance produced by a vaginal *Lactobacillus salivarius* strain. *Appl. Environ. Microbiol.* **65,** 5631–5635.
11. Ocaña, V. and Nader-Macias, M. E. (2001) Adhesion of *Lactobacillus* vaginal strains with probiotic properties to vaginal epithelial cells. *Biocell* **25,** 265–273.
12. Sanders, M. E. (1998) Development of consumer probiotics for the US market. *Br. J. Nutr.* **80,** S213–S218.
13. Chrisope, G. L. (Boulder, CO) (2000)Vaginal lactobacillus medicant. US Patent & Trademark Office, US Patent, GyneLogix, Inc., 093,394-6-21.
14. Larsen, B. (Huntington, WV) (1999) Vaginal pharmaceutical compositions. US Patent & Trademark Office, US Patent, Marshall University Research Corporation, Huntington, WV, 5,958,461-7-34.
15. Reid, G. (London, Canada), Bruce, A.W. (Toronto, Canada) (1997) *Lactobacillus* and skim milk compositions and methods for preventing microbial urogenital infections. US Patent & Trademark Office, US Patent, Research Corporation Technologies, Inc., Tucson, AZ, 5,643,830-1-1.
16. National Committee for Clinical and Laboratory Standards (NCCLS) (1997) Methods for dilution antimicrobial susceptibility testing for bacteria that graw aerobically. Approved Standard M 7-A 4. NCCLS, Wayne, PA.
17. Vintiñi, E., Alvarez, S., Medina, M., Medici, M., de Budeguer, M. V., and Perdigon G. (2000) Gut mucosal immunostimulation by lactic acid bacteria. *Biocell* **24,** 223–232.

38

Determination of Bacterial Adhesion to Intestinal Mucus

Carlos Gusils, Vilma Morata, and Silvia González

1. Introduction

The epithelial cells in the small intestine are covered by a relatively thick layer of mucus, secreted by specialized cells, which consists of mucin, many small associated proteins, glycoproteins, lipids, and glycolipids. The mucus contains receptors that recognize specific adhesion proteins *(1,2)*. Adhesion or close association of bacteria to the epithelial cells may further contribute to competitive exclusion. In addition, bacterial adhesion to intestinal mucus and epithelia seems to be important for individual stability of microbial flora *(3)*. Although the mucus layer covering the gastrointestinal tract has been recognized for many years, it has only recently been of interest in the study of the adhesion between mucus and bacteria. In this chapter, we describe a method for the study of colonization of the gastrointestinal mucus by bacteria and determine the possible effects on adhesion of pathogenic organisms.

2. Materials

2.1. Growth Media

1. de Man-Rogosa-Sharpe (MRS) broth *(4)*, instead of LAPTg broth, can be use for lactic acid bacteria growth. MRS broth: 10 g/L polypeptone, 10 g/L meat extract, 5 g/L yeast extract, 20 g/L glucose, 5 g/L sodium acetate, 2 g/L ammonium citrate, 2 g/L K_2HPO_4, 0.2 g/L $MgSO_4 \cdot 7H_2O$, 0.05 g/L $MnSO_4 \cdot 4H_2O$, and 1.08 g/L Tween-80. Sterilize in the autoclave at 121°C for 15 min (*see* **Note 1**).
2. Brain-heart infusion (BHI) broth is used for culturing enterobacteria. Composition: 200 g/L calf brains, 250 g/L infusion from beef heart, 10 g/L protease peptone, 2 g/L dextrose, 5 g/L sodium chloride, and 2.5 g/L disodium phosphate, pH 7.4 ± 0.2. Sterilize in autoclave at 121°C for 15 min.

From: *Methods in Molecular Biology, vol. 268: Public Health Microbiology: Methods and Protocols*
Edited by: J. F. T. Spencer and A. L. Ragout de Spencer © Humana Press Inc., Totowa, NJ

2.2. Mucus Collection

1. Newborn animals.
2. Sterile Petri dishes with two microscope covers.
3. HEPES-Hanks' buffer (Sigma).
4. Sterile tubes of centrifuge.
5. Refrigerated centrifuge.
6. 0.1 *M* Phosphate buffer pH 7.0:
   a. Solution A (0.2 *M* NaH_2PO_4): Weigh 27.6 g $NaH_2PO_4 \cdot H_2O$ and make to 100 mL with distilled water (dH_2O).
   b. Solution B (0.2 *M* Na_2HPO_4): Weigh 53.05 g of $Na_2HPO_4 \cdot 7H_2O$ and make to 100 mL with dH_2O.
   c. To prepare 1 L of 0.1 *M* phosphate buffer, pH 7.0, mix 195 mL of solution A and 305 mL of solution B and bring to 1000 mL with dH_2O.

2.3. Growth of Lactobacilli in Chicken Small Intestinal Mucus

1. An active culture (16 h at 37°C) grown in MRS broth.
2. HEPES-Hanks' buffer (Sigma).
3. UV lamp.
4. Physiological solution (PS): 8.5 g/L sodium chloride.
5. Rogosa SL agar (Merck): 10 g/L peptone from caseine, 5 g/L yeast extract, 20 g/L D(+) glucose, 6 g/L potassium dihydrogen phosphate, 2 g/L ammonium citrate, 1 g/L Tween-80, 15 g/L sodium acetate, 0.575 g/L magnesium sulfate, 0.034 g/L iron (II) sulfate, 0.12 g/L manganese sulfate, 15 g/L agar-agar, pH 5.5 ± 0.2. Suspend 74.5 g in 1 L of demineralized water by heating in a boiling water bath or in a current of steam; adjust the pH to 5.5 with acetic acid 96% (*see* **Note 2**).
6. Spectrophotometer.

2.4. In Vitro Adhesion Assay

1. An active culture grown in MRS broth for about 16 h at 37°C.
2. [Methyl, 1,2-^{3}H] thymidine (specific activity 117 Ci/m*M*; Sigma).
3. Sterile mucus.
4. HEPES-Hanks' buffer.
5. 50 mg/mL Bovine serum albumin (BSA; Sigma).
6. Sterile polystyrene tissue culture wells (Corning).
7. 5% Sodium dodecyl sulphate (SDS): weigh 5 g and make up to 100 mL in dH_2O with gentle stirring. Store at room temperature (*see* **Note 3**).
8. Scintillation liquid.

2.5. Characterization of Bacterial Adhesion to Mucus

1. An active bacterial culture grown in MRS broth for about 16 h at 37°C.
2. Sterile mucus.

2.5.1. Proteolytic Digestion

1. 1 mg/mL Trypsin (activity: 10,600 U/mg protein; Sigma).
2. 1 mg/mL Protease K (activity: 10–20 U/mg protein, Sigma).
3. Phosphate buffered saline (PBS), pH 7.0.

2.5.2. Characterization of Adhesion

1. Water bath at different temperatures (30, 37, and 42°C).
2. 0.1 *M* PBS at different pH values (6.0, 7.0, and 8.0).
3. 0.2 m*M* of different carbohydrate (Sigma) solutions (fucose, fructose, galactose, glucose, lactose, mannose, *N*-acetyl galactosamine, *N*-acetyl glucosamine, sialic acid, and sucrose).
4. 1 m*M* $CaCl_2$ (Anedra, Argentina) in PBS.
5. 0.15 m*M* EDTA (Sigma) in PBS.

2.6. Inhibition of Adhesion Assay

1. Pathogenic bacteria (for example, enterobacteria): Active cultures grown in BHI broth for about 16 h at 37°C.
2. An active bacterial culture grown in MRS broth for about 16 h at 37°C.
3. Sterile polystyrene tissue culture wells.
4. PBS, pH 7.0.
5. HEPES-Hanks' buffer.
6. Bacterial spent culture supernatants (SCS).

3. Methods

3.1. Mucus Collection

1. Obtain 5-cm samples from different sections of intestinal tract in asceptic conditions and put on a sterile microscope cover slip.
2. Isolate mucus from the sample walls by gently scraping with a sterile microscope cover slip, and put in sterile tubes of the centrifuge.
3. Dilute mucus twice with 10 *M* HEPES-Hank's buffer, pH 7.4, and shaken vigorously for 5 min.
4. Centrifuge once at 11,000*g* (10 min at 4°C) and once at 26,000*g* (15 min at 4°C) (*see* **Note 4**).
5. Store aliquots of the supernatant at –20°C.

3.2. Growth of Lactobacilli in Chicken Small Intestinal Mucus

1. Grow the strains in MRS broth at 37°C for 16 h.
2. Harvest the cells by centrifugation at 4000*g* for 5 min.
3. Wash the pellet obtained twice with PS (*see* **Note 5**).
4. Resuspend the cells in 5 mL PBS (0.1 *M*, pH 7.0).
5. Dilute the mucus in HEPES-Hanks' buffer (1:4).
6. Sterilize 3-mL aliquots of diluted mucus using UV light for 10 min.
7. Inoculate into 3 mL of UV-treated mucus an aliquot of 30 μL (10^{-3} dilution) of overnight bacterial culture.
8. Incubate at 37°C in a water bath.
9. Determine the optical density at 560 nm against blank (non-inoculated UV-treated mucus) and plated dilutions of inoculated and non-inoculated mucus on Rogosa SL agar, every hour for the first 8 h and after 12, 16, and 24 h of incubation.
10. Plot CFU/mL and OD values against incubation time.

3.3. In Vitro Adhesion Assay

1. Grow the strains of lactic acid bacteria in LAPTg broth at 37°C for 16 h.
2. Harvest the cells by centrifugation at 4000*g* for 5 min.

3. Wash the pellet obtained with PS twice (*see* **Note 5**).
4. Resuspend the cells in 5 mL PS.
5. Inoculate 50 µL in LAPTG broth with 1% [methyl, 1,2-^{3}H] thymidine (specific activity 117 Ci/m*M*, Sigma).
6. Incubate for 2 h at 42°C (*see* **Note 6**).
7. Collect cells by centrifugation at 4000*g* for 5 min.
8. Wash twice with HEPES-Hanks' buffer.
9. Resuspend the cells in HEPES-Hanks' buffer.
10. Add 300 µL of mucus and BSA in polystyrene tissue culture wells.
11. Incubate for 48 h at 4°C (*see* **Note 7**).
12. Wash twice with HEPES-Hanks' buffer to remove mucus or BSA excess.
13. Add 200 µL of bacterial suspension to the immobilized mucus and BSA coating.
14. Incubate tissue culture plates for 1 h at 42°C.
15. Wash twice with HEPES-Hanks' buffer to remove unbound bacteria.
16. Add 5 mL of 5% SDS to each well and incubate the plates overnight at 37°C to lyse the bacteria.
17. Add the lysed cells from each well to 2 mL of scintillation liquid.
18. Calculate the adhesion ratio (%) comparing the radioactivity of the original bacterial suspension with the final radioactivity from the lysed cells.

3.4. Characterization of Bacterial Adhesion to Mucus

1. Prepare the culture inoculum as described in **steps 1–9** from **Subheading 3.3.**

3.4.1. Proteolytic Digestion

1. Incubate mucus and mucus fractions with 1 mg/mL trypsin and 1 mg/mL protease K for 1 h at 37°C (*see* **Note 8**).
2. Wash twice with PBS (pH 7.0).
3. Determine adhesion capacity by method described in **steps 13–18** from **Subheading 3.3.**

3.4.2. Characterization of Adhesion

1. Prepare the culture inoculum and immobilized mucus as described in **Subheading 3.3.**
2. Incubate the plates at different temperatures (30, 37, and 42°C) and pH values (6.0, 7.0, and 8.0), and with 0.2 m*M* of different carbohydrate solutions, 1 m*M* $CaCl_2$, 0.15 m*M* EDTA.
3. Wash twice with PBS (pH 7.0).
4. Determine adhesion capacity by method described in **steps 13–18** from **Subheading 3.3.**

3.5. Inhibition of Adhesion Assay

1. Prepare the culture inoculum (pathogenic bacteria) and immobilized mucus as described in **Subheading 3.3.**
2. Incubate beneficial bacteria or spent culture supernatants (SCS; *see* **Note 9**) with mucus in polystyrene tissue culture wells for 1 h at 42°C.
3. Wash twice with PBS, pH 7.4.
4. Challenge labeled pathogenic bacteria for 1 h at 42°C.
5. Wash twice with HEPES-Hanks' buffer.
6. Determine adhesion capacity by the method described in **steps 13–18** from **Subheading 3.3.**

4. Notes

1. LAPTg broth *(5)*, instead of MRS broth, can be use to lactic acid bacteria growth. LAPTg broth: 15 g/L peptone, 10 g/L yeast extract, 10 g/L glucose, 10 g/L triptone, and 1% Tween-80, pH 6.4 ± 0.2. Sterilize in autoclave at 121°C for 15 min.
2. This media doesn't have to be autoclaved.
3. Solution may become cloudy at temperatures below 20°C, but warming to 30°C and mixing may restore clarity.
4. The centrifugation removes epithelial cells and large cellular debris, to eliminate contaminants.
5. Cell washes are important to minimize carryover of products from culture.
6. The incubation time can be longer (8 or 16 h).
7. This stage is necessary to immobilize the mucus in the plates.
8. Use equal volumes of mucus or fractions (without proteolytic enzymes) as control.
9. SCS are obtained from overnight cultures in LAPTg broth by centrifugation (3000*g*) and sterilization by filtration; they are then used in the inhibition assays.

References

1. Kim, Y. S., Morita, A., Miura, S., and Siddiqui, B. (1984). Structure of glycoconjugates of intestinal mucosal membranes. Role of bacterial adherence. In: *Attachment of organisms to the gut mucosa* (Boedecker, E. C., ed.). CRC, Boca Raton, FL, pp. 99–109.
2. Metcalfe, J. E., Krogfelt, H. C., Krivan, C., Cohen, P. S., and Laux, D. C. (1991) Characterization and identification of a porcine small intestine mucus receptor for K88ab fimbrial adhesion. *Infect. Immun.* **59,** 91–96.
3. Mikelsaar, M., Mändar, R., and Sepp, E. (1998). Lactic acid microflora in the human microbial ecosystem and its development. In: *Lactic Acid Bacteria, Microbiology and Functional Aspects, Food Science and Technology* (Salminen, S. and von Wright, A., eds.). Marcel Dekker, New York, pp. 279–342.
4. de Man, J. C., Rogosa, M., and Sharpe, E. (1960) A medium for the cultivation of lactobacilli. *J. Appl. Bacteriol.* **23,** 130–155.
5. Raibaud, P., Caulet, M., Galpin, J. V., and Mocquot, G. (1961) Studies on the bacterial flora of the alimentary tract of pigs II. Streptococci: selective enumeration and differentiation of the dominant group. *J. Appl. Bacteriol.* **24,** 285–291.

39

Animal Model for In Vivo Evaluation of Cholesterol Reduction by Lactic Acid Bacteria

María Pía Taranto, Gabriela Perdigón, Marta Médici, and Graciela Font de Valdez

1. Introduction

For many years, it has been recognized that elevated serum cholesterol is a risk factor associated with atherosclerosis and coronary heart disease, the latter being a major cause of death in Western countries. Numerous drugs that lower cholesterol have been used to treat hypocholesterolemic individuals *(1)*. However, the undesirable side effects of these compounds have caused concerns about their therapeutic use *(2)*. Ingestion of probiotic *(beneficial for health)* lactic acid bacteria (LAB) would possibly be a more natural method to decrease serum cholesterol in humans *(3)*, as has been was reported *(4,5)*.

Previous studies *(6,7)* have demonstrated that *Lactobacillus reuteri* administered in low doses has a hypocholesterolemic effect both therapeutically and preventively. One of the key studies in the development of a probiotic is to determine the minimal effective dose of live microorganisms that might be ingested without producing adverse effects (i.e., translocation) in the host.

In this chapter, we describe an animal model that allows us to evaluate reduction in hypercholesterolemia by LAB and, also to determine the minimal dose of the microorganism, a critical step in the development of a safe probiotic product.

2. Materials

2.1. Establishment of a Hypercholesterolemic Animal Model (see *Note 1*)

2.1.1. Test Group

1. Ten male Swiss albino mice aged 4–6 wk and weighing 20–25 g were used (*see* **Notes 2–5**). They were kept under a 12-h light/dark cycle at 22–26°C and a relative humidity of 50%.

From: *Methods in Molecular Biology, vol. 268: Public Health Microbiology: Methods and Protocols*
Edited by: J. F. T. Spencer and A. L. Ragout de Spencer © Humana Press Inc., Totowa, NJ

2. Cages (15 × 25 cm).
3. Fat-enriched liquid diet: 10% (w/v) sterile nonfat milk (NFM) supplemented with 10% cream. The composition is detailed in **Table 1** (*see* **Note 6**).
4. Glass containers with adapters to facilitate consumption by the mice of the fat-enriched liquid diet (*see* **Note 7**).
5. Solid commercial conventional diet: 32% protein, 5% fat, 2% fiber, and 61% nitrogen-free extract.
6. Balance.
7. Pasteur pipet.
8. Scissors.
9. Clamps.
10. Enzymatic reagent kit (Sigma, St. Louis, MO).

2.1.2. Control Group

1. The same as **items 1–2** of **Subheading 2.1.1.**
2. Tap water *ad libitum*.
3. The same as **items 4–10** of **Subheading 2.1.1.**

2.2. Lactobacilli Growth Medium

1. de-Man-Rogosa-Sharpe (MRS) broth *(8)*: 10 g/L polypeptone, 10 g/L meat extract, 5 g/L yeast extract, 20 g/L glucose, 5 g/L sodium acetate, 2 g/L ammonium citrate, 2 g/L KH_2PO_4, 0.25 g/L $MgSO_4 \cdot 7H_2O$, 0.058 g/L $MnSO_4 \cdot 4H_2O$, and 1.08 mL/L Tween-80, final pH 6.4 ± 0.2. Sterilize at 121°C for 20 min.

2.3. Media for Testing Bacterial Translocation

1. LBS agar broth.
2. MacConkey's agar: 20 g/L peptone from casein, 10 g/L lactose, 5 g/L Oxgall bile dried, and 0.01 g/L bromocresol purple.
3. Brain-heart infusion agar.

2.4. Testing Bacterial Translocation (see Note 8)

1. An active culture grown in MRS broth for about 16 h at 37°C (overnight culture).
2. Phosphate buffer.
   a. Solution A (0.2 *M* NaH_2PO_4): weigh 27.6 g $NaH_2PO_4 \cdot H_2O$ and make to 100 mL with distilled water (dH_2O).
   b. Solution B (0.2 *M* Na_2HPO_4): weigh 53.05 g $Na_2HPO_4 \cdot 7\ H_2O$ and make to 100 mL with dH_2O.
   c. To prepare 1 L of 0.1 *M* sodium phosphate buffer, pH 7.0, mix 195 mL of solution A and 305 mL of solution B, and bring to 1000 mL with dH_2O.
   d. Sterilize at 121°C for 20 min.
3. 1 g/L Peptone water: to prepare 1 L, weigh 1 g peptone and bring it to 1000 mL with dH_2O. Sterilize at 121°C for 20 min.
4. 8 g/L Saline solution: to prepare 1 L, weigh 8 g NaCl and bring it to 1000 mL with dH_2O. Sterilize at 121°C for 20 min.
5. LBS agar.
6. MacConkey's agar.

Table 1
Fat-Enriched Liquid Diet Composition

Component	Amount
Proteins	0.26 g
Carbohydrate	0.38 g
Lactic fat	0.26 g
Lecithine	0.002 g
Vitamin B_2	0.014 mg
Vitamin B_{12}	0.018 mg
Calcium	9.3 mg
Phosphorus	7.5 mg
Cholesterol	0.3 g
Lipids	0.2 g
Water	10 mL
Calories	10.5 kcal

7. Brain-heart infusion agar supplemented with blood.

2.5. Lactobacilli Treatment

2.5.1. Test Group

1. The same as **items 1–5** of **Subheading 2.1.1.**
2. An active culture grown in MRS broth for about 16 h at 37°C (overnight culture).
3. Phosphate buffer.
4. Nonfat milk reconstituted at 10%.
5. The same as **items 6–10** of **Subheading 2.1.1.**

2.5.2. Control Group

1. The same as items **1–6** of **Subheading 2.1.2.**
2. Nonfat milk reconstituted at 10%.
3. The same as **items 6–10** of **Subheading 2.1.1.**

3. Methods

3.1. Establish of the Hypercholesterolemic Model

3.1.1. Test Group

1. Place the mice in separate cages.
2. Feed the animals 5 mL of the fat-enriched liquid diet (to induce hypercholesterolemia) and the solid conventional diet *ad libitum* for 15 consecutive days (*see* **Note 9**).
3. Determine the weight of the animals daily and observe the characteristics of the feces (*see* **Note 10**).
4. Take blood samples carefully from the retro-orbital venous plexus with a Pasteur pipet and place them in perfectly clean glass tubes (*see* **Note 11**).
5. Centrifuge the blood (6000*g*, 10 min, at room temperature), separate the serum (supernatant), and determine total cholesterol, HDL cholesterol, LDL cholesterol, and serum triglycerides with enzymatic reagent kits following the instructions of the suppliers. Express

the results as mg/dL.

6. Dissect the animal at the level of the abdominal zone, cutting the skin and muscle with a scissors and avoiding damage to internal organs (*see* **Note 12**).
7. Remove the spleen and liver with clamps.
8. Weigh the removed organs and observe their anatomical characteristics (*see* **Note 13**).

3.1.2. Control Group

1. Place the mice in separate cages.
2. Feed the animals 5 mL of water and solid conventional diet *ad libitum* for 15 consecutive days.
3. Repeat **steps 3–6** of **Subheading 3.1.1.**

3.2. Bacterial Translocation

Lactic acid bacteria preparation is performed as follows:

1. Grow the probiotic strain in MRS broth at 37°C for 16 h (*see* **Note 14**).
2. Harvest the cells by centrifugation at 5000*g* for 10 min.
3. Wash the pellet obtained twice with phosphate buffer.
4. Suspend the pellet in 5 mL sterile 10% NFM at the following concentrations: 10^2–10^8 colony-forming units (CFU)/mL.
5. Feed the mice with the cell suspensions in NFM at 20% (v/v) in the drinking water for 2, 5, and 7 d.
6. Kill the mice by cervical dislocation after feeding each microbial suspension.
7. Remove and weigh the spleen and liver under aseptic conditions and homogenize the organs with saline solution
8. Plate in the selective agarized media the proper dilutions of the organs from 10-fold serial dilutions prepared in 1 g/L peptone water (*see* **Note 15**).
9. Incubate the plates in anaerobic jars (Oxoid Systems) at 37°C for 48 h, and count the resulting colonies. Results are expressed as CFU per mL per gram of organ tissue.
10. Select the minimum dose that does not produce translocation.

3.3. Evaluation of the Hypocholesterolemic Effect

3.1.1. Test Group

1. Place the mice in separate cages.
2. Repeat **steps 2–3** of **Subheading 3.1.1.**
3. Feed the animals 5 mL of the probiotic strain at the concentration established in **step 10** of **Subheading 3.2.** diluted to 20% (v/v) in the drinking water for seven consecutive days.
4. Repeat **steps 3–6** of **Subheading 3.1.1.**

3.1.2. Control Group

1. Place the mice in separate cages.
2. Repeat **steps 2–3** of **Subheading 3.1.1.**
3. Feed the animals with 5 mL of NFM (10%) diluted to 20% (v/v) in the drinking water for 7 consecutive days.
4. Repeat **steps 3–6** of **Subheading 3.1.1.**

4. Notes

1. Research using animal models should not be carried out until the protocol is reviewed by an appropriate animal care committee to ensure that the procedures are appropriate and humane.
2. Mice are selected because their physiological system is similar to that of humans and they are easy to obtain and handle. The animals used for this study belong to a closed colony. The Swiss albino mouse was chosen because it is extremely robust and is used for reproductive and physiological studies.
3. Young adult mice were chosen because they have an active cholesterol metabolism, guaranteeing a detectable reaction.
4. The sex of the mouse should be chosen with study aims in mind. In the hypercholesterolemic model, it is advisable to work with males because the hormonal system of the females can indirectly influence the lipid levels.
5. Each animal must be housed in a separate cage, for individual control of the normal consumption of food and other supplies corresponding to each treatment; it also allows evaluation of any secondary reaction. The number of mice for each group was established as 10 for statistical reasons. The experiment must be repeated at least three times.
6. Prepare the fat-enriched liquid diet using NFM sterilized at 115°C for 15 min and perfectly clean glass materials to ensure the best maintenance of the diet for long periods.
7. A suitable container was designed to administer both the liquid diet and the drinking water. This device ensures efficient food intake by the mice; being a closed container, it also allows better control of the amount ingested
8. It is very important to establish the minimum dose of probiotic bacterium that does not demonstrate undesirable effects such as a translocation ***(9)***. This phenomenon is known as the passage of the intestinal microbiota from the gut into such internal organs as the liver and spleen, which might later trigger septicemia. LAB are considered GRAS microorganisms (Generally Recognized as Safe); however, excessive proliferation of these bacteria in the gut may cause damage to the intestinal epithelial, generating cracks that favor the passage of undesirable enteric bacteria.
9. The fat-enriched liquid diet must be renewed every 24 h, and the containers must be perfectly washed before they are loaded with the recently prepared diet. These precautions will avoid diet spoilage owing to microbial contamination and prevent loss in cell viability of the probiotic strains during the treatment period.
10. Daily observation of the feces characteristics of each mouse during the test period is of utmost importance. This allows one to detect changes in color, consistency, and/or amount indicating either that the animal is not responding properly to the probiotic treatment or that it is necessary to reformulate the diet. Some mice within a treatment group may present negative symptoms and may need to be put aside. In case the number of mice discarded is above a significant value (more than three), it is advisable to reinitiate the experiments.
11. Harvesting of mouse blood must be made carefully to avoid hemolysis. This phenomenon releases the contents of the red cells, which may interfere with enzymatic determination of cholesterol and triglycerides, producing an overestimation.
12. Dissection procedure. For the dissection, perfectly clean and disinfected clamps and scissors as well as gloves are used. Kill the mouse, place it in a ventral position, and fix it with pins by the extremities. Cut the skin with the scissors, separate it carefully, and fix the skin with pins at both sides of the body. Cut the peritoneum, avoiding damage to the

internal organs. Remove carefully the organs of interest (liver, spleen, and small and large intestines) using clamps and scissors. Place the removed organs into Petri dishes containing sterile saline solution to avoid dehydration.

13. The spleen and liver should be observed post mortim. This allows one to detect any change in color, consistency, and/or size, which, as explained in **Note 10**, might indicate that the animal is responding negatively to the fat-enriched liquid diet. If negative changes are seen, it is advisable to redesign the assay conditions, taking into account the amount and composition of the diet and the period of administration.
14. The probiotic strains should be selected on the basis of in vitro cholesterol reduction and the rate of tolerance to bile and gastric juice ***(10)***.
15. The selective media used (LBS agar for the enumeration of lactobacilli, MacConkey's agar for enterobacteria, and blood-suplemented brain-heart infusion broth for anaerobes) were appropriate for selective enumeration (lactobacilli and enterobacteria) and total counts (anaerobes).

References

1. Suckling, K. E., Benson, G. M., Bond, B., et al. (1991) Cholesterol lowering and bile acid excretion in the hamster with cholestyramine treatment. *Atherosclerosis* **89,** 183–190.
2. Frick, M., Elo, O., and Haapa, K. (1987) Helsinki Heart Study: primary prevention trial with gemfibrozil in middle-aged men with dyslipemia. *N. Engl. J. Med.* **317,** 1237–1245.
3. Lin, S. Y., Ayres, J. W., Winkler, W., and Sandine, W. E. (1989) *Lactobacillus* effects on cholesterol: in vitro and in vivo results. *J. Dairy Sci.* **72,** 2885–2899.
4. Rodas, B. Z., Gilliland, S. E., and Maxwell, C. V. (1996) Hypocholesterolemic action of *Lactobacillus acidophilus* ATCC 43121 and calcium in swine with hypercholesterolemia induced by diet. *J. Dairy Sci.* **79,** 2121–2128.
5. Mc Namara, D. J., Lowell, A. M., and Sabb, J. E. (1989) Effect of yogurt intake on plasma lipid and lipoprotein levels in normolipidemic males. *Atherosclerosis* **79,** 167–171.
6. Taranto, M. P., Medici, M., Perdigón, G., Ruíz Holgado, A. P., and Valdez, G. F. (2000) Effect of *Lacobacillus reuteri* on the prevention of hypercholesterolemia in mice. *J. Dairy Sci.* **83,** 401–403.
7. Taranto, M. P., Medici, M., Perdigón, G., Ruíz Holgado, A. P., and Valdez, G. F. (1998) Evidence of hypocholesterolemic effect of *Lactobacillus reuteri* in hypercholesterolemic mice. *J. Dairy Sci.* **81,** 2336–2340
8. de Man, J. C., Rogosa, M., and Sharpe, M. E. (1960) A medium for the cultivation of lactobacilli. *J. Appl. Bacteriol.* **23,** 130–135
9. Berg, R. D. (1979) Bacterial translocation from the intestines. *Exp. Anim.* **34,** 1–16.
10. de Valdez, G. F. and Taranto, M. P. (2000) Probiotic properties of lactobacilli: cholesterol reduction and bile salt hydrolase activity. In: *Food Microbiology Protocols. Methods in Molecular Biology* (Walker, J., ed.). Humana, Totowa, NJ, pp. 173–181.

40

Assessing Survival of Dairy Propionibacteria in Gastrointestinal Conditions and Adherence to Intestinal Epithelia

Gabriela Zárate, Silvia González, and Adriana Pérez Chaia

1. Introduction

The genus *Propionibacterium* consists of two principal groups, *cutaneous* and *classical* or *dairy*. Cutaneous species are predominant members of the microbial population of human skin and have also been isolated from the feces of humans and other vertebrate animals. They are often considered opportunistic organisms and have been occasionally associated with infections in humans *(1)*.

Dairy propionibacteria are microorganisms extensively used in the industry for manufacture of Swiss-type cheeses and biological production of propionic acid and vitamin B_{12} *(2)*. They can be isolated from soil, vegetables, silage, raw milk, and dairy products such as kefir and different cheeses with "eyes."

In the last decade, several studies have demonstrated probiotic properties for members of the genus *Propionibacterium* *(3,4)*. The effects claimed are based on the production of bacteriocins *(5)*, vitamins, stimulation of growth of other colonic bacteria like bifidobacteria *(6)*, beneficial modification of the composition and metabolic activities of the intestinal microflora *(3,4)*, immunomodulation *(7)*, and antimutagenic activity *(8)*.

It is thought that to produce many of these health benefits, the probiotic microorganisms must be able to survive the transit through the hostile conditions of the gastrointestinal tract (GIT) and remain at high levels in the intestine, avoiding removal by peristaltic contractions of the gut. In this sense, microorganisms with a short generation time or the ability to adhere to the intestinal mucosa will survive for prolonged periods in the body of the host *(9)*. Therefore, two desirable properties for probiotic microorganisms are (1) resistance to gastric acidity, bile, and pancreatic enzymes; and (2) adhesion ability to mucosal surfaces.

From: *Methods in Molecular Biology, vol. 268: Public Health Microbiology: Methods and Protocols*
Edited by: J. F. T. Spencer and A. L. Ragout de Spencer © Humana Press Inc., Totowa, NJ

Dairy bacteria are traditionally not considered to persist as normal inhabitants of the human intestinal tract. Therefore, survival under GIT conditions and adherence are important properties to be considered, and tests to study them would be useful tools.

In the present chapter we describe the methods used in our laboratory to assess survival, metabolic activity, and adhesion of dairy propionibacteria to intestinal epithelial cells after gastrointestinal digestion ***(10–13)***.

The techniques used are described under the following headings:

1. *Gastrointestinal digestion in vitro*: as a screening procedure for selection of acid- and bile-tolerant strains. The method considers both stresses in a sequential way instead of either of these alone, as in conventional testing. Microbial counts and β-galactosidase activity are determined to evaluate damage produced by the digestion ***(10)***.
2. *Adhesion to intestinal epithelial cells in vitro*: as a useful tool to study the potential of strains to persist long enough in the intestine to exert beneficial effects ***(11,12)***.
3. *Intestinal transit rate*: to assess the time of permanence of the bacteria in each section of the GIT. This is important to determine the optimum sampling time in assays of adhesion to the mucosa. Spores of *Bacillus stearothermophilus* are commonly used as transit markers to determine the fate of bacteria in the bulk of the digestive tract ***(10)***.
4. *Adhesion to the intestinal mucosa in vivo*: as a useful tool to study the adhesion ability of selected strains in an animal model ***(11)***.
5. β-*galactosidase assay*: most of the industrial strains of *Propionibacterium* have the ability to metabolize lactose. A simple assay of β-galactosidase activity allows one to estimate the metabolic activity of the strains after gastrointestinal digestion, both in vitro and in vivo. Additional information about the ability of strains to improve lactose digestion within the intestine can be obtained ***(10,13–15)***.

2. Materials

2.1. Growth Media

1. de Man-Rogosa-Sharpe (MRS) broth ***(16)***: 10 g/L peptone, 10 g/L meat extract, 5 g/L yeast extract, 20 g/L glucose, 1.0 g/L Tween-80, 2 g/L K_2HPO_4, 5 g/L sodium acetate, 2 g/L ammonium citrate, 0.2 g/L $MgSO_4 \cdot 7H_2O$, 0.05 g/L $MnSO_4 \cdot 4H_2O$, pH 6.5. Autoclave at 121°C for 15 min and store at room temperature. Dehydrated MRS can be obtained from Difco (Detroit, MI).
2. YEL agar (sodium lactate agar): 10 g/L tryptone, 10 g/L yeast extract, 0.5 g/L L-cysteine, 0.5 g/L Tween-80, 0.25 g/L K_2HPO_4, 0.05 g/L $MnSO_4$, 14 g/L sodium lactate, 15 g/L agar, pH 6.5. Autoclave at 121°C for 15 min and store at room temperature. Tryptone and yeast extract were obtained from Difco, components were from Sigma (St. Louis, MO).
3. Lactose broth: The same composition as YEL without sodium lactate. Autoclave at 121°C for 15 min and then supplement with filter-sterilized lactose for a final concentration of 1%.
4. Lactose agar: Lactose broth with 1.5% agar.
5. BHI broth: 10.0 g/L brain-heart infusion, 8.5 g/L casein peptone, 8.5 g/L meat peptone, 2.5 g/L glucose, 2.5 g/L Na_2HPO_4, 5.0 g/L sodium chloride, pH 7.4. Autoclave at 121°C for 15 min and store prepared media at 2–8°C protected from light.
6. Plate count agar: 5 g/L casein peptone, 2.5 g/L yeast extract, 1 g/L glucose, 14 g/L agar, pH 7.0. Autoclave at 121°C for 15 min and store at room temperature.
7. Milk-yeast extract: 10 g/L skim milk, 10 g/L glucose, 5 g/L yeast extract. pH 6.0. Autoclave for 20 min at 115°C ($^3/_4$ atm) and store at refrigeration temperature.

2.2. Gastrointestinal Digestion In Vitro

1. Erlenmeyer flasks containing 50 mL of lactose broth.
2. Saline solution (0.9% NaCl). Autoclave at 121°C for 15 min and store at room temperature.
3. Artificial gastric fluid: 125 m*M* NaCl, 7 m*M* KCl, 45 m*M* $NaHCO_3$, 3 g/L pepsin, and HCl in variable concentration between 80 and 160 m*M* to adjust the final pH to 2.0, 3.0, and 4.0. Prepare control gastric juice by adding 5 *N* NaOH up to pH 7.0. Prepare just prior to use and sterilize by filtration.
4. Artificial intestinal fluid: 0.1% w/v Pancreatin USP (ICN Biomedicals) and 0.15 % w/v bile salts (oxgall, Difco) in water. Adjust the pH to 8.0 with 5 *N* NaOH (*see* **Note 1**). Prepare just prior to use and sterilize by filtration.
5. pHmeter (Digimeter IV, Luftman, Argentina).
6. Water bath at 37°C with shaker (Vicking S.R.L. Shaker, Dubnoff model, Industria Argentina).
7. Sterile screwtop glass tubes.
8. Refrigerated centrifuge.
9. Diluent medium: 0.1% peptone water sterilized by autoclaving at 121°C for 15 min.
10. Lactose agar medium.
11. Anaerobic incubator.

2.3. Animals (see Note 2)

1. Six to 8-wk-old Balb/c mice, each weighing 25–30 g, from a random-bred closed colony.
2. Metal or plastic cages to house animals.

2.4. Adhesion to Intestinal Epithelial Cells In Vitro

2.4.1. Preparation of Bacterial Suspension

1. Tubes with 5 mL of MRS broth.
2. Phosphate-buffered saline (PBS): 50 m*M* KH_2PO_4/Na_2HPO_4, containing 8 g/L NaCl and 2 g/L KCl, pH 7.2. Autoclave at 121°C for 15 min and store at room temperature.
3. Refrigerated centrifuge.
4. Spectrophotometer for glass tubes (Spectronic 20, Bausch & Lomb, Miltou Roy Company, USA).

2.4.2. Preparation of Intestinal Epithelial Cell Suspension

1. Three Balb/c mice.
2. Dissection board, pins, scissors, and dissection pliers.
3. Sterile microscope slides.
4. Sterile PBS, pH 7.2.
5. NCTC 135 medium (Sigma) supplemented with 2% fetal bovine serum, pH 7.5. Prepare according to the manufacturer's instructions (*see* **Note 3**).
6. Neubauer chamber for adjustment of the number of epithelial cells.
7. Refrigerated centrifuge
8. Phase-contrast microscope.

2.4.3. Adhesion Assay

1. Stove at 37°C with gas supply to provide a 5% CO_2/95% air atmosphere.
2. 50 m*M* PBS, pH 7.2.

3. Membrane filters of 8 µm pore size (filter membranes, nitrocellulose, cat. no. N-4021, Sigma).
4. Kit for Gram staining.
5. Phase-contrast microscope.

2.5. Intestinal Transit Rate

1. Spore suspension of *B. stearothermophilus* (Britania Laboratories, Argentina). Store at refrigeration temperature.
2. Tubes with 5 mL of BHI.
3. 40 Balb/c mice.
4. Dissection board, pins, scissors, and dissection pliers.
5. Sterile saline solution.
6. Diluent medium: 0.1% peptone water.
7. Plate count agar (PCA; Merck).
8. Sterile Petri dishes and screw-top glass tubes.
9. Precision digital balance.
10. Water bath at 80°C.
11. Stove at 65°C.

2.6. Adhesion to Intestinal Epithelial Cells In Vivo

1. Five Balb/c mice per treatment group.
2. 10% Nonfat milk.
3. Dissection board, pins, scissors, and dissection pliers.
4. 50 m*M* PBS, pH 7.2.
5. NCTC 135 medium.
6. Membrane filters of 8 µm pore size.
7. Kit for Gram staining.
8. Phase-contrast microscope.

2.7. Enzyme Activity (β-Galactosidase Assay)

1. Phosphate buffer: 50 m*M* KH_2PO_4/Na_2HPO_4, pH 7.0 containing 0.1% 2-mercaptoethanol (cat. no. M 6250, Sigma).
2. Substrate solution: 4.16 m*M* *o*-nitophenyl-β-D-galactopyranoside in water (ONPG; cat. no. N-1127, Sigma). Store prepared solution in a dark flask at 4°C for 1 mo or at –20°C for several months.
3. Stop solution: 0.5 *M* Na_2CO_3. Store at room temperature.
4. Calibration curve: 5 m*M* *o*-nitrophenol (ONP; cat. no. N-9256, Sigma) in phosphate buffer (**step 1**), as the product released by enzyme activity.
5. Spectrophotometer.

3. Methods

3.1. Propionibacteria Storage and Growth Conditions

1. For storage, harvest by centrifugation the cells from 10 mL of a late exponential phase culture (36–42 h). Wash once with saline solution and resuspend the pellet in 5 mL of milk-yeast extract. Transfer the suspension to 1-mL screw-top sterile cryovials and store at –20°C. Propionibacteria remain viable for several months.
2. Activate propionibacteria strains from stock cultures at –20°C in milk-yeast extract, by inoculating 100 µL in the selected medium and incubate statically at 37°C for 48 h. Make at least three successive transfers every 24 h before experimental use.

3.2. Gastrointestinal Digestion In Vitro

1. Inoculate four flasks containing 50 mL of lactose broth, at a level of 2% with a midexponential phase grown culture (24–30 h) of propionibacteria, and incubate statically at 37°C for 24 h.
2. Harvest the cells by centrifugation (20,000*g*, 10 min, 4°C), wash the pellet once, and resuspend each pellet in the 10th part of the original volume in sterile saline solution.
3. Add these cell suspensions to four flasks containing 50 mL of artificial gastric juice adjusted to pH 2.0, 3.0, 4.0, and 7, respectively. Mix with vortex and incubate the bacterial suspensions at 37°C with agitation at 200 opm to simulate peristalsis and the normal temperature of the human body.
4. Take samples of 5 mL in sterile screw-top glass tubes, for enumeration of viable microorganisms and β-galactosidase activity assay at 0, 30, 60, and 180 min (*see* **Note 4**).
5. After 180 min of gastric digestion, centrifuge 25 mL of each bacterial suspension at 20,000*g*, 10 min. Suspend the pellets to the original volume in intestinal fluid and incubate as before.
6. Take samples of 5 mL for enumeration of viable microorganisms and β-galactosidase activity at 0, 30, 60, and 180 min (*see* **Note 4**).
7. For survival of propionibacteria: Take 100 μL of each sample and dilute with 900 μL of sterile peptone water. Make serial 10-fold dilutions and inoculate 0.5 mL of each dilution in a Petri dish. Add melted lactose agar medium by homogenizing properly.
8. Once they are solidified, invert the plates and incubate at 35°C for 5 d in an anaerobic chamber (Forma Scientifica Anaerobic System, model 1024) with a gas mixture of 90% N_2 and 10% CO_2 (*see* **Note 5**).
9. For determination of β-galactosidase activity, *see* **Subheading 3.7.**

3.3. Adhesion to Intestinal Epithelial Cells In Vitro

3.3.1. Preparation of Bacterial Suspension

1. Inoculate 5 mL of MRS broth with an active culture of propionibacteria and incubate statically at 37°C up to the late exponential phase (42 h).
2. Harvest cells by centrifugation (10,000*g*, 10 min), wash twice with 50 m*M* PBS, pH 7.2, and finally resuspend in the same buffer.
3. Adjust the bacterial concentration in a spectrophotometer to approx 5×10^8 colony-forming units (CFU)/mL (OD_{560} = 0.8; *see* **Note 6**).

3.3.2. Preparation of Intestinal Epithelial Cells Suspension

1. The mice, maintained in metal cages with free access to food and water, must be fasted for 24 h before the performance of adhesion experiments.
2. Sacrifice mice by cervical dislocation and immobilize the body on the dissection board by fixing the extremities with metallic pins.
3. Cut the skin with scissors and open the abdominal cavity in aseptic conditions.
4. Remove the small bowel and place it in a sterile Petri dish.
5. Discard the intestinal contents by flushing PBS, pH 7.2, through the intestine with a glass syringe.
6. Cut the intestine longitudinally and open it in the Petri dish. Wash gut walls repeatedly with sterile PBS, pH 7.2.
7. Scrap the mucosa gently with the edge of a microscope slide and suspend the exfoliated cells in chilled PBS, pH 7.2.

8. Keep the suspended scrapings on ice for 15 min to allow the large debris to settle. Remove the fluid supernatant containing the intestinal epithelial cells (IECs) carefully and transfer to another sterile recipient.
9. Wash the IEC suspension twice with PBS by centrifuging at low speed (120*g*, 10 min) to avoid cell damage.
10. Resupend the IECs in NCTC 135 medium supplemented with 2% fetal bovine serum and adjust the suspension to a concentration of 5×10^5 cells/mL by counting cells in a Neubauer chamber (*see* **Note 7**).
11. Keep the IEC suspension on ice and use within 2 h for binding assay.

3.3.3. Adhesion Assay

1. Mix suspensions of propionibacteria and IECs at a 1:4 proportion and incubate at 37°C for 1 h in a 5% CO_2/95% air atmosphere. Make mixtures in duplicate.
2. After incubation, wash the mixtures once with PBS (120*g*, 5 min) and resuspend to the original volume in cell culture medium. Keep the mixtures on ice and determine adhesion indexes within 1 h.
3. Adhesion of propionibacteria to IECs is determined by examining 30 epithelial cells selected at random using phase-contrast microscopy ***(17)***. Two adhesion parameters can be determined: (1) the *adhesion percentage*, which indicates the number of intestinal cells containing attached bacteria per 100 IECs; and (2) the *adhesion index*, which gives the mean number of adherent bacteria per IEC. They are expressed as follows:
   a. Adhesion percentage: Number of IECs with adherent bacteria extrapolated to 100 IECs.
   b. Adhesion index: Average number of bacteria attached per IEC, only including IECs with bacteria.
4. To illustrate and identify adhesion patterns clearly, wash the duplicate reaction mixtures (120*g*, 5 min), resuspend in PBS, and filter through an 8-μm-pore membrane. Transfer filters retaining cells with adherent bacteria to albumin-coated microscope slides, fix with methanol, and Gram stain. Observe under the light microscope.

3.4. Adhesion to Intestinal Epithelial Cells After Digestion In Vitro

1. Repeat **steps 1** and **2** of **Subheading 3.3.1.** but resuspend the cells in the 10th part of the original volume.
2. Add the cell suspension to 5 mL of artificial gastric juice at pH 3. Mix with a vortex and incubate each bacterial suspension at 37°C with agitation at 200 opm (Dubnoff Metabolic Shaking Incubator, Precision CGA Corp.).
3. After 60 min of gastric digestion, centrifuge the bacterial suspension at 20,000*g* for 10 min. Suspend the pellet to the original volume in intestinal fluid and incubate as before.
4. After 60 min of intestinal digestion, harvest cells by centrifugation (10,000*g*, 10 min), wash twice with 50 m*M* PBS, pH 7.2, and finally resuspend in the same buffer.
5. Adjust the bacterial concentration in a spectrophotometer to approx 5×10^8 CFU/mL ($OD_{560} = 0.8$; *see* **Note 7**).
6. Carry out the assay as in **Subheading 3.3.3.**

3.5. Intestinal Transit Rate

1. Incubate one ampoule of *B. stearothermophilus* in a stove at 65°C for 24 h.
2. Open the ampoule and inoculate 100 μL of the grown culture in 5 mL of BHI broth. Incubate at 65°C for 24 h. Subculture twice in BHI (*see* **Note 8**).

3. Make a third subculture in duplicate and incubate at 65°C for 24 h. Let cells from one culture sporulate at room temperature for 7 d. With the other grown culture, make a calibration curve of absorbance against viable cells per mL by plating serial 10-fold dilutions of known OD_{560} into PCA with incubation at 65°C for 24 h.
4. Wash the sporulated stored culture three times with saline solution (2000*g*, 10 min) and suspend the cells in the appropriate volume of sterile deionized water in order to obtain 5 × 10^7 CFU/mL.
5. Give an oral dose of 1 × 10^6 spores of *B. stearothermophilus* in 50 µL of water to each of 39 mice fasted overnight.
6. Sacrifice three mice immediately and at 30-min intervals up to 360 min.
7. Obtain the stomach, small bowel, and cecum contents. Weigh the organs before and after removing the contents.
8. Heat the contents at 80°C for 10 min and prepare 10% homogenates with saline solution (*see* **Note 9**).
9. Make serial 10-fold dilutions in 0.1% peptone water. Pour different dilutions into PCA and incubate at 65°C for 24 h.
10. Calculate percentage of recovery in each section as: (no. of spores counted in each GI section/no. of spores orally delivered) × 100.

3.6. Adhesion to Intestinal Epithelial Cells In Vivo

1. Put mice at random into two treatment groups of five animals each.
2. Develop the selected *Propionibacterium* strain in MRS broth up to the late exponential phase (42 h). Wash by centrifugation (10,000*g*, 10 min) with sterile PBS, and suspend the cells in 10% nonfat milk to a concentration of 1 × 10^{10} CFU/mL.
3. Orally administer 50 µL of the bacterial suspension (5 × 10^8 CFU) to each mouse of each test group with a plastic-tipped pipet. Give an equal volume of sterile milk to control mice.
4. At the optimum time of sampling for maximal recovery of propionibacteria from the small bowel (determined from the intestinal transit rate experiment), sacrifice the animals and obtain IECs as described in **steps 1–10** of **Subheading 3.3.2.**
5. Determine the bacterial adhesion as in **steps 3** and **4** of **Subheading 3.3.3.** and compare results with controls obtained from animals receiving 50 µL of 10% nonfat milk administered orally.
6. Adhesion in vivo of propionibacteria can be also expressed as CFU/g of IECs by counting the bacteria in selective media *(11)* (*see* **Note 10**).

3.7. Enzyme Activity (β-Galactosidase Assay)

3.7.1. Determination of β-Galactosidase Activity of Bacterial Suspensions

1. Harvest cells by centrifugation (10,000*g*, 10 min, 4°C), wash twice with 0.05 *M* KH_2PO_4/Na_2HPO_4, pH 7.0 containing 0.1% 2-mercaptoethanol (phosphate buffer), suspend in the same buffer to adjust the OD_{560} to 0.5, and finally concentrate 20 times (approx 1 × 10^{10} cells/mL, or 3.85 ± 0.16 mg of dry weight).
2. Carry out the enzymatic reaction by mixing: 200 µL of cell suspension (*see* **Note 11**), 250 µL *o*-nitrophenyl β-D-galactopyranoside (1 m*M* final concentration of ONPG), and 550 µL phosphate buffer, pH 7.0 (*see* **Note 12**).
3. Prepare substrate and sample blanks by excluding sample and substrate, respectively, in the reaction mixtures.

4. Incubate all mixtures at 37°C for 15 min and stop with 800 μL of 0.5 *M* Na_2CO_3.
5. Remove bacterial cells by centrifugation (5000*g*, 10 min) and determine the absorbance at 440 nm of supernatants.

3.7.2. Determination of β-Galactosidase Activity of Intestinal Contents

1. Obtain the intestinal contents in a glass tube by flushing saline solution through the intestine with a glass syringe and a catheter.
2. Wash the contents with saline solution (1000*g*, 5 min) to eliminate residues of lactose that could compete with ONPG for β-galactosidase enzyme (*see* **Note 9**).
3. Make 10% homogenates with 0.05 M phosphate buffer, pH 7.0 (**step 1** of **Subheading 3.2.**).
4. Carry out the enzymatic reaction as in **steps 2** and **3** of **Subheading 3.7.1.** using the homogenates instead cell suspensions in the reaction mixtures. Use 200 μL of small bowel and 30 mL of large bowel contents (*see* **Note 9**).
5. Centrifuge at 10,000*g*, 10 min, 4°C and read the OD_{440} of supernatants (*see* **Note 9**).

3.7.3. Interpretation of β-Galactosidase Activity

1. The β-galactosidase activity is expressed as the rate of hydrolysis of *o*-nitrophenyl β-D-galactopyranoside. The enzyme activity is defined as nanomoles of ONP released from ONPG per milliliter of cell suspension or grams of intestinal contents and per minute of reaction under the assayed conditions.
2. Calculate activity with the following formula:

$$\frac{(OD_{440}\text{ sample} - OD_{440}\text{ higher blank}) \times F = \text{nanomoles OPN/min/mL/cell suspension}}{\text{reaction time (min)} \times \text{sample volume (mL)}}$$

$$\frac{(OD_{440}\text{ sample} - OD_{440}\text{ higher blank}) \times F = \text{nanomoles ONP/min/g/content}}{\text{reaction time (min)} \times \text{sample volume} \times 0.1\text{ g*}}$$

*For 10% homogenates (0.1 g of content in 1 mL of sample.

3 Factor *F* is the slope of a calibration curve of nanomoles of *o*-nitrophenyl (ONP) in function of absorbance at 440 nm. To obtain the standard curve include different concentrations of ONP, between 0 and 2 m*M*, in a final volume of 1.8 mL (which corresponds to the volume of stopped reaction mixtures), and read the OD_{440} of the different dilutions.

4. Notes

1. Pancreatin contains numerous enzymes from porcine pancreas with lipase, amylase, and protease activity. The main proteolytic enzymes are trypsin and chymotrypsin, with optimum activity at pH 8. Pancreatin activity could be replaced by individual enzymes at 1 mg/mL
2. Other murine strains or even other vertebrate animals (like chicken or rabbits) could be used for these tests.
3. Dissolve 9.4 g/L powdered medium plus 2.2 g/L sodium bicarbonate in water, and adjust the pH 0.3 units below the desired pH with 1 *N* HCl or 1 *N* NaOH, since it may rise during filtration. Sterilize by filtration using a 0.22-μm-pore membrane. Store prepared liquid medium at 2–8°C in the dark. Deterioration may be recognized by pH change, cloudy appearance, color change, or particulate matter throughout the solution.

4. Samples can be taken at small intervals and stored at 4°C to be processed all together after finishing sampling.
5. The plates can also be incubated in anaerobic jars with Anaerocult A (Merck, Germany) or GENbox Anaer Sachets (Biomeurieux).
6. The adjustment of the bacterial cell concentration could be made from a calibration curve of absorbance at 560 nm as a function of CFU/mL. From a concentrated suspension of the organism grown under the appropriate conditions, carry out a series of dilutions in PBS (or culture medium) in glass screw-top tubes and read their OD_{560}. Plate dilutions in YEL agar media, and incubate in anerobiosis for 5 d. Count colonies per mL at each dilution and plot OD_{560} against the corresponding CFU/mL.
7. Dilute 0.02 mL of bacterial sample in 0.38 mL of Turk´s solution. Fill the Neubauer chamber with the diluted sample using a plastic-tipped pipet. Count cells in the four large corner squares under a contrast-phase microscope. Calculate the number of cells by using the following formula:

   No. of cells × 1000/counted area (mm^2) × chamber depth (mm) × dilution = no. of cells/mL.

   where counted area four squares of 1 mm^2 = 4 mm^2; chamber depth = 0.1 mm; and dilution = 1:20.
8. Nutrient broth (3 g/L meat extract, 5 g/L peptone, 5 g/L yeast extract, 7×10^{-4} *M* $CaCl_2$, 1×10^{-3} *M* $MgCL_2$ and 5×10^{-5} *M* $MnCl_2$, pH 6.8) could also be used as a sporulation medium.
9. Calculate the weight of intestinal contents by the difference between the weight of the gut before and after extraction of the contents.
10. Calculate the weight of IECs by the difference between the weight of gut walls before and after extraction of the mucosa cells. Make dilutions of the IECs in sterile 0.1% peptone water and plate in a selective medium for propionibacteria. Incubate the plates anaerobically at 35°C for 5 d in an anaerobic chamber (Forma Scientifica Anaerobic System, model 1024) containing a gas mixture of 90% N_2 and 10% CO_2.
11. Cell suspensions of other bacterial species are commonly treated with toluene-acetone (1:9) to enhance β-galactosidase activity, but some strains of dairy propionibacteria loose a considerable portion of activity after this treatment. In consequence, sensitivity of the strains or other methods like treatments of whole cells with SDS, Triton X-100, bile, or sonication should be tested.
12. If you have many samples to test, add the substrate ONPG at 10-s intervals and the same for the stop solution.

References

1. Brook, I. (1994). Infection caused by *Propionibacterium* in children. *Clin. Pediatr. Phila.* **33,** 485–490.
2. Quesada Chanto, A., Afschar, A. S., and Wagner, F. (1994). Microbial production of propionic acid and vitamin B12 using molasses as sugar. *Appl. Microbiol. Biotechnol.* **41,** 378–383.
3. Pérez Chaia, A., Nader de Macías, M. E., and Oliver, G. (1995) Propionibacteria in the gut: effect on some metabolic activities of the host. *Lait* **75,** 435–445.
4. Pérez Chaia, A., Zárate, G., and Oliver, G. (1999) The probiotic properties of propionibacteria. *Lait* **79,** 175–185.

5. Grinstead, D. A. and Barefoot, S. F. (1992) Jenseniin G, a heat-stable bacteriocin produced by *Propionibacterium jensenii* P126. *Appl. Environ. Microbiol.* **58,** 215–220.
6. Roland, N., Bouglé, D., Lebeurrier, F., Arhan, P., and Maubois, J. L. (1998) *Propionibacterium freudenreichii* stimulate in vitro the growth of *Bifidobacterium bifidum* and increase faecal bifidobacteria in healthy human volunteers. *Int. Dairy J.* **8,** 587–588.
7. Roszkowski, W., Roszkowski, K., Ko, H. L., Beuth, J., and Jejaszweicz, J. (1990) Immunomodulation by propionibacteria. *Int. J. Med. Microbiol.* **274,** 289–298.
8. Vorobjeva, L., Lljasova, I., Khodjaev, E. Y., Ponomareva, G. M., and Variouhina, S. Y. (2001) Inhibition of induced mutagenesis in *Salmonella typhimurium* by the protein of *Propionibacterium freudenreichii* subsp. *sermanii. Anaerobe* **7,** 37–44.
9. Havenaar, R., Ten Brink, B., and Huis in't Veld. (1992) Selection of strains for probiotic use. In: *Probiotics, the Scientific Basis* (Fuller, R., ed.). Chapman & Hall, London, pp. 209–224.
10. Zárate, G., Perez Chaia, A., González, S., and Oliver, G. (2000) Viability and β-galactosidase activity of dairy propionibacteria subjected to digestion by artificial gastric and intestinal fluids. *J. Food Proteins* **63,** 1214–1221.
11. Zárate, G., Morata de Ambrosini, V., Pérez Chaia, A., and González, S. (2002) Adhesion of dairy propionibacteria to intestinal epithelial tissue In vitro and In vivo *J. Food Proteins* **65,** 534–539.
12. Zárate, G., Morata de Ambrosini, V., Pérez Chaia, A., and González, S. (2002) Some factors affecting the adherence of probiotic *Propionibacterium acidipropionici* CRL 1198 to intestinal epithelial cells. *Can. J. Microbiol.* **48,** 449–457.
13. Zárate, G., González, S., Pérez Chaia, A and Oliver, G. (2000) Effect of bile on the β-galactosidase activity of dairy propionibacteria. *Lait* **80,** 267–276.
14. Pochart, P., Dewit, O., Desjeux, J-F., and Bourlioux, P. (1989) Viable starter culture, β-galactosidase activity, and lactose in duodenum after yogurt ingestion in lactase deficient humans. *Am. J. Clin. Nutr.* **49,** 828–831.
15. Shah, N. and Jelen, P. (1990) Survival of lactic acid bacteria and their lactases under acidic conditions. *J. Food Sci.* **55,** 506–509.
16. de Man, J. C., Rogosa, M., and Sharpe, E. (1960) A medium for the cultivation of lactobacilli. *J. Appl. Bacteriol.* **23,** 130–135.
17. Morata de Ambrosini, V.I.; González, S.N., and Oliver, G. (1999) Study of adhesion of *Lactobacillus casei* CRL 431 to ileal intestinal cells of mice. *J. Food Proteins* **62,** 1430–1434.

VII

Special Methods

41

Bacterial Surface Characteristics Applied to Selection of Probiotic Microorganisms

María Claudia Otero, Virginia S. Ocaña,
and María Elena Nader-Macías

1. Introduction

For the study of probiotic microorganisms, the in vitro selection tests need to be based on a solid scientific foundation *(1)*. Surface characteristics, one of the in vitro properties are used to evaluate the potentially probiotic strains of lactobacilli *(2)*.

Bacterial surface properties have been associated with attachment to a variety of substrata. Bacterial adhesion to tissues is considered the first step, and such adhesion can also determine the colonization capability of a microorganism *(3)*. Through adhesion ability and colonization of tissues, probiotic microorganisms can prevent pathogen access by steric interactions or specific blockage on cell receptors *(4)*.

One of the main characteristics studied is the hydrophobic nature of the bacterial cell surface. To test this property, Rosenberg and Doyle *(3)* divided microbial cell hydrophobicity assays into two categories. The first includes contact angle measurements (CAMs), partitioning of cells into one or another liquid phase (TTP), and adsorption of individual hydrophobic molecular probes at the cell surface. The second category includes microbial adhesion to hydrocarbons (MATH), hydrophobic interaction chromatography (HIC), and adhesion to polystyrene and other hydrophobic solid surfaces. The tests included in the first category measure hydrophobic properties of the outer cell surface as a whole; those in the second measure hydrophobicity in terms of adhesion *(3)*. Finally, those bacterium classified as hydrophobic can be considered as able to mediate adhesion *(5,6)*.

The objective of this chapter is to describe three different methods applied in our laboratory for the study of bacterial surface properties. They can be used to screen characteristics of lactobacillus strains for probiotic purposes. They are:

- Microbial adhesion to hydrocarbons (MATH).
- Salt aggregation test (SAT).
- Hemagglutination (HA) reaction.

From: *Methods in Molecular Biology, vol. 268: Public Health Microbiology: Methods and Protocols*
Edited by: J. F. T. Spencer and A. L. Ragout de Spencer © Humana Press Inc., Totowa, NJ

MATH was assayed by mixing washed cell suspensions with the test hydrocarbons (*n*-hexadecane, *n*-octane, and *p*-xylene) for a given time and measuring adhesion simply as the decrease in turbidity in the aqueous phase *(7)*. This method was first described for bacteria isolated from the oral environment, more specifically, from dental plaque *(8)*.

The SAT was developed, among other reasons, to study the hydrophobicity of the bacterial cell surface *(9)*. Is based on the fact that highly hydrophobic bacteria are able to aggregate with characteristic patterns in the presence of different ammonium sulfate concentrations.

The HA reaction can be an important tool for monitoring bacterial adhesion, because the erythrocyte surface shares the ontogenetic origin of the tissue cells normally colonized by the organisms *(10)*. In particular, epithelial cells express outer molecules similar to the red blood cell membrane *(11)*.

2. Materials

2.1. Preparation of Microorganisms

2.1.1. Storage

1. LAPTg broth: 15 g/L peptone, 10 g/L tryptone, 10 g/L glucose, 10 g/L yeast extract, 1 mL/L Tween-80, pH 6.5, autoclaved at 121°C for 15 min, and stored at refrigeration temperature. The chemicals for LAPTg preparation were obtained from Britania Laboratories (Argentina).
2. Milk-yeast extract: 10% skim milk, 1% glucose, 0.5% yeast extract, pH 6.0, autoclaved for 20 min at 115°C ($^3/_4$ atm) and stored at refrigeration temperature.
3. Accessories: sterile screw-capped cryovials.

2.1.2. Culture Buildup for the Assay

1. LAPTg broth (*see* **Subheading 2.1.1., step 1**).
2. de man-Ragosa-Sharpe (MRS) broth: dehydrated MRS was obtained from Biokar Diagnostics (Beauvais, France). Composition: 10 g/L peptone, 10 g/L meat extract, 5 g/L yeast extract, 20 g/L glucose, 1.08 g/L Tween-80, 2 g/L dipotassium phosphate, 5 g/L sodium acetate, 2 g/L ammonium citrate, 0.2 g/L magnesium sulfate, 0.05 g/L manganese sulfate, pH 6.5, autoclaved at 121°C for 15 min and stored at refrigeration temperature.

2.2. Hydrophobic Partition Assay: MATH

1. Saline solution (0.85% NaCl), autoclaved at 121°C for 15 min and stored at refrigeration temperature.
2. Spectrophotometer (Spectronic 20, Bausch & Lomb, Miltou Roy, USA) for determinations of optical density (OD) in tubes.
3. Microorganisms: obtained as described in **Subheadings 3.1.** and **3.2.1.**
4. Hydrophobic solvents (all from Merck, Darmstadt): *n*-hexadecane, pro-analysis; *p*-xylene, pro-analysis; and toluene, pro-analysis. (*see* **Note 1**).
5. Accessories: Vortex (Vicking, Argentina), pro-pipets, screw-top glass tubes.
6. Determination of the degree of hydrophobicity.

$$\% \text{ hydrophobicity} = [(OD_b - OD_a)/OD_b] \times 100$$

OD_b = optical density before, obtained as described in **Subheading 3.2.1.**
OD_a = optical density after, obtained as described in **Subheading 3.2.3.**

2.3. Salt Aggregation Test

1. Microorganisms: obtained as described in **Subheading 3.1.3.**
2. Phosphate buffer: 0.02 *M*, pH 6.8, stored at room temperature.
3. Bacterial suspension: obtained as described in **Subheading 3.3.1.**
4. Ammonium sulphate ($[NH_4]_2SO_4$) solutions: 0.2–4 *M* in distilled water. The different concentration solutions are prepared by diluting the 4 *M* solution with distilled water. The solutions are stored at room temperature.

2.4. HA Characteristics

1. Microorganisms: obtained as described in **Subheading 3.1.3.**
2. Saline solution (*see* **Subheading 2.2., step 1**).
3. Spectrophotometer (*see* **Subheading 2.2., step 2**) and tubes.
4. Blood samples: from human or bovine sources, obtained by venipuncture.
5. Sodium citrate solution (3.8%), autoclaved at 121ºC for 15 min.
6. Sterile syringes and screw-top tubes.
7. Capillary test tubes used in hematology to determine red cell concentrations.
8. Accessories: round-bottomed microplates, automatic pipets, and laminar flow chamber (Clean Room Products, USA).

3. Methods

3.1. Preparation of Microorganisms

3.1.1. Microorganisms

Lactobacilli were isolated from vaginal swabs of women and cows under aseptical conditions. The microorganisms were identified by the methods described in *Bergey's Manual of Determinative Bacteriology* ***(12)***, standard techniques used in the laboratory, and API 50 CHL (Biomerieux-France).

3.1.2. Storage

1. Lactobacilli subcultivation: Screw-top glass tubes containing 5 mL LAPTg broth were inoculated with colonies of each of the isolated lactobacilli, and incubated at 37°C for 24 h. A 150-μL aliquot of these cultures was subcultured twice at 37°C for 12 h in 5 mL of the same medium.
2. Storage in milk-yeast extract: The third lactobacilli culture was centrifuged (10 min, 2000*g*), and the spent supernatant was discarded. The pellet was resuspended in 2 mL milk-yeast extract. This bacterial suspension was distributed in cryovials (1 mL) and stored at –20ºC.

3.1.3. Culture Buildup for the Experiments

1. Lactobacilli subcultivation: Stored lactobacilli were subcultured three times in LAPTg broth (5 mL were inoculated with 30 μL of lactobacilli stored at –20°C in milk-yeast extract), incubating the cultures between 12 and 14 h at 37°C. The third culture was used to prepare the bacterial cell suspension for each experiment.

3.2. Hydrophobic Partition Assay (MATH)

1. Bacterial-cell suspension preparation. The third culture was centrifuged (10 min, 2000*g*), and the spent supernatant was discarded. The pellet was washed three times with 3 mL

saline solution and then resuspended in the same solution to an OD_{600} from 0.4 to 0.6. The exact OD values were registered as OD_b values (OD before).

2. Hydrophobic partition assay. Aliquots of 3.6 mL of the bacterial suspensions were distributed in glass tubes. The solvents (0.6 mL) were added to each tube. The tubes were closed and vortexed vigorously for 1 min. They were kept still to allow the immiscible solvent and aqueous phase to separate.
3. Optical density measurements. After separation of the two phases, the aqueous layer was removed by using Pasteur pipets, and transferred to clean tubes. Absorbance was measured in this aqueous phase: OD values were registered as OD_a (OD after).
4. Calculation. The percent of hydrophobicity was obtained from the application of the following formula:

$$\% \text{ hydrophobicity} = [(OD_b - OD_a)/(OD_b] \times 100$$

The percentage of hydrophobicity was calculated for each strain with the three solvents. Each test was performed in triplicate, and the means of the data were calculated.

3.3. Salt Aggregation Test

1. Bacterial cell suspension preparation. The cultures were centrifuged at 2000*g* for 10 min, and the cells were washed with saline and resuspended in phosphate buffer (0.02 *M*, pH 6.8) to a final concentration of 1×10^9 CFU/mL (*see* **Note 2**).
2. Salt aggregation assay. The aggregating properties of lactobacilli in $(NH_4)_2SO_4$ were tested by mixing on a glass slide 25 µL of the bacterial suspension and the same volume of each one of the salt solutions. They were mixed with a circular movement for 2 min, and the formation of aggregates was observed.

   A *negative control* was prepared to compare the characteristics of the bacterial suspension with and without $(NH_4)_2SO_4$. The control was prepared by mixing 25 µL of the bacterial suspension with 25 µL of phosphate buffer 0.02 *M*, pH 6.8 (*see* **Note 3**). A *positive control* was prepared by mixing the same volume of bacterial suspension and 25 µL of the 4 *M* $(NH_4)_2SO_4$ solution. The results of the SAT were expressed as *positive* or *negative*, indicating the salt concentration. Bacteria that were able to form aggregates even after mixing with diluted $(NH_4)_2SO_4$ solutions were termed *highly aggregating*. Lactobacilli that were able to aggregate in the presence or absence of $(NH_4)_2SO_4$ were termed as *self-aggregating*.

3.4. Hemagglutination Ability

3.4.1. Bacterial Cell Suspensions Preparation

The suspension of 1×10^9 CFU/mL was serially diluted (1:2, 1:4, 1:8, and 1:16) in round-bottomed microplates (*see* **Note 2**).

3.4.2. Red Blood Cells Suspension

1. Fresh red blood cells: Blood samples obtained by venipuncture with sterile syringes were collected in sterile screw-top tubes by adding 9 vol of blood to 1 vol of sodium citrate solution (3.8%) (*see* **Notes 4** and **5**).
2. Washed red blood cells: The blood sample was centrifuged (10 min, 4500*g*), and the spent supernatant was discarded. The red cells were washed three times with 3 mL of saline solution and then resuspended in the same solution. Red cell concentration was determined by the microhematocrit method. This suspension was diluted to a 2% (vol/vol) concentration (*see* **Note 6–8).**

3.4.3. Hemagglutination Assay

1. In a round-bottomed microplate, 50 µL of bacterial cell suspension prepared as described in **Subheading 3.4.1.** were added with 50 µL of red blood cell suspension (2%) prepared as described in **Subheading 3.4.2.** Then they were mixed with a circular movement for 1 min and incubated first at 37°C for 30 min and later at 4°C for 6 h (*see* **Note 9**).
2. Controls of bacterial suspension and cell red blood suspension were always prepared (*see* **Note 10**).
3. The hemagglutination titer was visually determined, and a magnifying lens was used at the limit points. The results were expressed as the inverse of the highest dilution of bacteria producing agglutination.

4. Notes

1. **Caution**: Flammable product, irritating to eyes, respiratory system, and skin.
2. For determination of the numbers of microbial cells that should be used in each assays, a plot of OD vs number of viable microorganisms was calculated. The OD was determined by spectrophotometric measures and the number of microorganisms by the plate dilution method using peptone water for the dilutions and LAPTg agar for growing the microorganisms.
3. It was not possible to perform negative controls with self-aggregating lactobacilli, because they formed aggregates as soon as they were resuspended in phosphate buffer.
4. EDTA 2 mg/mL or sodium citrate solution (3.8%) may also be used.
5. Blood samples were used fresh or within 24 h of collection, stored at 4°C.
6. The red cell suspension must be carefully homogenized before loading the capillary tubes and centrifuging.
7. Human blood samples were serologically classified into groups A, B, and O, by using the respective antisera. Hemmaglutination was tested with each one.
8. Bovine blood samples: Red blood cells were centrifuged and washed. Then a pool of red cells was used.
9. This step and the steps in **Subheading 3.4.2.** were performed under sterile conditions (using a laminar flow chamber).
10. Both sets of controls must remain *negative*.

References

1. Morelli, L. (2000) In vitro selection of probiotic lactobacilli: a critical appraisal. *Curr. Issues Intest. Microbiol.* **1,** 59–67.
2. Reid, G. (1999) The scientific basis for probiotic strains of *Lactobacillus*. *Appl. Env. Microbiol.* **65,** 3763–3766.
3. Rosenberg, M. and Doyle, R. J. (1990) Microbial cell surface hydrophobicity: history, measurement, and significance. In: *Microbial Cell Surface Hydrophobicity* (Doyle, R. J. and Rosenberg, M., eds.). American Society for Microbiology, Washington, DC, pp. 1–37.
4. Mc.Groarty, J. (1993) Probiotic use of lactobacilli in the human female urogenital tract. *FEMS Inmunol. Med. Microbiol.* **6,** 251–264.
5. Marshall, K. C. and Cruickshank R. H. (1973) Cell surface hydrophobicity and the orientation of certain bacteria at interfaces. *Arch. Mikrobiol.* **91,** 29–40.
6. Marshall, K. C., Cruickshank, R. H., and Bushby, H. V. A. (1975) The orientation of certain root- nodule bacteria at interfaces, including root- hair surfaces. *J. Gen. Microbiol.* **91,** 198–200.

7. Rosenberg, M., Gutnick, D., and Rosenberg, E. (1980) Adherence of bacteria to hydrocarbons: a simple method for measuring cell-surface hydrophobicity. *FEMS Microbiol. Lett.* **9,** 29–33.
8. Rosenberg, M., Judes, H., and Weiss, E. (1983) Cell surface hydrophobicity of dental plaque microorganisms in situ. *Infect. Inmun.* **42,** 831–834.
9. Lindahl, M., Faris, A., Wadstrom, T., and Hjertén, S. (1981). A new test based on "salting out" to measure relative surface hydrophobicity of bacterial cells. *Biochim. Biophys. Acta* **677,** 471–476.
10. Ofek, I. and Doyle, R. (1994) Methods, models, and analyses of bacterial adhesion. In: *Bacterial Adhesion to Cells and Tissues.* Chapman and Hall, London, pp. 16–20.
11. Madigan, M., Martinko, J., and Parker, J. (1997) *Brock Biology of Microorganisms.* Prentice Hall, Upper Saddle River, NJ.
12. Kandler, O. and Weiss, N. (1986). The Genus Lactobacillus. In: *Bergey's Manual of Systematic Bacteriology*, Vol 2 (Sneath, P., ed.). Williams and Wilkins, Baltimore, MD, pp. 1209–1234.

42

Adhesion Ability of *Lactobacillus* to Vaginal Epithelial Cells

Study by Microbiological Methods

Virginia S. Ocaña and María Elena Nader-Macías

1. Introduction

Adhesion of lactobacilli to the epithelium has been described as the first step in the formation of a barrier to prevent undesirable microbial colonization ***(1)***; consequently, it has been defined as a characteristic of interest for selecting probiotic strains ***(2–4)***.

Several methods have been described to predict the adhesion ability of *Lactobacillus*. Early studies were phenomenological: it was useful to determine whether a particular bacterium could hemagglutinate, or bind to coated particles. Later, studies based on the adhesion of bacteria to epithelial cells in vitro were developed ***(1,5–10)***. These last assays were based on the numbers of bacteria attached to epithelial cells, which were determined by counting stained microorganisms under light microscopy or by measuring the radioactivity of previously radiolabeled bacteria. Because the microscope technique, had some disadvantages, a modification was developed in our laboratory. The technique described in this chapter is a modification of the Mardh and Weströn method ***(5,11)***. The method allows the study of the adhesion ability of bacteria, even if they are aggregating or if they are high-adherent bacteria covering a large area of the epithelial cell surface. Determination of the number of adherent bacteria by counting colonies grown in a selective media avoids the time-consuming, tedious, and hazardous counting under the light microscope and the use of radioactive methods.

2. Materials

2.1. Microorganisms

The microorganisms employed for this study were previously selected because of their ability to produce antimicrobial substances or because of their self-aggregating pattern, as shown by the salt aggregation test ***(12–15)***. They were: *L. crispatus* CRL 1266, *L. salivarius* subsp. *salivarius* CRL 1328, *L. acidophilus* CRL 1259, and *L. acido-*

From: *Methods in Molecular Biology, vol. 268: Public Health Microbiology: Methods and Protocols*
Edited by: J. F. T. Spencer and A. L. Ragout de Spencer © Humana Press Inc., Totowa, NJ

philus CRL 1294. *L. salivarius* S256 isolated from human saliva and *Lactobacillus* sp. CC18 isolated from chicken crop (kindly donated by María Ahumada and Carina Audisio, CERELA, Tucumán, Argentina) were also employed to obtain comparative results.

2.2. Culture and Conservation Media

1. LAPTg broth: 1.5% peptone, 1% tryptone, 1% glucose, 1% yeast extract, 0.1% Tween-80, pH 6.5. Sterilize at 121°C for 15 min (by autoclaving) and store under refrigeration.
2. LAPTg agar: 1.5% peptone, 1% tryptone, 1% glucose, 1% yeast extract, 0.1% Tween-80, 1.5% agar-agar, pH 6.5. Sterilize at 121°C for 15 min and store under refrigeration.
3. Milk-yeast extract (MYE): 10% low-fat milk, 0.5% yeast extract, 1% glucose. Sterilize at 121°C for 15 min and store at room temperature.
4. LBS agar: 1% peptone from casein, 0.5% yeast extract, 2% glucose, 0.6% potassium dihydrogen phosphate, 0.2% ammonium citrate, 0.1% Tween-80, 1.5% sodium acetate, 0.0575% magnesium sulfate, 0.0034% iron (II) sulfate, 0.012% manganese sulfate, 1.5% agar-agar, pH 5.5. Sterilize at 121°C for 15 min and store under refrigeration.
5. Eagle's minimal essential medium (EMEM), pH 7.0 (Gibco), used as prepared by the manufacturer. Store at 0–4°C until the expiration date (*see* **Note 1**).

2.3. Adherence Assay

1. Epithelial cells.
2. Cytobrush (Cytosoft, Medical Packaging) and Neubauer chamber.
3. Dulbecco's modified Eagle's medium (Gibco, cat. no. 11995-065). Store at 2–8°C and protect from light.
4. 2 *N* NaOH.
5. 2% Lactic acid.
6. Saline.
7. pH strips
8. Filters with 8-μm pore membranes (Millipore)
9. Shaker (for agitation at 37°C and 100 opm/min)
10. Peptone water: 1% peptone in distilled water. Store at room temperature.

3. Methods

3.1. Microorganisms

3.1.1. Conservation

1. Microorganisms isolated in LAPTg or LBS agar are grown in 5 mL of LAPTg for 12 h at 37°C, subcultured twice in the same media under identical conditions, and then separated by centrifugation at 2000*g* for 15 min.
2. The pellet is resuspended in 5 mL of MYE and stored at –20°C.

3.1.2. Bacterial Suspensions

1. Frozen lactobacilli (stored in MYE at –20°C) are subcultured three times in LAPTg broth ***(16)*** for 12–14 h at 37°C prior to adhesion studies. An inocula of 2% is used.
2. Overnight cultures grown in LAPTg broth are centrifuged at 2000 rpm for 15 min, and the supernatant was discarded.
3. Lactobacilli were washed twice with saline and once with EMEM, pH 7.0. They were resuspended in EMEM to obtain a final concentration of 10^7 bacteria/mL. An OD_{600} of 0.11 was determined previously by a calibration curve.

3.2. Vaginal Epithelial Cell Suspensions

1. Vaginal epithelial cells were collected by scraping the vaginal wall of healthy volunteers with a cytobrush and resuspended immediately in EMEM, pH 7.0.
2. Indigenous bacteria from epithelial cells were removed by washing three times with 10 mL of EMEM and performing a final filtration through an 8-μm-pore filter (Millipore) of the prewashed cells resuspended in 10 mL of EMEM. This filtration step allowed bacteria to pass through but retained epithelial cells (*see* **Note 2**).
3. Vaginal cells were resuspended in EMEM and adjusted to 10^5 cells/mL as determined in a Neubauer chamber under light microscopy. The OD_{600} of 10^5 cells/mL was approx 0.15. The OD of the cell suspension can be adjusted with EMEM to 0.15 before counting in the Neubauer chamber, which is done to confirm the epithelial cell concentration.

3.3. Adherence Assay

1. 1 mL of vaginal cell suspension is mixed with 1 mL of the bacterial suspension and incubated at 37°C for 1 h on a platform mixer (100 opm).
2. A control was prepared by substituting EMEM for the bacterial suspension.
3. Assays can be performed at pH 7.0 and pH 4.0, adjusting the pH with 2 *N* NaOH and 10% lactic acid, respectively.
4. Suspensions are centrifuged for 10 min at 500 rpm, the supernatant was discarded, and the pellet was resuspended in 10 mL of EMEM.
5. The suspension was passed again through an 8-μm pore membrane to remove nonadherent bacteria.
6. Membranes are washed with manual agitation for 1min in Petri dishes containing 5 mL of saline.
7. The remaining washing solution, containing the epithelial cells with the attached bacteria, was employed for counting viable adherent bacteria.

3.4. Evaluation of Adhesion

1. The number of viable bacteria is determined by the plate dilution method, using peptone water as diluent and LBS agar plates as growth media. The dilutions used are 1:10, 1:100, and 1:1000. Pure washing cell suspension (100- and 200-μL aliquots) were also added to LBS plates to detect poorly adherent bacteria (*see* **Note 3**).
2. LBS plates were incubated in a microaerophylic environment for 48 h at 37°C, and colonies were counted to calculate adhesion.
4. Adhesion can be expressed as adhesion percentage calculated as follows (*see* **Note 4**):

$$\frac{\%\ \text{Adhesion} = \text{Log UFC (adhered bacteria)} \times 100}{\text{Log of total number of bacteria}}$$

and as mean percentage of adhesion per epithelial cell:

$$\frac{\%\ \text{Adhesion/cell} = \text{Log UFC (adhered bacteria)} \times 100}{\text{Log epithelial cell number}}$$

4. Notes

1. Do not use if the pH indicator turns yellow or another color, or if there is any sign of contamination (turbidity).
2. Washing of the epithelial cells obtained from the vagina (*see* **Subheading 3.2.**, **step 1** is a very important step. By this procedure, indigenous bacteria are removed, as is the

mucus. The presence of mucus can block the membrane pores and affect the later filtration steps.

3. To verify that all t he bacteria and cells are removed form the filters after the washing step, the filter can be transferred to a new Petri dish and LBS agar can be added on top. The plate can be incubated in a microaerophilic environment for 48 h at 37°C, and the absence of colonies grown can be tested.
4. In the original Mardh and Weströn technique, the procedure applied to determine the number of attached bacteria is different. After the last filtration, which allows the separation of epithelial cells with the attached bacteria, cells retained on filters are transferred to slides by gently pressing the membranes against slides (that had previously been coated with a thin layer of albumin, as indicated in **Subheading 3.3., item 6**. The cells transferred to slides are fixed with 96° ethanol and Gram-stained. Then bacteria are counted under the light microscope (40X and 100X). With this method, self-aggregating bacteria cannot be individually counted. With the plate dilution method described in this chapter, it is possible to count all the bacteria by using the agitation procedure.

Acknowledgments

This work was supported by grants from CONICET (Consejo Nacional de Investigaciones Científicas y Técnicas), PID-BID 395, and Beca Carrillo Oñativia from the Ministerio de Salud Pública de la República Argentina.

References

1. Chan, R. V., Reid, G., Irvin, J., Bruce, A., and Costerton, W. (1985) Competitive exclusion of uropathogens from human uroepithelial cells by lactobacilli whole cells and cell wall fragments. *Infect. Immun.* **47,** 84–89.
2. Redondo-López, V., Cook, R. L., and Sobel, J. D. (1990) Emerging role of lactobacilli in the control and maintenance of the vaginal bacterial microflora. *Rev. Infect. Dis.* **12,** 856–872.
3. Havenaar, R., Brink, B. T., and Huis In´t Veld, J. H. J. (1992) Selection of strains for probiotics use. In: *Probiotics. The Scientific Basis* (Fuller, R., ed.). Chapman & Hall, London, pp. 209–223
4. Mc Goarty, J. (1993) Probiotic use of lactobacilli in the human female urogenital tract. *FEMS Immunol. Med. Microbiol.* **6,** 251–264.
5. Mardh, P. A. and Weströn, L. (1976) Adherence of bacteria to vaginal epithelial cells. *Infect. Immun.* **13,** 661–666.
6. Boris, S., Suárez, J. E., Vázquez, F., and Barbés, C. (1998) Adherence of human lactobacilli to vaginal epithelial cells and interaction with uropathogens. *Infect. Immun.* **66,** 1985–1989.
7. Kotarski, S. and Savage D. (1979) Models for study the specificity by which indigenous lactobacilli adhere to murine gastric epithelia. *Infect. Immun.* **26,** 966–975.
8. Sobel, J., Schneider, J., Kaye, D., and Levison, M. E. (1981a) Adherence of bacteria to vaginal epithelial cells at various times in the menstrual cycle. *Infect. Immun.* **32,** 194–197.
9. Sobel, J., Myers, P., Levison, M., and Kaye, D. (1981b) *Candida albicans* adherence to vaginal epithelial cells. *J Infect. Dis.* **143,** 76–82.
10. Wood, J. R., Sweet, R. L., Catena, A., Hadley, W. K., and Robbie, M. (1985) In vitro adherence of *Lactobacillus* species to vaginal epithelial cells. *Am. J. Obstet. Gynecol.* **153,** 740–743.
11. Ocaña, V. S. and Nader-Macías, M. E. (2001) Adhesion ability of vaginal *Lactobacillus* strains with probiotic properties. *Biocell* **25,** 253–273.

12. Ocaña, V. S., Ruiz Holgado, A. P., and Nader-Macías, M. E. (1999) Growth inhibition of *Staphylococcus aureus* by H_2O_2-producing *Lactobacillus paracasei* subsp. *paracasei* isolated from the human vagina. *FEMS Immunol. Med. Microbiol.* **23,** 87–92.
13. Ocaña, V. S., Ruiz Holgado, A. P. and Nader-Macías, M. E. (1999). Selection of vaginal H_2O_2-generating *Lactobacillus* species for probiotic use. *Curr. Microbiol.* **38,** 279–284.
14. Ocaña V. S., Bru, E., Ruiz Holgado, A. P., and Nader-Macías, M. E. Surface characteristics of lactobacilli isolated from human vagina. (1999) *J. Gen. Appl. Microbiol.* **45,** 203–212.
15. Ocaña V. S., Ruiz Holgado, A. P., and Nader-Macías, M. E. (1999). Characterization of a bacteriocin-like substance produced by a vaginal *Lactobacillus salivarius* strain. *Appl. Environ. Microbiol.* **65,** 5631–5635.
16. Raibaud, P., Galpin, J. V., Duclezeau, R., Mocquot, G., and Oliver, G. (1963) Le genre *Lactobacillus* dans le tube digestif du rat. II. Caracters de souches heterofermentaires isolates de rats. "holo" et "gnotoxeniques." *Ann. Microbiol.* (Ann. L'Instit. Pasteur) **124,** 2223–2235.

43

Hydroxylapatite Beads As an Experimental Model to Study the Adhesion of Lactic Acid Bacteria From the Oral Cavity to Hard Tissues

María del Carmen Ahumada Ostengo and María Elena Nader-Macías

1. Introduction

The oral environment contains many different types of microorganisms, including both Gram-negative and Gram-positive cocci, bacilli, and spirochetes. From the ecological point of view, the oral cavity is a perfect niche for certain bacteria such as lactobacilli *(1)* because they interact, forming different types of communities. Lactobacilli have been associated with the generation of caries in some reports, secondary to the cariogenic *Streptococcus (2)*.

In previous papers, the isolation and identification of 145 strains from healthy subjects and from subjects with active caries *(3–5)* were performed. Strains were characterized by their surface properties and also by the production of inhibitory substances *(3–5)*. From all the strains, one isolated from the teeth of healthy patients and another from a patient with caries, sharing some surface properties, were selected for further study of their adhesion properties in an experimental model by using hydroxylapatite beads.

Adhesion is the first step in the association of microorganisms with surfaces or mucous membranes. The first approach is a nonspecific interaction of both surfaces; later some other types of interactions can occur, involving more specific mediators of adhesions *(6,7)*.

Many different assays are available to study the adhesion phenomenon; some of them use predictive characteristics, and others use experimental models resembling the in vivo situation *(1)*. Two model systems predominate. The most widely used has been saliva-coated hydroxylapatite or hydroxylapatite coated with buffers, proteins, and other substances *(1)*.

From: *Methods in Molecular Biology, vol. 268: Public Health Microbiology: Methods and Protocols*
Edited by: J. F. T. Spencer and A. L. Ragout de Spencer © Humana Press Inc., Totowa, NJ

In an attempt to increase knowledge of the mechanisms of adhesion of oral lactobacilli with hard surfaces, this chapter describes an experimental model for studing the adhesion between lactobacilli and hard tissues represented by hydroxylapatite, the component most abundant in the teeth.

The following steps were performed:

1. Obtaining the microorganisms.
2. Preparation of the hydroxylapatite beads.
3. Adhesion assay.

2. Materials

2.1. Microorganisms and Culture Media

1. Probiotic microorganisms: The bacteria used in this study were *Lactobacillus salivarius* CRL (Centro de Referencia para Lactobacilos Culture Collection) 1414 and *L. plantarum* CRL 1356. They were isolated from the mouths of different groups of patients and characterized as described previously ***(2–4)***.
2. Storage and culture.
   a. Milk-yeast extract: 10% skim milk, 1% glucose, 0.5% yeast extract, pH 6.0.
   b. Sterilize by autoclaving.
   c. Store at refrigerated temperature.
   d. All the culture media and/or components were obtained from Britania-Argentina.
3. Lactobacilli subcultivation.
   a. LAPTg broth: 15 g/L peptone, 10 g/L tryptone, 10 g/L glucose, 10 g/L yeast extract, 1 mL Tween-80, pH 6.5).
   b. Sterilize by autoclaving.
   c. Store at refrigerated temperature.
4. Peptone water (diluent): 0.1% meat peptone.
5. Adhesion buffer: 50 m*M* KCl, 1 m*M* KH_2PO_4, 1 m*M* $CaCl_2$, 0.1 m*M* $MgCl_2$, pH 6.0. Store at refrigerated temperature. The prepared buffer can be stored under refrigeration for up to 6 mo, to control contamination.

2.2. Preparation of Hydroxylapatite Beads

1. Hydroxylapatite powder: fast flow (flow rate 115 mL/h/cm^2), mol wt = 1004 (ICN Biomedicals). Store at room temperature.
2. Powder and liquid glass ionomer filling material with type II dose dispenser (Klepp Universal Shade, Germany). Store at room temperature.
3. Adhesion buffer: as described in **Subheading 2.1., item 5**, pH 6.0. Store at refrigerated temperature.
4. Syringe restorative dispenser (3M Dental Products), with spatula and support glass.
5. Peptone water (diluent): 0.1% meat peptone distributed in glass tubes and stored at refrigeration temperature.
6. Other materials:
   a. Automatic pipets.
   b. Petri dishes.
   c. Tissue homogenizer with Teflon pestle (MSE, speed continuously variable).
   d. Glass tubes.

e. Glass slides.
f. Sterile tips for pipets.
g. Tube centrifuge (Rolco SRL model 2036, Argentina).

3. Methods

3.1. Preparation of Probiotic Bacteria Cell Suspension (see Note 1)

1. Lactobacilli were subcultured in LAPTg broth no more than three times, by using a 2% inoculum.
2. Each subculture was incubated for 12 h at 37°C.
3. The bacterial cells were collected by centrifugation at 2000*g* for 15 min and washed with the adhesion buffer.
4. Three different concentrations of microorganisms were established for each assay.
5. The Optical density was determined previously according to a calibration curve by which the number of colony-forming units (CFU)/mL was determined (by the successive dilution method and plating in agar media).
6. The three optical densities at 540 nm used in each assay were 0.1, 0.3, and 1.0, corresponding to 1×10^6 CFU/mL, 1.2×10^8 CFU/mL, and 1.3×10^9 CFU/mL for *L. platarum* CRL 1356 and 1×10^6 CFU/mL, 2×10^7 CFU/mL, and 2×10^8 CFU/mL for *L. salivarus* CRL 1414.

3.2. Preparation of Beads With Hydroxylapatite

1. The beads were prepared by mixing (with a dose dispenser) two parts of hydroxylapatite with two parts of powdered glass ionomer and 400 μL of liquid in a glass support.
2. The syringe was loaded with the mixture, and the material was spilled out, forming the beads on the same glass support.
3. The beads were allowed to rest at room temperature for 24 h.
4. Then the weight of the beads was determined, and the beads were distributed in Eppendorf tubes (approx 0.22 g of beads) and autoclaved.
5. The beads were stored at refrigeration temperatures and stabilized for 30 min at room temperature immersed in 1mL adhesion buffer before the experiment.

3.3. Adhesion Assay

1. Microorganisms.
   a. Before beginning the adhesion experiment, the number of initial microorganisms was determined by serial dilution in peptone water (dilutions 1:10 to 1:1,000,000).
   b. Aliquots were plated in LAPTg agar.
   c. The plates were incubated under microaerophilic conditions (Forma Scientific CO_2 incubator model 3185, water-jacketed) for 48 h at 37°C.
   d. This step was performed after the washing step described before.
2. Beads.
   a. Beads were stabilized at room temperature.
   b. After sterilization in autoclaving, they can be stored at refrigeration temperature, before being stabilized in the buffer at room temperature.
   c. Beads were added to 1 mL of adhesion buffer.
   d. The beads were always used at a similar weight, approx 0.22 g and stored in an Eppendorf tube.

3. Adhesion assay.
   a. Aliquots of 2.5 mL of microorganisms were placed in contact with the beads 1 mL of adhesion buffer.
   b. The mixture was taken to the shaker in a thermostatic water bath and incubated for 1 h at 37°C, 50 opm (*see* **Note 2**).
   c. After incubation, the beads were separated from the supernatant with a spatula in aseptic conditions.
   d. They were washed twice with adhesion buffer by gently moving the tubes for 5 min at room temperature.
   e. Disintegration of the beads was performed by using a Teflon pesttle connected to a tissue homogenizor (resuspended in 1 mL of water peptone).
   f. Aliquots of the supernatant were serially diluted in peptone water (as described before in **step 1a.**) and plated in LAPTg agar for determining the number of nonadherent microorganisms.
   g. The number of bacteria adhering to the beads was determined in the bead homogenates by serial dilution in peptone water and by plating aliquots in LAPTg agar.
   h. The plates were incubated in microaerophilic conditions (Forma Scientific CO_2 incubator, model 3185, water-jacketed) for 48 h at 37°C.
   i. Controls were represented by bacterial suspensions without beads.

3.4. Analysis of Data

Careful analyses of adhesion data can yield important insights into adhesion mechanisms. It is also useful to determine the numbers of combining sites (receptors) on a substratum, the affinity constant(s) describing the complex between an adhesin and its receptor, and the amount of an inhibitor required to cause a 50% reduction in adhesion. It is also important to determine the rate of an adhesion reaction and its corresponding desorption or dissociation. In a given system, the total (T) number of cells is equivalent to the combined bound (B) and unbound (U) or free cells. The knowledge of T and B data makes it possible to calculate U, or, conversely, knowledge of any two or the three entities provides a means of computing the third. In most adhesion experiments, T is known and B is determined.

The experimental protocol described in previous steps allowed us to obtain the following data:

3.4.1. Determination of the Number of Microorganisms (as CFU)

1. CFU/mL at the beginning of the experiment (Total [Bound + Free]).
2. CFU/mL adhering to the beads (Bound).
3. CFU nonadhering to the beads, after the adhesion steps (Free or Unbound).
4. Data from control tubes.

3.4.2. Plotting Procedure (see **Note 3**)

The general strategy in most properly designed adhesion experiments is to determine the extent of adhesion using several different bacterial cell densities but employing a constant amount of a receptor material ***(1)***. If the adhesion is directly proportional to the numbers of cells added, then Henry's law specifies that:

$$B = K \times U$$

In actual practice, it is preferable to plot B vs log T (or U).

Another approach to the study of adhesion phenomena is through application of the Langmuir adsorption isotherm, one common derivative of the law of mass action. It has found wide usage in treating adhesion data:

$$U/B = 1/KN + (U/N)$$

Plots for U/B vs U are convenient for computing N (1/slope) and $-1/K$ (the x-intercept). The N value represents the number of receptor sites, and K is the average association constant.

In the adhesion of microorganisms to surfaces, linearity may be observed only when reasonably narrow cells densities are employed.

The third approach to treat the adhesion data is through application of the Scatchard equation:

$$B/U = KN - KB$$

This method is probably of more value in microbial adhesion. A plot of B/U vs B yields a straight line with a slope of $-K$, and an ordinate intercepts yields $N \times K$. As with the Langmuir equation, linearity is achieved only with a system that possesses site-site homogeneity and is free of cooperative phenomenum .

The extent of cooperativity (positive or negative) may be assessed by use of the Hill equation:

$$B/N - B = KU^{n_H}$$

n_H is the Hill coefficient, or index of cooperativity. When $n_H > 1$, there is positive cooperativity. Similarly, negative cooperativity is indicated when $n_H < 1$.

Finally, adhesion data should be subjected to statistical analyses, such as the parametric (unpaired I) test or unpaired (Mann-Whitney U) test specifically when comparing bacteria and/or substrata.

4. Notes

1. Once the suspension of microorganisms is obtained, they can be stored under refrigeration for no more than 1 h for the adhesion assay.
2. It is recommended to carry out the assay in a thermostatized shaker to favor the contact of the bacteria with the inert support.
3. After the plotting step, these data are used as the basis to determine the effect of different substances on the degree of adhesion. The value of 50% adhesion is taken from that plot, meaning to the number of bacteria able to bind to the beads in a value of 50%. This is the first reference point when studying the effects of the different substances that can increase or decrease this value by promoting or diminishing the potential adhesion of the bacteria under study. Many different substances can be studied in terms of their effect on the adhesion pattern of each strain:
    a. Those related to the physical-chemical conditions of growth of the microorganisms i.e., culture media, pH, temperature.
    b. Other substances which are present in the oral cavity, such as compounds of saliva, proteins, ions, solubles, and insoluble glucans, glucosyltransferase, fibronectin, and others that can affect or interfere in the adhesion phenomenon.

c. Some chemical products (such as proteolytic enzymes, saliva amylases, or other compounds) that can be tested to study the nature of the adhesion determinants are used to determine the chemical components of the bacterial surface.

Acknowledgments

This work was supported by grants from the Consejo Nacional de Investigaciones Científicas y Técnicas de Argentina (CONICET), PIP 359.

References

1. Ofek I. and Doyle R. J. (1994) *Bacterial Adhesion to Cells and Tissues* Chapman & Hall, New York, pp. 16–37, 195–237.
2. Babaahmady, K. D., Challacombe, S. J., Marsh, P. D., and Newman, H. N. (1998) Ecological study of *Streptococcus mutans*, *Streptococcus sobrinus* and *Lactobacillus* spp. at sub-sites from approximal dental plaque from children. *Caries Res.* **32,** 51–58.
3. Ahumada, M. C., Colloca, M. E., López, M. E., Pesce de Ruiz Holgado, A., Nader-Macias M. E., (1999) Characterization of lactobacilli isolated from the tongue and gum. *Anaerobe* **5,** 129–135.
4. Colloca, M. E., Ahumada, M. C., López, M. E., and Nader Macias, M. E. (2000) Surface properties of lactobacilli isolated from healthy subjects. *Oral Dis.* **6,** 227–233.
5. Ahumada, M. C., López, M. E., Colloca, M. E., and Nader Macias, M. E. (2001) Lactobacilli isolation from dental plaque and saliva of a group of patients with caries and characterization of their surface properties. *Anaerobe* **7,** 71–77.
6. Marshall, K. C., Stout, R., and Mitchell, R. (1971) Mechanisms of initial events in the sorption of marine bacteria to surfaces. *J. Gen. Microbiol.* **68,** 337–348.
7. Marshall, K. C., Savage, D. C., and Fletcher, M. (eds.) (1985) Mechanisms of bacterial adhesion at solid-water interfaces. In: *Bacterial Adhesion. Mechanisms and Physiological Significance.* Plenum, New York, pp. 133–161.

44

Microtechnique for Identification of Lactic Acid Bacteria

Carlos Gusils, Adriana Pérez Chaia, Guillermo Oliver, and Silvia González

1. Introduction

Certain gut species are pathogens, but a number of other resident bacteria may be of some benefit to host health. Examples include enterococci, lactobacilli, propionibacteria, and bifidobacteria, which are present in the colon in significant numbers. Identification and classification are not identical. A group can be identified only after it has been classified, based on a pattern of properties shown by all the members of the group that other groups do not possess. The properties used in identification are often different from those used in classification. Biochemical, nutritional, and physiological characterization tests (usually carried out in bottles and tubes of solid and liquid media and on plates) have been developed and modified since the earliest days of bacteriology. Generally, the characteristics chosen for an identification plan should be easily determinable, whereas those used for classification (such as DNA homology) may be quite difficult to determine. Genera and species identification might not be based on only a few tests, but rather on the pattern given by a whole battery of tests. The members of the family Lactobacillaceae represent one example of this *(1)*. Some probiotic strains can be selected for their benefical properties as active antimicrobial agents against pathogenic microorganisms, hydrophobic ability, presence of substances with a capacity for adherence to epithelium, and so on. After isolation, identification is an important step before selecting probiotic strains.

These identifications can be facilitated through microtechniques. To alleviate the need to inoculate large numbers of tubes with media (conventional test), some rapid multitest systems have been devised and are commercially available (such as the API or Biolog systems). Although they are expensive for large studies and not always sufficiently versatile, these kits do offer the advantages of convenience, miniaturization, rapidity, and, above all, strict standardization. Other methods *(2)* such as ribotyping or randomly amplified polymorphic DNA do not take into account the phenotypic char-

From: *Methods in Molecular Biology, vol. 268: Public Health Microbiology: Methods and Protocols*
Edited by: J. F. T. Spencer and A. L. Ragout de Spencer © Humana Press Inc., Totowa, NJ

acteristics (biotyping). In this chapter, we describe a simple, rapid, and economical microplates technique to identify lactic acid bacteria and enterobacteria based on general metabolic characteristics, adapted for routine study of a large number of strains.

2. Materials

2.1. Growth Media

1. LAPTg broth *(3)*: This medium is used for culture of lactic acid bacteria. It contains 15 g/L peptone, 10 g/L yeast extract, 10 g/L glucose, 10 g/L triptone, and 1% Tween-80; pH 6.4 ± 0.2. Sterilize in an autoclave at 121°C for 15 min (*see* **Note 1**).

2.2. Study of Sugar Fermentations

1. Physiological solution (PS): 8.5 g/L sodium chloride. Sterilize in an autoclave at 121°C for 15 min (*see* **Note 2**).
2. Sugar solutions (20%, w/v): D-amygdaline, D-arabinose, D-cellobiose, D-erythritol, D-fructose, D-galactose, D-glucose, glycerol, inositol, D-lactose, maltose, D-mannitol, D-mannose, D-melibiose, D-melezitose, D-salicine, D-raffinose rhamnose, D-ribose, sorbitol, D-sucrose, D-trehalose, and D-xylose, sterilized by filtration (*see* **Note 3**).
3. Fermentation media.
   a. For lactobacilli, use MRS *(4)* broth: 10 g/L polypeptone, 10 g/L meat extract, 5 g/L yeast extract, 20 g/L glucose, 5 g/L sodium acetato, 2 g/L ammonium citrate, 2 g/L K_2HPO_4, 0.2 g/L $MgSO_4{\cdot}7H_2O$, 0.05 g/L $MnSO_4{\cdot}4H_2O$, and 1.08 g/L Tween-80; with 0.2 g/L of aggregated of clorophenol red, pH 6.5. Sterilize in an autoclave at 121°C for 15 min (*see* **Note 4**).
   b. For cocci, use base medium *(5)*: 1 g/L meat extract, 10 g/L protease peptone, 0.5 g/L sodium chloride, and 0.2 g/L phenol red, pH 7.6. Sterilize in an autoclave at 121°C for 15 min (*see* **Note 4**).
4. Sterile microplates (*see* **Note 5**).
5. Sterile Vaseline. Sterilize in autoclave at 121°C for 15 min.

2.3. Biochemical Assays (see *Note 6*)

1. Arginine.
   a. Lactobacilli: 3 g/L arginine, 4 g/L triptone, 4 g/L yeast extract, 1 g/L glucose, 6 g/L peptone, and 1 g/L Tween-80, pH 6.5.
   b. Cocci: 3 g/L arginine, 5 g/L triptone, 2.5 g/L yeast extract, 0.5 g/L glucose, 6 g/L peptone, and 2 g/L K_2HPO_4, pH 7.0.
   c. Sterilize in an autoclave at 121°C for 15 min.
   d. Test reagent: Nessler's reagent: A) 7 g potasium iodide, B) 19 g mercuric iodide, C) 10 g Potassium hydroxide, D) distilled water (to 100 mL).
      i. Dissolve A and B in 40 mL of distilled water and C in 50 mL of distilled water.
      ii. Mix the two solutions and add distilled water to 100 mL.
      iii. Allow the precipitate to settle, decant the clear supernatant liquid into a reagent bottle, and discard the precipitate.
2. Esculin.
   a. 1 g/L Esculin, 4 g/L triptone, 4 g/L yeast extract, 1 g/L glucose, 6 g/L peptone, and 1 g/L Tween-80, pH 6.5. Sterilize in autoclave at 121°C for 15 min.
   b. Test reagent: $FeCl_3$ 1% in 2% HCl.

3. Milk with methylene blue: 100 g/L skim milk, and 0.1 and 0.3% methylene blue. Sterilize in autoclave at 121°C for 15 min.
4. Indole.
    a. 10 g/L Triptone. Sterilize in autoclave at 121°C for 15 min.
    b. Test reagent: Gracian I: formaldehyde, H_2SO_4 conc. (1:1); Gracian II: 0.5 % $K_2Cr_2O_4$.
5. Nitrate.
    a. 10 g/L Peptone, 5 g/L sodium chloride, 1 g/L KNO_3, and inverted Durham tube. Sterilize in autoclave at 121°C for 15 min.
    b. Test reagent: Reagent 1: 1 g sulphanilic acid, 100 mL 5 N acetic acid; Reagent 2: 1 g α-naphtol, 100 mL ethanol.
6. Gluconate: 1.5 g/L triptone, 1 g/L yeast extract, 1 g/L K_2HPO_4, 40 g/L gluconate. Sterilize in autoclave at 121°C for 15 min.
7. Gibson's medium: 80 g/L skim milk, 2.5 g/L yeast extract, 50 g/L glucose, 10 mL 4% $MnSO_4$, 20 mL nutritive agar (*see* **Note 7**), pH 6.5. Sterilize in an autoclave at 121°C for 15 min.
8. Hippurate.
    a. 10 g/L Triptone, 3 g/L yeast extract, 2 g/L K_2HPO_4, 0.5 g/L glucose, and 20 g/L sodium hippurate. Sterilize in an autoclave at 121°C for 15 min.
    b. Test reagent: H_2SO_4 conc.
9. 2,3,5-Triphenyltetrazolium chloride (TTC): 10 mL skim milk and 0.01% 1mL 2,3,5-triphenyltetrazolium chloride. Sterilize in an autoclave at 121°C for 15 min.
10. LAPTg broth with different conditions: final pH of 9.2, 9.4, and 9.6; 0.02% of sodium azide; 1 and 4% bile salts; 4 and 6.5% sodium chloride.

3. Methods

3.1. Sugar Fermentation Assays

1. Grow the strains of lactic acid bacteria in LAPTg broth at 37°C for 16 h.
2. Harvest the cells by centrifugation at 4000*g* for 5 min.
3. Wash the pellet obtained with PS twice (*see* **Note 8**).
4. Resuspend the cells with the PS by vortexing to optical density $(OD)_{540\ nm} = 0.5$ (*see* **Note 9**).
5. Inoculate (1%) fermentation media (cocci or lactobacilli) with the bacterial suspension.
6. Dispense 10 μL of each sugar solution in wells of microplates under sterile conditions (*see* **Note 10**, and **Fig. 1**).
7. Add 200 μL of the specific inoculated medium and 50 μL of sterile Vaseline (*see* **Note 11**, and **Fig. 1**).
8. Incubate the microplates for 48 h at 37°C.
9. After 24 and 48 h of incubation, evaluate fermentation capacity (*see* **Note 12**).

3.2. Biochemicals Assays

1. Prepare the culture inoculum as described in **steps 1–4** from **Subheading 3.1.**
2. Dispense 200 μL of each sterile media in wells of microplates under sterile conditions (*see* **Note 13**).
3. Inoculate (1%) each well containing specific media with the bacterial suspension.
4. Incubate the microplates for 48 h at 37°C.

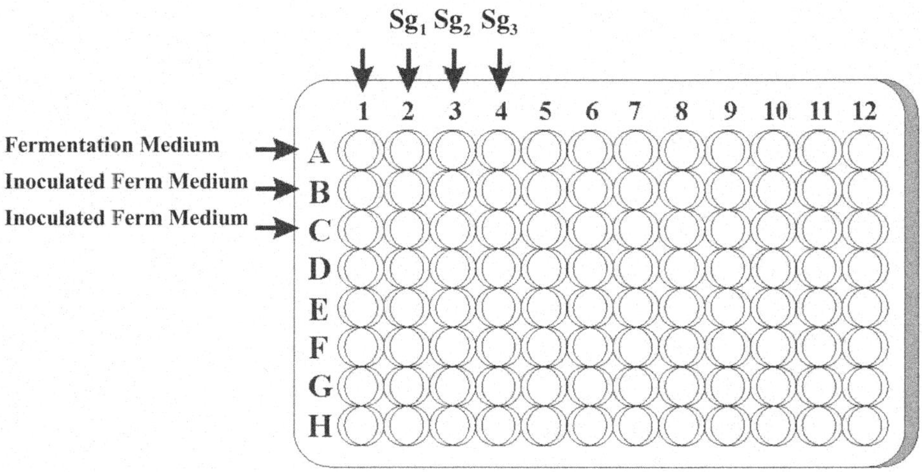

Fig. 1. Dispensation of sugar solution sin to wells of microplates. Column 2: sugar 1 (Sg1; example, arabinose); column 3, sugar 2 (example, glucose); file A, control sugars; file B, sugar fermentation (Ferm) by strain "A"; file C, sugar fermentation by strain "B."

5. Inoculate (1%) another three microplates containing LAPTg broth with the bacterial suspension. In wells with gluconate, add 50 μL of Vaseline-paraffin (1:1).
6. Incubate these microplates for 48 h at different temperatures (15, 45, and 50°C).
7. After 1, 2, 5, and 7 d of incubation, evaluate growth capacity, gas production, substance reduction, and other parameters in the media assays (*see* **Table 1**).
   a. Arginine: Add 0.1 mL bacterial culture and 0.1 mL Nessler's reagent.
   b. Esculin: Add 0.1 mL reagent and 0.1 mL bacterial culture.
   c. Indol: Add 0.1 mL Gracian I reagent and 0.1 mL bacterial culture, and then add 0.1 mL Gracian II reagent.
   d. Nitrate: Add 0.1 mL bacterial culture and 0.1 mL reagent 1, and then add 0.1 mL reagent 2. Negative reaction: Add 0.1 mL bacterial culture and add a small quantity of zinc.
   e. Hippurate: Add 0.1 mL bacterial culture and 0.1 mL reagent.

4. Notes

1. de Man-Ragosa-Sharpe (MRS) broth ***(4)***, instead of LAPTg broth, can be use to aid lactic acid bacteria growth. MRS broth: 10 g/L polypeptone, 10 g/L meat extract, 5 g/L yeast extract, 20 g/L glucose, 5 g/L sodium acetato, 2 g/L ammonium citrate, 2 g/L K_2HPO_4, 0.2 g/L $MgSO_4 \cdot 7H_2O$, 0.05 g/L $MnSO_4 \cdot 4H_2O$, and 1.08 g/L Tween-80. Sterilize in an autoclave at 121°C for 15 min.
2. Phosphate buffer, instead of physiological solution, can be use to wash cells
3. D-arabinose, D-erythritol, D-fructose, D-galactose, inositol, D-lactose, maltose, D-mannose, rhamnose, D-ribose, sorbitol, D-sucrose and D-xylose should be sterilized by filtration and another sugars by autoclaving at 121°C for 15 min.

Table 1
Tests, Test reagents and Results for Identification of Bacteria

Test	Test reagent	Result
Hydrolysis of arginine	Nessler's reagent	Development of an orange to brown color
Hydrolysis of esculine	1% $FeCl_3$ in 2% HCl	Black precipitated
Methylene blue milk		Color reduction and growth
Production of indole	Gracian I and Gracian II	Development of a deep red color
Reduction of nitrate		
Nitrite	Reagent 1 and reagent 2	Development of a red color
Residual nitrate	Small quantity of zinc	Development of a red color
Gluconate		Presence of gas
Gibson		Disruption of the medium
Hippurate	H_2SO_4 conc	White precipitated (5–10 min)
2,3,5-Triphenyltetrazolium (TCC)		Reduction of the colorless compound yields a red-colored compound
LAPTg broth		Bacterial growth
pH		
Sodium azide		
Bile salts		
Sodium chloride		

4. For determining the sugar fermentation capacity, it is important to use both pH indicators at 10-fold concentration, to obtain a better visualization of the change of color of the indicator.
5. Microplates can be sterilized by UV light for 1 h.
6. Studies of the biochemical properties of lactobacilli included hydrolysis of arginine, esculin growth in the presence of 0.1 and 0.3% methylene blue added to milk, indol, reduction of nitrate, gluconate, and Gibson's medium to study formation of carbon dioxide from glucose, and growth at different temperatures (15, 37, 45, and 50°C). Studies of the biochemical properties of cocci included arginine, esculin, hippurate, hydrolysis, 2,3,5-triphenyltetrazolium chloride, potassium tellurite, and nitrate reductions, growth in LAPTG at initial pH of 9.2, 9.4, and 9.6, growth at 4 and 6.5% sodium chloride, growth in the presence of 0.1 and 0.3% methylene blue added to milk, growth in 0.02% sodium azide, growth in the presence of 1 and 4% bile salts, and growth at different temperatures (15, 37, 45, and 50°C).
7. Nutritive agar: 1 g/L meat extract, 2 g/L yeast extract, 5 g/L peptone, 5 g/L sodium chloride, 3 g/L agar, pH 7.4.
8. Cell washes are important to minimize carryover of nutrient from culture.
9. These suspensions were used for inoculating biochemical studies or diluted 1:10 in adequate media for sugar fermentation.
10. The microplates with sugars can be stored for 60 d, but recent cell suspensions must be used.

11. Addition of the sterile Vaseline over fermentation media is very important to avoid contamination and to obtain an anaerobic atmosphere.
12. Fermentation is considered positive when the pH indicator in the medium becomes yellow after incubation; any shade of red is considered negative.
13. Studies of the biochemical properties of lactobacilli included arginine, esculin, methylene blue milk, indol, esculine, nitrate, gluconate, Gibson's medium, and growth at different temperatures (15, 37, 45, and 50°C). For cocci, studies included arginine, esculin, hippurate, tetrazolium chloride, potassium tellurite, nitrate, growth in LAPTg broth at initial pH of 9.2, 9.4, and 9.6; 4 and 6.5% sodium chloride; 0.1 and 0.3% methylene blue milk; 0.02% sodium azide; 1 and 4% bile salts; and at different temperatures (15, 37, 45, and 50°C).

References

1. Fagnant, J. E., Sanders, C. C., and Sanders, E. (1982) Development and evaluation of a biochemical scheme for identification of endocervical lactobacilli. *J. Clin. Microbiol.* **16,** 926–934.
2. Du Plessis, E. M. and Dicks, L. M. T. (1995) Evaluation of random amplified polymorphic DNA (RAPD)-PCR as a method to differentiate *Lactobacillus acidophilus, L. crispatus, L. amyglovorus, L. gallinarum, L. grasseri* and *L. johnsonii. Curr. Microbiol.* **31,** 114–118.
3. Raibaud, P., Caulet, M., Galpin, J. V., and Mocquot, G. (1961) Studies on the bacterial flora of the alimentary tract of pigs II. Streptococci: selective enumeration and differentiation of the dominant group. *J. Appl. Bacteriol.* **24,** 285–291.
4. De Man, J. C., Rogosa, M., and Sharpe, E. (1960) A medium for the cultivation of lactobacilli. *J. Appl. Bacteriol.* **23,** 130–155.
5. Larpent, J. P. and Larpent-Gourgaud, M. (1975) Mémento Technique de Microbiologie Technique et Documentation, Paris, pp. 1–200.

45

Differentiation of Lactic Acid Bacteria Strains by Postelectrophoretic Detection of Esterases

Roxana Beatriz Medina, Marta Beatriz Katz, and Silvia González

1. Introduction

Lactic acid bacteria (LAB) comprise a diverse group of Gram-positive, non-spore-forming microorganisms. These bacteria are widely used in food technology. The species identification of LAB depends mainly on physiological and biochemical criteria. The esterolytic systems of LAB remain poorly characterized. Esterases (EC 3.1.1.3) represent a diverse group of hydrolases catalyzing the cleavage and formation of esters bonds *(1)* Screening of esterases is usually performed either by employing chromophoric substances (e.g., α- or β-naphthyl esters of short-chain fatty acids). The post-electrophoretic detection of esterases is a sensitive technique applied in bacterial systems, that mainly provides information on the similarity of strains within the same species or subspecies according to their esterase patterns. This technique is principally used to determine the number and substrate specificity of esterases and lipases, revealing the complexity of lipase and esterase systems *(2,3)*. The present chapter describes the technique of polyacrylamide gel electrophoresis (PAGE; in the absence of sodium dodecyl sulfate [SDS]), in non-denaturing conditions, to find intracellular fractions for strain typing of LAB.

2. Materials

2.1. Growth Media

1. de-Man-Rogosa-Sharpe (MRS) broth *(4)*: 10 g/L polypeptone, 10 g/L meat extract, 5 g/L yeast extract, 20 g/L glucose, 5 g/L sodium acetate, 1 g/L Tween-80, 2 g/L ammonium citrate, 2 g/L K_2HPO_4, 0.2 g /L $MgSO_4 \cdot 7H_2O$, and 0.05 g/L $MnSO_4 \cdot 4H_2O$, pH 6.4 ± 0.2. Sterilize in an autoclave at 121°C for 15 min.

From: *Methods in Molecular Biology, vol. 268: Public Health Microbiology: Methods and Protocols*
Edited by: J. F. T. Spencer and A. L. Ragout de Spencer © Humana Press Inc., Totowa, NJ

2.2. Postelectrophoretic Detection of Esterases

1. Phosphate buffer.
   a. Solution A (0.2 M NaH_2PO_4): weigh 27.6 g $NaH2PO_4 \cdot H_2O$ and make to 100 mL with distilled water (dH_2O).
   b. Solution B (0.2 M Na_2HPO_4): weigh 53.05 g of $Na_2HPO_4 \cdot 7H_2O$ and make to 100 mL with dH_2O.
   c. To prepare 1 L of 0.1 M phosphate buffer (pH 7.0), mix 195 mL of solution A and 305 mL of solution B and bring to 1000 mL with dH_2O.
2. Acrylamide solution.
   a. Weigh 29.2 g acrylamide and 0.8 g *N,N'*-methylene-*bis*-acrylamide, add about 50 mL of dH_2O, stir until disolved, and dilute to 100 mL.
   b. Remove insoluble material by filtration.
   c. Store at 4°C in a dark bottle (30 d maximum to avoid hydrolysis of acrylamide to acrylic acid).
   d. **Caution:** acrylamide is neurotoxic; repeated skin contact or inhalation is dangerous.
3. Separation gel buffer: dissolve 18.15 g Tris in 80 mL dH_2O, adjust to pH 8.8 with HCl, and make up to 100 mL with dH_2O. The solution is stable for at least 2 wk when stored at 4°C.
4. Stacking gel buffer: dissolve 6 g Tris in 80 mL dH_2O, adjust to pH 6.8 with HCl, and make up to 100 mL with dH_2O. The solution is stable for at least 2 wk when stored at 4°C.
5. 10% (w/v) Ammonium persulfate (APS): dissolve 100 mg APS in 1 mL dH_2O. Always use a freshly prepared solution.
6. Reservoir buffer: prepare solution by combining 6.05 g Tris-HCl and 28.8 g glycine. Dissolve with dH_2O to a final volume of 2 L. Final pH should be about 8.3. Store at 4°C; warm to 30°C before use if precipitation occurs.
7. *N,N,N',N'*-Tetramethylethyenediamine (TEMED).
8. Sample buffer: prepare solution by combining 4 mL dH_2O, 1 mL stacking gel buffer, 0.8 mL glycerol, and 0.002% bromophenol blue. Dilute the sample 1:1 with sample buffer.
9. 1% (w/v) Substrate solution: dissolve 100 mg α- and β-naphthyl derivate of fatty acids of C2–C10 carbon atoms in 1 mL acetone. Dissolve 100 mg α- and β-naphthyl derivate of fatty acid of C12 carbon atoms in propanol-2. Always use a freshly prepared solution.
10. Solution substrate for developing color: dissolve 5 mg of Fast Red TR (Sigma) in 10 mL phosphate buffer 0.1 *M*, pH 7.0, and add 200 µL substrate solution.
11. Glass beads no. 31/14 (diameter, 0.1–0.11 mm; B. Braun Biotech International, Germany).
12. Cell disruptor (B. Braun Melsungen, Germany).
13. Mini-Protean II electrophoresis cell (Bio-Rad).

3. Methods

3.1. Cultivation of Bacteria

1. Inoculate 2% of the microorganisms in 200 mL MRS broth.
2. Incubate overnight at 30°C (lactococci, leuconostocs, and mesophilic lactobacilli) or 37°C (enterococci and thermophilic lactobacilli) (*see* **Note 1**)
3. Collect cells by centrifugation at 10,000*g* for 10 min at 4°C.
4. Wash the pellet twice with 10 mL of 0.1 *M* sodium phosphate buffer, pH 7.0.
5. Resuspend at 40% w/v washed cells in the afore-mentioned buffer by vortexing. Be sure that the pellet is completely resuspended before proceeding.

3.2. Preparation of Intracellular Fraction

1. Add 1–2 g of glass beads to the resuspended cells.
2. Disrupt the cells five times for 1 min under CO_2 atmosphere.
3. Cool the samples on ice for 5 min between each mixing sequence.
4. Centrifuge the sample at 20,000*g* for 30 min at 4°C to remove glass beads and unbroken cells (*see* **Note 2**).
5. Transfer the supernatant (intracellular fraction) to a clean tube by carefully decanting from the pellet and maintain on ice (*see* **Note 3**).

3.3. Postelectrophoretic Detection of Esterases

In this method the enzyme extracts (intracellular fraction) are first subjected to a protein separation (electrophoresis in non-denaturing conditions [without SDS]), and then lipases and esterases are revealed using α- and β-naphthyl derivates of fatty acids. Released naphthol is coupled with diazotized salts (Fast Red TR salt or Fast Garnet GBC salt) to give a colored product. The PAGE procedure considered here is based on the technique described by Laemmli *(5)*.

3.3.1. Preparation of Separation Gel (12% Acrylamide)

1. Place a comb into the assembled gel sandwich. With a pen, place a mark on the glass plate 2 cm below the teeth of the comb. This will be the level to which the separating gel is poured.
2. To a clean flask, add 3.45 mL of dH_2O, 2.5 mL separation gel buffer, 4.0 mL acrylamide solution, 50 μL of 10% APS, and 10 μL TEMED (*see* **Notes 4** and **5**).
3. Mix thoroughly by stirring. Pour the solution smoothly to the mark and avoid the inclusion of air bubbles.
4. Immediately overlay the solution with 1.5 mL water-satured isobutanol to obtain a flat surface.
5. Allow the gel to polymerize for 30 min to 1 h at 25°C. The acrylamide is gelled when a clear line can be observed 1–2 mm below the water-satured isobutanol.
6. Rinse off the overlaying solution four times with dH_2O.

3.3.2. Preparation of Stacking Gel (4% Acrylamide)

1. To a clean flask, add 6.2 mL of dH_2O, 2.5 mL stacking gel buffer, 1.33 mL acrylamide solution, 50 μL 10% APS, and 10 μL TEMED. Swirl the solution gently but thoroughly.
2. Rinse the top of the gel with approx 1 mL of stacking gel solution, pour off, and fill the cassettes immediately with stacking gel solution.
3. Insert the comb into cassettes; avoid trapping air bubbles.
4. Allow the gel to polymerize at least 1 h at 20°C
5. The gel can be used after 12 h or stored for 24 h at 4°C in a closed plastic bag to avoid evaporation.

3.3.3. Loading the Samples

1. After polymerization, fill the upper and lower reservoirs with reservoir buffer. Remove any air bubbles from the bottom of the gel so that good electric contact is achieved.
2. Insert the samples through the reservoir buffer and into the wells with a Hamilton syringe. The volume depends on the concentration of proteins in the cellular fraction. Apply approx 10 μg of protein from each fraction. Adjust the volume to 15 μL with phosphate buffer and mix with 15 μL sample buffer. Load 20 μL of these samples in each well (*see* **Note 6**).

3.3.4. Running the Gel

1. Attach the electrical leads to a suitable power supply with the proper polarity. Be sure to connect the positive lead to the lower chamber.
2. Run the gel at constant current 60 mA per gel until the tracking dye reaches 1 cm of the botton of the gel (*see* **Note 7**).
3. Turn off the power supply and disassemble the glass plates.

3.3.5. Detection of Esterase or Lipase Activity

1. Immerse the gel in a container and wash it twice with 10 mL phosphate buffer 0.1 *M*, pH 7.0, cooled at 10°C for 10 min.
2. Incubate the gel in 10 mL solution to detect esterase activity at 37°C for 8 h with shaking gentle
3. Wash the gel with dH_2O to eliminate the complex color precipitated (*see* **Note 8**).
4. Activities were identified as colored bands on the gels (*see* **Notes 9** and **10**).

3.3.6. Storage

Gels may be stored between cellophane sheets or dried onto mylar sheets using a gel dryer (Bio-Rad).

3.3.7. Interpretation of Bacterial Esterase Activity Profiles

Visual comparison is the most frequently used method for interpretation of bacterial esterase electrophoretic patterns. The M_F value (distance moved by the esterase band as a percentage of the distance moved by the dye front) is used for comparison only. Electrophoretic mobility variants are numbered for each esterase in decreasing order of migration (*see* **Note 11**).

4. Notes

1. Maximum esterase production is observed at the early stationary phase of growth.
2. Other methods to lyse cells can be used, such as lysozyme treatment with sonication, or pressure techniques.
3. The esterase system is intracellular in LAB
4. The electrophoretic reagents must be of high quality.
5. The high-molecular-weight proteins (>100 kDa) are better resolved on 7–8% acrylamide gels. We obtain better results with 12% acrylamide gels.
6. Band resolution is improved if small volumes of samples are applied to the gels.
7. Electrophoresis can be run at 4°C to avoid enzyme denaturalization.
8. The disadvantage of this technique is that a substrate of α-NA fatty acid derivatives longer than propionate may not be easily accessibile through the net of the electrophoresis gel, lending one to the false conclusion that there is no activity *(6)*.
9. Fast Red TR forms a complex red color that precipitates, but the bands of activity can be noticable.
10. Fast Blue can be used as a substrate to develop color. In this case a complex dark blue color is formed.
11. The electrophoretic protein patterns provide a reliable way to differentiate strains and reveal the complexity of esterase systems *(7)*.

References

1. Bornscheuer, U. T. (2002) Microbial carboxyl esterases: classification, properties and application in biocatalysis. *FEMS Microbiol. Rev.* **26,** 73–81.
2. Tsakalidou, E., Zoidou, E., and Kalantzopoulos, G. (1992) Esterase activities of cell-free extracts from strains of Lactococcus lactis subsp. lactis isolated from traditional Greek cheese. *J. Dairy Res.* **59,** 111–113.
3. Vafopoulou-Mastrojiannaki, A., Litopoulou-Tzanetaki, E., and Tzanetakis, N. (1996) Esterase activities of cell-free extracts from wild strains of leuconostoc and heterofermentative lactobacilli isolated from traditional Greek cheese. *Lett. Appl. Microbiol.* **23,** 367–370.
4. De Man, J. C., Rogosa, M., and Sharpe, M. E. (1960) A medium for the cultivation of lactobacilli. *J. Appl. Bacteriol.* **23,** 130–135.
5. Laemmli, U. K. (1970) Cleavage of structural proteins during the assembly of the head of bacteriophage T_4. *Nature* **277,** 680–685.
6. Morichi, T., Sharpe, M. E., and Reiter, D. (1968). Esterase and other soluble proteins of some lactic acid bacteria. *J. Gen. Microbiol.* **53,** 405–414
7. Katz, M., Medina, R., Gonzalez, S., and Oliver, G. (2002) Esterolytic and lipolytic activities of lactic acid bacteria isolated from ewe´s milk and cheese. *J. Food Prot.* **65,** 1997–2001.

46

Determination of Esterolytic and Lipolytic Activities of Lactic Acid Bacteria

Roxana Beatriz Medina, Marta Beatriz Katz, Silvia González, and Guillermo Oliver

1. Introduction

Lactic acid bacteria (LAB) are considered weakly lipolytic compared with many other groups of bacteria (e.g., *Pseudomonas*, *Bacillus*, and *Achromobacter*). The esterolytic and lipolytic systems of dairy LAB remain poorly characterized. Esterases from lactic acid bacteria, yeasts, and *Pseudomonas* organisms may be involved in the development of fruity flavors in foods *(1)*, and pregastric lipase and esterases are essential for the development of typical flavor in Italian cheese *(2)*. Microbial lipases and esterases may improve quality or accelerate the maturation of cheeses, cured bacon, and fermented sausages *(3)*.

Lipases are defined as glycerol ester hydrolases (EC 3.1.1.3) that hydrolyze tri-, di-, and monoglycerides present at an oil-water interface. Esterases (EC 3.1.1.6) hydrolyze esters in solution and may also hydrolyze tri- and especially di- and monoglycerides containing short-chain fatty acids *(4)*.

Some probiotic strains of LAB can hydrolyze the triglycerides, releasing most short and medium chain, and essential fatty acids, which are valuable to today's health-conscious consumer. Medium chain fatty acids (C6–C14), in particular, have become accepted treatment for patients with malabsorption symptoms, a variety of metabolic disorders, cholesterol problems, and infant malnutrition. These probiotic bacteria could alleviate lipase deficiency in the digestive tract during digestion (steatorrhea) *(5)*.

In this chapter, we describe different methods routinely used in our laboratory to determine the esterolytic and lipolytic activity of LAB. These techniques include the use of α- and β-naphthyl derivatives of fatty acids (chromogenic method), the *p*-nitrophenyl (pNP) derivative of fatty acids (chromogenic method), and triglycerides (agar-well assay technique and titrimetric test) as substrates.

From: *Methods in Molecular Biology, vol. 268: Public Health Microbiology: Methods and Protocols*
Edited by: J. F. T. Spencer and A. L. Ragout de Spencer © Humana Press Inc., Totowa, NJ

2. Materials

2.1. Growth Media

1. de Man-Rogosa-Sharpe (MRS) broth ***(6)***: 10 g/L polypeptone, 10 g/L meat extract, 5 g/L yeast extract, 20 g/L glucose, 5 g/L sodium acetate, 1 g/L Tween-80, 2 g/L ammonium citrate, 2 g/L K_2HPO_4, 0.2 g /L $MgSO_4{\cdot}7H_2O$, and 0.05 g/L $MnSO_4{\cdot}4H_2O$, pH 6.4 ± 0.2. Sterilize in an autoclave at 121°C for 15 min.

2.2. Chromogenic Method

2.2.1. Substrates: α- and β-Naphthyl Derivatives of Fatty Acids

1. 0.1 *M* Sodium phosphate buffer, pH 7.0.
   a. Solution A (0.2 *M* NaH_2PO_4: Weigh 27.6 g $NaH_2PO_4{\cdot}H_2O$ and make to 100 mL with distilled water (dH_2O).
   b. Solution B (0.2 *M* Na_2HPO_4): Weigh 53.05 g of $Na_2HPO_4{\cdot}7H_2O$ and make to 100 mL with dH_2O.
   c. To prepare 1 L of 0.1 *M* phosphate buffer (pH 7.0), mix 195 mL of solution A and 305 mL of solution B and bring to 1000 mL with dH_2O.
2. 66 m*M* α- and β-naphthyl substrate in ethanol/water (1:1). Always use a freshly prepared solution.
3. 10% (w/v) Sodium dodecyl sulfate (SDS): dissolve 10 g SDS in 100 mL dH_2O with gentle stirring. Store at room temperature. Solution may become cloudy at temperatures below 20°C, but warming to 30°C and mixing may restore clarity.
4. 10% Fast Garnet GBC solution: dissolved 5 mg of Fast Garnet GBC (Sigma) in 1 mL 10% SDS. Always use a freshly prepared solution.

2.2.2. Substrates: p-Nitrophenyl Derivatives of Fatty Acids

1. 0.15 *M* Sodium phosphate buffer, pH 7.0.
   a. Solution A (0.2 *M* NaH_2PO_4): Weigh 27.6 g $NaH_2PO_4{\cdot}H_2O$ and make to 100 mL with distilled water (dH_2O).
   b. Solution B (0.2 *M* Na_2HPO_4): Weigh 53.05 g of $Na_2HPO_4{\cdot}7H_2O$ and make to 100 mL with dH_2O.
   c. To prepare 1 L of 0.15 *M* phosphate buffer (pH 7.0), mix 293 mL of solution A and 457 mL of solution B and bring to 1000 mL with dH_2O.
2. Solvent mixture: acetone/water at a concentration of 1:1
3. Substrate solution: 22.5 m*M* pNP derivate of fatty acids in solvent mixture. Always use a freshly prepared solution.

2.3. Triglyceride Assay

2.3.1. Agar-Well Assay Technique

1. 0.2 *M* Phosphate buffer, pH 7.0, prepared as described in **Subheading 2.2.1.**
2. Substrate solution: dissolve 0.25 mL triglyceride, 25 mg sodium azide, and 100 mL 0.2 *M* phosphate buffer, pH 7.0.
3. 1.5% Agar-Agar ultra pure (Merck).

2.3.2. Titrimetric Test

1. 10 % (w/v) Gum arabic: dissolve 10 g gum arabic in 100 mL dH_2O.
2. 10% (v/v) Triglycerides (substrate solution): Dissolve 10 mL triglycerides in 10 mL gum arabic.

3. 10 m*M* $CaCl_2$/1.0 *M* Tris-HCl buffer, pH 7.0: dissolve 12.11 g Tris-HCl and 0.11 g $CaCl_2$ in 80 mL dH_2O, adjust to pH 7.0 with HCl, and make up to 100 mL with dH_2O. The solution is stable for at least 2 wk when stored at 4°C.
4. 1.0 *M* NaCl solution: weigh 5.85 g NaCl, add about 50 mL of dH_2O, stir until dissolved, and dilute to 100 mL.
5. 0.2 *N* H_2SO_4/2.0 *M* NaCl solution: weigh 11.7 g NaCl and make up 50 mL with H_2O, stir until dissolved, add 0.53 mL H_2SO_4, and dilute to 100 mL.
6. Extraction mixture: light petroleum ether/diethyl ether (1: 2.75 [v/v]).
7. 10 m*M* NaOH: dissolve 0.04 g NaOH in 100 mL ethanol.
8. 1% (w/v) Phenolphthalein: weigh 0.1 g phenolphthalein and make up 10 mL with ethanol.

3. Methods

3.1. Cultivation of Bacteria

1. Inoculate 2% of the microorganisms in 200 mL MRS broth (*see* **Note 1**).
2. Incubate overnight at 30°C (mesophilic bacteria) or 37°C (thermophilic bacteria) (*see* **Note 2**).
3. Collect cells by centrifugation at 10,000*g* for 10 min at 4°C.
4. Wash the pellet twice with 10 mL of 0.1 *M* sodium phosphate buffer, pH 7.0.
5. Resuspend at 40% w/v washed cells in the afore-mentioned buffer by vortexing. Be sure that the pellet is completely resuspended before proceeding.

3.2. Preparation of Cellular Fraction

3.2.1. Extracellular Fraction

1. The supernatant of the growth medium is considered the extracellular fraction.
2. Neutralize the supernatant with 5 *M* NaOH.
3. Sterilize by filtration using a 0.22 μm-pore filter, white GSWP, 25 mm (Millipore).
4. Keep on ice to preserve the enzyme activity.

3.2.2. Intracellular Fraction

1. Add 1–2 g of glass beads to the resuspended cells.
2. Disrupt the cells five times for 1 min under CO_2 atmosphere (*see* **Note 3**).
3. Cool the samples in ice for 5 min between each mixing sequence.
4. Centrifuge the sample at 20,000*g* for 30 min at 4°C to remove glass beads and unbroken cells.
5. Transfer the supernatant (intracellular fraction) to a clean tube by carefully decanting from the pellet.
6. Keep on ice to preserve the enzyme activity.

3.3. Chromogenic Method

3.3.1. Substrate: α- or β-Naphthyl Derivatives of Fatty Acids

In this method, the substrate naphthyl derivatives are hydrolyzed, releasing α- or β-naphthol, which is coupled with Fast Garnet GBC salt (diazotized salts) to give a colored product.

The method is as follows:

1. To 100 μL of intracellular or extracellular fraction, add 160 μL of 100 m*M* sodium phosphate buffer, pH 7.0, and 20 μL of α- and β-naphthyl substrate (*see* **Note 4**).

2. Mix gently and incubate at 30°C (mesophilic bacteria) or 37°C (thermophilic bacteria) for 1 h.
3. Add 0.6 mL of Fast Garnet GBC solution for color development of (red).
4. Incubate at room temperature for 15 min (*see* **Note 5**).
5. Measure the absorbance at 560 nm.
6. The concentration of color complex formed can be calculated using a standard α- or β-naphthol curve.

One unit of enzyme activity is defined as the amount of enzyme that released 1 μmol of α- and β-naphthol per min under the conditions of the assay.

3.3.2. Substrate: p-*Nitrophenyl Derivatives of Fatty Acids*

In this method, the amount of pNP released from the pNP derivatives of fatty acids by the action of LAB esterases or lipases is measured at 410 nm.

The method is as follows:

1. To 200 μL of the intracellular or extracellular fraction, add 400 μL of 150 m*M* sodium phosphate buffer, pH 7.0, 775 μL of solvent mixture, and 125 μL of the pNP derivative substrate (*see* **Notes 4** and **6**).
2. Mix gently and incubate at 30°C (mesophilic bacteria) or 37°C (thermophilic bacteria) for 1 h.
3. Stop the reaction by addition of 175 μL of 80% (v/v) acetic acid.
4. Centrifuge for 5 min at 13,000*g*.
5. Measure the release of pNP at 410 nm.
6. The concentration of pNP released can be calculated from the derived value of molar adsorption of pNP at 410 nm:

$$\mu M \text{ pNP} = \varepsilon \Delta A_{410} F \times 10^3$$

where ΔA_{410} is the experimentally observed change of absorbance at 410 nm using a 1-cm light path, and F is the dilution factor corresponding to the assay procedure. One unit of enzyme activity is defined as the amount of enzyme that released 1 μmol of pNP per min under conditions of the assay.

3.4. Triglyceride Assay

3.4.1. Agar-Well Assay Technique

This technique is generally used for screening and is based on diffusion assays in solid medium. The triglycerides emulsified are hydrolyzed by lipases. Lipase activity is monitored by measuring the clearing zone surrounding the wells. The disadvantages of this technique are as follows: (1) tributyrin is not always a good substrate for lipase activity; and (2) the sensitivity of this technique depends on triglyceride concentration. The advantages of this method are as follows: (1) synthetic medium is a reproducible substrate; and (2) it is more adapted for weak lipolytic microorganisms

The method is as follows:

1. Emulsify 10 mL of substrate solution at maximal amplitude for 3 min. During sonic treatment, keep cold in an ice-water bath to prevent excessive heating of the emulsion (*see* **Note 7**).

2. Add to emulsified substrate solution 1.5 g Agar-Agar ultra pure.
3. Heat emulsion to dissolve the agar
4. Add 15 mL hot emulsion to sterile Petri dishes and solidify.
5. Cut 9-mm-diameter wells in the agar bed.
6. Place 100-μL cellular fractions in the well.
7. Incubate at 30°C (mesophilic bacteria) or 37°C (thermophilic bacteria) for 72 h (*see* **Note 8**).
8. The clearing zone surrounding the wells indicates positive lipase activity (*see* **Notes 9** and **10**).

3.4.2. Titrimetric Test

In this method, triglycerides are used as the substrate (natural ones such as olive oil or butter fat or synthetic ones such as like triacetin, tributyrin, triolein, and others). Thus, they are often involved in the characterization of lipase, except when triacetin and (less frequently) tributyrin are used. Hydrolysis of triglycerides leads to the production of free fatty acids (FFAs), which can be titrated. Such techniques can be divided into three groups: (1) titration of FFA after extraction; (2) direct titration; and (3) continuous titration. We describe here the titration of FFAs after extraction.

The method is as follows:

1. Emulsify 1.5 mL of substrate solution at maximal amplitude for 3 min. During sonic treatment, cool in an ice-water bath to prevent excessive heating of the emulsion
2. Add the emulsified substrate solution to 1 mL cellular fraction, 0.5 mL 10 m*M* $CaCl_2$/1.0 *M* Tris-HCl buffer, pH 7.0, 0.5 mL 1.0 *M* NaCl, and 2 mL dH_2O (reaction mixture) (*see* **Note 4**).
3. Mix gently and incubate at 30°C (mesophilic bacteria) or 37°C (thermophilic bacteria) for 48 h in shaker (*see* **Notes 11** and **12**).
4. Remove 1 mL of reaction mixture and add 1 mL 0.2 *N* H_2SO_4/2.0 *M* NaCl solution for the stop reaction.
5. Add 6 mL extraction mixture, shake vigorously for 20 s, and let rest for 20 min (*see* **Note 13**).
6. Remove 2 mL of the upper phase, place into an Erlenmeyer flask, add 6 drops of phenolphthalein, and mix gently.
7. Titrate with 10 m*M* NaOH.
8. One unit of enzyme activity is defined as the amount of enzyme that released 1 μmol FFA per min under the conditions of the assay (*see* **Notes 14** and **15**).

4. Notes

1. Rich media are recommended to obtain good growth of the microorganisms.
2. Maximum esterase and lipase production is observed at the early stationary phase of growth.
3. Other methods to lyse cells can be used, such as lysozyme treatment with sonication or pressure techniques.
4. Include blanks containing substrate without enzyme, boiled enzyme, and enzyme alone.
5. This time is recommended for formation and stabilization of the color complex.
6. When the intracellular fraction is obtained by the procedure described in **Subheading 3.2.2.**, color is observed in the blank containing substrate without enzyme. When the pressure technique is used, this blank has no color. This method is recommended for purified enzymes.

7. The best sensitivity is obtained with 0.25% triglyceride.
8. Observe the presence of lipase activity at 12, 24, 48, and 72 h.
9. To quantify, measure the clearing-zone size surrounding the wells and express in mm.
10. Activity can be expressed as diffusion units, defined as the minimum fold dilution of the enzyme preparation that failed to give a detectable zone of hydrolysis *(7)*.
11. Titrate samples at 0, 16, 24, and 48 h.
12. In this technique, the lipid substrate and the cells are constantly mixed, resulting in more direct contact between the substrate and the enzyme. However, in the agar-well assay technique, the substrate is essentially immobile, and the enzyme or cells must diffuse or spread within the agar to reach the substrate, a process that is relatively slow.
13. Extraction efficiency is not always high, depending on the nature of the solvents and the FFAs. Calibrations must be done. Measurements do not take the acid molecules into account.
14. Make a standard curve with butyric acid.
15. All assays are performed in duplicate.

References

1. McKay, A. M. (1993) Microbial carboxylic ester hydrolases (EC 3.1.1) in food biotechnology. *Lett. Appl. Microbiol.* **16,** 1–6
2. Fox, P. F. and Stepaniak, L. (1993) Application of enzymes in cheese technology. *Int. Dairy J.* **3,** 509–536.
3. Lee, S. Y. and Lee, B. H. (1990) Esterolytic and lipolytic activities of *Lactobacillus casei* subsp. *casei* LLG. *J. Food Sci.* **55,** 119–122.
4. Jaeger, K. E., Ransac, S., Dijkstra, B. W., Colson, C., van Huevel, M., and Misset, O. (1994) Bacterial lipases. *FEMS Microbiol. Rev.* **15,** 29–64.
5. Layer, P. and Keller, J. (1999) Pancreatic enzymes: secretion a nd luminal nutrient digestion in health and disease. *J. Clin. Gastroenterol.* **28,** 3–10.
6. De Man, J. C., Rogosa, M., and Sharpe, M. E. (1960) A medium for the cultivation of lactobacilli. *J. Appl. Bacteriol.* **23,** 130–135.
7. Fox, P. and Stepaniak, L. (1983) Isolation and some properties of extracellular heat-stable lipases from *Pseudomonas fluorescens* strain AFT 36. *J. Dairy Res.* **50,** 77–89.

47

Meat-Model System Development for Proteolytic Activity Determination

Graciela Vignolo, Silvina Fadda, Patricia Castellano, and Cecilia Fontana

1. Introduction

Many technological parameters that affect the nature and functional properties of proteins are involved in the preparation of meat products. The dry-curing process is quite complex because of the coexistence of enzymes from endogenous and bacterial origins. The protein breakdown that takes place during the ripening of dry fermented sausages leads to an increase in the concentration of peptides and free amino acids *(1,2)*. The proteolytic events have been thoroughly investigated not only because of their physiological significance but also for their technological connotations in terms of texture and flavor development *(3,4)*. Lactic acid bacteria and *Staphylococcus* or *Kocuria* are used as starter cultures in fermented meat products. In recent years, the proteolytic system of lactobacilli involved in meat fermentation became the focus of an increasing number of studies because of the technological roles of these organisms *(5–7)*. Although results obtained from broth systems show proteolytic activity, studies involving food systems must be done to confirm their effectiveness. A simplified soluble muscle extract to determine the capacity to degrade meat proteins by lactic acid bacteria was developed using a spectrophotometric method *(8)* based on the reaction of the α-amino groups released by hydrolysis with *o*-phtaldialdehyde and β-mercaptoethanol to form an adduct (1-thioalkyl-2-alkylisoindoles) that is enhanced at basic pH and absorbs strongly at 360 nm.

2. Materials

2.1. Soluble Muscle Extract

1. Stomacher 400 blender (London, UK) and stomacher bags.
2. Bovine semimembranosus muscle or porcine longissimus dorsi muscle after 24 h of exsanguinization.

From: *Methods in Molecular Biology, vol. 268: Public Health Microbiology: Methods and Protocols*
Edited by: J. F. T. Spencer and A. L. Ragout de Spencer © Humana Press Inc., Totowa, NJ

3. Distilled water (dH_2O).
4. Centrifuge.
5. Whatman no. 5 paper.
6. Sterifilter system (Millipore); 0.22-μm membrane filter.
7. 0.1 *N* NaOH.
8. 1% (w/v) Glucose.
9. 0.1% (v/v) Sterile Tween-80.
10. Plate count agar (PCA).

2.2. Culture Media

1. de Man-Rogosa-Sharpe (MRS) broth culture medium for lactic acid bacteria: 10 g/L peptone, 10 g/L meat extract, 5 g/L yeast extract, 20 g/L glucose, 5 g/L sodium acetate, 1.08 g/L Tween-80, 2 g/L ammonium citrate, 2 g/L K_2HPO_4, $MgSO_4·7H_2O$ (0.2g/L), 0.05 g/L $MnSO_4·4H_2O$, pH 6.5. Sterilize in an autoclave at 121°C for 15 min.
2. MC broth culture medium for *Staphylococcus* sp.: 10 g/L meat extract, 5 g/L yeast extract, 5 g/L NaCL, 1 g/L Na_2HPO_4, pH 7.0. Sterilize in an autoclave at 121°C for 15 min.
3. Baird-Parker broth culture medium for *Kocuria* sp.: 10 g/L peptone, 5 g/L meat extract, 1 g/L yeast extract, 10 g/L sodium pyruvate, 12 g/L glycine, 5 g/L lithium chloride. Sterilize in an autoclave at 121°C for 15 min.

2.3. Ortho-Phthaldialdehyde Method (OPA)

1. 0.1 *M* Phosphate buffer, pH 7.0.
   a. Solution A (0.2 *M* NaH_2PO_4): Weigh 27.6 g $NaH_2PO_4·H_2O$ and make up to 100 mL with dH_2O.
   b. Solution B (0.2 *M* Na_2HPO_4): Weigh 53.05 g $Na_2HPO_4·7H_2O$ and make up to 100 mL with dH_2O.
   c. To prepare 1 L of 0.1 *M* phosphate buffer, pH 7.0, mix 195 mL solution A and 305 mL solution B and bring to 1000 mL with dH_2O.
2. Sodium dodecyl sulfate (SDS; 20% w/v): Dissolve 20 g SDS in 100 mL dH_2O with gentle stirring. Store at room temperature (*see* **Note 1**).
3. 100 m*M* sodium tetraborate (borax): weigh 3.81 g of $Na_2B_4O_7·10H_2O$ and make up to 100 mL with dH_2O. This solution is stable for at least 1 mo at 4°C.
4. OPA solution (working reagent): mix 25 mL of 100 m*M* borax, 2.5 mL 20% SDS, 40 mg OPA (dissolved in 1 mL methanol), and 100 μL β-mercaptoethanol. Dilute to a final 50 mL with dH_2O. This reagent must be prepared daily.
5. 0.75 *N* Trichloroacetic acid (TCA): weigh 61.24 g TCA and dilute to 500 mL with dH_2O. The solution is stable for at least 2 wk when stored at 4°C.

3. Methods

3.1. Soluble Muscle Extract

1. Weigh 10 g of lean muscle in a stomacher bag (*see* **Note 2**).
2. Add 90 mL of 20 m*M* phosphate buffer, pH 7.
3. Homogenize in a stomacher 400 blender for 3 min.
4. Centrifuge the protein solution (12,000*g* at 4°C for 20 min).
5. Adjust the pH to 6.5 with 1 *N* NaOH.
6. Add 1% (w/v) of glucose (*see* **Note 3**).
7. Filter the supernatant containing the proteins through Whatman paper.

8. Filter-sterilize this solution by using a 250-mL capacity filter (Bio-Rad) with a vacuum pump (*see* **Note 4**).
9. Add 0.1% Tween-80, previously sterilized (*see* **Note 5**).

3.1.1. Sterility Control

1. To quantify the total aerobic organisms, inoculate them in duplicate Petri dishes with 0.5 mL of protein extract.
2. Add 10 mL of the melted PCA and homogenize appropriately.
3. The growth of microorganisms must be negligible, with colony-forming units (CFU) below 1×10^2 CFU/mL.

3.2. Proteolytic Activity of Lactic Acid Bacteria in the Soluble Muscle Extract

3.2.1. Culture Conditions of the Strains

1. Propagate *Lactobacillus* and *Staphylococcus/Kocuria* sp. strains in the specific media and incubate at 30°C.
2. Harvest the cells at logarithmic phase (overnight culture) and wash twice with 20 m*M* phosphate buffer (pH 7.0).

3.2.2. Growth in the Soluble Muscle Extract

1. Inoculate 30 mL of the soluble muscle extract medium with an overnight culture to yield an initial number of 10^5 CFU/mL corresponding to an OD_{680} of 0.15 (*see* **Note 6**).
2. Incubate at 30°C for 96 h.
3. Take samples every 24 h for pH, bacterial growth, and proteolytic activity analysis.
4. Measure the pH values with a pH meter using 2 mL of the sample.
5. Determine bacterial cell counts using 0.1% peptone water as the diluent.
6. Plate 0.5 mL of each dilution in duplicate in Petri dishes adding the medium corresponding to each organism.
7. Incubate at 30°C for 48 h.

3.2.3. OPA Spectrophotometric Assay

1. Add 500 μL of 0.75 *N* TCA to 250 μL of homogenized sample.
2. Vortex vigorously and allow to stand for 15 min at 4°C.
3. Centrifuge the solution at 13,000*g* for 10 min and transfer the supernatant to a clean Eppendorf tube with a micropipet (*see* **Note 7**).
4. Add 1 mL of OPA reagent to 50 μL supernatant in a 1-mL quartz cuvet.
5. Mix briefly by inversion and incubate for 5 min at room temperature.
6. Measure the absorbance at 340 nm against the blank (sterile soluble muscle extract).
7. Calculate the millimoles per liter of the α-amino acid released (m*M*) from the following relationship: $mM = \varepsilon \Delta A_{340} F$, where ΔA_{340} is the experimentally observed change of absorbance at 340 nm using a 1-cm light path, F is the dilution factor corresponding to the assay procedure, and ε is the molar absorption coefficient (6000 *M*/cm).

4. Notes

1. The solution may become cloudy at temperatures below 20°C but may restored by warming at 30°C and mixing.
2. Before mincing the beef/pork muscle, remove the fat and connective tissue in aseptic conditions, cut the lean muscle into small cubes (3 cm), and keep at –30°C until use.

3. To ensure the development of lactic acid bacteria in this system, it is necessary to add a carbon source.
4. This experimental system contain approx 4.5 g/L of protein.
5. The concentration of this growth factor should be determined for each lactic acid bacterium.
6. To avoid interference with the sarcoplasmic system, an optical density of 680 nm was selected.
7. After the TCA treatment, samples can be frozen at –20°C for further use.

References

1. Aristoy, M. C. and Toldrá F. (1991) Deproteinization techniques for HPLC amino acid analysis in fresh pork muscle and dry cured ham. *J. Agric. Food Chem.* **39,** 1792–1795.
2. De Masi, T. M., Wardlaw, F. B., Dick, R. L., and Ancton, J. C. (1990) Non protein nitrogen (NPN) and free amino acid contents of dry fermented sausages. *Meat Sci.* **27,** 1–12.
3. Fadda, S., Sanz, Y., Vignolo, G., Aristoy, M-C., Oliver, G., and Toldrá, F. (1999) Characterization of muscle sarcoplasmic and myofibrillar protein hydrolysis caused by *Lactobacillus plantarum. Appl. Environ Microbiol.* **65,** 3540–3546.
4. Hammes, W. P., Bantleon, A., and Min, S. (1990) Lactic acid bacteria in meat fermentation. *FEMS Microbiol. Rev.* **87,** 165–174.
5. Johansson, G., Berdagué, J. L., Larsson, M., Tran, N., and Borch, E. (1994) Lipolysis, proteolysis and formation of volatile components during ripening of a fermented sausage with *Pediococcus pentosaceus* and *Staphylococcus xylosus* as starter cultures. *Meat Sci.* **38,** 203–218.
6. Montel, M. C., Seronine, M. P., Talon, R., and Hebraud, M. (1995) Purification and characterization of a dipeptidase from *Lactobacillus sake. Appl. Environ. Microbiol.* **61,** 837–839.
7. Sanz, Y., Fadda, S., Vignolo, G., Aristoy, M-C., Oliver, G., and Toldrá, F. (1999) Hydrolytic action of *Lactobacillus casei* CRL705 on pork muscle sarcoplasmic and myofibrillar proteins. *J. Agric. Food Chem.* **47,** 3441–3448.
8. Church, F. C., Swaisgood, H. E., Porter, H. D., and Castignani, G. L. (1983) Spectrophotometric assay using *o*-phaldialdehyde for determination of proteolysis in milk and isolated milk proteins. *J. Dairy Sci.* **66,** 1219–1227.

48

β-Glucuronidase Method to Determine Mastitis Levels in Goat Milk

Rubén Oliszewski, Martha Núñez de Kairúz, Silvia González, and Guillermo Oliver

1. Introduction

Mastitis is a general term that refers to the inflammation of the mammary gland. It is the most common illness in dairy farms and it has different causes, mainly a great number of germs that infect the gland.

These infectious diseases induce gross variations in milk composition, reflected by physical, chemical, and bacteriological changes. They produce milk jellification, a decrease in important components such as lactose, casein, and fats and minerals such as calcium, phosphorus, and potassium and increases in other unimportant technological components, such as serum proteins and chlorides; all these affect the cheese efficiency and the starter culture action *(1)*.

Assuming that cheese making is the principal use of goat milk in industry, an evaluation of the quality of the milk used as the raw material is of fundamental importance. It is impossible to obtain quality products by using milk with an anomalous chemical composition *(2)*.

Somatic cell count (SCC) is the indicator most used for mastitis detection. These cells, which are contained in milk, can be grouped into three types: epithelial cells, blood cells, and cytoplasmatic particles. During an attack of mastitis, the immune defenses of the udder are activated, polynucleated leukocytes pass from the blood toward the mammary gland in large numbers, and the number of somatic cells in the milk increases *(2)*.

The level of somatic cells in goat milk is characterized by great variability between different countries and between regions of the same country. Different authors show averages between 750,000 and 5,400,000 cells/mL *(1–4)*. These values differ greatly between cow and goat milk, mainly because normally nonleukocytic cell-like particles can be found as a result of the particular apocrine secretion process in the goat

From: *Methods in Molecular Biology, vol. 268: Public Health Microbiology: Methods and Protocols*
Edited by: J. F. T. Spencer and A. L. Ragout de Spencer © Humana Press Inc., Totowa, NJ

mammary gland. These particles are large fragments of cytoplasm originating from the distal portion of alveolar secretory cells and are of similar size (5–30 μm in diameter) to milk leukocytes. They contain abundant RNA-positive granular material (associated with dilated cisternae of the rough endoplasmic reticulum), large amounts of protein, and some lipids, but no DNA ***(5,6)***. Thus it is important to use techniques that disregard these other substances and allow only a count of somatic cells ***(7,8)***.

The afflux of polymorphonuclear leukocytes and macrophages is one of the essential body defenses against clinical and subclinical mastitis. These cells are used as a base for the SCC method. However, their secretion products can also be used. These cells begin the phagocytic process with the secretion of hydrolytic enzymes, which, together with other factors, participate in the degradation of tissue constituents in the inflammatory process. One of these enzymes is β-glucuronidase ***(7)***. It has proved to be the most significant selectively released enzyme in this process ***(9,10)***.

A β-glucuronidase test was recently standardized for goat milk and was shown to be reliable, measuring only the presence of the enzyme in milk, without the other substances that are picked up by the SCC method ***(1)***. In this chapter, we describe the procedure used in our laboratory to determine β-glucuronidase levels in milk.

2. Materials

2.1. β-Glucuronidase Method

2.1.1. Standard Curve

1. 1 m*M* *p*-Nitrophenol solution: 13.9 mg *p*-nitrophenol in 100 mL acetate buffer, to prepare a standard curve.
2. 0.5 *M* Na_2CO_3.
3. Spectrophotometer at 410 nm, for optical density (OD) assay to determine the extinction coefficient.

2.1.2. Enzyme Assay

1. Milk samples collected and immediately cooled to 4°C.
2. 40 m*M* pnPG: 1.5 g *p*-nitrophenol β-D-glucuronide in 100 mL of distilled water.
3. 1 *M* Acetate buffer, pH 4.0.
4. 0.5 *M* Carbonate buffer, pH 10.0.

2.2. Somatic Cell Count

1. Water bath to 30–40°C.
2. Clean microscope slides.
3. Micropipet, with appropriate tips, suitable for rapid and convenient transfer of 0.01 mL of milk.
4. Slide drier: heat source regulated at 40–45°C.
5. Forceps or slide holder: required for dipping and holding slides.
6. Slide storage: clean, dust-free insect-proof boxes, cases, or files.
7. Microscope: for computing single strip factor (SSF).
8. Immersion oil.
9. Carnoy's fixative: prepared by combining 60 mL chloroform, 20 mL glacial acetic acid, and 120 mL 100% ethyl alcohol.

10. Pyronin Y-methyl green stain (PYMG) for goat milk: prepared by combining 1.0 g pyronin Y, 0.56 g methyl green, and 196 mL water. Filter through Whatman no. 1 paper before use. Stain is light-sensitive; store in brown bottle.

3. Methods

3.1. β-Glucuconidase Method

3.1.1. Standard Curve

1. To 1 mL of whole milk add 0.01–0.25 mL standard solution of 1 m*M* *p*-nitrophenol.
2. Made up to a final volume of 4 mL with 0.5 *M* Na_2CO_3.
3. Centrifuge twice for 20 min at 3000*g*.
4. Determine the extinction coefficient at 410 nm and plot the OD against n*M* of *p*-nitrophenol.
5. One unit of β-glucuronidase is defined as n*M* of *p*-nitrophenol released from the substrate per mL of milk per min.

3.1.2. Enzyme Assay Procedure

1. Remove milk fat from the samples by cool centrifugation at 0°C and 10,000*g* (*see* **Note 1**).
2. To 0.4 mL of milk add 0.2 mL of 40 m*M* pnPG (1.5 g *p*-nitrophenol β-D-glucuronide in 100 mL of distilled water) and 0.4 mL of 1 **M** acetate buffer, pH 4.0.
3. Incubate the mixture for 4 h at 50°C.
4. Stop the reaction with 4 mL of 0.5 *M* carbonate buffer, pH 10.0.
5. Centrifuge twice for 20 min at 3000*g*. Measure the *p*-nitrophenol liberated by the substrate in the supernatant against a blank with nonincubated milk (instead of the substrate solution) in a spectrophotometer at 410 nm (*see* **Note 2**).

3.2. Direct Microscopic Somatic Cell Count

3.2.1. Slide Identification and Sample Preparation

1. Identify each sample area legibly and indelibly on the margin of slide.
2. Mix samples with a vortex 25 times.
3. Heat samples in a water bath to 30–40°C.
4. Vortex the samples again.
5. Mix them carefully and cool to the temperature at which the microsyringe has been calibrated.

3.2.2. Sample Measurement and Smear Preparation

1. Place 0.01 mL of milk sample on a clean slide in a 1-cm^2 area.
2. Holding the microsyringe vertical, place the tip near the center of the area to receive the smear, touch the slide with the tip, and expel the test portion.
3. With the plunger still fully depressed, touch once against a dry spot.
4. Release the plunger only after touching and removing the tip from the slide.

3.2.3. Staining Films

1. Dry the sample in a heat source regulated at 40–45°C and fix it in Carnoy's fixative solution for 5 min; then was hydrate the film for 1 min each in 50% ethanol, 30% ethanol, and distilled water.
2. Stain for 6 min in PYMG stain (*see* **Note 3**).
3. Dry the film completely, immerse in butanol and distilled water for 1 min each, and dry again.

3.2.4. Examination

1. Examine the smear microscopically.
2. Adjust the microscope lamp to provide maximal optical resolution. Place one drop of immersion oil on the smear, carefully lower the oil immersion lens, focus, locate the center of the microscope area, and begin counting cells. Cells stain blue or blue-green; RNA and background stain pink.
3. Count all cells in a field-wide strip across the diameter of a single smear, focusing up and down as necessary. Cells are generally 8 μm or larger; do not count cells less than 4 μm, and count fragments only if more than 50% of the nuclear material is visible. Clusters of cells must be counted as 1 unless nuclear unit(s) are clearly separated; focus up and down to ensure that there are no bridges connecting nuclear masses. Count cells touching only the top or bottom half of strip. If in doubt, do not count.
4. Based on values taken from different researchers ***(1,8,12–15)***, a count between 1.0 and 1.3×10^6 cells/mL can be accepted as the limit between infected and noninfected milk.
5. Each analyst measures the field diameter and calculates the SSF, rounding to three significant figures.

3.3. Interpretation of β-Glucuronidase Test

When this technique is used for the first time, the values obtained must be compared with those obtained by the SCC method, fixing the somatic cell level for infected and noninfected animals and between different degrees of infection (*see* **Note 4**). Based on this comparison, the β-glucuronidase activity levels can then also be established for the same categories described above. Values between 0 and 100 U of β-glucuronidase are determined and a value of 15 U is fixed as the limit between infected and noninfected milk ***(1)***.

Next, the false-negative and -positive values obtained by comparing the β-glucuronidase method with the SCC method must be identified. Thus false positives of 6% were established for goat milk (samples showing low SCC levels but higher β-glucuronidase levels than the established limit owing to false-positive results or other factors; *see* **Note 5**) ***(1)***. On the other hand, false negatives of 12% were thus established (samples showing high SCC levels but with β-glucuronidase levels lower than the established limit) ***(1)***.

This technique will allow identification of clinically or subclinically infected animals. Thus interference and errors produced by counting the normal nonleukocytic cell-like particles characteristic of goat milk will be eliminated.

4. Notes

1. It is advisable to remove milk fat from the milk samples before analysis by cool centrifugation. The grease remains in the upper part of the tube and can be easily taken away with a clean spatula. If not removed, it will interfere with the spectrophotometer measure.
2. It is convenient to include a blank for each sample of analyzed milk that contains the same milk sample but is not incubated, because each sample has a different fat content and therefore the fat taken away from each one will be different.
3. Use disposable plastic gloves to handle stains.

4. The California Mastitis Test can also be used for comparing with the β-glucuronidase method. It is faster but less reliable.
5. The number of somatic cells in milk may be increased when goats are in different lactation stage, are subjected to stress, or have undergone recent parturition.

References

1. Oliszewski, R., Núñez de Kairúz, M. S., González, S. N., and Oliver, G. (2002) Assessment of β-glucuronidase levels in goat's milk as an indicator of mastitis: comparison with other mastitis detection methods. *J. Food Prot.* **65,** 864–866.
2. Parisi, A., Sanna, A., and Caria, A. (2000) Raw materials: quality and pricing. Proceedings of the Conference on Development Strategy for the Sheep and Goat Sector. Bulletin of the International Dairy Federation 354/2000, Brussels, Belgium.
3. Shearer, J.K. and Harris, B. Jr. (1992) Mastitis in dairy goats. University of Florida, Gainesville, DS **85**.
4. Dulin, A. M., Paape, M. J., Schultze, W. D., and Weinland, B. T. (1983) Effect of parity, stage of lactation and intramammary infection on concentration of somatic cells and cytoplasmic particles in goat milk. *J. Dairy Sci.* **66,** 2426
5. Dulin, A. M., Paape, M. J., and Wergin, W. P. (1982) Differentiation and enumeration of somatic cells in goat milk. *J. Food Prot.* **45,** 435–439.
6. Wooding, F. B. P., Peaker, M., and Linzell, J. L. (1970) Theories of milk secretion: evidence from the electron microscopic examination of milk. *Nature* **226,** 762–764.
7. Perdigon, G., Medici, M., Oliver, G., and Pesce de Ruiz Holgado, A. A. (1988) Relationship between β-glucuronidase levels and cells counts in bulk milk. *Milchwissenschaft* **43,** 25–26.
8. Zeng, S. S., Popham, T., and Escobar, E. N. (1999) Seasonal variation of somatic cell count and chemical composition in bulk tank goat milk. *Dairy Food Environ. Sanitation* **19,** 685–690.
9. Schnyder, D. and Bagglioni, M. (1978) Secretion of lysosomal hydrolases by stimulated and non-stimulated macrophages. *J. Exp. Med.* **148,** 435–450.
10. Schorlemmer, H. U., Burger, R., Hylton, W., and Allison, A. C. (1977) Induction of lysosomal enzyme release from culture macrophages by dextran sulfate. *Clin. Immunol. Immunopathol.* **7,** 88.
11. Csapó, J., Csapó-Kiss, Z., and Stefler, J. (1995) Influence of mastitis on D-amino content of milk. J. Dairy Sci. 78, 2375–2381.
12. Páez, R. and Maggio, A. (1997) Leche de cabra: un camino hacia la calidad. *Infortambo* **104,** 132–134.
13. Tietze, M. and Majewski, T. (1995) Somatic cell count in cow, ewe and goat milk. Annales Univesitatis Mariae Curie-Sklodowska. *Sectio EE Zootechnica* **13,** 145–149.
14. Fahr, R. D., Schulz, J., Finn, G., Lengerken, G. V. and Walther, R. (1999) Total and differential somatic cell counts in goat milk. Variability and influencing factors. *Ausgabe G. Grosstiere/Nutztiere* **27,** 99–106.
15. Poutrel, B. and Lerondelle, C. (1983) Cell content of goat milk. California mastitis test, Coulter counter and Fossomatic for predicting half infection. *J. Dairy Sci.* **66,** 2575–2579.

49

Differences Between Biogenic Amine Detection by HPLC Methods Using OPA and Dansyl Derivates

María R. Alberto, Mario E. Arena, and María C. Manca de Nadra

1. Introduction

Biogenic amines can be formed and degraded as a result of normal metabolic activity in animals, plants, and microorganisms and are usually produced by the decarboxylation of amino acids *(1)*. Recent trends in food security are promoting an increasing search for trace compounds that can affect human health. Although they are present in fermented foods and beverages in low quantities, they exhibit interactions with normal human metabolism (e.g., having vasoactive or psychoactive properties) that justify the research on their presence in foods and the possible related toxicological effects that they may cause.

Estimation of the biogenic amines histamine, tyramine, agmatine, putrescine, and cadaverine is important not only from the point of view of their toxicity, but also because they can be used as indicators of the degree of freshness or spoilage of food. Until recently, because of the difficulty in detecting and quantifying amines reliably we have had insufficient information about their occurrence in different types of foods and beverages. These problems are related to matrix interference (e.g., the presence of free amino acids) and the low levels at which the amines are found.

Early techniques for the determination of biogenic amines in foods were based on thin-layer chromatography. More modern analytical techniques have since been developed that allow the acquisition of reliable quantitative data and better separation/resolution of various amines. The quantitative determination of biogenic amines is generally accomplished by overpressure-layer chromatography, high-performance liquid chromatography (HPLC), and gas chromatography *(1)*.

From: *Methods in Molecular Biology, vol. 268: Public Health Microbiology: Methods and Protocols*
Edited by: J. F. T. Spencer and A. L. Ragout de Spencer © Humana Press Inc., Totowa, NJ

The use of reverse-phase column and precolumn derivatization was more efficient and faster than the conventional ion-exchange techniques *(2)*. This study was conducted to evaluate two HPLC derivatization methods for quantitative determination of biogenic amines: the method described by Gonzales de Llano et al. *(3)* for amino acid analysis and the method described by Eerola et al. *(4)*.

2. Materials

2.1. Chemical

The biogenic amine standards agmatine, cadaverine, histamine, putrescine, tyramine and 1,7-diaminoheptane were from Sigma. All solvents used in the derivatization process and in the chromatographic separation were HPLC quality.

2.2. OPA Method

1. Reverse-phase (RP)-HPLC.
   a. An ISCO HPLC system (Lincoln, NE).
   b. Model 121 fluorimeter (340 nm excitation filter and 425 nm emission filter).
   c. A Waters Novapack C18 column, 3.9 × 150 mn, 4-μm particle size for the stationary phase, with a flowrate of 1.5 mL/min (*see* **Note 1**).
2. Solvents.
   a. For derivatization: 200 mg *o*-phthaldialdehyde (OPA) in 9 mL methanol, 1 mL 0.4 *M* sodium borate, pH 10, and 160 μL 2-mercaptoethanol (MCE; *see* **Note 2**).
   b. For separation: A: methanol, 10 m*M* sodium phosphate buffer, pH 7.3, and tetrahydrofuran (19:80:1); B: methanol and 10 m*M* sodium phosphate buffer, pH 7.3 (80:20).

2.3. Dansyl Method

1. Gilson system connected to a Gilson 118 UV detector at 254 nm.
2. A reversed-phase Phenomenex ODS2 column 4.6 × 300 mn id, 4 mm particle size, was used for the stationary phase, with a flowrate of 1.0 mL/min.
3. For deivatization dansyl chloride solution: 5 mg dansyl chloride in 0.5 mL acetone.
4. For separation: Solvent A: 0.1 *M* ammonium acetate; solvent B: acetonitrile.

3. Methods

3.1. OPA Method (see *Note 3*)

3.1.1. Derivatization

1. A standard solution of biogenic amines was prepared by dissolving each amine into a 0.1 *N* HCl solution to reach a concentration of 2.5 μmol/mL.
2. 50, 100, 200, and 500 μL of these solutions were adjusted to 25 mL with 0.4 *M* borate buffer, pH 10, and filtered through a 0.45-μm filter.
3. Standards were derivatized prior to column injection as follows: 50 μL of sample were reacted with 50 μL OPA/MCE reagent for exactly 1 min, and 25 μL of this solution were immediately injected (*see* **Note 4**).
4. Solvent gradient conditions were as follows: 8 min (20% B); 8 min (30% B); 12 min (40% B); 16 min (80% B); 6 min (100% B) and 12 min (20% B). *See* **Note 5**.

3.1.2. Quantification

The external standard method was used (*see* **Figs. 1** and **2** and **Note 6**).

3.2. Dansyl Method (see Note 7)

3.2.1. Derivatization

1. Standards were derivatized prior to column injection as follows: dansyl derivatization was performed with the addition of 100 μL 2 *N* sodium hydroxide solution, 150 μL saturated sodium bicarbonate and dansyl chloride solution.
2. The reaction mixture was incubated at 40°C for 45 min.
3. After incubation the residual dansyl chloride was removed by addition of 50 μL ammonia.
4. After 20 min the sample was adjusted to 2.5 mL with acetonitrile.
5. The sample was filtered with a 0.22-μm filter.
6. The precolumn derivatization and the column apparatus were at 40°C.
7. Standard curves were prepared with pure compounds at different concentrations (*see* **Note 8**).
8. Solvent gradient conditions began with 50% B and ended with 90% B over 19 min (*see* **Note 9**).

3.2.2. Quantification

The internal standard (1,7-diaminoheptane) method was used (*see* **Fig. 3** and **Note 10**).

3.3. Statistical Analysis

To validate the methods, the MINITAB Student's test was used. Five replicate determinations were carried out.

4. Notes

1. The precolumn derivatization and the column apparatus were at room temperature.
2. OPA/MCE reagent was not used within 24 h of preparation.
3. In OPA method, to achieve a better separation between amines and between amines and acidic and neutral amino acids, the time of analysis was sacrificed (50 min). The acidic and neutral amino acids are those present at higher levels than amines eluted first and into the chromatogram. However, the simple preparation of the sample and the rapid derivatization of the amines considerably reduce time and effort.
4. The formation of *o*-phthaldialdehyde derivatives was performed automatically with reactant solution 100–150-folds higher in concentration than the amino nitrogen. The determination of amino nitrogen was performed using the method described previously *(5)*.
5. The entire gradient cycle lasted for 62 min, including the time necessary for stabilization of the column after each injection. The analysis time was only 50 min in the OPA method.
6. Quantification by the external standard method, used in the OPA method, is based on the linearity of the detector response: a double concentration of product led to a peak of double area. The fluorescence exhibited a linear correlation with the concentration in the range 1–10 mg/L. A good linear regression between peak area and concentration for each biogenic amine was obtained, with correlation coefficients ranging from 0.985 to 0.998

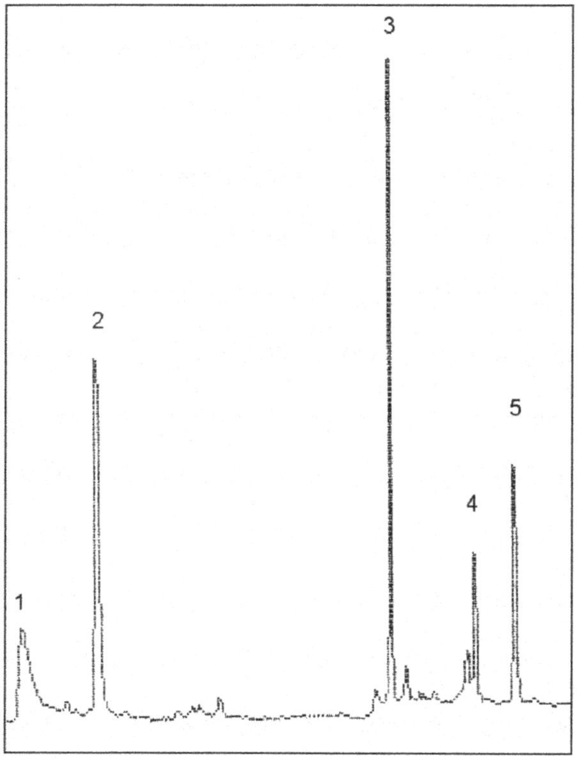

Fig. 1. HPLC profile of amines in standard solution (OPA method). 1, agmatine; 2, histamine; 3, tyramine; 4, putrescine; and 5, cadaverine. The elution time was between 11.23 and 42.27 min. Agmatine was the first amine eluted, followed by histamine, tyramine, putrescine, and cadaverine. A relative standard deviation (RSD) for retention time was less than 4.00%. An RSD of less than 6.00% for the response factor was obtained. Agmatine and putrescine had an RSD for the response factor greater than average (2.99%) (*see* **Notes 11–13**).

for the OPA method. This confirms the accuracy of this method for determining biogenic amine content.

7. Some modifications of this procedure were carried out: inclusion of agmatine, omission of centrifugation by filtration using a filter of 0.22 μm to avoid microbial spoilage, increase of column length (higher plate number), and use of a UV detector.
8. In the Dansyl method the linearity was tested by dansylating the standard solutions at different concentrations. Dansylated amines showed linear responses in the concentration range 1–10 mg/mL, with correlation coefficients ranging from 0.980 to 0.998. The greatest variation in results was observed in the case of cadaverine.
9. The total run time was 35 min including the washing time. Washing was essential to maintain column performance in the Dansyl method.

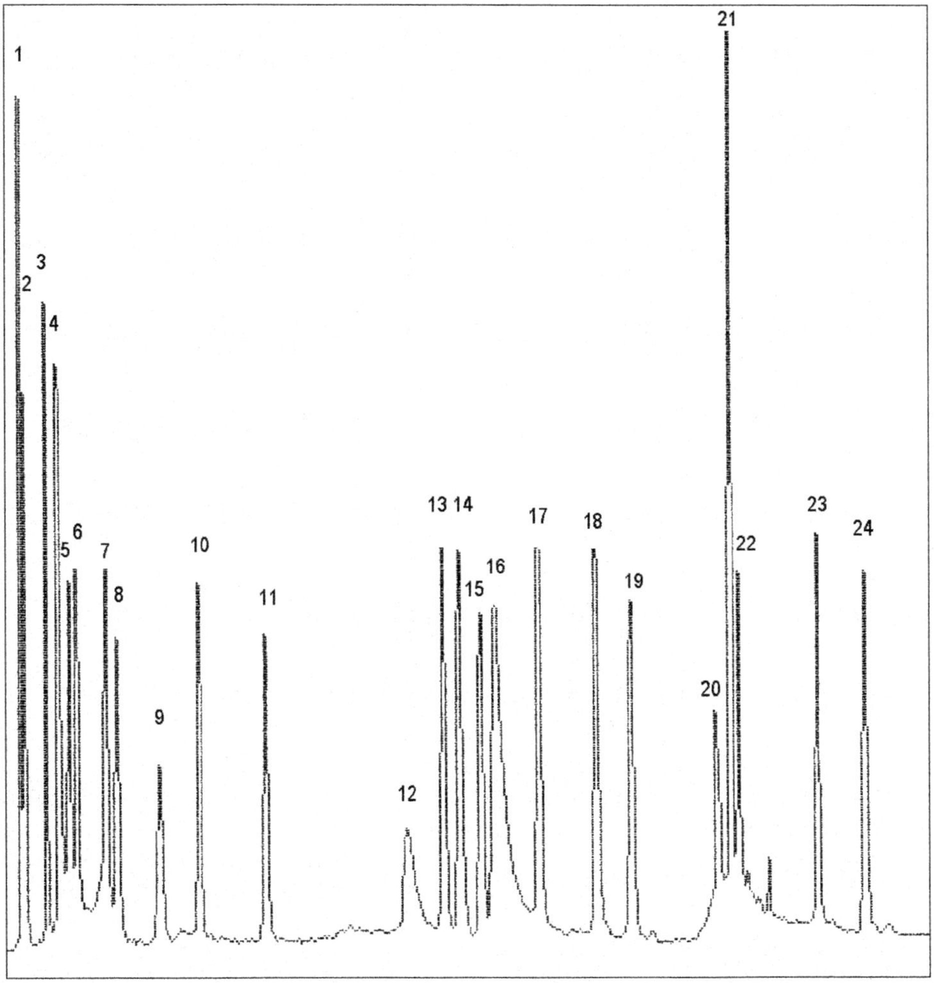

Fig. 2. HPLC profile of amines and amino acids (OPA method). 1, aspartic acid; 2, glutamic acid; 3, asparagines; 4, serine; 5, glutamine; 6, histidine; 7, glycine; 8, threonine; 9, arginine; 10, alanine; 11, tyrosine; 12, agmatine; 13, methionine; 14, valine; 15, tryptophan; 16, histamine; 17, phenylalanine; 18, isoleucine; 19, leucine; 20, ornithine; 21, tyramine; 22, lysine; 23, putrescine; 24, cadaverine. Using the OPA method and beginning the gradient with 6 min (0% B) and 11 min (15% B), it is possible to separate and quantify amines and amino acids from a sample in a single chromatogram (*see* **Note 14**).

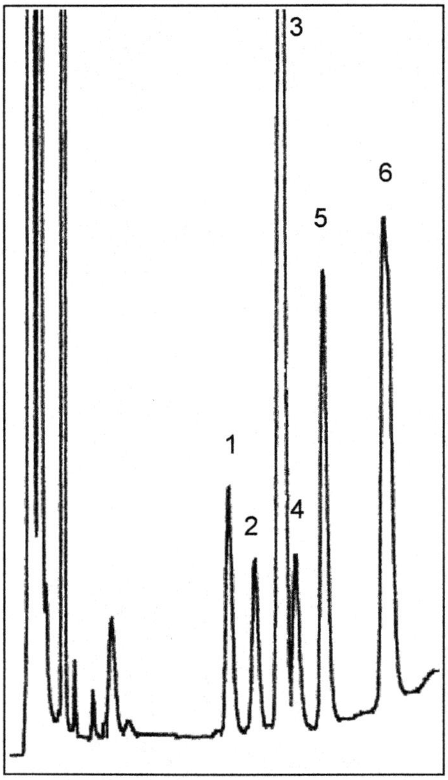

Fig. 3. HPLC profile of amines in standard solution (Dansyl method). 1, putrescine; 2, cadaverine; 3, agmatine; 4, histamine; 5, 1,7 diaminoheptane; and 6, tyramine. The elution time lay within the range of 12.90 to 17.24 min. The elution order was putrescine, cadaverine, agmatine, histamine, and tyramine. The internal standard 1,7 diaminoheptane eluted between histamine and tyramine. Values for the relative standard deviation (RSD) ranged from 2.08 to 5.30%. An RSD ranging from 1.30 to 3.50% for the response factor was obtained. Putrescine and cadaverine had an RSD greater than the average (2.30%; *see* **Notes 11–13**).

10. In the Dansyl method, an internal standard was used. It involves a compound not present in the sample as an internal standard. Internal standardization compensates for variations in conditions during sampling and derivatization, as well as for variations in injection volumes and retention times during the chromatographic run.
11. The peaks of the biogenic amines were satisfactorily resolved, and there were no interfering peaks, so simple observation of the chromatograms suggests the presence of these compounds.
12. Without derivatization the amines would elute as broad peaks shown by fluorescence detection. UV absorbance detection is only possible for the heterocyclic and aromatic amines; therefore derivatization is necessary for the detection of aliphatic amines and for increased sensitivity.

13. Programmed elution with a polarity gradient was necessary to obtain optimum separation and quantification of the corresponding amines owing to the wide range of polarities of these molecules.
14. Our results using the OPA method point out the possibility of obtaining good resolution for concentrated samples of biogenic amines and amino acids, without the interference in the analysis from the other less concentrated biogenic amines, which are next to these compounds in the chromatogram.

References

1. Halász, A., Baráth Á., Simon-Sarkadi, L., and Holzapfel, W. (1994) Biogenic amines and their production by microorganisms in food. *Trends Food Sci. Technol.* **5,** 42–49.
2. Marcé, R. M., Callull, M., Guasch, J., and Borrull, F. (1989) Determination of free amino acids in wine by HPLC using precolumn derivatization with phenylisothiocyanate. *Am. J. Enol. Vitic.* **40,** 194–198.
3. Gonzalez de Llano, D., Polo, M. C., and Ramos, M. (1991) Production, isolation and identification of low molecular mass peptides from blue cheese by high performance liquid chromatography. *J. Dairy Res.* **58,** 363–372
4. Eerola, S., Hinkkanen, R., Lindfors, E., and Hurvi, T. (1993) Liquid chromatographic determination of biogenic amines in dry sausages. *J. AOAC Int.* **76,** 575–577.
5. Doi, E., Shibata, D., and Matoba, T. (1981) Modified colorimetric ninhydrin methods for peptidase assay. *Anal. Biochem.* **118,** 173–184.

VIII

Reviews

50

Microflora of the Gastrointestinal Tract

A Review

Wei-Long Hao and Yuan-Kun Lee

1. Introduction

The mucosal surface of the human gastrointestinal (GI) tract is about 200–300 m^2 and is colonized by 10^{13-14} bacteria of 400 different species and subspecies. Savage *(1)* has defined and categorized the gastrointestinal microflora into two types, autochthonous flora (indigenous flora) and allochthonous flora (transient flora). Autochthonous microorganisms colonize particular habitats, i.e., physical spaces in the GI tract, whereas allochthonous microorganisms cannot colonize particular habitats except under abnormal conditions. Most pathogens are allochthonous microorganisms; nevertheless, some pathogens can be autochthonous to the ecosystem and normally live in harmony with the host, except when the system is disturbed *(2)*.

The prevalence of bacteria in different parts of the GI tract appears to be dependent on several factors, such as pH, peristalsis, redox potential, bacterial adhesion, bacterial cooperation, mucin secretion, nutrient availability, diet, and bacterial antagonism. Because of the low pH of the stomach and the relatively swift peristalsis through the stomach and the small bowel, the stomach and the upper two-thirds of the small intestine (duodenum and jejunum) contain only low numbers of microorganisms, which range from 10^3 to 10^4 bacteria/mL of the gastric or intestinal contents, mainly acid-tolerant lactobacilli and streptococci. In the distal small intestine (ileum), the microflora begin to resemble those of the colon, with around 10^7–10^8 bacteria/mL of the intestinal contents. With decreased peristalsis, acidity, and lower oxidation-reduction potentials, the ileum maintains a more diverse microflora and a higher bacterial population *(3)*. Probably because of slow intestinal motility and very low oxidation-reduction potentials, the colon is the primary site of microbial colonization in humans. The colon harbors tremendous numbers and species of bacteria. However, 99.9% of colonic microflora are obligate anaerobes.

From: *Methods in Molecular Biology, vol. 268: Public Health Microbiology: Methods and Protocols*
Edited by: J. F. T. Spencer and A. L. Ragout de Spencer © Humana Press Inc., Totowa, NJ

2. Effects of Gastrointestinal Microflora

The most important beneficial effect of the indigenous microflora is to make it more difficult for exogenous pathogenic bacteria to colonize the GI tract and cause disease, a phenomenon known as colonization resistance *(4)*. Indigenous microflora create this barrier effect by occupying available habitats (adhesion sites) at the mucosal level, competing for metabolic substrates and the production of regulatory factors such as short-chain fatty acids and bacteriocins.

On the other hand, indigenous microflora may have potentially deleterious influences on the host's health. There are strong links among dietary factors, the metabolic activities of the indigenous GI microflora, and bowel cancer *(5)*. In fact, it is well known that indigenous GI bacteria transform certain dietary substances into precarcinogens or carcinogens.

Colonization of bacteria is a prerequisite to affect the host, and the process of colonization is directly dependent on the ability of a resident to adhere to the substratum.

3. Adhesion

Adhesion or adherence is defined as the measurable union between a bacterium and a substratum. A bacterium is said to have adhered to a substratum when energy is required to separate the bacterium from the substratum *(6)*.

Adhesion of bacteria to intestinal mucosa is often recognized as a prerequisite for colonization of the human GI tract. Therefore, adhesion properties are important for both pathogenic bacteria and bacteria belonging to the normal human gut microflora.

The study of adhesion will give us better insight into the nature of the interaction between bacteria and cell surfaces. One particular goal in adhesion studies is to be able to inhibit or promote adhesion according to a particular need. Detailed knowledge of the characteristics of adhesins and receptors will provide new approaches to the prevention of serious bacterial infections by interfering with the adhesion process. For example, the bacterial lectin that serves as an adhesin (i.e., carbohydrate-binding adhesin) can be inhibited either by antibodies or by relatively high concentrations of soluble carbohydrates specific for the lectin *(7)*. Moreover, one can screen and characterize certain types of probiotic to exclude specific pathogens or groups of pathogens based on adhesin and receptor properties *(8)*.

3.1. Animal Cell Surface

All animal cell membranes have common compositional and organizational features. Receptors for bacterial adhesions are found in all three classes of membrane constituents, namely, integral, peripheral, and cell surface coat components.

Integral membrane constituents include glycolipids and glycoproteins. Chemically, the glycolipids are either neutral or acidic. The acidic glycolipids contain sialic acids (i.e., gangliosides), which contribute significantly to the net negative charge of the animal cell surface. Virtually all proteins of animal membranes are glycosylated and are similar to soluble glycoproteins *(9)*. There are no significant differences in the overall amino acid composition and carbohydrate content. The monosaccharide constituents usually include the hexoses D-galactose and D-mannose and the methylpentose

L-fuctose, *N*-acetylhexosamines, and the sialic acids. In addition, two types of carbohydrate-peptide linkages are common to both groups of glycoproteins: the *N*-glycosyl bond between *N*-acetyl-D-glucosamine and asparagines (*N*-acetyl-D-glucosaminyl-asparagine) and the *O*-glycosidic bond between *N*-acetyl-D-galactosamine and a serine or threonine of the protein polypeptide backbone.

The peripheral proteins and glycoproteins are anchored to the surface of the membrane by weak ionic interactions or by hydrogen bonding with integral constituents of the cell membranes. Fibronectin is a component of peripheral membrane constituents. The structure of fibronectin is characterized by the presence of several distinct domains with binding sites for specific ligands, such as heparan sulfate, collagen, and hyaluronic acid ***(10)***. Numerous bacteria are able to bind to membrane-associated fibronectin.

The cell coat is a substantial layer of carbohydrate-containing materials of variable thickness outside but in close association with the plasma membrane. In healthy human beings, mucosal cells of the respiratory, GI, and urinary tracts are coated with a layer rich in highly sialylated, high molecular weight glycoproteins known as mucins. For a bacterium to colonize tissues, it must either bind to or penetrate the cell coat.

3.2. Bacterial Cell Surfaces

The surface composition of bacteria may vary with the stage of the bacteria, the medium composition, and the presence of antimicrobial agents, all of which may influence adhesin function directly or indirectly. In most cases, the bacterial adhesions are assembled and must dock or anchor on the bacterial surface before they can participate in adhesive processes.

There are four major cell "compartments" that serve to anchor adhesins onto the bacterial surface. Three compartments, namely, the cytoplasmic membrane, peptidoglycan, and the S layer, are found in both Gram-positive and Gram-negative bacteria, and one compartment, the outer membrane, is found only in Gram-negative bacteria.

Bacteria adhere to host cell targets by means of adhesins, which are proteins that recognize a defined carbohydrate sequence on host cell glycoproteins, glycolipids, or, less often, a defined protein structure. In Gram-negative bacteria, adhesins are often located on different types of appendages protruding from the bacterial surface: fimbrial, fibrillar, or curli ***(11,12)***. In Gram-positive bacteria, adhesins are usually located in the cell wall or surface coat, although fimbrial structures have been demonstrated on vaginal lactobacilli ***(13)***. A bacterium may carry one or several adhesins. The synthesis of fimbriae and adhesins is switched on and off by the bacteria, depending on the environmental conditions, a process called phase variation.

In general, bacterial adhesion to substrata involves two types of mechanisms, specific adhesin-receptor interactions and nonspecific interactions.

3.3. Specific Adhesion

Specific adhesion was defined by Ofek and Doyle ***(6)*** as an association between the bacterium and substratum that requires rigid stereochemical constraints. Specific adhesion may require hydrogen bond formation, ion-ion pairing, or the hydrophobic effect. In its simplest form, it requires the participation of two factors: a receptor and an adhesin.

A typical example of stereospecific interaction in bacterial adhesion is that between a lectin (carbohydrate-binding lectin adhesin) and a carbohydrate (on the intestinal surface) ***(14)***. A well-studied example is the type I fimbriae of *Escherichia coli*, which recognize D-mannose as the receptor site on the host mucosal surface. These adhesin-receptor interactions will lead to irreversible adhesion and may be blocked by specific carbohydrate analogs of the receptor site at the host cell surface ***(15)***. The adhesin molecules can be regulated by the bacteria under different condition. Many bacteria have evolved protein adhesins whose specificity limits their choice of econiches. The selective expression of a specific adhesion in response to given environmental demands is an important mechanism that allows the organism to proliferate and survive ***(16)***.

The facultative anaerobes have lower adhesion to Caco-2 cells after incubation in the anaerobic condition compared with the aerobic condition. It has been shown that in the aerobic condition more adhesins are expressed than in the anaerobic condition (**Fig. 1**). A recent study has shown that *Bacteroides thetaiotaomicron* can induce matrilysin production is mammalian cells, but it was not detected in germ-free mice. Matrilysin is known for its function in the repair of epithelium and adhesion modification ***(17)***.

3.3.1. Fucose and Mannose As Modulators for Adhesion

Fucose suppressed the adhesion of lactobacilli but enhanced the adhesion of the other GI bacteria on Caco-2 cells (**Fig. 2**).

We also found that the effect of fucose on the adhesion of *Lactobacillus casei shirota* on Caco-2 was owing to a decrease in the number of adhesions on the bacterial surface but an increase in the affinity of the adhesions for the receptors (data not shown). The effect of fucose on *E. coli* 11775 was owing to an increase in the number of adhesins on the bacterial cell surface. An earlier study demonstrated that the occurrence of *B. thetaiotaomicron* ***(18)*** induces the production of host GDP-L-fucose (β-D-galactoside 2-α-L-fucosyltransferase) in the ileum of germ-free mice. Although the effect and roles of fucose on intestinal bacteria are not clear ***(19)***, its effect on adhesion is obvious.

Mannose strongly enhanced (increase of 220%) the adhesion of *Bacteroides* organisms to Caco-2 cells and had no effect on lactobacilli (**Fig. 3**). The study suggests that free fucose and mannose on the intestinal mucosal surface could serve as modulators of intercellular communication between the bacteria and the host for regulation of the bacteria population on the intestinal surface.

In germ-free mice, preculturing both the *E. coli* 11775 and *B. fragilis* in mannose altered their adhesion on the small intestinal surface but not on the colon surface (Lee et al., unpublished data). This observation suggests that the properties of bacterial surfaces adhesins are different, the adhesins to small intestine surface receptors being mannose-dependent, and those for the colon being mannose-independent. It also may be possible that the free mannose had bound to the adhesins, preventing them from binding onto the small intestinal surface receptors, and thus resulting in lower concentrations of adherent mannose-cultured *E. coli.* Klemm and Schembri ***(20)*** classify the

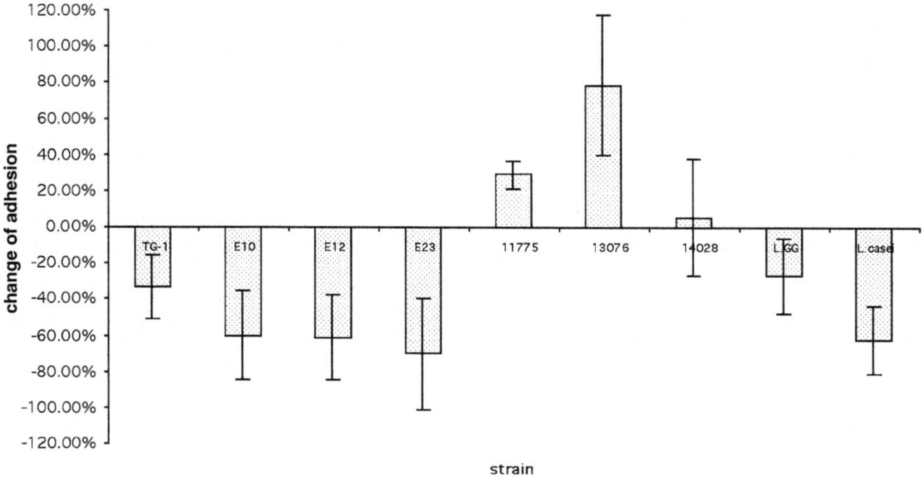

Fig. 1. Percent changes in adhesion to Caco-2 cells of *Lactobacillus*, *E. coli*, and *Salmonella* organisms incubated in the anaerobic compared with aerobic condition. TG1, *E. coli*; E10, *S. typhimurium*; E12, *S. typhimurium*; E23, *S. bellurup*; 11775, *E. coli*; 13076, *S. choleraesuis* subsp. *choleraesuis* serotype *enteritidis*; 14028, *S. typhimurium*; LGG, *L. rhamnosus*; *L. casei*, *L. casei*.

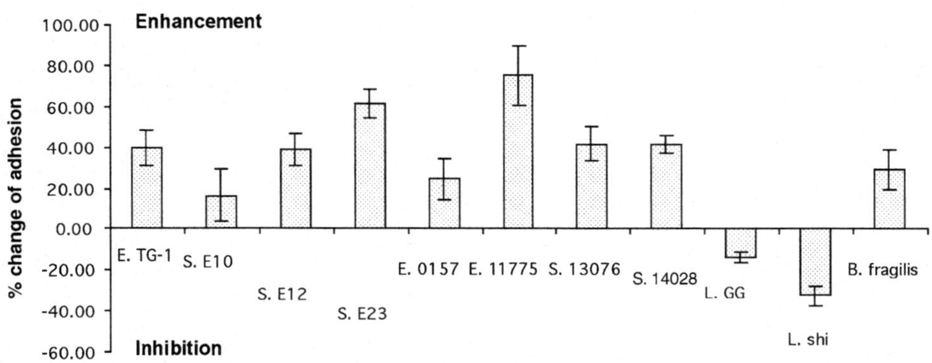

Fig. 2. Percent changes in adhesion to Caco-2 cells of lactobacilli and GI bacteria cultured with fucose (0.5% w/v) in media. ETG1, *Escherichia coli*; SE10, *Salmonella typhimurium*; SE12, *S. typhimurium*; SE23, *S. bellurup*; EO157, *E. coli*; E11775, *E. coli*; S13076, *S. choleraesuis* subsp. *choleraesuis* serotype *enteritidis*, S14028, *S. typhimurium*; LGG, *L. rhamnosus*; *L. shi*, *L. casei*; *B. fragilis*, *Bacteroides fragilis*.

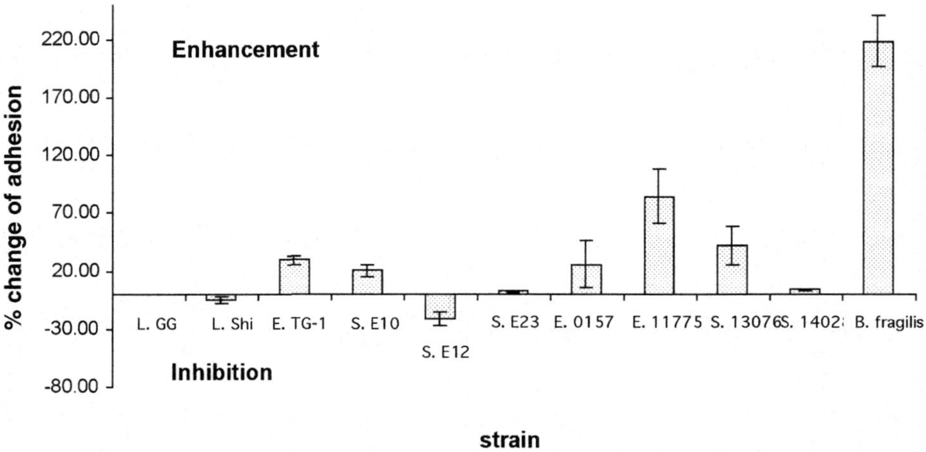

Fig. 3. Percent changes in adhesion to Caco-2 cells of lactobacilli and GI bacteria cultured with mannose (0.5% w/v). LGG, *Lactobacillus rhamnosus*; *L. shi*, *L. casei*; ETG1, *Escherichia coli*; SE10, *Salmonella typhimurium*; SE12, *S. typhimurium*; SE23, *S. bellurup*; EO157, *E. coli*; E11775, *E. coli*; S13076, *S. choleraesuis* subsp. *choleraesuis* serotype *enteritidis*; S14028, *S. typhimurium*; *B. fragilis*, *Bacteroides fragilis*.

adhesins of *E. coli* into two groups: mannose-sensitive (which involved FimH of type 1 fimbriae) and mannose-resistant. Our study found that the mannose sensitivity of adhesion depends on the site of the GI tract.

3.4. Nonspecific Adhesion

An association between a bacterium and a substratum that involves the same forces does not require a precise stereochemical fit. Nonspecific interaction basically involves the physicochemical forces, i.e., van der Waals and electrostatic forces, hydrogen bonding, and hydrophobic bonding.

4. In Vitro Models for the Study of Adhesion

Adhesion is difficult to study in vivo, and therefore many in vitro methods have been developed.

4.1. Human Intestinal Cell Lines

Cultured cell lines from human intestinal origin are commonly used to study bacterial adhesion, because the cell lines mimic the epithelium of the intestine. Caco-2 and HT-29 are the most used cell lines; they are known to display a typical enterocytic differentiation. Although the cell lines originate from the colon, the cells are able to differentiate into enterocytes *(21)*. The differentiation and polarization of Caco-2 cells is spontaneous, without the requirement of inducers, but the differentiation of the HT-29 cell line only occurs by replacing the glucose in the medium with galactose *(21)*. Another good choice of a human intestinal cell line is HT-29-MTX, a stable mucus-secreting subpopulation of HT-29 cells created by adapting the cell line to methotrexate *(22)*. The time-course for this cell line to differentiate is much longer, 30 d. These valuable tools share the morphological and physiological characteristics of normal enterocytes and permit studies dealing with adhesion of enteropathogens.

4.2. Human Intestinal Mucus

In healthy human beings, the intestinal epithelial surface is covered by a layer of mucus, comprised mainly of mucin (mucus glycoprotein). Mucus glycoproteins are synthesized and secreted from the salivary glands, esophagus, stomach, small and large intestine, gallbladder, and pancreatic ducts.

The mucins are composed of glycoprotein monomers linked through disulfide bridges. Thus the mucins are characterized by their high molecular weight and consist of a number of carbohydrate side chains, composed of *N*-acetylgalactosamine, *N*-acetylglucosamine, galactose, fucose, and sialic acid, attached to a protein core *(23)*. The presence of sialic acids and esters of sulfate makes the mucus more viscous and less susceptible to bacterial attack. There is a core region containing *N*-acetylgalactosamine linking the oligosaccharide side chain to the protein core; a backbone region often branched, of repeating D-galactose and *N*-acetylglucosamine; and a peripheral region at the nonreducing end where the terminal sugar is responsible for the antigenicity of the mucin *(24)*. The ratio of mucin constituent protein to carbohydrate is about 1:4 by weight.

Woods and coworkers *(25)* found that pili mediated the attachment of *Pseudomanas aeruginosae* to buccal epithelial cells. Pili have protein subunits, with a characteristic *N*-methyl-phenylalanine (NMetPhe) residue, that are common to *Vibrio cholera*, *Neisseria gonorrhoea*, and *B. nodosus*. The same authors suggested that outer menbrane proteins and flagellae account for non-pilus-mediated adhesion, although specific receptors have not been defined.

Using mucin as in vitro model, He and coworkers *(26)* reported that the mucosal adhesive property of bifidobacteria was reduced with the aging of the host, and it was suggested that reduced adhesive ability causes the low levels of bifidobacteria found in aging people. An earlier study showed that infant-type bacteria adhered better to adult mucin than to infant mucin and that adult-type bacteria adhered better to infant mucin than to adult mucin. This finding suggested that there are many other factors such as immunity that control the intestinal flora *(27)*.

Immobilized human intestinal mucus glycoproteins have been used as substrata for lactobacilli and GI pathogen adherence *(28)*. Both mucus glycoproteins isolated from feces and ileostomy glycoproteins from ileostoma patients have been used as a model for intestinal mucus and typical small intestinal mucus, respectively *(29,30)*.

5. In Vivo Model for Study of Bacterial Interaction

The most widely used models are germ-free mice and antibiotic-treated mice. Although the bacterial species in the human GI tract are different from those in mice, it is likely that the principles involved in bacterial competition within the mouse gut hold for interactions in the human gut as well. Freter et al. *(31)* established a continuous-flow (CF) culture, i.e., inoculating the mouse flora into a CF culture within an anaerobic chamber, and maintaining the entire ecosystem of the culture in equilibrium for months. The CF culture was composed of anaerobic species at about the same proportions as in the conventional mouse. The CF culture model has been used to investigate the mechanisms of interaction between *Clostridium difficile* and the colonic flora *(32)*. Similar models have been used to study human flora *(33)*. Minekus et al. *(34)* established a computer-controlled system to simulate conditions in the large intestine. This system showed a stable microflora population after inoculation with human fecal samples, but it could not mimic colonization in the intestine.

6. Interaction Between the GI Tract and Lactobacilli

In vitro studies have shown that lactobacilli were able to compete with many pathogenic bacteria for adhesion. (**Fig. 4**),

In addition, lactobacilli produce antagonistic substances including organic acids such as lactic acid, low molecular weight antimicrobial substances such as reuterin, and high molecular weight bacteriocins. All these substances are able to suppress growth and interfere with the adhesion of pathogenic strains of bacteria. Antimicrobial activity against *Salmonella enterica* serovar *typhimurium* (*S. typhimurium*) adhesion and invasion of the Caco-2 cell line by strains of *Lactobacillus* GG, *L. acidophilus* LB, and *L. acidophilus* LA1 was documented *(35)*.

Heinemann and co-workers *(36)* noted that a surface binding protein from *L. fermentum* RC-14 inhibited the adhesion of *Enterococcus faecalis* 1131. Interestingly, Mack and co-workers *(37)* found that the ability of *L. plantarum* 299v to inhibit adherence of pathogenic *E. coli* to HT-29 cells was mediated through their ability to increase expression of MUC2 and MUC3 intestinal mucins.

Other than the adhesion-inhibiting activity mentioned above, it was also found that *Lactobacillus* whole cells and cell wall fragments were able to exclude pathogens competitively *(38)*. In addition, Coconnier et al. *(39)* reported that *L. acidophilus* could prevent, by steric hindrance, the attachment of enterogenic *E. coli* to enterocyte-like Caco-2 cells. It was suggested that the strong antiadhesion activity of *L. crispatus* JCM 8779 against *E. faecalis* appears to be a combined effect of both bactericidal activity and competition for attachment site. Osset et al. *(40)* showed that strains of *Lactobacillus* (especially hemagglutination groups III) were able to block (by exclusion, competition, and displacement) uropathogen adherence, although this was strain-dependent, indicating that both lactobacilli and uropathogens compete for receptors on vaginal epithelial cells. Using mucins extracted from human feces as a model for intestinal mucus, the adhesion of *S. typhimurium* was significantly inhibited by *L. johnsonii* LJ1 and *L. casei shirota (41)*. Lactobacilli have been demonstrated to possess surface adhesins similar to those on bacterial pathogens and are thus compete for the specific receptors on the mucosal surface. For example, Neeser et al. *(42)* reported that *L. johnsonii* La1 shares carbohydrate binding specificities with several enteropathogenic bacteria, and Mukai et al. *(43)* proposed that *L. reuteri* strains share glycolipid specificity with *Helicobacter pylori*. These findings confirm that the bacteria itself can potentially block the adhesion of pathogens to the intestinal mucosal surface.

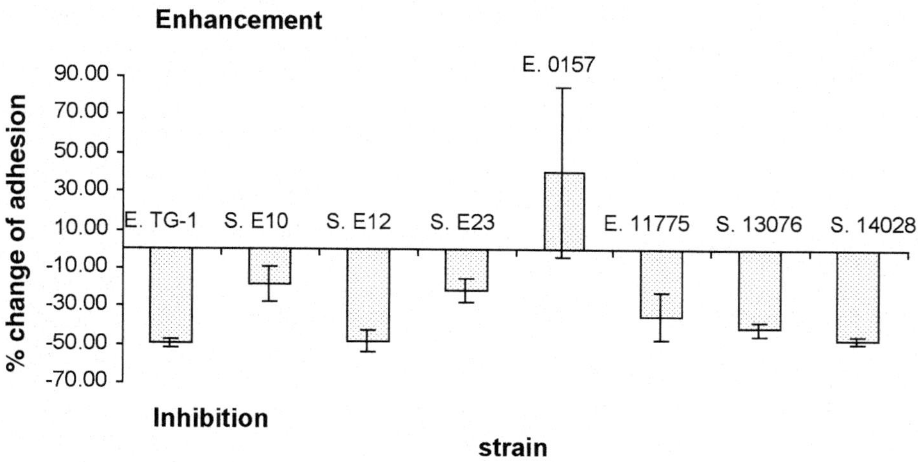

Fig. 4. Competition for adhesion between *Lactobacillus rhamnosus* GG and GI bacteria on Caco-2 cells. ETG1, *Escherichia coli*; SE10, *Salmonella typhimurium*; SE12, *S. typhimurium*; SE23, *S. bellurup*; EO157, *E. coli*; E11775, *E. coli*; S13076, *S. choleraesuis* subsp. *choleraesuis* serotype *enteritidis*; S14028, *S. typhimurium*.

Lactobacilli have been shown to inhibit bladder cancer growth in vivo ***(44)*** and to prevent recurrence of human superficial bladder cancer ***(45)***. In recent studies lactobacilli have been shown to induce production of interleukin-6 (IL-6), granulolyte macrophage colony-stimulating factor, and IL-8, which are cytokines that mediate potent immune responses ***(46)***. Our study shows that *L. casei shirota* can help in the recovery of the total population of microbes in ampicillin-treated mice. The total population in the duodenum and ileum increases steadily 1 d after introduction of *L. casei shirota*. The pathogen *E. coli* O157 did not have this property, however.

References

1. Savage, D. C. (1977) Interactions between the host and its microbes. In: *Microbial Ecology of the Gut* (Clark, R. T. J. and Bauchop, T., eds.). Academic, San Diego, pp. 277–310.
2. Trenschel, R., Peceny, R., Runde, V., et al. (2000) Fungal colonization and invasive fungal infections following allogeneic BMT using metronidazole, ciprofloxacin and fluconazole or ciprofloxacin and fluconazole as intestinal decontamination. *Bone Marrow Transplant* **26,** 993–997.
3. Tannock, G. W. (1983) Effect of dietary and environmental stress on the gastrointestinal microbiota. In: *Human Intestinal Microflora in Health and Disease* (Hentges, D. J., ed.). Academic, London, p. 517.
4. Brassart, D. and Schiffrin, E. J. (1997) The use of probiotics to reinforce mucosal defence mechanisms. *Trends Food Sci. Technol.* **8,** 321–326.
5. Gorbach, S. L. and Goldin, B. R. (1990) The intestinal microflora and the colon cancer connection. *Rev. Infect. Dis.* **12(suppl 2),** S252–S261.
6. Ofek, I. and Doyle, R. J. (1994) Principles of bacterial adhesion. In: *Bacterial Adhesion to Cells and Tissues* (Ofek, I. and Doyle, R. J., eds.). Chapman & Hall, New York, pp. 1–16.

7. Mouricout, M., Petit, J. M., Carias, J. R., and Julien, R. (1990) Glycoprotein glycans that inhibit adhesion of *Escherichia coli* mediated by K99 fimbriae: treatment of experimental colibacillosis. *Infect. Immun.* **58,** 98–106.
8. Lee, Y. K. and Puong, K. Y. (2002) Competition for adhesion between probiotics and human gastrointestinal pathogens in the presence of carbohydrate. *Br. J. Nutr.* **88,** S1–S8.
9. Sharon, N. and Lis, H. (1981) Glycoproteins: research booming on long-ignored, ubiquitous compounds. *Chem. Engr. News* **59,** 21–24.
10. Hynes, R. O. and Yamada, K. M. (1982) Fibronectins: multifunctional molecular glycoproteins. *J. Cell Biol.* **95,** 369–377.
11. Klemm, P. (1985) Fimbrial adhesion of *Escherichia coli. Rev. Infect. Dis.* **7,**321–340.
12. Hulgren, S. J., Abraham, S., Caparon, M., Falk, P., St. Geme, J. W., and Normark, S. (1993) Pilus and nonpilus bacterial adhesion: assembly and function in cell recognition. *Cell* **73,** 887–901.
13. McGrcarty, J. P. (1994) Cell surface appendages of lactobacilli. *FEMS Microbiol. Lett.* **124,** 405–410.
14. Yamamoto, K., Miwa, T., Taniguchi, H., et al. (1996) Binding specificity of *Lactobacillus* to glycolipids. *Biochem. Biophys. Res. Commun.* **228,** 148–152.
15. Neeser, J. R., Granato, D., Rouvet, M., Servin, A., Teneberg, S., and Karlsson, K. A. (2000) *Lactobacillus johnsonii* La1 shares carbohydrate-binding specificities with several enteropathogenic bacteria. *Glycobiology* **10,** 1193–1199.
16. Prince, A. (1996) *Pseudomonas aeruginosa*: versatile attachment mechanisms. In: *Bacterial Adhesion* (Fletcher, M., ed.). Wiley-Liss, New York, pp. 183–199.
17. Lopez-Boado, Y. S., Wilson, C. L., Hooper, L. V., Gordon, J. I., and Hultgren, S. J. (2000) Bacterial exposure induces and activates matrilysin in mucosal epithelial cells. *J. Cell Biol.* **148,** 1305–1315.
18. Bry, L., Falk, P. G., Midtvedt, T., and Gordon, J. I. (1996) A model of host-microbial interactions in the an open mammalian ecosystem. *Science* **273,** 1380–1383.
19. Hooper, L. V. and Gordon, J. I. (2001) Commensal host-bacterial relationships in the gut. *Science* **292,** 1115–1118.
20. Klemm, P. and Schembri, M. A. (2000). Bacterial adhesions: function and structure. *Int. J. Med. Microbiol.* **290,** 27–35.
21. Pinto, M., Robine-Leon, S., Appay, M. D., et al. (1983) Enterocyte-like differentiation and polarization of the human colon carcinoma cell line Caco-2 in culture. *Biol. Cell* **47,** 323–330.
22. Lesuffleur, T., Barbat, A., Dussaulx, E., and Zweibaum, A. (1990) Growth adaptation to methotrexate of HT-29 human colon carcinoma cells associated with their ability to differentiate into columnar absorptive and mucus-secreting cells. *Cancer Res.* **50,** 6334–6343.
23. Janet, F. F. and Gordon, G. F. (1987) Gastrointestinal mucus. In: *Physiology of the Gastrointestinal Tract* (Johnson, L. R., ed.). Raven, New York, pp. 1255–1284.
24. Mantle, M. (1996) The anti-adherent role of intestinal mucus: mechanisms and physiopathology. *Mucus Dialogue On-line* **2,** 1–6.
25. Woods, D. E., Straus, D. C., Johanson, W. G., Berry, V. K., and Bass, J. A. (1980): Role of pili in adherence of a *Pseudomonas aeruginosa* to mammalian buccal epithelial cells. *Infect. Immun.* **29,** 1146–1151.
26. He, F., Ouwehand, A. C., Isolauri, E., et al. (2001) Differences in composition and mucosal adhesion of bifidobacteria isolated from healthy adults and healthy seniors. *Curr. Microbiol.* **43,** 351–354.

27. Matsumoto, M., Tani, H., Ono, H., Ohishi, H., and Benno, Y. (2002) Adhesive property of *Bifidobacterium lactis* LYM 512 and predominant bacteria of intestinal microflora to human intestinal mucin. *Curr. Microbiol.* **44,** 212–215.
28. Ouwehand, A. C., Tuomola, E. M., Lee, Y. K., and Salminen, S. (2001) Microbial interactions to intestinal mucosal models. *Methods Enzymol.* **337,** 200–212.
29. Ouwehand, A. C., Conway, P. L., and Salminen, S. J. (1995) Inhibition of S-fimbria-mediated adhesion to human ileostomy glycoproteins by a protein isolated from bovine colostrums. *Infect. Immun.* **63,** 4917–4920.
30. Apostolou, E., Pelto, L., Kirjavainen, P.V., Isolauri E., Salminen, S. J., and Gibson, G. R. (2001) Differences in the gut bacterial flora of healthy and milk-hypersensitive adults, as measured by fluorescence in situ hybridization. *FEMS Immunol. Med. Microbiol.* **30,** 217–221.
31. Freter, R., Stauffer, E., and Cleven, D. (1983) Continuous-flow cultures as in vitro models of the ecology of large intestinal flora. *Infect Immun.* **36,** 666.
32. Wilson, K. H. and Perini, F. (1988) Role of competition for nutrients in suppression of *Clostridum difficile* by the colonic microflora. *Infect Immun.* **56,** 2610.
33. Miller, T. L. and Wolin, M. J. (1981) Fermentation by the human large intestine microbial community in an semicontiuous culture system. *Appl. Environ. Microbiol.* **42,** 400.
34. Minekus, M., Smeets-peeters, M., Bernakier, A., et al. (1999) A computer-controlled system to simulate conditions of the large intestine with peristaltic mixing, water absorption and adsorption of fermentation products. *Appl. Microbiol. Biotechnol.* **53,** 108–114.
35. Coconnier, M. H., Liévin, V., Lorrot, M., and Servin, A. L. (2000) Antagonistic activity of *Lactobacillus acidophilus* LB against intracellular *Salmonella enterica* serovar *typhimurium* infecting human enterocyte-like Caco-2/TC-7 cells. *Appl. Environ. Microbiol.* **66,** 1152–1157.
36. Heinemann, C., van Hylckama Vlieg, J. E. T., Janssen, D. B., Busscher, H. J., van der Mei, H. C., and Reid, G. (2000) Purification and characterization of a surface-binding protein from *Lactobacillus fermentum* RC-14 that inhibits adhesion of *Enterococcus faecalis* 1131. *FEMS Microbiol. Lett.* **190,** 177–180.
37. Mack, D. R., Michail, S., Wei, S., McDougall, L., and Hollingsworth, M. A. (1999) Probiotics inhibit enteropathogenic *E. coli* adherence in vitro by inducing intestinal mucin gene expression. *Am. J. Physiol.* **276,** G941–G950.
38. Chan, R. C. Y., Reid, G., Irvin, R. T., Bruce, A. W., and Costerton, J. W. (1985) Competitive exclusion of uropathogens from human uroepithelial cells by *Lactobacillus* whole cells and cell wall fragments. *Infect. Immun.* **47,** 84–89.
39. Coconnier, M.-H., Bernet, M.-F., Kerneis, S., Chauviere, G., Fourniat, J., and Servin, A. L. (1993) Inhibition of adhesion of enteroinvasive pathogens to human intestinal Caco-2 cells by *Lactobacillus acidophilus* strain LB decreases bacterial invasion. *FEMS Microbiol. Lett.* **110,** 299–306.
40. Osset, J., Bartolomé, R. M., Garcia, E., and Andreu, A. (2001) Assessment of the capacity of *Lactobacillus* to inhibit the growth of uropathogens and block their adhesion to vaginal epithelial cells. *J. Infect. Dis.* **183,** 485–491.
41. Tuomola, E. M., Ouwehand, A. C., and Salminen, S. J. (1999) The effect of probiotic bacteria on the adhesion of pathogens to human intestinal mucus. *FEMS Immunol. Med. Microbiol.* **26,** 137–142.
42. Neeser, J. R., Granato, D., Rouvet, M., Servin, A., Teneberg, S., and Karlsson, K. A. (2000) *Lactobacillus johnsonii* La1 shares carbohydrate-binding specificities with several enteropathogenic bacteria. *Glycobiology* **10,** 1193–1199.

43. Mukai, T., Asasaka, T., Sato, E., Mori, K., Matsumoto, M., and Ohori, H. (2002) Inhibition of binding of *Helicobacter pylori* to the glycolipid receptors by probiotic *Lactobacillus reuteri. FEMS Immunol. Med. Microbiol.* **32,** 105–110.
44. Lim, B. K., Mahendran, R., and Lee, Y. K. (2002) Chemopreventive effect of *Lactobacillus rhamnosus* on growth of a subcutaneously implanted bladder cancer cell line in the mouse. *Jpn. J. Cancer Res.* **93,** 36–51.
45. Aso, Y., Akaza, H., and Kotake, T. (1995) Preventive effect of a *Lactobacillus casei* preparation on the recurrence of superficial bladder cancer in a double blind clinical trial. *Eur. Urol.* **27,** 104–109.
46. Yong, J. Y., Mahendran, R., Lee, Y. K., and Bay BH. *Lactobacillus* spp. potentiate GM-CSF and IL-8 production in malignant urothelial cells. Submitted.

51

Public Health Implications Related to Spread of Pathogens in Manure From Livestock and Poultry Operations

J. Lloyd Spencer and Jiewen Guan

1. Introduction

During the 20th century, food animal agriculture grew from small operations, where livestock (cattle, sheep, and swine) and poultry (chickens and turkeys) had access to free range, to large operations where animals and poultry were concentrated and confined to feed lots or buildings. The quantity of manure produced by confinement animals in the United States has been estimated to be in excess of 61 million tons of dry matter per year *(1)*, and another report states that 1.2 billion tons of manure are produced by cattle annually in the United States (US Senate Agricultural Committee, 1998). As urban developments have come closer to livestock operations, there has been increasing public concern for the impact of the latter on public health and the environment.

Although management practices for livestock production have increased in efficiency, insufficient attention has been given to managing and utilizing wastes so that they benefit rather than pollute the environment. Animal manure includes urine and various bodily secretions such as those from the nose, vagina, and mammary glands. Dust from animals and manure may be blown from buildings by powerful fans, and manure is often piled near the animal quarters or is spread on land in solid or liquid form. Public concerns associated with disposal of animal manure include objectionable odors, flies, excessive levels of phosphorous and nitrogen, and the potential for spread of human pathogens. It has been observed that despite linkages between outbreaks of gastroenteritis in humans and livestock operations, the importance of animal manure in the spread of infectious agents tends to be underestimated *(2)*.

Slogans such as "farm to fork" and "plough to plate" have been used by the food industry to emphasize that food safety must begin on the farm. Technological advances aimed at the control of potential foodborne pathogens include the development of antimicrobial products, competitive exclusion products, and vaccines, along with improved methods for detecting infection or contamination in animals and their environment *(3–5)*.

From: *Methods in Molecular Biology, vol. 268: Public Health Microbiology: Methods and Protocols*
Edited by: J. F. T. Spencer and A. L. Ragout de Spencer

However, despite these advances, insufficient progress has been made in reducing the prevalence of potential foodborne pathogens on the farm. Poultry and other farm animals sold for slaughter frequently carry pathogens of human health concern to the killing floor of packing plants. Hazard analysis critical control point (HACCP) procedures have been implemented in killing and processing plants in an effort to prevent contamination of carcasses and meat with *Salmonella* and other foodborne pathogens. Although these measures have proven beneficial for reducing *Salmonella* contamination *(6)*, once the equipment in the killing plant becomes contaminated, it is virtually impossible to prevent further contamination from occurring. In addition to contamination of meat, there is a concern for contamination of vegetables, fruits, spices, and water on the farm. Thus, to be successful in protecting food from contamination with pathogens, HACCP principles must also be more fully implemented on the farm.

Biosecurity is the term used with respect to livestock and poultry production to include measures that can or should be taken to prevent potential pathogens from gaining entrance to production facilities *(7)*. However, despite the implementation of these measures, a high percentage of poultry and animal production facilities are contaminated with food- and waterborne pathogens. Many of these infectious agents persist on farms, and the practice of spreading raw manure on land is no doubt a contributing factor. Crops produced for animal feed as well as mice, flies, wild birds, dust, and water may become contaminated with pathogens in the environment and serve as vehicles to carry these agents back into animal quarters. Thus, biosecurity should include measures to prevent pathogens from exiting as well as entering animal facilities.

The following sections give further information on the relevance and spread of specific pathogens found in animal manure that could impact on human health. Measures that could be taken on farms to sanitize wastes and prevent environmental contamination are also considered.

2. Communicable Diseases Related to Management of Animal Manure

Animal manure has been found to be the source of more than 100 pathogens, including bacteria, parasites, and viruses, that could be transmitted from animals to humans *(8)*. Many of these pathogens survive well in the environment and frequently persist in livestock operations. Infection of livestock and poultry with such pathogens may not produce clinical disease. In fact, animals act as asymptomatic carriers in amplifying the pathogenic agent that goes back into the environment in a cyclic manner. In most cases the producer would be unaware that the manure or bedding from animal quarters contains these pathogens. Since there is no official program in place to monitor animal manure and bedding routinely for pathogens of concern to public health, there is little incentive on the part of livestock producers to implement management practices to eliminate them from animal manure before it is spread on land.

2.1. Salmonella *spp.*

Salmonella serotypes of food-safety concern are often referred to as paratyphoid Salmonellae. More than 2300 serovars have been described, and many infect warm-blooded animals *(3,9)*. Some of the serotypes of *Salmonella* frequently isolated from humans with symptoms of enterocolitis have been commonly detected in poultry meat.

In a US study, *Salmonella enterica* were isolated from 22.8% of the broiler chickens and from 1.8% of the beef carcasses ***(10)***. Salmonellosis may affect 1 to 5 million people each year in the United States ***(3)***. Rarely, *Salmonella* infection in humans leads to chronic conditions such as reactive arthritis, osteomyelitis, cardiac inflammation or neural disorders ***(11,12)***.

Young livestock and poultry are most susceptible to *Salmonella* infection, but usually the infected stock do not develop symptoms of disease. Infected stock, whether asymptomatic or sick, may shed high numbers of *Salmonella* organisms into the feces. It is not difficult to understand then that wastes from poultry and other animals, if not properly treated, can contribute to the spread of *Salmonella* ***(13)***. Surveys reported in 1991 found that the prevalence of *Salmonella* spp. was 52.9% in Canada's registered layer flocks and 76.9% in registered broiler flocks ***(14,15)***. *Salmonella* spp. were found in the manure from about 70% of the layer flocks surveyed in Southern California ***(16)***. In a survey in New York State, *Salmonella* organisms were isolated from poultry houses on all 28 farms included in the study ***(17)***. In one poultry house, *S. enteritidis* phage type 2 was isolated from chickens, manure, and mice. Other studies have shown that mice play a particularly important role in the epizootiology of *Salmonella* ***(18)***. In studies in Denmark, ***(19)*** *S. typhimurium* DT12 clone was found to have persisted in a piggery. The organism was isolated from waste slurry and from agricultural soil 14 d after slurry had been applied. It was suggested that contamination of soil could have been a factor in the persistence of the organism in the herd environment. A review article reported that *Salmonella* survived for 87 d in tap water, 115 d in pond water, 120 d in pasture soil, 280 d in garden soil, and 28 mo in naturally infected avian feces ***(20)***. These findings were in agreement with another review indicating that *Salmonella* survived for 9 mo in soil and for 3 yr in animal feces ***(21)***. *Salmonella* organisms have survived for 10 mo in slurry tanks of cattle manure and for 3 mo in soil after spreading of manure, plowing, and seeding ***(22)***. The fact that *Salmonella* organisms survive in the environment makes this pathogen a constant concern for poultry and livestock operations.

There is also growing concern over the emergence of antibiotic-resistant strains of *Salmonella*, and it is believed that widespread use of antibiotics in livestock production, for both preventative and treatment purposes, has contributed to the development of this problem. Although it is not known how the multiple antibiotic-resistant strain *S. typhimurium* DT 104 evolved, 37% of 323 isolations of salmonellae from swine in the United Kingdom in 1998 were this organism ***(21)***.

In addition to poultry and red meat, other sources of *Salmonella* are brought into domestic kitchens. *Salmonella* has been isolated from lettuce and other fresh vegetables, and large outbreaks of disease have been traced to eating cantaloupes and other fruits. It was suggested that the fruit may have become contaminated by irrigation with polluted water or by distributing products under poorly hygenic conditions ***(11)***. In experiments in which tomato plants were inoculated with salmonellae during the flowering stage, the organisms were found to survive in or on the developing fruits ***(23)***. This capacity to invade plant tissues indicates that simply washing and disinfecting produce may not be sufficient to eliminate *Salmonella*.

2.2. Campylobacter *spp.*

Campylobacter jejuni and *C. coli* are the species of *Campylobacter* most frequently implicated in foodborne infections. Infection in humans usually results in symptoms related to an enteritis, and about 2 million cases occur annually in the United States ***(24)***. A relationship has been found between intestinal infection with *C. jejuni* and the development of severe postrecovery autoimmune conditions including Guillain-Barré syndrome and Reiter's syndrome, a nonpurulent arthritis. The annual economic impact of Guillain-Barré syndrome in the United States was estimated to be \$0.2–1.8 billion US, and about one-third of these cases were believed to be initiated by *C. jejuni* infection ***(25)***.

As is the case with *Salmonella*, food animals, especially poultry, are usually silent carriers of *Campylobacter*. Broiler flocks often shed the organism in feces from 2 to 3 wk of age onward. Thus it is not surprising that the organism has been isolated from a variety of environmental samples including rodents, wild birds, flies, feed, farmer's boots, and chicken feathers ***(26)***. Prevalence rates of *Campylobacter* infection among flocks have been reported to range from 18% in Norway to 90% in the United States ***(25)***. These wide differences in prevalence of infection between countries may be caused by the manure management practices and biosecurity measures that are practiced on farms ***(25,26)***. *Campylobacter* organisms can survive for several months in an aquatic environment but are susceptible to drying. There is a concern that some environmental stresses on the organisms may lead to viable nonculturable forms ***(25)***.

The high prevalence of infection with *Campylobacter* in poultry flocks has important public health implications. Humans may become infected by consumption of contaminated water or milk and meat products. Infection with *C. jejuni* has been an occupational hazard for poultry plant workers ***(25)***. The emergence of plasmid-mediated, antibiotic-resistant strains of *Campylobacter* has added to the health concerns ***(27)***.

2.3. *Verotoxogenic* Escherichia coli

Escherichia coli is a natural inhabitant of the intestinal tract of mammals and birds, and large numbers are shed in feces. In environmental monitoring programs, the number of *E. coli* in water is taken as an indicator of the level of recent fecal contamination and thus the potential presence of enteric pathogens. Although most of the *E. coli* strains found in environmental specimens are not long lived and are not pathogens, the cycling of this bacterium in animals and the environment could promote changes in the bacterial genome that could potentially increase virulence ***(28)***. In the early 1980s, *E. coli* O157:H7 emerged as a pathogen that could cause debilitating human illnesses and deaths. Syndromes in affected humans may include hemorrhagic colitis, hemolytic uremic syndrome, and thrombocytopenic purpura. Disease outbreaks in humans have usually been associated with ingestion of undercooked ground beef and raw milk. However, drinking water contaminated with *E. coli* O157:H7 has also been responsible for major outbreaks of disease ***(28)***.

Fruits, vegetables, and spices may also become contaminated with the organism on farms ***(2,29)***. Although contamination may be superficial, *E. coli* O157:H7 has been found to penetrate apple tissue ***(30)***. Also, it was found to enter lettuce plants through the root system and to survive in the plant tissues ***(31)***, thereby increasing the possibility for its persistence in environmental niches.

Dairy and beef cattle are the most important reservoirs of *E. coli* O157:H7, and large numbers of the pathogen may be found in bovine feces ***(21)***. In one study ***(32)*** in which *E. coli* O157:H7 was added to feces, the organism survived for up to 56 d at 22°C and for up to 70 d at 5°C. During these survival studies, isolates retained the ability to produce toxins that cause disease. Thus, unlike other strains of *E. coli* that are short lived, *E. coli* O157:H7 can survive in adverse environments and has been found to be very tolerant to changes in pH ***(21)***. The practice of spreading cow manure slurry on pastures has created an environment that favors the persistence of *E. coli* O157:H7 ***(33)*** and could be a source of infection for animals.

2.4. Listeria

Sheep, cattle, pigs, and chickens and many other species of animals and birds are susceptible to infection with *Listeria monocytogenes* and shed the organism in their feces. *L. monocytogenes* is known for its ability to survive well in the environment, partly because of its ability to grow at temperatures as low as 4°C, an attribute not shared with the aforementioned enteric pathogens. The organism has been shown to be more resistant to changes in pH and to accumulations of ammonia in manure than were *S. typhimurium* and *E. coli* O157:H7, thus enhancing its environmental persistence ***(34)***. *L. monocytogenes* has been found in high numbers in poorly fermented silage, and this feedstuff has frequently been the source of the pathogen in on-farm outbreaks ***(21)***. In one such report, sheep fed corn silage developed listerial encephalitis, and losses from the disease stopped when feeding of the silage ceased ***(35)***.

On dairy farms, milking practices were shown to influence the odds of detecting *L. monocytogenes* in milk, and there was evidence that manure was the source of the contamination ***(36,37)***. On one Japanese farm ***(37)***, the organism was frequently isolated from milk and was also detected in manure and in soil on which the fodder plants were grown. Application of manure to the land may have led to contamination of fodder plants and subsequently to infection of cattle.

Humans have become infected with *L. monocytogenes* by eating raw vegetables that had been fertilized with sheep manure ***(2)***. Perinatal infections with this pathogen may cause a high rate of fetal mortality, and infections in immunologically compromised individuals may result in meningitis and death ***(38)***.

2.5. Cryptosporidium parvum

Cryptosporidium parvum is a protozoan parasite that infects cattle, sheep, horses, and pigs. Oocysts are the infectious stage of the parasite found outside the host; they may remain viable for several months ***(39)***. Since the oocysts are shed in feces, manure management practices have been a factor in spread of infection. A nationwide survey in the United States found that more than half of the dairy farms studied had calves that tested positive for infection with *C. parvum*, indicating a widespread distribution of infection. Although the infection was usually self-limiting, it was noted that infected neonatal calves could excrete 30 billion oocytes over a period of 2 wk

(40,41). Although animals acquired resistance with age, apparently healthy cows in one study were found to shed 18,000 oocysts/g of feces ***(42)***. The parasite has been shown to be resistant to a number of commercial disinfectants including the routine levels of chlorine used for treatment of public drinking water ***(43)***.

Humans become infected by drinking water or by swimming in recreational water contaminated with the pathogen. Usually, those infected develop a diarrheal illness, but in immunocompromised individuals, a chronic infection can develop and may be fatal. In a major outbreak in Wisconsin, 400,000 people, most of whom were healthy, became ill after they drank contaminated public water ***(44)***.

2.6. Influenza

Influenza viruses are not considered to be foodborne pathogens, but the feces shed by infected birds may be a source of infection for humans ***(45)***. Influenza viruses belong to the Orthomyxoviridae family and are classified as type A, B, or C based on antigenic differences in nucleoproteins and matrix proteins. They are further subtyped based on the antigenicity of two surface glycoproteins known as hemagglutinin and neuraminidase. Wild aquatic birds are silent carriers of influenza viruses and are a source of infection to other animal species. As the influenza viruses spread among birds and animals, there is a possibility for mutations and for exchanging of genetic material among these viruses that may lead to emergence of strains that are pathogenic for humans. The H1N1 virus, which caused the great pandemic of 1918–19, known as Spanish influenza, resulted in more than 20 million human deaths ***(46)***.

Viruses that cycle in avian populations are of type A and are usually no threat to human health. However, in 1997, an H5N1 virus that had circulated in Hong Kong's poultry population caused 18 cases of severe respiratory illness and 6 fatalities in humans. Previously, H5 viruses had only been isolated from the avian species. In the outbreak, all internal genes, including those that encoded the surface glycoproteins, were of avian origin. This was evidence that the H5N1 virus from poultry had crossed the species barrier ***(47)***. However, it has been suggested that H9N2 virus may have provided replicating genes for H5N1 ***(48)***. H9N2 viruses had been isolated in the Hong Kong area from poultry, pigs, and humans. It is important to note that small farms that sell poultry through live bird markets tend to keep multiple species of birds and animals. The movement of stock, crates, and vehicles to and from the markets would favor spread, multiplication, and mixing of influenza viruses. Thus live bird market operations are a potential reservoir for influenza viruses of concern to animal and public health.

In the 1997 Hong Kong outbreak, the human index case occurred in May, and there were no further cases until November of that year. Presumably there was a buildup of virulent virus in the poultry population between those months. It has been suggested that slaughter of all poultry in Hong Kong may have averted the continued spread of the disease to humans. What happened in Hong Kong may be analogous to what has occurred in outbreaks of highly pathogenic avian influenza in large commercial poultry operations. Typically, in those outbreaks low virulent viruses circulate in the population and then a highly virulent mutant emerges. This was the case in an outbreak in

Pennsylvania in 1983. In April of that year an avirulent H5N2 influenza virus appeared in poultry populations, and then in October of the same year, a highly virulent virus emerged that caused more than 80% mortality in some flocks. In the eradication effort, over 17 million chickens and turkeys were destroyed at a cost of over $61 million. Analyses indicated that genes of the virulent virus were derived from an avirulent virus ***(46)***.

Influenza viruses multiply in the respiratory and intestinal tract, and large numbers of virus particles may be excreted in the feces of infected birds. It was noteworthy that the H5N1 virus that caused disease in humans had been detected in bird feces in poultry markets in Hong Kong. Studies on stability showed that infectivity of H5N1 was lost within 1 d in feces dried at room temperature. However, when feces was kept moist at 25°C, the virus was detectable after 4 d, and at 4°C there was no detectable loss in moist feces after 40 d ***(45)***. Although airborne transmission of influenza viruses occurs, it has been shown that the fecal-oral transmission of H5N1 was more common ***(48)***. Fecal contamination of water would be expected to be important in the perpetuation of infection in waterfowl and shore birds and may also contribute to the spread of disease to humans. In addition, fecal contamination of poultry meat would also be a potential source of infection for humans. All these observations on the stability and spread of influenza viruses in manure indicate that farm waste management practices should be an important consideration in the prevention and control of this disease.

3. Composting for Elimination of Pathogens From Animal Wastes

As noted above, many pathogens of public health concern may be found in large numbers in animal manure. The more microorganisms such as *E. coli* and influenza viruses pass from animal to animal and from species to species, the greater is the likelihood for mutants to emerge with undesirable properties such as increased virulence and, in the case of bacteria, increased resistance to antimicrobial products ***(28,48)***. Thus, implementing manure management practices, in order to eliminate pathogens from animal wastes before land application, could greatly reduce the spread of disease to animals and humans.

In crisis situations such as outbreaks of *S. enteritidis* or influenza, the decision might be made to depopulate the animal premises. In such cases, measures should be taken to dispose of animals and manure safely. However, in practice, the importance of microbes in manure appears to be given lower priority than those in carcasses ***(2)***. In some cases carcasses may be burned to ashes to ensure complete destruction of pathogens, whereas manure from these animals, laden with the same pathogen, may be spread directly on land. The recommendation to plow the field within 24 h of applying the contaminated manure, in order to discourage birds and other scavengers ***(49)***, seems inadequate.

In addition to crisis situations, waste management continues to be a concern in routine livestock production. With the intensification of management practices, many producers have adopted slurry systems for the disposal of manure. A study in Sweden ***(22)*** reported that the spread of salmonellae in herds of cattle was greater on farms using slurry systems than on those using solid manure systems. This was attributed in part to the killing of salmonellae by heat generated in piles of manure. The traditional manure pile, consisting of a mixture of moist manure and bedding, is a crude compost pile, but conditions are not suitable for thoroughly sanitizing the wastes. The interior of such piles may reach temperatures sufficient to kill pathogens, but the exterior may be close to ambient temperatures and may favor survival of *Salmonella* and other pathogens ***(16,49)***.

A properly designed and managed composting operation has the potential for eliminating animal and human pathogens from the entire compost pile consisting of manure, animal carcasses, and other organic wastes. Composting offers a safe, economical, and environmentally acceptable alternative for disposal of animal wastes ***(50,51)***. Systems can be designed to dispose of manure and carcasses together, and the end product can be spread on land as a soil amendment. Although it is possible to compost liquid manure ***(52)***, large quantities of bulking agent must be mixed in to bring the moisture content to an acceptable level. Temperatures of 55–60°C for 3 d have been sufficient for killing most pathogens in compost ***(21,53)***. In addition to heat, enzymatic activity and ammonia production may also contribute to the sanitizing process during composting, and it was suggested that the action of composting lowered the temperature required to kill certain plant pathogens ***(54)***. Thermotolerant mutants of *Salmonella* and *E. coli* have been developed under laboratory conditions ***(55)***, but there appears to be no evidence that such mutants have emerged and caused problems in the environment during millenia of composting.

Chicken manure has been composted in high-rise cage layer barns in an effort to control flies ***(56)***. In those studies, manure was mixed with wheat straw, and the biomass reached temperatures of at least 43°C. When the biomass was turned, the temperatures rose to about 60°C and then rapidly declined. Although temperatures achieved were sufficient to control flies, temperatures of 55–60°C were not maintained long enough to ensure the killing of pathogens. This suggests the need for an improved composting design to conserve heat.

Like manure piles, some of the methods that have been developed for composting animal carcasses may not ensure complete sanitization of wastes. Sheep carcasses have been composted in bins that were filled with layers of barley straw, cured compost, dead sheep, and manure ***(57)***. In other studies, animal carcasses were buried in sawdust ***(50,58)***. In these operations the outer surfaces of the mass in the primary bins would not reach temperatures sufficient to kill pathogens. However, if contaminated materials were placed toward the center of the piles, the risk of spreading pathogens would be reduced. In addition, when the temperature declines in primary bins, transfer of the mass to secondary bins would cause a resurgence of heat production that would further reduce levels of viable pathogens. Measures that can be taken to make composting of farm animals a safe and acceptable alternative to burning or burying have been reported ***(51)***.

A static pile, passive aeration system has been used to compost manure and fish wastes ***(52,59)***. Compost piles were built over aeration pipes and remained static until the compost had matured. Since turning of the compost was not necessary to provide aeration, the system could have advantages for disease control. Using this system, compost piles were prepared that were composed of chicken litter contaminated with *S. typhimurium*. The *Salmonella* was eliminated toward the center of the piles within 10 d of composting but persisted at the surface. In subsequent studies, piles were covered with a vapor barrier fabric and then insulation was laid over the fabric. The fabric conserved moisture needed for microbial activity within the piles and protected the overlying insulation from becoming contaminated. Sufficient heat was generated throughout the entire compost pile to kill *Salmonella* and other pathogens (Spencer, unpublished data).

Anaerobic digestion of animal manure has been used for biogas production, and it appears that toxic substances produced during this process may kill pathogens. It is also possible to compost the anaerobically digested materials to complete the sanitization process ***(54)***.

4. Air Management

Air quality inside animal facilities is influenced by such factors as temperature, humidity, air circulation, and manure management practices. A buildup of dust or ammonia and other odorous chemicals in animal quarters can render animals more susceptible to disease. Many infectious agents are carried on dust particles and can travel several miles through the air. For example, in areas of high-density poultry production, virtually all flocks must be vaccinated against a number of important virus diseases, yet nonvaccinated chicks can be kept free of these and other diseases if they are raised in buildings where all air entering the facility passes through HEPA filters ***(60,61)***. These so-called filtered air positive pressure (FAPP) houses are now routinely used for production of the specific pathogen-free chickens required for research and production of biologicals. Although FAPP facilities have been considered too costly to be used by the commercial poultry industry, their effectiveness in preventing entrance of disease-causing organisms has shown the importance of air in the transmission of disease.

Biofilters have been designed for removing odors from air exiting livestock facilities ***(62)***, and it is likely that these filters also remove some pathogens. Research is needed to develop biofilters or other systems to remove pathogens efficiently from air exiting animal production buildings.

In conclusion, public health problems associated with the food- and waterborne pathogens appear to have grown with the poultry and livestock industries. However, the notion that salmonellae are ubiquitous in the natural environment should be qualified ***(11)***. Large poultry breeding companies have demonstrated that chickens can be raised totally free of *Salmonella* and other zoonotic pathogens ***(4)***. The added costs to producers of developing and implementing composting and other practices to prevent pathogens from exiting and entering animal quarters may be more than compensated for by reduced losses from diseases. In addition, public health would benefit. Proactive measures to prevent environmental pollution from agricultural operations could make it more acceptable for large production operations to be developed closer to urban centers, resulting in further economic benefits.

References

1. Anon. (1996) *Integrated Animal Waste Management*. Task Force Report no. 128. Council for Agricultural Science and Technology, Ames, IA.
2. Pell, A. N. (1997) Manure and microbes: public and animal health problem? *J. Dairy Sci.* **80,** 2673–2681.
3. Gast, R. K. (1997) *Salmonella* infections. In: Diseases of Poultry, 10th ed. (Calnek, B. W., Barnes, H. J., Beard, C. W., McDougald, L. R., and Saif, Y. M., eds.). Iowa State University Press, Ames, pp. 81–82.
4. Barrow, P. A. (2000) The paratyphoid salmonellae. Diseases of poultry: world trade and public health implications. Office International des Épizooties. *Sci. Tech. Rev.* **19,** 351–375.
5. Kurowski, P. B., Traub-Dargatz, J. L., Morley, P. S., and Gentry-Weeks, C. R. (2002) Detection of *Salmonella* spp in fecal specimens by use of real-time polymerase chain reaction assay. *Am. J. Vet. Res.* **63,** 1265–1268.
6. Olsen, S. J., Bishop, R., Brenner, F. W., et al. (2001) The changing epidemiology of *Salmonella*: trends in serotypes isolated from humans in the United States, 1987–1997. *J. Infect. Dis.* 183:753–761.
7. Zander, D. V., Bermudez, A. J., and Mallinson, E. T. (1997) Principles of disease prevention: diagnosis and control. In: *Diseases of Poultry*, 10th ed. (Calnek, B. W., Barnes, H. J., Beard, C. W., McDougald, L. R., and Saif, Y. M., eds.). Iowa State University Press, Ames, pp. 3–45.
8. Jones, P. W. (1982) A short survey of the most important pathogens found in farm animal wastes. In: *Communicable Diseases Resulting from Storage, Handling, Transport and Landspreading of Manures* (Walton, J. R. and White, E. G., eds.). Commission of the European Communities, Director-General of Information on Markets and Innovations, Bâtiment Jean Monnet, Luxembourg, pp. 139–147.
9. Brenner, F. W., Villar, R. G., Angula, F. J., Tauxe, R. V., and Swaminathan, B. (2002) *Salmonella* nomenclature. *J. Clin. Microbiol.* **38,** 2465–2467.
10. Sarwari, A. R., Magder, L. S., Levine, P., et al. (2001) Serotype distribution of *Salmonella* isolates from food animals after slaughter differs from that of isolates found in humans. *J. Infect. Dis.* **183,** 1295–1299.
11. D'Aoust, J-Y. (1994) Review paper. *Salmonella* and the international food trade. *Int. J. Food Microbiol.* **24,** 11–31.
12. D'Aoust, J-Y. (2001) Foodborne salmonellosis: current international concerns. *Food Safety Mag.* **7,** 10–17, 51.
13. Kinde, H., Read, D. H., Ardans, A., et al. (1996) Sewage effluent: likely source of *Salmonella enteritidis*, phage type 4 infection in a commercial chicken layer flock in Southern California. *Avian Dis.* **40,** 672–676.
14. Poppe, C., Irwin, R. J., Forsberg, C. S., Clarke, R. C., and Oggel, J. (1991) The prevalence of *Salmonella enteritidis* and other *Salmonella* spp. among Canadian registered commercial layer flocks. *Epidemiol. Infect.* **106,** 259–270.
15. Poppe, C., Irwin, R. J., Messier, S., Finley, G. G., and Oggel, J. (1991) The prevalence of *Salmonella enteritidis* and other *Salmonella* spp. among Canadian registered commercial chicken broiler flocks. *Epidemiol. Infect.* **107,** 201–211.
16. Riemann, H., Himathongkham, S., Willoughby, D., Tarbell, R., and Breitmeyer, R. (1998) A survey for *Salmonella* by drag swabbing manure piles in California egg ranches. *Avian Dis.* **42,** 67–71.

17. Mutalib, A., McDonough, P., Shin, S., Patten, V., and Lein, D. (1992) *Salmonella enteritidis* in commercial layer farms in New York state; environmental survey results and significance of available monitoring tests. *J. Vet. Diagn. Invest.* **4,** 416–418.
18. Henzler, D. J. and Opitz, H. M. (1992) The role of mice in the epizootiology of *Salmonella enteritidis* infection on chicken layer farms. *Avian Dis.* **36,** 625–631.
19. Baloda, S. J., Christensen, L., and Trajcevska, S. (2001) Persistence of Salmonella enterica serovar *typhimurium* DT12 clone in a piggery and in agricultural soil amended with *Salmonella*-contaminated slurry. *Appl. Environ. Microbiol.* **67,** 2859–2862.
20. Morse, E. V. and Duncan, M. A. (1974) Salmonellosis—an environmental health problem. *J. Am. Vet. Med. Assoc.* **165,** 1015–1019.
21. Nicholson, F. A., Hutchison, M. L., Smith, K. A., Keevil, C. W., Chambers, B. J., and Moore, A. (2000) A study on farm manure applications to agricultural land and an assessment of the risks of pathogen transfer into the food-chain. Report to The Ministry of Agriculture, Fisheries and Food, UK.
22. Thunegard, E. (1975) On the persistence of bacteria in manure. *Acta Vet. Scand. Suppl.* **56,** 1–86.
23. Guo, X., Chen, J., Brackett, R. E., and Beuchat, L. R. (2001) Survival of salmonellae on and in tomato plants from the time of inoculation at flowering and early stages of fruit development through fruit ripening. *Appl. Environ. Microbiol.* **67,** 4760–4764.
24. Mead, P. S., Slutsker, L., Dietz, V., et al. (1999) Food-related illness and death in the United States. *Emerg. Infect. Dis.* **5,** 607–625.
25. Shane, S. M. (2000) *Campylobacter* infection of commercial poultry. Diseases of poultry: world trade and public health implications. Office International des Épizooties. *Sci. Tech. Rev.* **19,** 376–387.
26. Stern, N .J., Myszewski, M. A., Barnhart, H. M., and Dreesen, D.W. (1997) Flagellin A gene restriction fragment length polymorphism patterns of *Campylobacter* spp. isolates from broiler production sources. *Avian Dis.* **41,** 899–905
27. Bradbury, W. C. and Monroe, D. L. G. (1985) Occurrence of plasmids and antibiotic resistance among *Campylobacter jejuni* and *Campylobacter coli* isolated from healthy and diarrheic animals. *J. Clin. Microbiol.* **22,** 339–346.
28. Souza, V., Castillo, A., and Eguiarte, L. E. The evolutionary ecology of *Escherichia coli*. *Am. Sci.* **90,** 332–341.
29. Hilborn, E. D., Mermin, J. H., Mshar, P. A., et al. (1999) A multistate outbreak of *Escherichia coli* O157:H7 infections associated with consumption of mesclun lettuce. *Arch. Intern. Med.* **159,** 1758–1764.
30. Burnett, S. L., Chen, J., and Beuchat, L. R. (2000) Attachment of *Escherichia coli* O157:H7 to the surfaces and internal structures of apples as detected by confocal scanning laser microscopy. *Appl. Environ. Microbiol.* **66,** 4679–4687.
31. Solomon, E. B., Yaron, S., and Mathews, K. R. (2002) Transmission of *E. coli* O157, H7 from contaminated manure and irrigation water to lettuce plant tissue and its subsequent internalization. *Appl. Environ. Microbiol.* **68,** 397–400.
32. Wang, G., Zhao, T., and Doyle, M. P. (1996) Fate of enterohemorrhagic *Escherichia coli* O157:H7 in bovine feces. *Appl. Environ. Microbiol.* **62,** 2567–2570.
33. Hancock, D. D., Besser, T. E., Kinsel, M. L., Tarr, P. I., Rice, D. H., and Paros, M. G. (1994) The prevalence of *E. coli* O157:H7 in dairy and beef cattle in Washington State. *Epidemiol. Infect.* **113,** 199–207.

34. Himathongkham, S. and Riemann, H. (1999) Destruction of *Salmonella typhimurium*, *Escherichia coli* O157:H7 and *Listeria monocytogenes* in chicken manure by drying and/or gassing with ammonia. *FEMS Microbiol. Lett.* **171,** 179–182.
35. Wiedmann, M., Arvic, T., Bruce, J. L., et al. (1997) Investigation of a listeriosis epizootic in sheep in New York State. *Am. J. Vet. Res.* **58,** 733–737.
36. Hassan, L., Mohammed, H. O., and McDonough, P. L. (2001) Farm-management and milking practices associated with the presence of *Listeria monocytogenes* in New York State dairy herds. *Prev. Vet. Med.* **51,** 63–73.
37. Yoshida, T., Kato, Y., Sato, M., and Hirai, K. (1998) Sources and routes of contamination of raw milk with *Listeria monocytogenes* and its control. *J. Vet. Med. Sci.* **60,** 1165–1168.
38. Siegman-Igra, Y., Levin, R., Weinberger, M., et al. (2002) *Listeria monocytogenes* infection in Israel and review of cases world-wide. *Emerg. Infect. Dis.* **8,** 305–310.
39. Fayer, R., Graczyk, T. K., Lewis, E. J., Trout, J. M., and Farley, C. A. (1998) Survival of infectious *Cryptosporidium parvum* in seawater and eastern oysters (*Crassostrea virginica*) in the Chesapeake Bay. *Appl. Environ. Microbiol.* **64,** 1070–1074.
40. Garber, L. P., Salman, M. D., Hurd, H. S., Keele, T., and Schlater, J. L. (1994) Potential risk factors for Cryptosporidium infection in dairy calves. *J. Am. Vet. Med. Assoc.* **205,** 86–91.
41. Kuczynska, E. and Shelton, D. R. (1999) Method for detection and enumeration of *Cryptosporidium parvum* oocysts in feces, manures, and soil. *Appl. Environ. Microbiol.* **65,** 2820–2826.
42. Scott, C. A. Smith, H. V., and Gibbs, H. A. (1994) Excretion of *Cryptosporidium parvum* oocysts by a herd of beef suckler cows. *Vet. Rec.* **134,** 172.
43. Weir, S. C., Pokorny, N. J., Carreno, R. A., Trevors, J. T., and Lee, H. (2002) Efficacy of common laboratory disinfectants on the infectivity of *Cryptosporidium parvum* oocysts in cell culture. *Appl. Environ. Microbiol.* **68,** 2576–2579.
44. MacKenzie, W. R., Hoxie, N. J., Proctor, M. E., et al. (1994) A massive outbreak of Cryptosporidium infection transmitted through the public water supply. *N. Engl. J. Med.* **331,** 161–167.
45. Shortridge, K. F., Zhou, N. N., Guan, Y., et al. (1998) Characterization of avian H5N1 influenza viruses from poultry in Hong Kong. *Virology* **252,** 331–342.
46. Horimoto, T. and Kawaoka, Y. (2001) Pandemic threat posed by avian influenza viruses. *Clin. Microbiol. Rev.* **14,** 129–149.
47. Lu, X., Tumpey, T. M., Morken, T., Zaki, S. R., Cox, N. J., and Katz, J. M. (1999) A mouse model for the evaluation of pathogenesis and immunity to influenza A (H5N1) viruses isolated from humans. *J. Virol.* **73,** 5903–5911.
48. Shortridge, K. F., Gao, P., Guan, Y., et al. (2000) Interspecies transmission of influenza viruses: H5N1 virus and a Hong Kong SAR perspective. *Vet. Microbiol.* **74,** 141–147.
49. McDaniel, H. A. (1991) Environmental protection during animal disease eradication programmes. Office International des Epizsoties *Rev. Sci. Tech.* **10,** 867–884.
50. Brodie, H. L and Carr, L. E. Composting farm animal mortalities. Fact Sheet 717. Cooperative Extension Service, University of Maryland, College Park, MD.
51. Elwell, D. L., Keener, H. M., Brown, T. J., and Monnin, M. J. (2000) Composting farm animal mortalities in Ohio: legal and practical considerations. In: *Proceedings of the 8th International Symposium on Animal, Agricultural and Food Processing Wastes*, October 9–11, Des Moines, IA (Moore, J. A., ed.). American Society of Agricultural Engineers, 2950 Niles Rd., St. Joseph, MI, 49085-9659, pp. 329–335.

52. Sartaj, M., Fernandes, L., and Patni, N. K. (1995) Influence of aeration pipes and temperature variations in passively aerated composting. *Am. Soc. Agric. Eng.* **38,** 1835–1841.
53. Hussong, D., Burge, W. D., and Enriki, N. K. (1985) Occurrence, growth and suppression of Salmonella in composted sewage sludge. *Appl. Environ. Microbiol.* **50,** 887–893.
54. Ryckeboer, J., Cops, S., and Coosemans, J. (2002) The fate of plant pathogens and seeds during anaerobic composting of source separated household wastes. *Compost Sci. Utilization* **10,** 204–216.
55. Brinton, Jr. W. F. and Droffner, M. W. (1994) Microbial approaches to characterization of composting processes. *Compost Sci. Utilization* **Summer,** 12–17.
56. Miner, Jr. F. D., Koenig, R. T., and Miller, B. E. (2001) The influence of bulking material type and volume on in-house composting in high-rise caged layer facilities. *Compost Sci. Utilization* **9,** 50–59.
57. Stanford, K., Larney, F. J., Olson, A. F., Yanke, L. J., and McKenzie, R. H. (2000) Composting as a means of disposal of sheep mortalities. *Compost Sci. Utilization* **8,** 135–146.
58. Morris, J. R., O'Connor, T., Kains, F., and Fraser, H. (1997) Composting livestock mortalities. Factsheet 97-001. Ontario Ministry of Agriculture, Food and Rural Affairs, Agdex 725/400.
59. Mathur, S. P., Daigle, J.-Y., Lévesque, M., and Dinel, H. (1986) The feasibility of producing high quality composts from fish scrap and peat with seaweeds and crab wastes. *Biol. Agric. Horticulture* **4,** 27–38.
60. Drury, L. N., Patterson, W. C., and Beard, C. W. (1969) Ventilating poultry houses with filtered air under positive pressure to prevent airborne diseases. *Poultry Sci.* **48,** 1640–1646.
61. Grunder, A. A., Gavora, J. S., Spencer, J. L., and Turnbull, J. E. (1975) Prevention of Marek's disease using a filtered air positive pressure house. *Poultry Sci.* **54,** 1189–1192.
62. Nicolai, R. E. and Janni, K. A. (2000) Designing biofilters for livestock facilities. In: *Proceedings of the 2nd international conference on Air Pollution from Agricultural Operations*, October 9–11, Des Moins, IA. American Society of Agricultural Engineers, 2950 Niles Rd., St. Joseph, MI, 49085-9659, pp. 376–383.

52

Molecular Aspects of Disease Pathogenesis in the Transmissible Spongiform Encephalopathies

Suzette A. Priola and Ina Vorberg

1. Introduction

The transmissible spongiform encephalopathy (TSE) diseases are a group of rare, fatal, and transmissible neurodegenerative diseases that include kuru and Creutzfeldt-Jakob disease (CJD) in humans, scrapie in sheep, transmissible mink encephalopathy (TME; *82*), and chronic wasting disease (CWD) in mule deer ***(141)*** and elk ***(142)***. Over the last 20 yr, they have gone from a fascinating but relatively obscure group of diseases to one that is a major agricultural and economic problem as well as a threat to human health. The shift in the relative impact of the TSE diseases began in the late 1970s when the United Kingdom altered the process by which animal carcasses were rendered to provide a protein supplement (i.e., meat and bone meal) to sheep, cattle, and other livestock. Several years later a new disease was recognized in the British cattle population. The pathological and immunohistochemical characteristics of the disease clearly placed it among the TSEs. The new disease was named bovine spongiform encephalopathy (BSE) by the scientific community ***(2,16,138)*** and "mad cow disease" by the less-than-scientific press. At its peak in the UK, several thousand cattle a year were diagnosed with BSE, and millions of cattle were slaughtered. Introduction of the specified offals ban as well as banning the practice of feeding ruminants to other ruminants has led to a drastic decrease in the number of yearly BSE cases in the UK (less than 500 in 2003), and the epidemic is clearly on the wane. However, BSE has now spread throughout the rest of Europe, as well as to Japan, Russia, Canada, and Israel and thus remains a worldwide problem.

A primary concern following the identification of BSE in 1985 was that it might cross species barriers to infect humans. Initially, it was thought that transmission of BSE to humans was unlikely, given that humans appeared to be resistant to scrapie, an

From: *Methods in Molecular Biology, vol. 268: Public Health Microbiology: Methods and Protocols*
Edited by: J. F. T. Spencer and A. L. Ragout de Spencer © Humana Press Inc., Totowa, NJ

animal TSE that had been endemic in British sheep for centuries. However, a few years after BSE was first recognized, a previously unknown form of CJD (variant CJD or vCJD) was identified in young people in Great Britain ***(140)***. The hypothesis that vCJD was the consequence of exposure of humans to BSE ***(140)*** has now been supported by several different studies ***(20,35,36,111)***, and over 140 cases of vCJD have been confirmed.

Variant CJD has increased public awareness in low-risk BSE countries of the dangers of exposure of humans to animal TSE diseases. For example, in the United States the presence of CWD in wild populations of deer and elk has raised concerns about the possible transmission of CWD to humans through consumption of venison. Indeed, recent in vitro studies have suggested that humans might be susceptible to CWD ***(110)***. There is also concern that CWD might cross species barriers to infect livestock that share range land with infected deer and elk and thus be transmitted to humans via infected cattle or sheep. In either case the end result could be reminiscent of what has happened in the UK: the occurrence of a new form of TSE disease in humans.

The impact of BSE has thus spread far beyond the agricultural arena and is now considered a significant threat to human health. Variant CJD has made it clear that exposure of humans to TSE-contaminated products can lead to disease. This in turn has fueled concerns about the possible spread of vCJD via contaminated blood products that has led to stringent new blood donor criteria in the United States and abroad. The only way to eliminate these concerns is to develop sensitive and specific diagnostic tests for TSE infection as well as effective postexposure therapeutics. Currently these are two very active areas of TSE research, although the unique nature and long-term pathogenesis of these diseases has made development of TSE diagnostics and therapeutics especially difficult. However, new methodologies have increased our understanding of the biochemistry of these diseases. This new understanding may in turn eventually lead to novel approaches in TSE diagnostics and therapeutics.

2. Prion Protein

One of the most intriguing aspects of the TSEs is that no virus or bacteria has ever been associated with infectivity. This, coupled with the extreme resistance of TSE infectivity to inactivation by a variety of harsh treatments, led to the hypothesis that the infectious agent in these diseases was an infectious, self-replicating protein ***(1,51)***. In the early 1980s a protein was found that was closely associated with TSE infectivity, and this protein was termed prion protein or PrP ***(11)***. Further analysis revealed that PrP was in fact a normal host cellular protein ***(11,41,78)*** that accumulates during TSE infection as an abnormal isoform. This accumulation occurs primarily in nervous system tissues but often also in tissues of the lymphoreticular system. It was this abnormal form of PrP that was first identified as scrapie-associated fibrils from scrapie-infected brain tissue ***(84,85)***. The fact that abnormal PrP is only found in TSE-infected tissues and is very closely associated with tissue preparations enriched for infectivity has led to the proposal that abnormal PrP itself is the infectious protein agent ***(100)***. Although this hypothesis remains a point of contention in TSE research, the key role of PrP in disease pathogenesis is undeniable ***(17,21)***.

2.1. PrP

Normal PrP, an approx 253-amino acid glycoprotein that is expressed on the cell surface in a wide variety of tissues, is both detergent soluble and sensitive to digestion with proteinase K (PrP-sen)*. PrP-sen is a cell surface glycoprotein containing two N-linked glycosylation sites *(26,52)*. Variable glycosylation of these two sites results in both PrP-sen and PrP-res migrating as three discrete bands consisting of unglycosylated, partially glycosylated, and fully glycosylated PrP (**Fig. 1**). A glycophosphotidyl-inositol (GPI) membrane anchor is also added post-translationally to PrP-sen, allowing it to be inserted into the plasma membrane *(10,57,58)*. By contrast, the abnormal form of PrP associated with disease (PrP-res) forms insoluble aggregates and is partially resistant to digestion with proteinase K such that approx 60–70 N-terminal amino acids are cleaved. This leads to a size shift characteristic for the PrP-res molecule (**Fig. 1**). PrP-res is post-translationally derived from mature PrP-sen *(12,27)*. Since both PrP-sen and PrP-res are central to the pathogenesis of the TSE diseases, an understanding of their basic biosynthesis is critical.

The normal function of PrP is unclear. Two different strains of mice in which PrP has been ablated develop normally but demonstrate subtle nervous system abnormalities *(37,133)*, whereas a third strain of PrP knockout mice showed a loss of Purkinje cells in the brain and developed a cerebellar dysfunction *(119)*. PrP-sen has been reported to be involved in mitogen-induced activation in lymphocytes *(23)*. Recent studies have provided compelling evidence that normal PrP may be involved in fyn kinase signaling *(86)* as well as pro- and antiapoptotic mechanisms *(15,87)*. These latter observations are particularly intriguing since conversion of PrP to its abnormal form could conceivably lead to a loss of function that may help to explain the mechanism of neurodegeneration during TSE infection. At present it is difficult to determine which, if any, of these observations reflect the true function of PrP.

2.2. Cell Biology of PrP-res Formation

Historically, it has been quite difficult to infect cells in vitro with a TSE agent, hindering attempts to study the cellular biology of PrP-res formation. Despite the almost ubiquitous expression of PrP-sen in many different cell types throughout the body, PrP-res accumulation in vivo appears to be limited to cells of the nervous and lymphoreticular systems. Similarly, even though most cell lines express PrP-sen, the susceptibility of tissue culture cells to TSE infection is also restricted. Susceptibility of cells to infection is probably dependent not only on the cell type and sequence of the cellular PrP-sen protein (**Fig. 2**) but also on the strain and infectious titer of the TSE agent. Occasionally, even when cells had been successfully infected with a TSE agent, the low percentage of infected cells meant that PrP-res was not even detectable

*The normal form of PrP has been referred to as either PrP^{C} for PrP-cellular or PrP-sen, which designates its sensitivity to proteinase K. The abnormal form of PrP has been referred to as either PrP^{Sc} for PrP-scrapie or PrP-res, which designates its partial resistance to proteinase K. Since the terms PrP-sen and PrP-res reflect a general operational definition rather than a disease-specific phenotype, they will be used throughout this discussion.

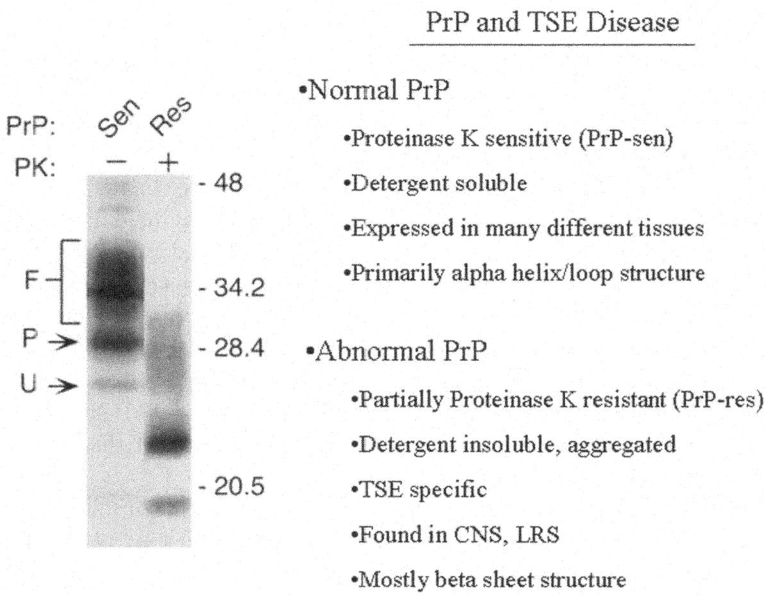

Fig. 1. Properties of normal and disease-associated prion protein (PrP). The panel on the left is a Western blot of normal PrP (PrP-sen) and abnormal PrP (PrP-res) isolated from mouse scrapie-infected neuroblastoma cells ***(107)***. PrP is a cell surface glycoprotein that migrates as three distinct glycosylated bands: a fully glycosylated band (F) representing complex glycans at both N-linked glycosylation sites, a partially glycosylated band (P) representing partially processed glycans at one or both N-linked glycosylation sites, and an unglycosylated band (U). Following digestion with proteinase K (PK), PrP-res shifts 6–7 kDa in size compared with PrP-sen. This shift is diagnostic for PrP-res and TSE diseases in general. Molecular mass markers in kilodaltons are indicated on the right. CNS, central nervous system; LRS, lymphoreticular system; TSE, transmissible spongiform encephalopathy.

(**Fig. 3**) until after the cells were cloned ***(106)***. Despite these technical difficulties, a limited number of mouse scrapie strains have been successfully used to infect mouse and rat cells ***(22,107,114,120)***. Additionally, although persistent attempts to infect cells with BSE and CJD have not been successful, rabbit epithelial cells overexpressing the sheep PrP protein have recently been infected with sheep scrapie ***(134)***. All these infected cell lines make PrP-sen, accumulate PrP-res, and replicate scrapie infectivity. As a result, the cell biology of PrP-res formation has now been studied in some detail.

In mouse scrapie-infected mouse neuroblastoma (Sc^+-MNB) cells, in which PrP-sen is the precursor to PrP-res, the biosynthesis of PrP-res differs dramatically from that of PrP-sen ***(12,26,27,30,128)***, and only a small proportion of the available PrP-sen is converted to PrP-res ***(27,30)***. The conversion of PrP from the protease-sensitive to the protease-resistant form probably occurs either on the plasma membrane or along an endocytic pathway ***(4,13,27,30,83)***, possibly in a cellular compartment containing

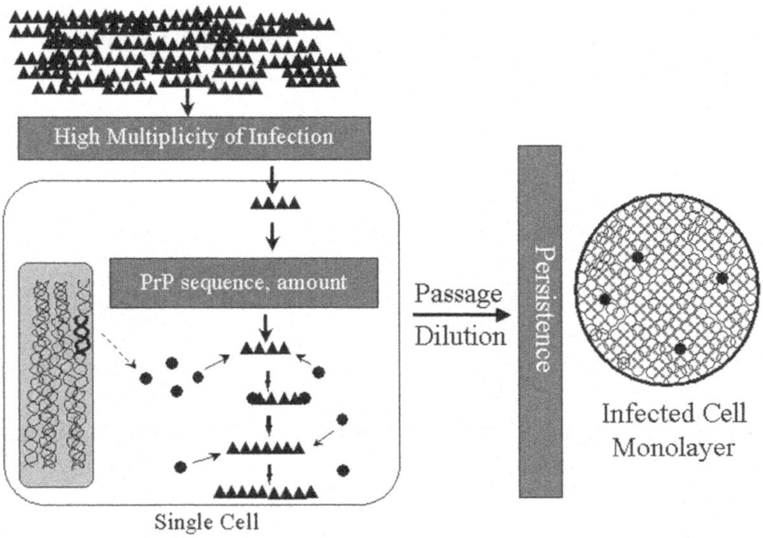

Fig. 2. Scrapie infection of tissue culture cells. A single cell (large square) and its DNA genome (light gray rectangle) is pictured. The major obstacles to successfully infecting cells in vitro are represented by the dark gray rectangles. A prion protein (PrP)-res (black triangles) positive, high-titer brain homogenate from a scrapie-infected animal is layered onto cells that encode PrP-sen (black circles). Following an incubation period of several hours, medium is added, and the cells are cultured until confluent and then passaged many times to remove any remaining brain homogenate. At different passages post infection, the cells are then assayed for scrapie infectivity in vivo and/or PrP-res formation in vitro ***(107)***. Several factors appear to influence the likelihood of obtaining persistently infected cells including the scrapie titer of the brain homogenate and multiplicity of infection, the PrP sequence, the expression level of PrP-sen, the scrapie strain, and the cell type. Persistently infected cells may account for less than 1% ***(106)*** or up to 10% ***(134)*** of the total cell population. When low numbers of cells are persistently infected, PrP-res can only be detected after the cells have been cloned (*see* **Fig. 3**).

cholesterol-rich membranes ***(129)***. In Sc^+-MNB cells, PrP-res eventually accumulates in the secondary lysosomes ***(27,30)*** with little, if any, present on the cell surface ***(12,25,83)***. In vivo, PrP-res can also accumulate in the extracellular spaces as amorphous deposits or more organized fibrils or amyloid plaques ***(64–67)***. Despite its exposure to endolysosomal hydrolases, PrP-res has a half-life of greater than 48 h ***(12,27)***. This metabolic stability of PrP-res could explain its accumulation in vivo, especially in the nondividing cells of the central nervous system.

Both the normal and abnormal forms of PrP are derived from the same gene and have the same amino acid sequence. No post-translational modifications have been identified that can account for the different biochemical properties of PrP-sen and PrP-res. Thus, it is likely that the difference between the two molecules is a conformational one. Analysis of PrP-res by infrared spectroscopy ***(29,32,89)*** and the recent

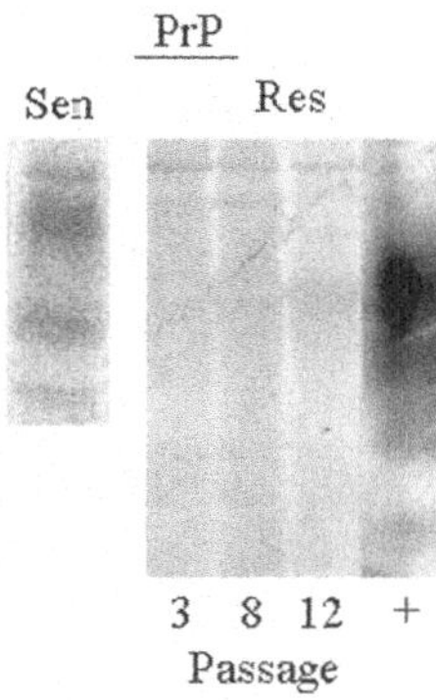

Cell Passage (PI)	Incubation time
Passage 3	8/8 (192 ± 20)[a]
Passage 8	8/8 (198 ± 14)
Passage 12	8/8 (192 ± 10)

a=scrapie positive/total animals, (days to death ± standard deviation)

Fig. 3. Uncloned scrapie-infected tissue culture cells do not necessarily have detectable levels of PrP-res. The left side of the figure shows a Western blot of PrP-sen expressed in mouse neuroblastoma cells infected with scrapie compared with PrP-res expression in these cells at passages 3, 8, and 12 post infection. For the bioassay summarized in the table on the right, cell lysates were injected intracranially into mice susceptible to scrapie, and the animals were monitored for disease ***(107)***. Note that although no PrP-res was detectable by Western blot, the fact that all mice at each passage tested succumbed to scrapie clearly demonstrates that the cells are persistently infected with mouse scrapie. These data illustrate the higher sensitivity of the in vivo bioassay versus Western blotting when monitoring scrapie infection. (+), PrP-res from scrapie-infected brain.

determination of the structure of PrP-sen by nuclear magnetic resonance (NMR) ***(42,112,113,143)*** and X-ray crystallography ***(71)*** support this hypothesis and demonstrate that PrP-res has a much higher β-sheet content than PrP-sen (**Fig. 4**). A critical question within the TSE diseases that remains unresolved is how the conformational conversion of PrP-sen to PrP-res occurs.

2.3. Conversion of PrP-sen to PrP-res

The conversion of PrP-sen to PrP-res is a process that is critical to disease pathogenesis, especially if PrP-res is by itself the infectious agent. Two primary models have been used to describe the self-propagation of PrP-res. The heterodimer model ***(9,51,101)*** proposes that a monomer of PrP-res interacts with a monomer of PrP-sen and converts it to more PrP-res by inducing a conformational change. The two molecules, now both PrP-res, separate, and the process continues, leading to accumulation of PrP-res and disease. In the seeded polymerization model ***(45,63)***, a "seed" of PrP-res composed of an organized array of multiple PrP-res molecules interacts with PrP-sen and converts it to PrP-res by inducing a conformational change. The newly formed PrP-res molecule remains a part of the polymer, and the process repeats, eventually leading to accumulation of PrP-res and disease. Recent data suggest that only aggregates of PrP-res can induce the conversion of radiolabeled PrP-sen to PrP-res ***(31)***. In

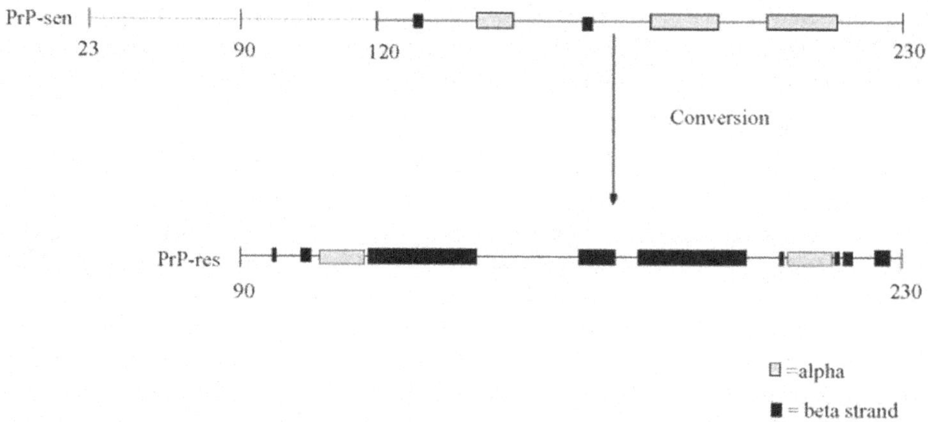

Fig. 4. PrP-res formation involves a conformational change in PrP-sen. During disease pathogenesis, PrP-sen (upper panel) undergoes a conformational change from a primarily α-helical structure (gray boxes) to a primarily β-sheet structure (black boxes) following conversion to PrP-res (lower panel). A linear representation of the NMR structure of processed PrP-sen from amino acid residues 23–230 is shown ***(113)***. Solid lines indicate loop structures; a dashed line represents disordered structure. Following digestion with proteinase K, which removes the N-terminal 60–70 amino acid resides, protease-resistant PrP-res lacks most of the disordered N-terminal region. The structure of PrP-res is theoretical and is based on predictions from infrared spectroscopy ***(32)***. Note that significant regions of β-sheet are predicted in PrP-res downstream of the small regions of β-strand present in PrP-sen.

yeast, another organism in which prion-like activity has been found ***(139)***, only aggregated forms of yeast prions can induce the formation of new yeast prions ***(47,91,124,130)***. Dissociation of PrP-res to a soluble form leads not only to a loss of its "converting activity" but also to a loss of infectivity ***(31,46,117)***. Finally, monomeric PrP-res as resistant to proteinase K as aggregated PrP-res has never been detected, although this does not rule out the possibility that less proteinase K-resistant monomeric PrP-res is present and active at low levels. Therefore, most of the currently available evidence supports the seeded polymerization model.

3. The Role of PrP in TSE Diseases

Any model of PrP-res formation must be able to explain at a molecular level (1) how PrP-res could induce its own formation (i.e., replicate); (2) how PrP can control not only disease incubation time but also species barriers to infection; (3) the occurrence of different scrapie strains in animals with one PrP genotype; and finally (4) human familial TSE diseases that are both infectious and heritable. Fortunately, several recently developed assays including a polymerase chain reaction (PCR)-like assay of PrP-res formation ***(116)***, two different cell-free conversion systems ***(72,137)***, persistently infected and acutely exposed Sc^{+}-MNB cells ***(107,137)***, as well as transgenic mice ***(21,108,121)***, have allowed researchers to study PrP-res formation in greater detail than ever before and have led to valuable new insights into the conversion process.

3.1. "Replication" of PrP-res

The most direct evidence that PrP-res can induce its own formation comes from an experimental system in which radiolabeled tissue culture-derived PrP-sen can be converted into a proteinase K-resistant form using PrP-res isolated from TSE-infected brains ***(72–74)***. In this cell-free conversion system (**Fig. 5A**), the primary components are immunoprecipitated PrP-sen and enriched preparations of PrP-res. A similar system using less purified components and PrP-sen tagged with a unique antibody epitope instead of radioactively labeled PrP-sen, has also recently been developed ***(137)*** (**Fig. 5B**). Via an interaction that most likely involves the C-terminal portion of PrP ***(59,73)***, PrP-res binds selectively to PrP-sen, and conversion of PrP-sen into PrP-res occurs. Some evidence suggests that binding and conversion are two distinct events ***(40,61,99)***, a hypothesis that is supported by data suggesting that initial PrP-PrP interactions may map to a different region of PrP than the region that appears to influence conversion ***(60,74,97,99,123)***. The fact that PrP-res can induce PrP-sen to convert to the abnormal form in the absence of a living cell provides compelling evidence that PrP-res can "replicate" itself in the absence of any metabolic activity and demonstrates that a viable virus or bacteria is not needed for this process to occur. However, it is still unclear whether or not this newly formed PrP-res is infectious.

Fig. 5. *(opposite page)* Formation of PrP-res in vitro. (**A**) The cell-free assay for PrP-res formation in vitro ***(72)***. Radiolabeled PrP-sen (^{35S}PrP-sen) is isolated by immunoprecipitation from eukaryotic tissue culture cells and mixed with PrP-res isolated from an animal infected with a TSE disease. Following a 2-d incubation at 37°C, the mixture is treated with proteinase K (PK) and assayed by autoradiography for PK-resistant, radiolabeled PrP-res that has shifted in molecular weight the expected 6–7 kDa (right panel). This assay can be performed using any species of PrP-sen or PrP-res as long as the necessary tissue culture cells and infected brain homogenates are available. (**B**) A cell lysate-based assay for PrP-res formation ***(137)***. The principle is similar to that of the cell-free conversion assay, except that radioactivity is not used and purified PrP-sen and PrP-res are not needed. A TSE-infected brain homogenate containing both PrP-sen and PrP-res (gray circles and squares, respectively) is mixed with a cell lysate containing an epitope-tagged PrP-sen molecule (flagged black circle) at a ratio of 1:8. The epitope in PrP-sen must be recognized by an antibody that does not recognize the PrP in the infected brain homogenate. Following a 4-d incubation at 37°C and treatment with PK, formation of new PrP-res (flagged black square) is assayed using an antibody that will only recognize epitope-tagged PrP

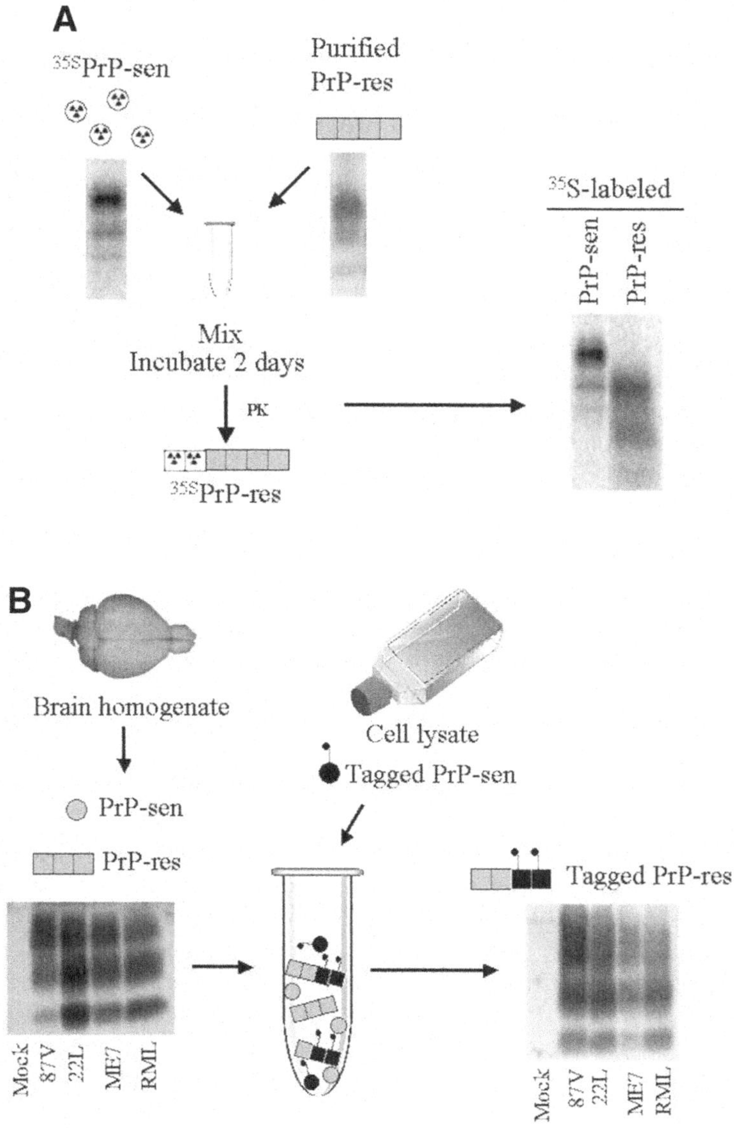

Fig. 5. *(continued)* molecules. A Western blot of the input PrP-res from an uninfected (mock) brain homogenate or from brain homogenates from mice infected with four different scrapie strains (87V, 22L, ME7, and RML) is shown on the left. Note that each strain has a distinctive pattern of three glycosylated PrP-res bands. A Western blot of newly formed epitope-tagged PrP-res for each of the four strains is shown on the right. It was developed using an antibody that does not recognize the PrP-res molecules shown in the left panel.

3.1.1. Cofactors in PrP-res Formation

Although these cell-free conversion systems could be seen as demonstrating that only PrP-sen and PrP-res are needed for conversion to occur, the fact that the preparations used are not absolutely pure precludes this interpretation. If other molecules are involved they might only need to be present in very low levels to have an effect. The role of other molecules in the conversion process is unclear. Glycosaminoglycans (GAGs) have been proposed as one possible cofactor (see ref. ***94*** for review). GAGs bind PrP ***(24,44)***, inhibit formation of PrP-res ***(28)***, are often associated with PrP plaques in the brain of infected individuals ***(126)***, and can inhibit TSE disease in vivo ***(70)***. Chaperones have been shown to enhance the formation of PrP-res in the cell-free conversion system, although none of the chaperones tested are found in cellular compartments where PrP-res formation most likely occurs ***(40)***. Other as yet unidentified cell factors have also been proposed as potential cofactors in the conversion process ***(132)***. It is certainly possible that any or all of the proposed cofactors could play a role in formation of PrP-res, but further studies are necessary before any definite conclusions can be drawn.

3.2. PrP-res and TSE Species Barriers

Species barriers in the TSE are defined as the resistance to TSE disease by one species following infection with the TSE agent of another species (see ref. ***93*** for review). This resistance is reflected by extremely long incubation times upon first passage of a TSE agent into a new host species. The barrier to infection upon first exposure can be very strong and can lead to incubation times that exceed the lifetime of the host. However, as the agent is serially passaged through the new host, adaptation occurs and incubation times shorten until a stable incubation time is reached.

The onset of the BSE and CWD epidemics in the UK and US, respectively, and the fact that BSE has been linked to vCJD in humans has made understanding species barriers to TSE infection of paramount importance. The fact that PrP-res binds PrP-sen suggests that the PrP amino acid sequence may be involved in the conversion process and that homology between the two molecules is important if conversion is to occur. Molecular control of species barriers could therefore depend on homology between the incoming PrP-res and the host PrP-sen. Homologous PrP molecules would form PrP-res more efficiently than less homologous PrP molecules, leading to a more rapid accumulation of PrP-res to pathogenic levels and thus to clinical disease. Conversely, heterologous PrP molecules with mismatches at key amino acid positions might have a reduced potential to interact and form additional PrP-res, thus slowing the disease process.

3.2.1. The Mouse–Hamster Scrapie Species Barrier

To study TSE species barriers, researchers have taken advantage of a strong barrier to infection that exists between hamsters and mice. Although mice are fully susceptible to mouse scrapie, they are resistant to infection with a particular strain of hamster scrapie. Mice infected with this strain of hamster scrapie do not become clinically ill within the lifetime of the animal. However, when transgenic mice were engineered that overexpressed hamster PrP-sen, they became fully susceptible to hamster scrapie ***(121)***. This was good evidence suggesting that the primary amino acid sequence of PrP could influence TSE species barriers. Subsequent studies demonstrated that the hamster/mouse species barrier could be crossed even when hamster PrP-sen expression was restricted to neurons ***(108)*** or astrocytes ***(109)***. Thus, homologous PrP molecules can act as a type of susceptibility factor in cross-species transmission of TSE.

The corollary to homologous PrP acting as a susceptibility factor is that heterologous PrP molecules might act as a type of host resistance factor. Interestingly, in transgenic mice that express mouse PrP-sen and overexpress hamster PrP-sen, mouse scrapie disease incubation times are increased ***(108,121)***. Furthermore, hamster PrP-expressing transgenic mice in which the mouse PrP gene has been ablated develop hamster scrapie more rapidly than hamster PrP transgenic mice which still express mouse PrP ***(102,109)***. These observations suggest that the presence of heterologous PrP molecules interferes with disease and that this interference is dependent on expression level. In fact, studies in Sc$^+$-MNB cells have shown that heterologous PrP molecules interfere with PrP-res formation in an expression level dependent manner ***(56,95,135,144)***, whereas in vivo both amino acid sequence and PrP expression levels are important in mouse models of scrapie ***(80,121)***. Thus, inefficient formation of PrP-res owing to heterologous PrP molecules could provide an explanation for the maintenance of species barriers in the TSE.

Although TSE species barriers are protective, it is important to note that they are not absolute. Even the very strong experimental species barrier between mice and hamsters can be broken. Hamster infectivity can be sequestered in infected mice ***(103)*** and can even replicate, albeit quite slowly ***(55,105)***. Furthermore, these mice exhibit no clinical signs of scrapie ***(55,103)***. Thus, a subclinical carrier state can be maintained following infection with a TSE agent ***(105)*** that can only be detected following serial in vivo passage of the brain of a subclinical animal ***(104)***. The implication of these data is that subclinical carrier states may also exist in natural forms of TSE such as BSE, scrapie, and CWD, leading to an undetectable reservoir of infected animals.

3.2.2. Regions of PrP Involved in the Species-Specific Formation of PrP-res

In vivo, transgenic studies have mapped the region of PrP important in TSE species barriers to an area encompassing the middle portion of PrP (**Fig. 6**) ***(122,132)***. Sc$^+$-MNB cells have also been used to assay the species-specific conversion of PrP-sen into PrP-res. Since mouse PrP-sen is converted into mouse PrP-res, but hamster PrP-sen is not converted into PrP-res in Sc$^+$-MNB cells ***(97,123)***, these cells provide an in vitro model for studying the in vivo hamster/mouse species barrier. Expression of epitope-tagged mouse and hamster PrP-sen or chimeric hamster/mouse PrP-sen molecules in Sc$^+$-MNB cells demonstrated that species-specific PrP-res formation could be controlled by a region of PrP-sen encompassing residues 112–187 of the PrP gene (**Fig. 6**) ***(97,123)***. These studies also demonstrated that a single amino acid residue mismatch at position 138 of mouse PrP could control mouse PrP-res formation ***(97)***, illustrating the exquisite sequence specificity of PrP-res formation. Further experiments have shown that critical amino residues involved in the species-specific forma-

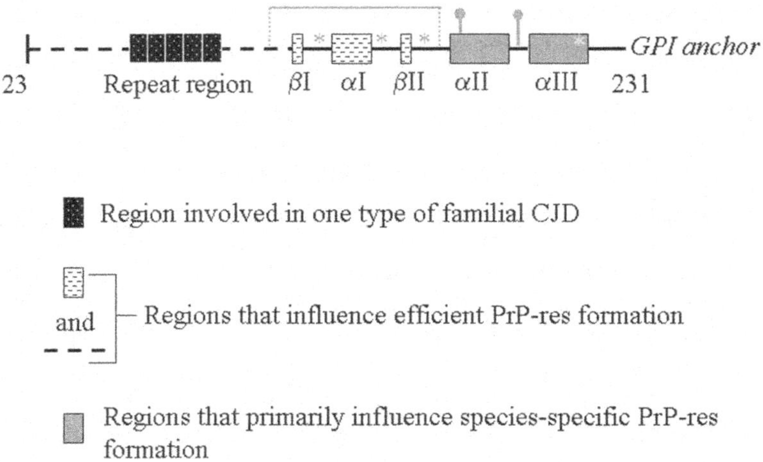

Fig. 6. Regions of prion protein (PrP) important in PrP-res formation. A linear representation of the mouse PrP-sen NMR structure is shown. The processed form of PrP-sen is depicted with the N-terminal signal peptide sequence (residues 1–22) removed as well as the C-terminal glycophosphotidylinositol (GPI) anchor addition site (residues 232–254). The dashed line indicates the disordered PrP N-terminus that is important in PrP-res formation. Solid lines indicate loop regions; the areas of α-helix and β-strand are labeled and indicated by boxes. Asterisks indicate the position of amino acid residues that are important in species-specific PrP-res formation. Areas shaded gray indicate other structures involved in formation of species-specific PrP-res. The two grey lollipop structures indicate the two N-linked glycosylation sites in PrP. CJD, Creutzfeldt-Jakob disease.

tion of PrP-res can differ between species ***(14,96,111,136)***. For example, homology at PrP amino acid residue 155 is necessary for the efficient formation of hamster PrP-res ***(96)***, whereas multiple amino acid residues within the rabbit PrP protein appear to prevent formation of abnormal rabbit PrP, possibly explaining the apparent resistance of rabbits to scrapie infection ***(136)***. In the context of species-specific amino acid mismatches between PrP-sen and PrP-res, even PrP glycosylation can influence species-specific PrP-res formation by influencing the first step in the conversion process: the binding of PrP-sen to PrP-res ***(99)***.

Taken together, these studies suggest that species-specific PrP-res formation can be influenced at the level of PrP-sen/PrP-res binding by PrP glycosylation, and at the level of conversion of PrP-sen to PrP-res by mismatches at critical amino acid residues ***(99)***. Thus, PrP amino acid residue mismatches have the potential to affect cross-species transmission of TSE disease profoundly. Indeed, there is ample evidence that mismatches between PrP-sen and PrP-res at a single amino acid residue can effect disease. Heterologous amino acids at residue 101 can alter disease incubation times in transgenic mice across species barriers ***(3,81)***. A mismatch at residue 138 (142 in goat PrP) has been associated with the resistance of some goats to sheep scrapie and BSE ***(50)***. The resistance of hamsters to BSE might also map to this position (139 in hamster PrP) ***(111)***. Resistance of sheep to scrapie has been correlated with amino acid residue 171 ***(49)***, whereas in humans homozygosity at codon 129 has been associated with susceptibility to CJD ***(88)***. All these single amino acid polymorphisms reside within the loop regions of PrP-sen (**Fig. 6**), suggesting that mutations within these regions of PrP have a significant effect on species-specific PrP-res formation.

The dependence of TSE species barriers on the amino acid sequence of PrP suggests one possible explanation for how BSE could cross a species barrier to infect humans. Human and bovine PrP differ at 22 positions, but only 5 of these differences are within the 75-amino acid residues of the PrP molecule that most often influence the species-specific formation of PrP. If none of these differences was sufficient to prevent bovine PrP-res from converting human PrP-sen, the implication would be that humans exposed to BSE-contaminated materials might be at risk of infection. Studies in the cell-free conversion system, which at the level of formation of PrP-res have generally correlated with the in vivo transmissibility of several different TSE agents ***(14,74,111)***, have demonstrated that bovine PrP-res can in fact convert human PrP-sen to human PrP-res at low levels ***(111)***, as can cervid (i.e., elk and deer) PrP-res ***(110)***. The amino acid sequence differences among bovine, cervid, and human PrP were therefore not sufficient to prevent bovine or cervid PrP-res from interacting with human PrP-sen. This could be at least part of the reason that BSE has been able to cross species barriers to cause vCJD in humans and raises concerns that CWD may do the same.

3.2.3. Role of PrP Secondary Structure in PrP-res Formation Within a Single Species

Since the conversion of PrP-sen to PrP-res most likely involves a conformational change, identifying the regions of PrP involved in the general formation of PrP-res is of great interest. Studies in transgenic mice have shown that part of the N-terminal unstructured region of PrP (**Fig. 4**) is important in susceptibility to scrapie infection ***(43,127)***. Other studies also indicate that residues between 108 and 124 in this region are important in PrP-res formation ***(75,95,136)***. The short stretches of β-strand present in PrP-sen (**Fig. 4**) have been hypothesized to act as a "nucleation site" for the conformational change to PrP-res. Involvement of the β-strands in PrP-res formation is suggested by recent studies showing that peptides to this region can inhibit PrP-res formation ***(33,59)***. Indeed, PrP-sen molecules with the first β-strand deleted (residues 128–131), although processed normally, cannot be converted to PrP-res ***(135)***. Deletion of the second β-strand (residues 161–164) also abolishes PrP-res formation, although this may be because this molecule is processed aberrantly ***(135)***. Even deletion of the first α-helix (residues 144–154) is sufficient to prevent PrP-res formation ***(135)***. Taken together, the data suggest that it is the overall tertiary structure of PrP-sen, rather than any single secondary structure, that is important in PrP-res formation ***(136)***.

3.3. PrP-res and TSE Strains

Strains of TSE agent have been described for CJD *(90)*, scrapie *(68,69)*, and hamster-adapted TME *(7)*. Strains in the TSE are defined by differences in brain pathology, disease incubation time, clinical disease, species tropism, and PrP-res properties (for review, see ref. *19*). It is especially challenging to explain TSE strains in terms of PrP-res formation since multiple strains have been described in species with one PrP genotype. It is difficult to imagine how PrP-res with the same amino acid sequence could cause several different disease phenotypes. Since the TSE diseases appear to be diseases of protein folding, however, the answer may reside not in the amino acid sequence but rather in the conformation of the protein. This hypothesis is supported by data from studies using protease digestion of hamster-adapted TME strains *(6)*. If PrP-res can adopt several stable variations in structure, it might be able to pass on its own strain-specific characteristics to PrP-sen via direct PrP-PrP interactions. In fact, recent structural data using infrared spectroscopy *(29)* or a conformation-dependent immunoassay *(118)* show that PrP-res from different mouse and hamster TSE strains have different conformations.

3.3.1. PrP-res Conformation and Strains

Two strains of hamster scrapie originally derived from mink infected with TME *(7)* have provided the best studied example of strain-specific differences in PrP-res conformation. In both strains, PrP-res is derived from the same hamster PrP-sen, but PrP-res derived from hamster brains infected with the "drowsy" strain is approx 1 kDa smaller than PrP-res derived from hamster brains infected with the "hyper" strain *(7)*. When drowsy or hyper PrP-res were mixed with the same radiolabeled PrP-sen precursor in the cell-free conversion system (**Fig. 5A**), the strain-specific size difference could be detected in the newly made protease-resistant PrP. Thus, protease-resistant PrP generated by drowsy-derived PrP-res was approx 1 kDa smaller than that generated from hyper PrP-res *(5,8)*. In vivo studies using transgenic mice infected with different strains of human CJD also imply that TSE strains may be dependent on the conformation of PrP-res *(36,131)*. Thus, there is evidence to support the hypothesis that self-propagation of PrP-res may proceed via distinctive three-dimensional structures, which in turn could provide some molecular basis for scrapie strains.

3.3.2. PrP-res Glycosylation and Strains

It has also been suggested that PrP-res glycosylation, which can differ between strains (**Fig. 5B**), might also encode strain-specific phenotypes. However, in vivo studies have suggested that PrP glycosylation does not encode strain characteristics. Strain-specific characteristics are maintained upon transmission of infectivity from one animal to another even when the infected tissues contain PrP-res molecules with very different glycosylation patterns *(54,115)*. The most recent evidence suggests that strain-specific PrP-res glycosylation does not encode strain-specific phenotypes but rather may be indicative of the cellular compartment where the majority of strain-specific PrP-res formation occurs *(137)*.

3.4. PrP-res and Familial TSE

Although there is no correlation with exposure to an infectious agent, TSE agent infectivity is detectable in the brains of individuals afflicted with familial TSE. Intriguingly, PrP-res is also detectable in most familial forms of human TSE, although its size may vary. The question then is how exactly abnormal PrP is generated in familial TSE in the absence of exposure to PrP-res from an exogenous source. Since these diseases are heritable, genetics must play a role. Given the critical role that PrP has in infectious forms of TSE, the most likely genetic component is the PrP gene itself. In fact, familial TSE diseases have been correlated with particular mutations in the PrP gene (**Fig. 7**) and in most families association of the PrP genotype with onset of disease (i.e. penetrance) is 100%. As seen in **Fig. 7**, these mutations occur throughout most of the PrP molecule, although many are clustered toward the C-terminus. The current hypothesis is that mutations in the PrP gene lead to a PrP molecule that is more likely to convert spontaneously to the abnormal form. The difference between this type of "spontaneous" conversion and the "induced" conversion associated with exposure to infectivity is that both PrP-sen and PrP-res are derived from the host.

It is now quite clear that mutant PrP molecules have properties different from wild-type PrP and that these altered properties lead to a molecule with properties reminiscent of PrP-res. Insertion of extra copies of an octapeptide repeat in PrP (**Fig. 7**) has been associated with CJD in several different families and is probably the best studied familial TSE mutant. These mutants are processed differently ***(39,76)*** and somewhat resemble PrP-res in that as the number of octapeptide repeats increases they have a greater tendency to aggregate as well as an increased resistance to proteinase K ***(77,98)***. Transgenic mice overexpressing one of these mutants even develop a neurological disease ***(34)***. Other PrP familial mutations demonstrate different phenotypes. PrP with the Gerstmann-Sträussler-Scheinker (GSS)-associated aspartic acid to asparagine mutation at codon 178 shows an altered cell surface expression ***(92)***. Mice overexpressing the GSS-associated proline to leucine mutation at codon 102 develop a neurodegenerative disease ***(62)***, although transgenic mice expressing wild-type levels of this mutant PrP protein do not ***(81)***. Thus, these data suggest that overexpression of the mutant protein may be required for disease. Altered PrP processing has also been linked to a PrP mutation linked to GSS. Mutation of an alanine to a valine at codon 117 results in an increase in transmembrane forms of PrP ***(79)***, and transgenic mice overexpressing this PrP mutant develop a neurological disease and are more susceptible to scrapie infection ***(53)***.

Although all these studies show that mutant PrP molecules behave differently in vitro and in vivo, the nature of their role in pathogenesis remains unclear. Although most of these PrP mutants are slightly more resistant to proteinase K, their level of resistance is several hundred fold less than that of PrP-res isolated from the brain. In fact, when tested in the cell-free conversion system, PrP mutants containing extra copies of the octapeptide repeat region were unable to form PrP-res spontaneously and were no different from nonmutant PrP in their ability to induce PrP-res formation ***(98)***. Thus, despite all these recent advances, the molecular basis for familial TSE remains an open question and may in fact differ for different familial TSEs.

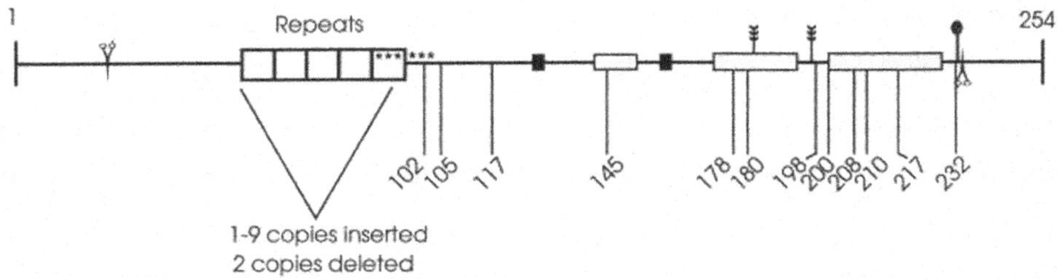

Fig. 7. Mutations in human PrP associated with familial TSE. The structure of human PrP is shown *(143)*. The positions of the signal peptide cleavage at the N-terminus and the GPI anchor addition site at the C-terminus are indicated by the scissors. The open boxes labeled "Repeats" represent an octapeptide sequence that is tandemly repeated five times in normal human PrP. The triple-headed arrows represent the two N-linked glycosylation sites. The asterisks indicate the site of proteinase K cleavage, which can vary somewhat in PrP-res isolated from familial TSE cases. The locations of single amino acid mutations and insertion or deletion mutants associated with familial TSE are indicated (see ref. ***48*** for review).

4. Conclusions

In the study of TSE diseases at both the molecular and pathogenic level, a great many questions remain. For example, how could different conformations of PrP-res lead to different lesion profiles in the brain? Are other molecules involved in PrP-res formation and how could these potential cofactors influence disease? How does the infectious agent enter a cell? For that matter, the question of whether or not PrP-res is actually the infectious agent remains unresolved largely because, owing to the difficulty in purifying or generating new PrP-res, it has never been shown that PrP-res alone can induce disease.

Independent of our understanding the mechanisms of TSE disease, and within the context of the current situation in Great Britain and elsewhere, the most important areas of TSE research are early diagnosis of the disease and development of anti-TSE prophylactic and therapeutic agents. It is still impossible to diagnose CJD in humans prior to the onset of clinical signs, and thus the disease is always fatal. New tests to detect a protein (14-3-3) that is associated with neurological damage ***(38)***, the potential for using tonsil biopsy or urinalysis to detect PrP-res prior to the onset of clinical signs ***(54,125)***, and a technique to amplify PrP-res ***(116)*** are all promising. Identifying effective inhibitors of TSE disease is also of the highest priority. Many different compounds have already been shown to inhibit PrP-res formation and/or delay disease (see ref. ***18*** for review). Although most of these compounds can delay disease onset, none can actually prevent disease if given just a few days after infection nor can any be used after the onset of clinical signs. The search for rapid diagnostic tests and effective anti-TSE drugs is ongoing and critically important if the threat of TSE as a re-emerging disease is to be overcome.

References

1. Alper, T., Cramp, W. A., Haig, D. A., and Clarke, M. C. (1967) Does the agent of scrapie replicate without nucleic acid? *Nature* **214,** 764–766.
2. Anderson, R. M., Donnelly, C. A., Ferguson, N. M., et al. (1996) Transmission dynamics and epidemiology of BSE in British cattle. *Nature* **382,** 779–788.
3. Barron, R. M., Thomson, V., Jamieson, E., et al. (2001) Changing a single amino acid in the N-terminus of murine PrP alters TSE incubation time across three species barriers. *EMBO J.* **20,** 5070–5078.
4. Beranger, F., Mange, A., Goud, B., and Lehmann, S. (2002) Stimulation of PrP(C) retrograde transport toward the endoplasmic reticulum increases accumulation of PrP(Sc) in prion-infected cells. *J. Biol. Chem.* **277,** 38,972–38,977.
5. Bessen, R. A., Kocisko, D. A., Raymond, G. J., Nandan, S., Lansbury, P. T. Jr., and Caughey, B. (1995) Nongenetic propagation of strain-specific phenotypes of scrapie prion protein. Nature 375, 698–700.
6. Bessen, R. A. and Marsh, R. F. (1992) Biochemical and physical properties of the prion protein from two strains of transmissible mink encephalopathy agent. *J. Virol.* **66,** 2096–2101.
7. Bessen, R. A. and Marsh, R. F. (1994) Distinct PrP properties suggest the molecular basis of strain variation in transmissible mink encephalopathy. *J. Virol.* **68,** 7859–7868.
8. Bessen, R. A., Raymond, G. J., and Caughey, B. (1997) *In situ* formation of protease-resistant prion protein in transmissible spongiform encephalopathy-infected brain slices. *J. Biol. Chem.* **272,** 15,227–15,231.
9. Bolton, D. C. and Bendheim, P. E. (1988). A modified host protein model of scrapie. In: *Novel Infectious Agents and the Central Nervous System* (Bock, G. and Marsh, J., eds.). John Wiley & Sons, Chichester, pp. 164–181.
10. Bolton, D. C., Bendheim, P. E., Marmostein, A. D., and Potempska, A. (1987). Isolation and structural studies of the intact scrapie agent protein. *Arch. Biochem. Biophys.* **258,** 579–590.
11. Bolton, D. C., McKinley, M. P., and Prusiner, S. B. (1982). Identification of a protein that purifies with the scrapie prion. *Science* **218,** 1309–1311.
12. Borchelt, D. R., Scott, M., Taraboulos, A., Stahl, N., and Prusiner, S. B. (1990). Scrapie and cellular prion proteins differ in the kinetics of synthesis and topology in cultured cells. *J. Cell Biol.* **110,** 743–752.
13. Borchelt, D. R., Taraboulos, A., and Prusiner, S. B. (1992). Evidence for synthesis of scrapie prion protein in the endocytic pathway. *J. Biol. Chem.* **267,** 16188–16199.
14. Bossers, A., Belt, P. B. G. M., Raymond, G. J., Caughey, B., de Vries, R., and Smits, M. A. (1997). Scrapie susceptibility-linked polymorphisms modulate the *in vitro* conversion of sheep prion protein to protease-resistant forms. *Proc. Natl. Acad. Sci. USA* **94,** 4931–4936.
15. Bounhar, Y., Zhang, Y., Goodyer, C. G., and LeBlanc, A. (2001). Prion protein protects human neurons against Bax-mediated apoptosis. *J. Biol. Chem.* **276,** 39145–39149.
16. Bradley, R. and Wilesmith, J. W. (1993). Epidemiology and control of bovine spongiform encephalopathy (BSE). *Br. Med. Bull.* **49,** 932–959.
17. Brandner, S., Isenmann, S., Raeber, A., et al. (1996) Normal host prion protein necessary for scrapie-induced neurotoxicity. *Nature* **379,** 339–343.
18. Brown, P. (2002) Drug therapy in human and experimental transmissible spongiform encephalopathy. *Neurology* **58,** 1720–1725.

19. Bruce, M. E. (1996) Strain typing studies of scrapie and BSE. In: *Methods in Molecular Medicine: Prion Diseases* (Baker, H. and Ridley, R. M., eds.). Humana Press, Totowa, NJ, pp. 223–236.
20. Bruce, M. E., Will, R. G., Ironside, J. W., et al. (1997) Transmissions to mice indicate that 'new variant' CJD is caused by the BSE agent. *Nature* **389,** 498–501.
21. Bueler, H., Aguzzi, A., Sailer, A., et al. (1993) Mice devoid of PrP are resistant to scrapie. *Cell* **73,** 1339–1347.
22. Butler, D. A., Scott, M. R. D., Bockman, J. M., et al. (1988) Scrapie-infected murine neuroblastoma cells produce protease-resistant prion proteins. *J. Virol.* **62,** 1558–1564.
23. Cashman, N. R., Loertscher, R., Nalbantoglu, J., et al. (1990) Cellular isoform of the scrapie agent protein participates in lymphocyte activation. Cell 61, 185–192.
24. Caughey, B., Brown, K., Raymond, G. J., Katzenstien, G. E., and Thresher, W. (1994) Binding of the protease-sensitive form of PrP (prion protein) to sulfated glycosaminoglycan and Congo red. *J. Virol.* **68,** 2135–2141.
25. Caughey, B., Neary, K., Buller, R., et al. (1990) Normal and scrapie-associated forms of prion protein differ in their sensitivities to phospholipase and proteases in intact neuroblastoma cells. *J. Virol. 64,* 1093–1101.
26. Caughey, B., Race, R. E., Ernst, D., Buchmeier, M. J., and Chesebro, B. (1989) Prion protein (PrP) biosynthesis in scrapie-infected and uninfected neuroblastoma cells. *J. Virol.* **63,** 175–181.
27. Caughey, B. and Raymond, G. J. (1991) The scrapie-associated form of PrP is made from a cell surface precursor that is both protease- and phospholipase-sensitive. *J. Biol. Chem.* **266,** 18,217–18,223.
28. Caughey, B. and Raymond, G. J. (1993) Sulfated polyanion inhibition of scrapie-associated PrP accumulation in cultured cells. *J. Virol.* **67,** 643–650.
29. Caughey, B., Raymond, G. J., and Bessen, R. A. (1998) Strain-dependent differences in beta-sheet conformations of abnormal prion protein. *J. Biol. Chem.* **273,** 32,230–32,235.
30. Caughey, B., Raymond, G. J., Ernst, D., and Race, R. E. (1991) N-terminal truncation of the scrapie-associated form of PrP by lysosomal protease(s): implications regarding the site of conversion of PrP to the protease-resistant state. *J. Virol.* **65,** 6597–6603.
31. Caughey, B., Raymond, G. J., Kocisko, D. A., and Lansbury, P. T., Jr. (1997) Scrapie infectivity correlates with converting activity, protease resistance, and aggregation of scrapie-associated prion protein in guanidine denaturation studies. *J. Virol.* **71,** 4107–4110.
32. Caughey, B. W., Dong, A., Bhat, K. S., Ernst, D., Hayes, S. F., and Caughey, W. S. (1991) Secondary structure analysis of the scrapie-associated protein PrP 27-30 in water by infrared spectroscopy. *Biochemistry* **30,** 7672–7680.
33. Chabry, J., Caughey, B., and Chesebro, B. (1998) Specific inhibition of in vitro formation of protease-resistant prion protein by synthetic peptides. *J. Biol. Chem.* **273,** 13,203–13,207.
34. Chiesa, R., Piccardo, P., Ghetti, B., and Harris, D. A. (1998) Neurological illness in transgenic mice expressing a prion protein with an insertional mutation. Neuron 21, 1339–1351.
35. Collinge, J., Palmer, M. S., Sidle, K. C., et al. (1995) Unaltered susceptibility to BSE in transgenic mice expressing human prion protein. *Nature* **378,** 779–783.
36. Collinge, J., Sidle, K. C. L., Meads, J., Ironside, J., and Hill, A. F. (1996) Molecular analysis of prion strain variation and the aetiology of "new variant" CJD. *Nature* **383,** 685–690.
37. Collinge, J., Whittington, M. A., Sidle, K. C., et al. (1994) Prion protein is necessary for normal synaptic function. *Nature* **370,** 295–297.

38. Collins, S., Boyd, A., Fletcher, A., et al. (2000) Creutzfeldt-Jakob disease: diagnostic utility of 14-3-3 protein immunodetection in cerebrospinal fluid. *J. Clin. Neurosci.* **7,** 203–208.
39. Daude, N., Lehmann, S., and Harris, D. A. (1997) Identification of intermediate steps in the conversion of a mutant prion protein to a scrapie-like form in cultured cells. *J. Biol. Chem.* **272,** 11,604–11,612.
40. DebBurman, S. K., Raymond, G. J., Caughey, B., and Lindquist, S. (1997) Chaperone-supervised conversion of prion protein to its protease-resistant form. *Proc. Natl. Acad. Sci. USA* **94,** 13,938–13,943.
41. Diringer, H., Gelderblom, H., Hilmert, H., Ozel, M., Edelbluth, C., and Kimberlin, R. H. (1983) Scrapie infectivity, fibrils and low molecular weight protein. *Nature* **306,** 476–478.
42. Donne, D. G., Viles, J. H., Groth, D., et al. (1997) Structure of the recombinant full-length hamster prion protein PrP(29-231): the N terminus is highly flexible. *Proc. Natl. Acad. Sci. USA* **94,** 13,452–13,457.
43. Fischer, M., Rulicke, T., Raeber, A., et al. (1996) Prion protein (PrP) with amino-proximal deletions restoring susceptibility of PrP knockout mice to scrapie. *EMBO J.* **15,** 1255–1264.
44. Gabizon, R., Meiner, Z., Halimi, M., and Bensasson, S. A. (1993) Heparin-like molecules bind differentially to prion proteins and change their intracellular metabolic-fate. *J. Cell. Physiol.* **157,** 319–325.
45. Gadjusek, D. C. (1988) Transmissible and nontransmissible amyloidoses: autocatalytic post-translational conversion of host precursor proteins to beta-pleated configurations. *J. Neuroimmunol.* **20,** 95–110.
46. Gasset, M., Baldwin, M. A., Fletterick, R. J., and Prusiner, S. B. (1993) Perturbation of the secondary structure of the scrapie prion protein under conditions that alter infectivity. *Proc. Natl. Acad. Sci. USA* **90,** 1–5.
47. Glover, J. R., Kowal, A. S., Schirmer, E. C., Patino, M. M., Liu, J., and Lindquist, S. (1997) Self-seeded fibers formed by sup35, the protein determinant of [PSI^+], a heritable prion-like factor of S. cerevisiae. *Cell* **89,** 811–819.
48. Goldfarb, L. G., Cervenakova, L., Brown, P., and Gajdusek, D. C. (1996) Genotype-phenotype correlations in familial spongiform encephalopathies associated with insert mutations. In: *Transmissible Subacute Spongiform Encephalopathies: Prion Diseases* (Court, L. and Dodet, B., eds.). Elsevier, Paris, pp. 425–431.
49. Goldmann, W., Hunter, N., Smith, G., Foster, J., and Hope, J. (1994) PrP genotype and agent effects in scrapie: change in allelic interaction with different isolates of agent in sheep, a natural host of scrapie. *J. Gen. Virol.* **75,** 989–995.
50. Goldmann, W., Martin, T., Foster, J., ett al. (1996) Novel polymorphisms in the caprine PrP gene: a codon 142 mutation associated with scrapie incubation period. *J. Gen. Virol.* **77,** 2885–2891.
51. Griffith, J. S. (1967) Self-replication and scrapie. *Nature* **215,** 1043–1044.
52. Haraguchi, T., Fisher, S., Olofsson, S., et al. (1989) Asparagine-linked glycosylation of the scrapie and cellular prion proteins. *Arch. Biochem. Biophys.* **274,** 1–13.
53. Hegde, R. S., Mastrianni, J. A., Scott, M. R., et al. (1998) A transmembrane form of the prion protein in neurodegenerative disease. *Science* **279,** 827–834.
54. Hill, A. F., Butterworth, R. J., Joiner, S., et al. (1999) Investigation of variant Creutzfeldt-Jakob disease and other human prion diseases with tonsil biopsy samples. *Lancet* **353,** 183–189.
55. Hill, A. F., Joiner, S., Linehan, J., Desbruslais, M., Lantos, P. L., and Collinge, J. (2000) Species-barrier-independent prion replication in apparently resistant species. *Proc. Natl. Acad. Sci. USA* **97,** 10248–10253.

56. Holscher, C., Delius, J., and Burkle, A. (1998) Overexpression of nonconvertible PrP^C del114-121 in scrapie-infected mouse neuroblastoma cells leads to *trans*-dominant inhibition of wild-type PrP^{Sc} accumulation. *J. Virol.* **72,** 1153–1159.
57. Hope, J., Morton, L. J. D., Farquhar, C. F., Multhaup, G., Beyreuther, K., and Kimberlin, R. H. (1986) The major polypeptide of scrapie-associated fibrils (SAF) has the same size, charge distribution and N-terminal protein sequence as predicted for the normal brain protein (PrP). *EMBO J.* **5,** 2591–2597.
58. Hope, J., Multhaup, G., Reekie, L. J. D., Kimberlin, R. H., and Beyreuther, K. (1988) Molecular pathology of scrapie-associated fibril protein (PrP) in mouse brain affected by the ME7 strain of scrapie. *Eur. J. Biochem.* **172,** 271–277.
59. Horiuchi, M., Baron, G. S., Xiong, L. W., and Caughey, B. (2001) Inhibition of interactions and interconversions of prion protein isoforms by peptide fragments from the C-terminal folded domain. *J. Biol. Chem.* **276,** 15,489–15,497.
60. Horiuchi, M., Chabry, J., and Caughey, B. (1999) Specific binding of normal prion protein to the scrapie form via a localized domain initiates its conversion to the protease-resistant state. *EMBO J.* **18,** 3193–3203.
61. Horiuchi, M., Priola, S. A., Chabry, J., and Caughey, B. (2000) Interactions between heterologous forms of prion protein: binding, inhibition of conversion, and species barriers. *Proc. Natl. Acad. Sci. USA* **97,** 5836–5841.
62. Hsiao, K. K., Scott, M., Foster, D., Groth, D. F., DeArmond, S. J., and Prusiner, S. B. (1990) Spontaneous neurodegeneration in transgenic mice with mutant prion protein. *Science* **250,** 1587–1590.
63. Jarrett, J. T. and Lansbury, P. T., Jr. (1993) Seeding "one-dimensional crystallization" of amyloid: a pathogenic mechanism in Alzheimer's disease and scrapie? *Cell* **73,** 1055–1058.
64. Jeffrey, M., Goodsir, C. M., Bruce, M., McBride, P. A., Scott, J. R., and Halliday, W. G. (1994) Correlative light and electron microscopy studies of PrP localisation in 87V scrapie. *Brain Res.* **656,** 329–343.
65. Jeffrey, M., Goodsir, C. M., Bruce, M. E., McBride, P. A., Fowler, N., and Scott, J. R. (1994) Murine scrapie-infected neurons in vivo release excess prion protein into the extracellular space. *Neurosci. Lett.* **174,** 39–42.
66. Jeffrey, M., Goodsir, C. M., Holliman, A., ett al. (1998) Determination of the frequency and distribution of vascular and parenchymal amyloid with polyclonal and N-terminal-specific PrP antibodies in scrapie-affected sheep and mice. *Vet. Rec.* **142,** 534–537.
67. Jeffrey, M., Martin, S., Barr, J., Chong, A., and Fraser, J. R. (2001) Onset of accumulation of PrP^{RES} in murine ME7 scrapie in relation to pathological and PrP immunohistochemical changes. *J. Comp. Path.* **124,** 20–28.
68. Kascsak, R. J., Rubenstein, R., Merz, P. A., et al. (1986) Immunological comparison of scrapie-associated fibrils isolated from animals infected with four different scrapie strains. *J. Virol.* **59,** 676–683.
69. Kascsak, R. J., Rubenstein, R., Merz, P. A., Carp, R. I., Wisniewski, H. M., and Diringer, H. (1985) Biochemical differences among scrapie-associated fibrils support the biological diversity of scrapie agents. *J. Gen. Virol.* **66,** 1715–1722.
70. Kimberlin, R. H. and Walker, C. A. (1986) Suppression of scrapie infection in mice by heteropolyanion 23, dextran sulfate, and some other polyanions. *Antimicrob. Agents Chemother.* **30,** 409–413.
71. Knaus, K. J., Morillas, M., Swietnicki, W., Malone, M., Surewicz, W. K., and Yee, V. C. (2001) Crystal structure of the human prion protein reveals a mechanism for oligomerization. *Nat. Struct. Biol.* **8,** 770–774.

72. Kocisko, D. A., Come, J. H., Priola, S. A., et al. (1994) Cell-free formation of protease-resistant prion protein. *Nature* **370,** 471–474.
73. Kocisko, D. A., Lansbury, P. T., Jr., and Caughey, B. (1996) Partial unfolding and refolding of scrapie-associated prion protein: evidence for a critical 16-kDa C-terminal domain. *Biochemistry* **35,** 13,434–13,442.
74. Kocisko, D. A., Priola, S. A., Raymond, G. J., Chesebro, B., Lansbury, P. T., Jr., and Caughey, B. (1995) Species specificity in the cell-free conversion of prion protein to protease-resistant forms: a model for the scrapie species barrier. *Proc. Natl. Acad. Sci. USA* **92,** 3923–3927.
75. Lawson, V. A., Priola, S. A., Wehrly, K., and Chesebro, B. (2001) N-terminal truncation of prion protein affects both formation and conformation of abnormal protease-resistant prion protein generated in vitro. *J. Biol. Chem.* **276,** 35,265–35,271.
76. Lehmann, S. and Harris, D. A. (1995) A mutant prion protein displays an aberrant membrane association when expressed in cultured cells. *J. Biol. Chem.* **270,** 24,589–24,597.
77. Lehmann, S. and Harris, D. A. (1996) Mutant and infectious prion proteins display common biochemical properties in cultured cells. *J. Biol. Chem.* **271,** 1633–1637.
78. Locht, C., Chesebro, B., Race, R., and Keith, J. M. (1986) Molecular cloning and complete sequence of prion protein cDNA from mouse brain infected with the scrapie agent. *Proc. Natl. Acad. Sci. USA* **83,** 6372–6376.
79. Lopez, C. D., Yost, C. S., Prusiner, S. B., Myers, R. M., and Lingappa, V. R. (1990) Unusual topogenic sequence directs prion protein biogenesis. *Science* **248,** 226–229.
80. Manson, J. C., Clarke, A. R., McBride, P. A., McConnell, I., and Hope, J. (1994) PrP gene dosage determines the timing but not the final intensity or distribution of lesions in scrapie pathology. *Neurodegeneration* **3,** 331–340.
81. Manson, J. C., Jamieson, E., Baybutt, H., et al. (1999) A single amino acid alteration (101L) introduced into murine PrP dramatically alters incubation time of transmissible spongiform encephalopathy. *EMBO J.* **18,**6855–6864.
82. Marsh, R. F. and Kimberlin, R. H. (1975) Comparison of scrapie and transmissible mink encephalopathy in hamsters. II. Clinical signs, pathology, and pathogenesis. *J. Infect. Dis.* **131,** 104–110.
83. McKinley, M. P., Taraboulos, A., Kenaga, L., et al. (1991) Ultrastructural localization of scrapie prion proteins in cytoplasmic vesicles of infected cultured cells. *Lab. Invest.* **65,** 622–630.
84. Merz, P. A., Somerville, R. A., Wisniewski, H. M., and Iqbal, K. (1981) Abnormal fibrils from scrapie-infected brain. *Acta Neuropathol.* **54,** 63–74.
85. Merz, P. A., Somerville, R. A., Wisniewski, H. M., Manuelidis, L., and Manuelidis, E. E. (1983) Scrapie associated fibrils in Creutzfeldt-Jakob disease. *Nature* **306,** 474–476.
86. Mouillet-Richard, S., Ermonval, M., Chebassier, C., et al. (2000) Signal transduction through prion protein. *Science* **289,** 1925–1928.
87. Paitel, E., Fahraeus, R., and Checler, F. (2003) Cellular prion protein sensitizes neurons to apoptotic stimuli through Mdm2-regulated and p53-dependent caspase 3-like activation. *J. Biol. Chem.* **278,** 10,061–10,066.
88. Palmer, M. S., Dryden, A. J., Hughes, J. T., and Collinge, J. (1991) Homozygous prion protein genotype predisposes to sporadic Creutzfeldt-Jakob disease. *Nature* **352,** 340–342.
89. Pan, K.-M., Baldwin, M., Nguyen, J., et al. (1993) Conversion of alpha-helices into beta-sheets features in the formation of the scrapie prion protein. *Proc. Natl. Acad. Sci. USA* **90,** 10,962–10,966.

90. Parchi, P., Castellani, R., Capellari, S., et al. (1996) Molecular basis of phenotypic variability in sporadic Creutzfeldt-Jakob disease. *Ann. Neurol.* **39,** 767–778.
91. Paushkin, S. V., Kushnirov, V. V., Smirnov, V. N., and Ter-Avanesyan, M. D. (1997) In vitro propagation of the prion-like state of yeast Sup35 protein. *Science* **277,** 381–383.
92. Petersen, R. B., Parchi, P., Richardson, S. L., Urig, C. B., and Gambetti, P. (1996) Effect of the D178N mutation and the codon 129 polymorphism on the metabolism of the prion protein. *J. Biol. Chem.* **271,** 12,661–12,668.
93. Priola, S. A. (1999) Prion protein and species barriers in the transmissible spongiform encephalopathies. *Biomed. Pharmacother.* **53,** 27–33.
94. Priola, S. A. and Caughey, B. (1994) Inhibition of scrapie-associated PrP accumulation: Probing the role of glycosaminoglycans in amyloidogenesis. *Mol. Neurobiol.* **8,** 113–120.
95. Priola, S. A., Caughey, B., Race, R. E., and Chesebro, B. (1994) Heterologous PrP molecules interfere with accumulation of protease-resistant PrP in scrapie-infected murine neuroblastoma cells. *J. Virol.* **68,** 4873–4878.
96. Priola, S. A., Chabry, J., and Chan, K. (2001) Efficient conversion of normal prion protein (PrP) by abnormal hamster PrP is determined by homology at amino acid residue 155. *J. Virol.* **75,** 4673–4680.
97. Priola, S. A. and Chesebro, B. (1995) A single hamster amino acid blocks conversion to protease-resistant PrP in scrapie-infected mouse neuroblastoma cells. *J. Virol.* **69,** 7754–7758.
98. Priola, S. A. and Chesebro, B. (1998) Abnormal properties of prion protein with insertional mutations in different cell types. *J. Biol. Chem.* **273,** 11,980–11,985.
99. Priola, S. A. and Lawson, V. A. (2001) Glycosylation influences cross-species formation of protease-resistant prion protein. *EMBO J.* **20,** 6692–6699.
100. Prusiner, S. B. (1982) Novel proteinaceous infectious particles cause scrapie. *Science* **216,** 136–144.
101. Prusiner, S. B. (1991) Molecular biology of prion diseases. *Science* **252,** 1515–1522.
102. Prusiner, S. B., Scott, M., Foster, D., et al. (1990) Transgenetic studies implicate interactions between homologous PrP isoforms in scrapie prion replication. *Cell* **63,** 673–686.
103. Race, R. and Chesebro, B. (1998) Scrapie infectivity found in resistant species. *Nature* **392,** 770.
104. Race, R., Meade-White, K., Raines, A., Raymond, G. J., Caughey, B., and Chesebro, B. (2002) Subclinical scrapie infection in a resistant species: persistence, replication, and adaptation of infectivity during four passages. *J. Infect. Dis.* **186(suppl 2),** S166–S170.
105. Race, R., Raines, A., Raymond, G. J., Caughey, B., and Chesebro, B. (2001) Long-term subclinical carrier state precedes scrapie replication and adaptation in a resistant species: analogies to bovine spongiform encephalopathy and variant Creutzfeldt-Jakob disease in humans. *J. Virol.* **75,** 10,106–10,112.
106. Race, R. E., Caughey, B., Graham, K., Ernst, D., and Chesebro, B. (1988) Analyses of frequency of infection, specific infectivity, and prion protein biosynthesis in scrapie-infected neuroblastoma cell clones. *J. Virol.* **62,** 2845–2849.
107. Race, R. E., Fadness, L. H., and Chesebro, B. (1987) Characterization of scrapie infection in mouse neuroblastoma cells. *J. Gen. Virol.* **68,** 1391–1399.
108. Race, R. E., Priola, S. A., Bessen, R. A., et al. (1995) Neuron-specific expression of a hamster prion protein minigene in transgenic mice induces susceptibility to hamster scrapie agent. *Neuron* **15,** 1183–1191.

109. Raeber, A. J., Race, R. E., Brandner, S., et al. (1997) Astrocyte-specific expression of hamster prion protein (PrP) renders PrP knockout mice susceptible to hamster scrapie. *EMBO J.* **16,** 6057–6065.
110. Raymond, G. J., Bossers, A., Raymond, L. D., et al. (2000) Evidence of a molecular barrier limiting susceptibility of humans, cattle and sheep to chronic wasting disease. *EMBO J.* **19,** 4425–4430.
111. Raymond, G. J., Hope, J., Kocisko, D. A., et al. (1997) Molecular assessment of the transmissibilities of BSE and scrapie to humans. *Nature* **388,** 285–288.
112. Riek, R., Hornemann, S., Wider, G., Billeter, M., Glockshuber, R., and Wuthrich, K. (1996) NMR structure of the mouse prion protein domain PrP(121-231) *Nature* **382,** 180–182.
113. Riek, R., Hornemann, S., Wider, G., Glockshuber, R., and Wuthrich, K. (1997) NMR characterization of the full-length recombinant murine prion protein, mPrP(23-21). *FEBS Lett.* **413,** 282–288.
114. Rubenstein, R., Carp, R. I., and Callahan, S. M. (1984) *In vitro* replication of scrapie agent in a neuronal model: infection of PC12 cells. *J. Gen. Virol.* **65,** 2191–2198.
115. Rubenstein, R., Merz, P. A., Kascsak, R. J., et al. (1991) Scrapie-infected spleens: analysis of infectivity, scrapie-associated fibrils, and protease-resistant proteins. *J. Infect. Dis.* **164,** 29–35.
116. Saborio, G. P., Permanne, B., and Soto, C. (2001) Sensitive detection of pathological prion protein by cyclic amplification of protein misfolding. *Nature* **411,** 810–813.
117. Safar, J., Roller, P. P., Gajdusek, D. C., and Gibbs, C. J. Jr. (1993) Thermal stability and conformational transitions of scrapie amyloid (prion) protein correlate with infectivity. *Protein Sci.* **2,** 2206–2216.
118. Safar, J., Wille, H., Itri, V., et al. (1998) Eight prion strains have PrP(Sc) molecules with different conformations [see comments]. *Nat. Med.* **4,** 1157–1165.
119. Sakaguchi, S., Katamine, S., Nishida, N., et al. (1996) Loss of cerebellar Purkinje cells in aged mice homozygous for a disrupted PrP gene. *Nature* **380,** 528–531.
120. Schatzl, H. M., Laszlo, L., Holtzman, D. M., et al. (1997) A hypothalamic neuronal cell line persistently infected with scrapie prions exhibits apoptosis. *J. Virol.* **71,** 8821–8831.
121. Scott, M., Foster, D., Mirenda, C., et al. (1989) Transgenic mice expressing hamster prion protein produce species-specific scrapie infectivity and amyloid plaques. *Cell* **59,** 847–857.
122. Scott, M., Groth, D., Foster, D., et al. (1993) Propagation of prions with artificial properties in transgenic mice expressing chimeric PrP genes. *Cell* **73,** 979–988.
123. Scott, M. R., Kohler, R., Foster, D., and Prusiner, S. B. (1992) Chimeric prion protein expression in cultured cells and transgenic mice. *Protein Sci.* **1,** 986–997.
124. Serio, T. R., Cashikar, A. G., Kowal, A. S., et al. (2000) Nucleated conformational conversion and the replication of conformational information by a prion determinant. *Science* **289,** 1317–1321.
125. Shaked, G. M., Shaked, Y., Kariv, Z., Halimi, M., Avraham, I., and Gabizon, R. (2001) A protease resistant PrP isoform is present in urine of animals and humans affected with prion diseases. *J. Biol. Chem.* **276,** 31,479–31,482.
126. Snow, A. D., Kisilevsky, R., Willmer, J., Prusiner, S. B., and DeArmond, S. J. (1989) Sulfated glycosaminoglycans in amyloid plaques of prion diseases. *Acta Neuropathol.* **77,** 337–342.

127. Supattapone, S., Bosque, P., Muramoto, T., et al. (1999) Prion protein of 106 residues creates an artifical transmission barrier for prion replication in transgenic mice. *Cell* **96,** 869–878.
128. Taraboulos, A., Raeber, A. J., Borchelt, D. R., Serban, D., and Prusiner, S. B. (1992) Synthesis and trafficking of prion proteins in cultured cells. *Mol. Biol. Cell.* **3,** 851–863.
129. Taraboulos, A., Scott, M., Semenov, A., Avraham, D., Laszlo, L., and Prusiner, S. B. (1995) Cholesterol depletion and modification of COOH-terminal targeting sequence of the prion protein inhibit formation of the scrapie isoform. *J. Cell. Biol.* **129,** 121–132.
130. Taylor, K. L., Cheng, N., Williams, R. W., Steven, A. C., and Wickner, R. B. (1999) Prion domain initiation of amyloid formation in vitro from native Ure2p. *Science* **283,** 1339–1343.
131. Telling, G. C., Parchi, P., DeArmond, S. J., et al. (1996) Evidence for the conformation of the pathologic isoform of the prion protein enciphering and propagating prion diversity. *Science* **274,** 2079–2082.
132. Telling, G. C., Scott, M., Mastrianni, J., et al. (1995) Prion propagation in mice expressing human and chimeric PrP transgenes implicates the interaction of cellular PrP with another protein. *Cell* **83,** 79–90.
133. Tobler, I., Gaus, S. E., Deboer, T., et al. (1996) Altered circadian activity rhythms and sleep in mice devoid of prion protein. *Nature* **380,** 639–642.
134. Vilette, D., Andreoletti, O., Archer, F., et al. (2001) Ex vivo propagation of infectious sheep scrapie agent in heterologous epithelial cells expressing ovine prion protein. *Proc. Natl. Acad. Sci. USA* **98,** 4055–4059.
135. Vorberg, I., Chan, K., and Priola, S. A. (2001) Deletion of beta-strand and alpha-helix secondary structure in normal prion protein inhibits formation of its protease-resistant isoform. *J. Virol.* **75,** 10024–10032.
136. Vorberg, I., Groschup, M. H., Pfaff, E., and Priola, S. A. (2003) Multiple amino acid residues within the rabbit prion protein inhibit formation of its abnormal isoform. *J. Virol.* **77,** 2003–2009.
137. Vorberg, I. and Priola, S. A. (2002) Molecular basis of scrapie strain glycoform variation. *J. Biol. Chem.* **277,** 36,775–36,781.
138. Wells, G. A. H., Scott, A. C., Johnson, C. T., et al. (1987) A novel progressive spongiform encephalopathy in cattle. *Vet. Rec.* **121,** 419–420.
139. Wickner, R. B. (1994) [URE3] as an altered URE2 protein: evidence for a prion analog in *Saccharomyces cerevisiae. Science* **264,** 566–569.
140. Will, R. G., Ironside, J. W., Zeidler, M., et al. (1996) A new variant of Creutzfeldt-Jakob disease in the UK. *Lancet* **347,** 921–925.
141. Williams, E. S. and Young, S. (1980) Chronic wasting disease of captive mule deer: a spongiform encephalopathy. *J. Wildl. Dis.* **16,** 89–98.
142. Williams, E. S. and Young, S. (1982) Spongiform encephalopathy of Rocky Mountain elk. *J. Wildl. Dis.* **18,** 465–471.
143. Zahn, R., Liu, A., Luhrs, T., et al. (2000) NMR solution structure of human prion protein. *Proc. Natl. Acad. Sci. USA* **97,** 145–150.
144. Zulianello, L., Kaneko, K., Scott, M., et al. (2000) Dominant-negative inhibition of prion formation diminished by deletion mutagenesis of the prion protein. *J. Virol.* **74,** 4351–4360.

Index

E

F

CPSIA information can be obtained
at www.ICGtesting.com
Printed in the USA
LVOW01*1614040216
473698LV00014B/175/P

9 781588 291172